Equine Reproductive Physiology, Breeding and Stud Management, 3rd Edition

To Mum and Dad

Equine Reproductive Physiology, Breeding and Stud Management, 3rd Edition

Mina C.G. Davies Morel
Institute of Rural Studies
University of Wales, Aberystwyth
UK

www.cabi.org

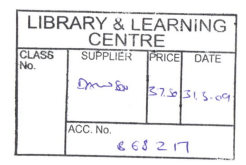
CABI is a trading name of CAB International

CABI Head Office
Nosworthy Way
Wallingford
Oxfordshire OX10 8DE
UK

Tel: +44 (0)1491 832111
Fax: +44 (0)1491 833508
E-mail: cabi@cabi.org
Website: www.cabi.org

CABI North American Office
875 Massachusetts Avenue
7th Floor
Cambridge, MA 02139
USA

Tel: +1 617 395 4056
Fax: +1 617 354 6875
E-mail: cabi-nao@cabi.org

A catalogue record for this book is available from the British Library, London, UK.

Library of Congress Cataloging-in-Publication Data

Davies Morel, Mina C.G.
  Equine reproductive physiology, breeding, and stud management/M.C.G. Davies Morel – 3rd ed.
    p. cm.
  Includes bibliographical references and index.
  ISBN 978-1-84593-450-7 (alk. paper)
1. Horses – Reproductive. 2. Horses – Breeding. I. Title.

SF768.2.H67D39 2008
636.1'082–dc22

2008008422

ISBN-13: 978 1 84593 450 7

Typeset by SPi, Pondicherry, India.
Printed and bound in the UK by Cambridge University Press, Cambridge.

Contents

1 The Reproductive Anatomy of the Mare

1.1. Introduction

This chapter details the anatomy and function of the mare's reproductive system. Further accounts may be found in other texts, such as Ashdown and Done (1987), Frandson and Spurgen (1992), Ginther (1992), Kainer (1993), Dyce *et al.* (1996), Bone (1998), Senger (1999), Bergfelt (2000), Hafez and Hafez (2000) and Le Blare *et al.* (2004). The reproductive tract of the mare may be considered as a Y-shaped tubular organ with a series of constrictions along its length. The perineum, vulva, vagina and cervix can be considered as the outer protective structures, providing protection for the inner, more delicate structures, the uterus, Fallopian tubes and ovaries, which are responsible for fertilization and embryo development. Figure 1.1, taken after slaughter, shows the reproductive structures of the mare, and Figs 1.2 and 1.3 illustrate these diagrammatically. Each of these structures will be dealt with in turn in the following account.

1.2. The Vulva

The vulva (Fig. 1.4) is the external area of the mare's reproductive system, protecting the entrance to the vagina. The outer area is pigmented skin with the normal sebaceous and sweat glands along with the nerve and blood supply normally associated with the skin of the mare. The inner area is lined by mucous membrane and is continuous with the vagina. The upper limit (the dorsal commissure) is situated approximately 7 cm below the anus. Below the entrance to the vagina, in the lower part of the vulva (the ventral commissure), lie the clitoris and the three clitoral sinuses (ventral, medial and lateral; Fig. 1.5). These sinuses are of importance in the mare as they provide an ideal environment for the harbouring of many venereal disease (VD) bacteria, such as *Taylorella equigenitalis* (causal agent for contagious equine metritis (CEM)), *Klebsiella pneumoniae* and *Pseudomonas aeroginosa*. Hence,

this area is regularly swabbed in mares prior to covering, and, indeed, in the Thoroughbred industry such swabbing is compulsory (McAllister and Sack, 1990; Ginther, 1992; Horse Race Betting Levy Board, 2008). Within the walls of the vulva lies the vulva constrictor muscle, running along either side of the length of the vulval lips. This muscle acts to maintain the vulval seal and to invert and expose the clitoral area during oestrus, known as winking (Ashdown and Done, 1987; Kainer, 1993).

1.3. The Perineum

The perineum is a rather loosely defined area in the mare, but includes the outer vulva and adjacent skin along with the anus and the surrounding area. In the mare, the conformation of this area is of clinical importance, due to its role in the protection of the genital tract from the entrance of air. Malconformation in this area predisposes the mare to a condition known as pneumovagina or vaginal wind-sucking, in which air is sucked in and out of the vagina through the open vulva. Along with this

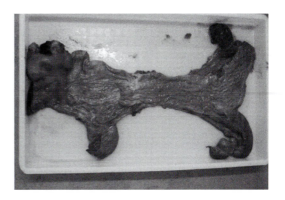

Fig. 1.1. The mare's reproductive tract after slaughter and dissection (see also Fig. 1.3).

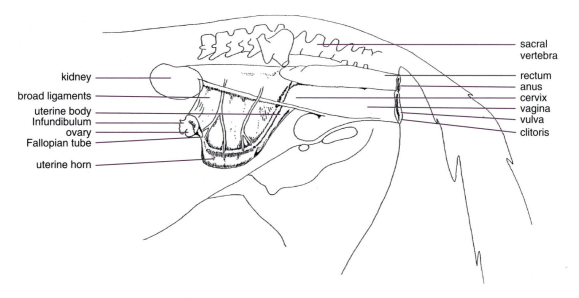

kidney

broad ligaments

uterine body
Infundibulum
ovary
Fallopian tube

uterine horn

sacral
vertebra

rectum
anus
cervix
vagina
vulva
clitoris

Fig. 1.2. A lateral view (from the side) of the mare's reproductive tract.

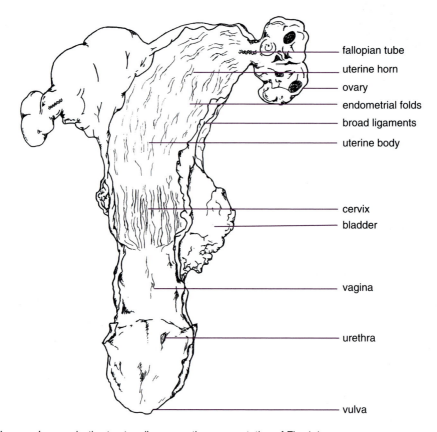

fallopian tube

uterine horn

ovary

endometrial folds

broad ligaments

uterine body

cervix

bladder

vagina

urethra

vulva

Fig. 1.3. The mare's reproductive tract: a diagrammatic representation of Fig. 1.1.

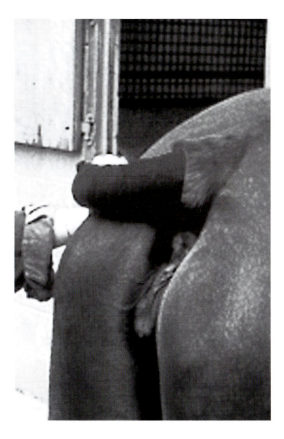

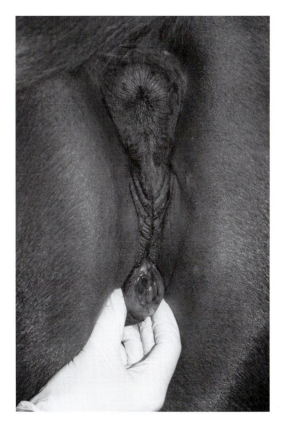

Fig. 1.4. The vulval area of the mare: in this instance, the conformation of the perineal area is poor with the anus sunken cranially, opening up the vulva to faecal contamination.

Fig. 1.5. The vulva of the mare showing the ventral commissure within which lie the clitoris and clitoral sinuses on either side.

passage of air also go bacteria, which bombard the cervix, exposing it to unacceptably high levels of contamination, which it is often unable to cope with, especially during oestrus when it is less competent. Passage of bacteria into the higher, more susceptible parts of the mare's tract may result in bacterial infections, such as CEM, and other VDs leading to endometritis (uterine infection). Chapter 19 (this volume) gives further details on the causes of VD infection in the mare, all of which adversely affect fertilization rates (Ginther, 1992; Easley, 1993; Kainer, 1993).

1.3.1. Protection of the genital tract

Adequate protection of the genital tract is essential to prevent the adverse effects of pneumovagina.

There are three seals within the tract: the vulval seal, the vestibular or vaginal seal and the cervix; these are illustrated in Fig. 1.6.

The perineal area plus the vulva constrictor muscle in the walls of the vulva form the vulval seal. The vestibular seal is formed by the natural collapsing and apposition of the walls of the posterior vagina, where it sits above the floor of the pelvic girdle plus the hymen, if still present. The tight muscle ring within the cervix forms the cervical seal. This series of seals is affected by the conformation of an individual and also by the stage of the oestrous cycle (Figs 1.7 and 1.9).

The ideal conformation is achieved if 80% of the vulva lies below the pelvic floor. A simple test can be performed to assess this. If a sterile plastic tube is inserted through the vulva into the vagina

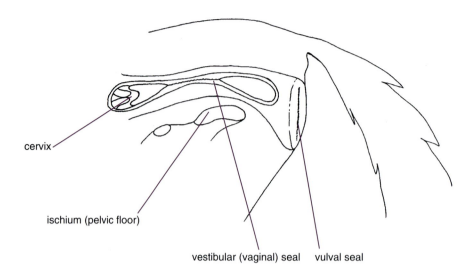

cervix

ischium (pelvic floor)

vestibular (vaginal) seal   vulval seal

Fig. 1.6. The seals of the mare's reproductive tract during dioestrus.

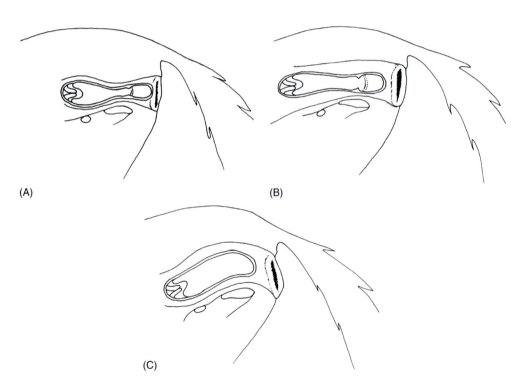

(A)

(B)

(C)

Fig. 1.7. The effect of conformation on the competence of the vulval, vestibular and cervical seals in the mare: (A) a low ischium (pelvic floor) results in an incompetent vestibular seal – in this case, the vulval seal is still competent; therefore, infection risk is limited; (B) a low ischium results in an incompetent vestibular seal – in this case, the vulval seal is also incompetent; therefore, infection risk is increased; and (C) an incompetent vestibular and vulval seal plus a sloping perineal area result in a significant infection risk, especially from faecal contamination.

and allowed to rest horizontally on the vagina floor, the amount of vulva lying below this tube should be approximately 80% in well-conformed mares. This technique is illustrated diagrammatically in Fig. 11.4.

If the ischium of the pelvis is too low, the vulva tends to fall towards the horizontal plane as seen in Fig. 1.7. This opens up the vulva to contamination by faeces, increasing the risk of uterine infection due to pneumovagina. Additionally, a low pelvis causes the vagina to slope inwards, preventing the natural drainage of urine at urination leading to urinovagina, which further increases the risk of uterine infection. Pascoe (1979a) suggested that mares should be allocated a Caslick index derived by multiplying the angle of inclination of the vulva with the distance from the ischium to the dorsal commissure. This index can then be used to classify mares into three types and so predict the likely occurrence of endometritis (Fig. 1.8).

The effect of poor conformation of the perineum area may be alleviated by a Caslick's vulvoplasty operation, developed by Dr Caslick in 1937 (Caslick, 1937). The lips on either side of the upper vulva are cut, and the two sides are then sutured together. The two raw edges heal together, as in the healing of an open wound, and hence seal the upper part of the vulva. The hole left at the ventral commissure is adequate for urination but prevents the passage of faeces into the vagina (Fig. 1.9).

The chance that a mare requiring a Caslick's operation will pass on the trait to her offspring is reasonably high. This, coupled with the fact that the operation site has to be cut to allow mating and foaling, casts doubt on whether such mares should be bred. Mares that have been repeatedly cut and resutured become increasingly hard to perform a Caslick's operation on, as the lips of the vulva become progressively more fibrous and therefore difficult to suture. In such cases, a procedure termed a Pouret may be carried out (Pouret, 1982). This is a more major operation and involves the realignment of the anus as well as the vulva (Knottenbelt and Pascoe, 2003).

Perineal malformation is particularly prevalent in Thoroughbred mares. It is causal to both pneumovagina (collection of air within the vagina) and urinovagina (collection of urine within the vagina), both conditions being precursors for endometritis and hence infertility. The condition tends to be exacerbated in mares with a low body condition score and also in multiparous, aged mares and those in fit athletic condition. Its continued existence is largely due to the selection of horses for athletic performance rather than reproductive competence (Caslick, 1937; Pascoe, 1979a; Pouret, 1982; Le Blanc, 1991; Easley, 1993).

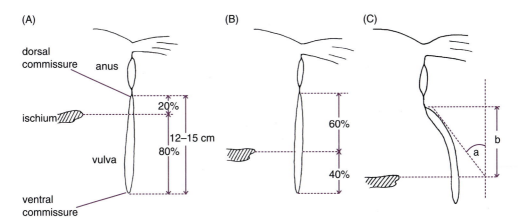

Fig. 1.8. A lateral view of the relationship between the anus, vulva and ischium, indicating: (A) type I mare with good conformation, Caslick index <50 (b = 2–3 cm, a < 10°) – no Caslick required; (B) type II mare with poor conformation, predisposing to type III in later life, Caslick index 50–100 (b = 6–7 cm, a = 10–20°) – no immediate need for a Caslick but likely in later life; and (C) type III mare with very poor conformation, including vulva lips in a horizontal plane, Caslick index >150 (b = 5–9 cm, a ≥ 30°) – Caslick required immediately, significant chance of endometritis and a reduction in reproductive success.

(A)

(B)

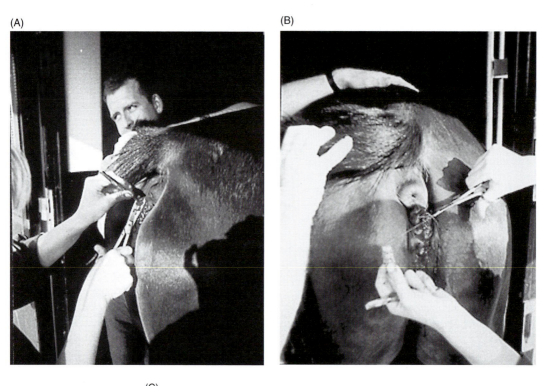

(C)

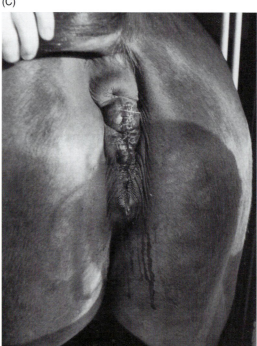

Fig. 1.9. A Caslick operation in the mare showing (A) the cutting of the vulval lips; (B) suturing; and (C) the finished job.

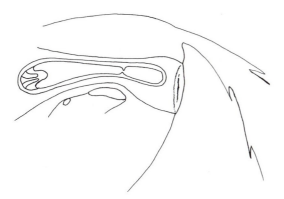

Fig. 1.10. The effect of oestrus on the competence of the vulval, vestibular and cervical seals in the mare: oestrus causes a relaxation of the seals and, therefore, an increase in infection risk.

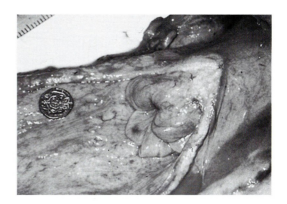

Fig. 1.11. The internal surface of the mare's vagina (coin measures 21 mm in diameter).

The oestrous cycle also has an effect on the competence of the three seals. Further details of the effect of the oestrous cycle on the reproductive tract are given in Chapter 3 (this volume). However, in summary, oestrus results in the slackening of all three seals, due to a relaxation of the muscles associated with the reproductive tract, especially the cervix (Fig. 1.10). This allows intromission at covering but also decreases the competence of the reproductive tract seals and so increases the chance of bacterial invasion. In part, this is compensated for by elevated oestradiol levels characteristic of oestrus, which enhance the mare's immunological response, thus reducing the chance of uterine infection, despite the increased chance of bacterial invasion. Recent work by Causey (2007) indicates that the uterus may also be adapted to provide mucociliary clearance of bacteria, a further defence against uterine bacterial invasion (see Section 1.6).

1.4. The Vagina

The vagina of the mare is on average 18–23 cm long and 10–15 cm in diameter. In the well-conformed mare the floor of the vagina should rest upon the ischium of the pelvis, and the walls are normally collapsed and apposed, forming the vestibular seal. The hymen, if present, is also associated with this seal and divides the vagina into anterior (cranial) and posterior (caudal) sections. In some texts the posterior vagina is referred to as the vestibule. The urethra, from the bladder, opens just caudal to the

hymen. Within the body cavity, the vagina is mainly covered by the peritoneum and is surrounded by loose connective tissue, fat and blood vessels. The walls of the vagina are muscular with a mucous lining; the elasticity conferred by the muscle layer allows the major stretching required at parturition (Figs. 1.11 and 1.16).

The vagina acts as the first protector and cleaner of the system. It is aglandular but contains acidic to neutral secretions, originating from the cervix and small glands situated just cranial to the vulval lips. These acidic secretions are bacteriocidal, but also have the disadvantage of being spermicidal and of attacking the epithelial cell lining of the vagina, necessitating the secretion of mucus by the cells lining the vagina, in order to provide a protective mucous layer. Thus, at ejaculation, sperm is deposited into the top of the cervix and/or bottom of the uterus, to avoid the detrimental effect of the acidic conditions within the vagina. The exact composition of vaginal secretion is controlled by the cyclical hormonal changes of the mare's reproductive cycle (Ginther, 1992; Kainer, 1993).

1.5. The Cervix

The cervix lies at the entrance to the uterus. It is a tight, thick-walled, sphincter muscle, acting as the final protector of the system. In the sexually inactive, dioestrous state, it is tightly contracted, white in colour and measures on average 6–8 cm long and 4–5 cm in diameter; cervical secretion is minimal and thick in consistency. The muscle tone and, therefore,

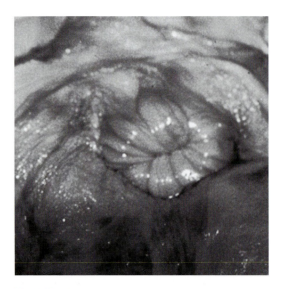

Fig. 1.12. The oestrous cervix protrudes (flowers) into the vagina of the mare.

Fig. 1.13. The internal surface of the cervix and uterus illustrating the cervical endometrial folds and uterine.

cervix size, along with its secretion are again governed by cyclic hormonal changes. Muscle tone relaxes during oestrus and there is an increase in secretion, easing the passage of the penis into the entrance of the cervix. The oestrous cervix appears pink in colour and may be seen protruding or 'flowering' into the vagina (Fig. 1.12; Lieux, 1970).

The lining of the cervix consists of a series of folds or crypts, as shown in Fig. 1.13. These crypts are continual with the folds in the uterine endometrium and enable the significant expansion of the cervix required at parturition (Ginther, 1992; Kainer, 1993).

1.6. The Uterus

The uterus of the mare is a hollow muscular organ joining the cervix and the Fallopian tubes (Figs 1.1 and 1.3). This upper part of the tract including the uterus is attached to the lumbar region of the mare by two broad ligaments, outfoldings of the peritoneum, on either side of the vertebral column. The broad ligaments provide the major support for the reproductive tract (Fig. 1.14) and can be divided into three areas: the mesometrium, attached to the uterus; the mesosalpinx, attached to the Fallopian tubes; and the mesovarium, attached to the ovaries (Ginther, 1992).

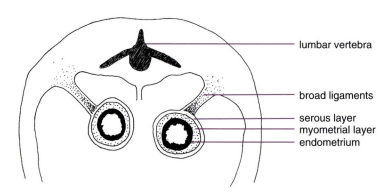

Fig. 1.14. Cross section through the abdomen of the mare illustrating the reproductive tract support provided by the broad ligaments.

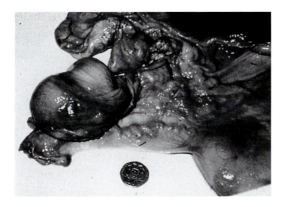

Fig. 1.15. The convoluted Fallopian tube running through the mesovarian section of the broad ligaments, from the uterine horn on the right to the ovary on the left.

The uterus is divided into two areas: the body and the horns. The body of the uterus normally measures 18–20 cm long and 8–12 cm in diameter, and divides into two uterine horns which are approximately 25 cm long and which reduce in diameter from 4–6 cm to 1–2 cm as they approach the Fallopian tubes (Fig. 1.15). The size of the uterus is affected by age and parity, older multipa-

rous mares tending to have larger uteri. The uterus of the mare is termed a simplex bipartitus, due to the relatively large size of the uterine body compared to the uterine horns (60:40 split). This differs from that in other farm livestock, where the uterine horns are the more predominant feature. The lack of a septum dividing the uterine body is also notable (Hafez and Hafez, 2000). *In situ* the uterine walls are flaccid and intermingle with the intestine, the only lumen present being that formed between the endometrial folds.

The uterine wall (Fig. 1.16) consists of three layers: an outer serous layer (perimetrium) continuous with the broad ligaments; a central muscular layer (myometrium); and an inner mucous membrane lining (endometrium). The central myometrial layer consists of external longitudinal muscle fibres, a central vascular layer and internal circular muscle fibres. It is this central myometrial layer that allows the considerable expansion of the uterus during pregnancy and provides the force for parturition. The inner endometrium is arranged in 12–15 longitudinal folds (Figs 1.3 and 1.13) and comprises luminal epithelial cells, stroma of connective tissue or lamina propria and associated endometrial glands and ducts (Fig. 1.16). Collagenous connective tissue cores support these folds. The activity and, therefore, appearance of

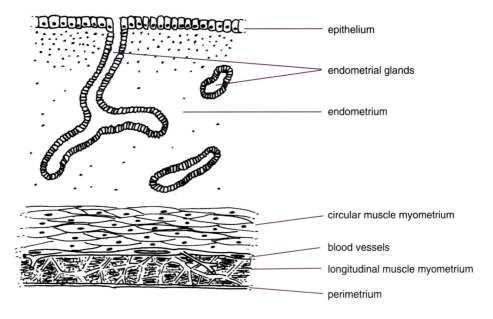

Fig. 1.16. Cross section through the uterine wall.

these endometrial glands are dependent on the cyclical hormonal changes. It is this endometrial layer that is largely responsible for supporting the developing conceptus and for placental attachment and development (Ashdown and Done, 1987; Ginther, 1992, 1995; Kainer, 1993; Sertich, 1998). New work by Causey (2007) suggests that the epithelium of the uterine endometrium exhibits mucus-secreting and ciliated cells, supporting a mucopolysaccharide blanket. These features are also found in the respiratory tract to aid the clearance of foreign material/infection. Hence, Causey (2007) suggests that the presence of such cells in the mare's uterus is indicative of a similar mucociliary clearance mechanism to help eliminate bacteria, and so provides a further defence against uterine bacterial invasion.

1.7. The Utero-tubular Junction

The utero-tubular junction is a constriction or sphincter formed by a high concentration of muscle cells from the circular myometrium of the Fallopian tube. The junction, which appears as a papilla in the endometrium, separates the end of the uterine horns from the beginning of the Fallopian tubes (Fig. 1.17). Fertilization takes place in the Fallopian tubes, and only fertilized ova can pass through this junction and on to the uterus for implantation and further development. Fertilized ova appear to actively control their own passage, possibly via a localized secretion of prostaglandin E (PGE; see Section 5.3; Ball and Brinsko, 1992), leaving the unfertilized ova on the Fallopian tube side of the junction. These then gradually degenerate (Ginther, 1992; Kainer, 1993; Fig. 1.17).

1.8. The Fallopian Tubes

The mare has two Fallopian tubes or oviducts of 25–30 cm length, which are continuous with the uterine horns (Fig. 1.15). The diameter of these tubes varies slightly along their length, being 2–5 mm at the isthmus end, nearest the uterine horn, and gradually increasing to 5–10 mm at the ampulla, nearest the ovary. The division of the Fallopian tube between the isthmus and ampulla is approximately equal. The Fallopian tubes lie within peritoneal folds, which form the mesosalphinx part of the broad ligaments. They have walls very similar in structure to the uterus, but thinner,

Fig. 1.17. The utero-tubular junction in the mare, as seen from the uterine horn side (the dark colour of the uterine endometrium is not natural but serves to allow easier identification of the utero-tubular junction).

composed of three layers: the outer fibrous serous layer, continuous with the mesosalpinx; a central myometrial layer of circular and longitudinal muscles fibres; and an inner mucous membrane. Fertilization takes place in the ampulla, a region lined with fimbrae (hair-like projections), which act to waft unfertilized ova into the ampulla to await the sperm and to waft fertilized ova out of the ampulla and on towards the utero-tubular junction. The ampulla of each Fallopian tube ends in the infundibulum, a funnel-like opening close to the ovary (Sisson, 1975).

The infundibulum in the mare is closely associated with a specific part of the ovary termed the ovulation fossa, which is unique to the mare and is the only site of ova release; in other mammals ovulation may occur over the whole surface of the ovary. The infundibulum is, therefore, relatively hard to distinguish in the mare, not being so evident as a funnel-shaped structure surrounding the whole ovary. The infundibulum is lined, like the ampulla, by fimbrae, which attract and catch the ova guiding them towards the entrance of the Fallopian tubes (Ginther, 1992; Kainer, 1993).

1.9. The Ovaries

The ovaries of the mare are both cytogenic and endocrine in function, producing gametes (ova) and hormones. They are evident as two bean-shaped structures situated ventrally to (below) the fourth and fifth lumbar vertebrae and supported by the mesovarium part of the broad ligaments. They make the total length of the reproductive tract in the mare in the region of 50–60 cm. In the sexually inactive stage, i.e. during the non-breeding season, the mare's ovaries measure 2–4 cm in length and 2–3 cm in width and are hard to the touch due to the absence of developing follicles. During the sexually active stage when the mare is in season, they increase in size to 6–8 cm in length and 3–4 cm in width; they are also softer to the touch due to the development of fluid-filled follicles (Fig. 1.18). Older, multiparous mares tend to show larger ovaries which can be up to 10 cm in length.

The convex outer surface or border of the ovary is attached to the mesovarian section of the broad ligaments (Figs 1.15 and 1.18) and is the entry point for blood and nerve supply; the concave inner surface is free from attachment and is the location of the ovulation fossa. The whole ovary is contained within a thick protective layer, the tunica albuginea, except for the ovulation fossa. The tissue of the ovary in the mare is arranged as the inner cortex (active gamete-producing tissue) and the outer medulla (supporting tissue). Ova release at ovulation occurs only through the ovulation fossa, and all follicular and corpora lutea (CL) development occurs internally, within the cortex of the ovary (Witherspoon, 1975). The mare differs in these aspects from other mammals, in which the

medulla and cortex are reversed, ovulation occurring over the surface of the ovary and all follicular and CL development occurring on the outer borders. Rectal palpation, as a clinical aid to assess reproductive function in the mare, is not, therefore, as easy to perform as it is in other farm livestock, for example, the cow. However, with the advent of ultrasound, assessment of ovarian characteristics in the mare is now quite accurate (see Sections 11.2.6.3 and 11.2.6.5; Ginther, 1992, 1995; Kainer, 1993; Sertich, 1998; Hafez and Hafez, 2000).

1.9.1. Folliculogenesis (follicular development) and ovulation

The ovary is made of two basic cell types: interstitial cells (stroma), which provide support; and germinal cells, which provide a reservoir from which all future ova are produced. The number of potential ova contained within the female ovary is dictated prior to birth; subsequently, no addition to that pool of ova can be made. These very immature ova are termed oogonia, and there are many more than an individual will use within her reproductive lifetime. These oogonia, with their full complement of chromosomes (64) and surrounding a single layer of epithelial cells, are termed primordial follicles. At birth, the ovary contains many thousands of these primordial follicles which, at varying rates, start to undergo development (folliculogenesis) to form ova mature enough to be fertilized. Folliculogenesis can be divided into two phases (Del Campo *et al.*, 1990; Ginther, 1992; Pierson, 1993; Hafez and Hafez, 2000). The first phase can start to occur anytime after birth and develops these oogonia within their primordial follicles into primary oocytes. These primary oocytes, surrounded by their epithelial or granulosa cells, undergo the first stages of meiosis. This first phase of folliculogenesis is not dependent upon gonadotrophic hormones (reproductive hormones produced by the anterior pituitary; see Section 3.2.2); hence, why some oogonia can start the first phase of folliculogenesis before the onset of puberty. These partially developed primary oocytes, within what are now termed primary follicles, then await puberty, when hormones secreted from the anterior pituitary drive their further development.

From puberty onwards, the second phase of folliculogenesis can occur where primary oocytes develop within their primary follicles and complete the final stages of meiosis in waves, designed to ensure that a regular supply of developed follicles

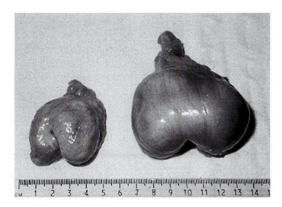

Fig. 1.18. The ovaries of the mare. Note the difference in size between the ovary on the left, which is inactive, and the one on the right, which is active. The concave surface (position of the ovulation fossa) and the convex surface (the hilus, entry point for blood and nerve supply) of the ovary are clearly seen.

is available for ovulation every 21 days during the breeding season. This second phase of folliculogenesis is gonadotrophic hormone-dependent and so is linked to the mare's 21-day cycle. However, not all primary follicles go on to ovulate, many are wasted along the way, degenerating and becoming atretic; in monovular species, such as the mare, normally just one reaches the stage ready for ovulation (Davies Morel and O'Sullivan, 2001).

The length of the second phase of folliculogenesis is unclear in the mare but may be as long as 21 days. Whatever, the waves of second-phase folliculogenesis occur continually and, if they coincide with elevating hormone levels towards the end of the mare's 21-day cycle, will result in a pre-ovulatory or graafian follicle(s) (Ginther *et al.*, 2001, 2003). As the primary follicle is driven by these hormones it develops, its surrounding epithelial cells differentiate into follicular epithelial cells which secrete follicular fluid, filling the cavity surrounding the oocyte. The follicle grows in size as fluid accumulation increases. The primary oocyte itself now also increases in size and develops a thick, outer jelly-like layer, the zona pellucida; it is now termed a secondary oocyte and has a haploid number of chromosomes (32). The secondary oocyte becomes associated with one inner edge of the follicle and lies on a mound of follicular cells

called cumulus oophorus. The epithelial cells surrounding the follicle become organized into two cell populations: the theca membrane, the inner layer of which is vascularized, whereas the outer layer is not; and immediately inside this, the granulosa layer. The follicles continue to develop and are termed graafian follicles (Fig. 1.19).

During this second phase of folliculogenesis the follicle develops hormone receptors, initially follicle-stimulating hormone (FSH) receptors and then luteinizing hormone (LH) receptors. These receptors allow it to develop in synchrony with the oestrous cycle (Chapter 3, this volume). There are three stages in the final development of graafian follicles: recruitment (recruitment of follicles from the pool of available primary follicles); selection or emergence (the selection or emergence of a few as potential pre-ovulatory follicles that undergo further development); and dominance (identification of one, possibly two, follicles that will go on to ovulate and that now start to suppress the development of other follicles). The identification of a dominant follicle is also sometimes termed divergence, as from this stage onwards the dominant follicle (normally 3 cm or greater in diameter) maintains its growth rate while the growth rate of other follicles (termed subordinate follicles) slows and they start to regress (Gastal *et al.*, 2004). The suc-

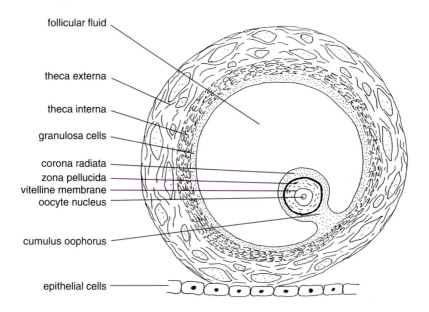

follicular fluid

theca externa

theca interna

granulosa cells

corona radiata
zona pellucida
vitelline membrane
oocyte nucleus

cumulus oophorus

epithelial cells

Fig. 1.19. The equine graafian follicle.

cessful development of follicles through these three stages depends on their ability to react to increasing FSH, LH and oestradiol levels, though the exact mechanisms are unclear (Fay and Douglas, 1987; Roy and Greenwald, 1987; Gastal *et al.*, 1999). In the mare when follicles reach 3 cm in diameter this appears to be a critical stage and follicles that develop beyond this size are very likely to become the dominant follicles destined for ovulation and start to inhibit the growth of subordinate follicles (Ginther *et al.*, 2002). The number that develop to a stage appropriate for ovulation depends on a number of factors including breed. In native ponies it is very rare for more than one follicle to develop to a stage appropriate for ovulation. However, in up to 25% of Thoroughbreds two or more dominant follicles may develop and ovulate, resulting in multiple ovulation (Davies Morel and O'Sullivan, 2001; Section 3.2.5).

In those follicles destined to ovulate, follicular diameter increases and at the same time the follicles appear to move within the stroma of the ovary and orientate themselves to await ovulation through the ovulation fossa. Ovulation of the mature follicle occurs in two stages, which normally (99% of occasions) occur concurrently (Ginther, 1992;

Pierson, 1993; Hafez and Hafez, 2000). The two stages are follicular collapse and ova release. The whole process may take from a matter of seconds up to a few hours, with ova release occurring at the later stages of ovulation (J. Newcombe, Wales, 2001, personal communication). The ova and follicular fluid are released through the ovulation fossa to be caught by the infundibulum and passed down the Fallopian tube for potential fertilization. Ovulation of follicles of diameter less than or greater than 3 cm does occur but this is the exception (Sirosis *et al.*, 1989; Ginther and Bergfeldt, 1993).

After the release of the ova and follicular fluid, the old follicle collapses and the theca membrane and remaining follicular epithelial cells become folded into the old follicular cavity. Bleeding from the theca interna occurs into the centre of this cavity, forming a clot. This clot, the theca cells and any remaining follicular epithelial cells make up the corpus luteum (CL or yellow body). Blood capillaries and fibroblasts then invade the CL. It is initially a reddish-purple colour. As the CL ages it becomes browner in colour and, if the mare is not pregnant, regresses and shrinks to yellow then white (corpora albucans) as it becomes non-functional. The luteal

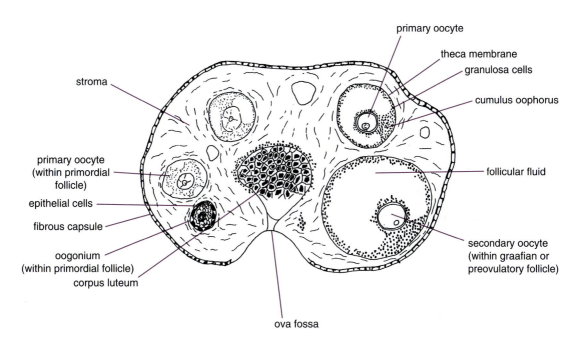

Fig. 1.20. Diagrammatic representation of follicular development and ovulation within the ovary.

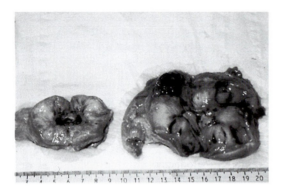

Fig. 1.21. A cross section taken through the two ovaries pictured in Fig. 1.18. Note in the active ovary the CL (dark mass at top left) and the large follicle (hollow or space at the top right).

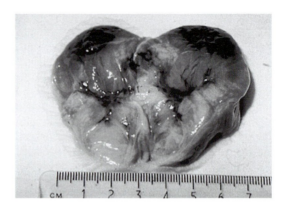

Fig. 1.23. A cross section taken through an active ovary illustrating a large CL at the top of the ovary.

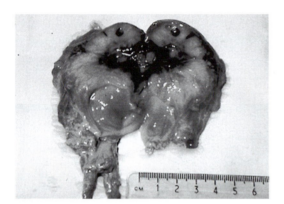

Fig. 1.22. A cross section taken through an active ovary illustrating the presence of a large pre-ovulatory follicle (3 cm in diameter) at the bottom of the ovary.

tissue is then gradually replaced with scar tissue (Fig. 1.20; Vogelsang *et al.*, 1987; Del Campo *et al.*, 1990; Kainer, 1993; Pierson, 1993; Ginther, 1995; Sertich, 1998; Hafez and Hafez, 2000). Figures 1.20–1.23 show sections through a mare's ovary, illustrating the presence of developing follicles and CL.

1.10. Conclusion

It can be concluded that the reproductive tract of the mare is a remarkable system designed not only to maximize the chance of fertilization and subsequent maintenance of the resulting conceptus in a sterile environment, but also to expel that conceptus successfully at term.

2 The Reproductive Anatomy of the Stallion

2.1. Introduction

This chapter details the anatomy and function of the stallion's reproductive system. Further, more detailed accounts may be found in other texts such as Ashdown and Done (1987), Sack (1991), Ginther (1995), Dyce *et al.* (1996), Samper (1997), Bone (1998), Turner (1998), Davies Morel (1999), Chenier (2000) and Hafez and Hafez (2000). Figure 2.1 illustrates the main structures of the stallion's reproductive tract. Figure 2.2 shows the reproductive system of the stallion after slaughter and Fig. 2.3 provides a diagrammatic representation of this plate. The reproductive organs of the stallion will be discussed in turn in the following account.

2.2. The Penis

The penis of the stallion may be divided into the glans penis, the body or shaft and the roots. In the resting position it lies retracted and hence protected within its sheath, or prepuce, out of sight; it is held in this position by muscles, including the retractor muscle. The prepuce is a double-folded covering to the penis, that folds back on itself to give a twofold protection.

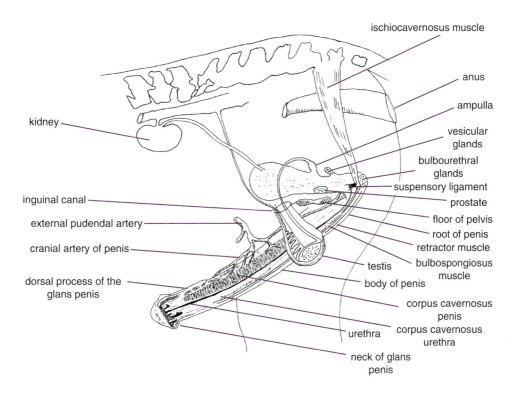

Fig. 2.1. Diagrammatic representation of the reproductive system of the stallion.

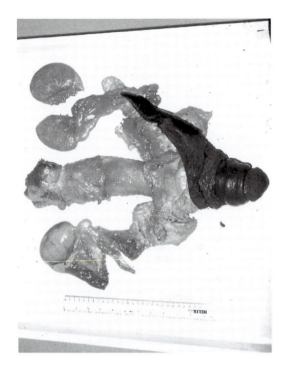

Fig. 2.2. The stallion's reproductive tract after slaughter and dissection; the accessory glands are not included (see also Fig. 2.3).

Within the inner fold lies the end of the penis, the glans penis (or rose), giving this sensitive area additional protection. Protruding, by 5 mm, from the centre of the glans penis lies the exit of the urethra. Around this protrusion lies the urethral fossa and below it a dorsal diverticulum, both of which are often filled with smegma, a red-brown secretion of the prepubital glands lining the prepuce, plus epithelial debris. These areas provide an ideal environment for bacteria, often harbouring venereal disease (VD) bacteria such as *Klebsiella pneumoniae*, *Taylorella equigenitalis* and *Pseudomonas aeroginosa*.

The penis of the stallion is attached by its two roots to the lower part of the pelvic bone by the ischiocarvenosus muscles. Here at its origin, the penis is also held in position by the suspensory ligament attachment to the pelvis. The urethra, running from the bladder, connects with the vas deferens and runs between the two roots before entering the body of the penis. The body of the stallion's penis contains a large percentage of erectile, as opposed to fibrous, tissue and as such it is termed haemodynamic (reacts to increasing blood pressure). Figure 2.4 illustrates a cross-sectional view through the stallion's penis.

Figure 2.4A shows that the main body of the penis is divided into two sections: the lower corpus

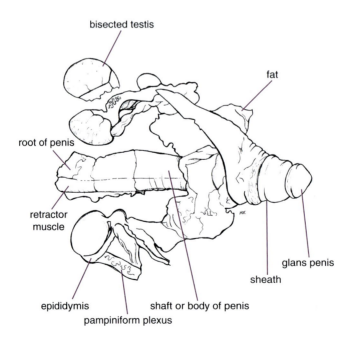

Fig. 2.3. The stallion's reproductive tract; a diagrammatic representation of Fig. 2.2.

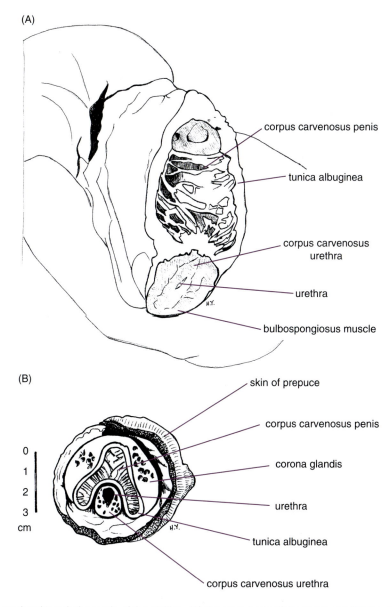

(A)

corpus carvenosus penis

tunica albuginea

corpus carvenosus urethra

urethra

bulbospongiosus muscle

(B)

skin of prepuce

corpus carvenosus penis

corona glandis

urethra

tunica albuginea

corpus carvenosus urethra

0
1
2
3
cm

Fig. 2.4. Cross section through the penis of the stallion: (A) the main body of the penis and (B) the glans penis.

carvenosus urethra and the upper corpus carvenosus penis. Through the corpus carvenosus urethra runs the urethra, surrounded by some trabeculae (sheets of connective tissue) enclosing small areas of erectile tissue, all enclosed within the bulbospongiosus muscle. The corpus carvenosus penis, the largest part of the penis, contains a dense network of trabeculae, associated muscle tissue and scattered cavities, and makes up the major erectile tissue of

the penis. The corpus carvenosus penis is contained within the tunica albuginea, a fibro-elastic capsule or sheet, which maintains the integrity of the penis but still allows the doubling in size seen at erection. Finally, running along the bottom of the penis is a retractor muscle, contraction of which returns the penis to within the prepuce. The major erectile tissue of the glans penis is a continuation of the corpus carvenosus urethra, the corona glandis (Fig. 2.4B).

This large amount of erectile tissue, which is not confined by the fibro-elastic capsule of the corpus carvenosus penis, allows the greater expansion (up to three times) of this area at ejaculation.

2.2.1. Erection, ejaculation and emission

Deposition of semen within the mare involves three stages: erection, emission and ejaculation. The control of these stages is discussed in detail in Section 4.2.1.6. Erection is the first reaction to a sexual stimulus and causes relaxation of the penile muscles and retractor muscle, allowing the penis to extrude from its sheath. This is followed by blood engorgement of the erectile tissue of the penis resulting in an initial turgid pressure. In order for intromission (entry into the mare) to be successful blood pressure has to further increase to intromission pressure, this is achieved by partial occlusion of the venous return from the penis, along with an increase in heart rate and general increase in circulating blood pressure. It is essential that intromission pressure is achieved before the stallion enters the mare; if not, it may cause permanent damage to the stallion or at least reduce his enthusiasm for future covering.

Emission and ejaculation are the culmination of erection. Both are the result of contraction of the muscle walls of the epididymis, vas deferens, accessory glands, ischiocarvenosus muscle and penile muscles causing sperm to be forced from the epididymis up the vas deferens to be mixed with seminal plasma (emission) and from there passed out into the mare (ejaculation). Ejaculation is in the form of 6–9 jets of semen, the initial three jets containing the majority of the biochemical components, forming the sperm-rich part of the ejaculate. In addition to the sperm-rich fraction, pre-sperm and post-sperm fractions are evident, both of which are smaller in volume and do not contain viable sperm (Kosiniak, 1975).

At full erection, the penis doubles in size to 80–90 cm in length and 10 cm in width. At ejaculation, the glans penis triples in size, forcing open the cervix to allow sperm deposition directly into the uterus; it may also have a role in preventing initial leakage of semen from the mare. The glans penis has to return to near normal size before the stallion is able to leave the mare. Failure to allow time for the glans penis to return to normal (detumescence) can cause damage to the mare and/or stallion. Problems may be encountered in overzealous stallions that demonstrate enlargement of the glans penis prior to entry into the mare; in such cases, intromission is not safe until the glans penis has returned to its normal size. Such problems are especially evident in stallions of high libido and young stallions, particularly if they have only a limited workload (Ginther, 1995; Turner, 1998; Davies Morel, 1999).

2.3. The Accessory Glands

The accessory glands are a series of four glands; ampulla, vesicular, prostate and bulbourethral glands (some authors consider there to be three glands, excluding the ampulla), situated between the end of the vas deferens and the roots of the penis. Collectively these glands are responsible for the secretion of seminal plasma.

2.3.1. Seminal plasma

Seminal plasma is the major fluid fraction of semen. Seminal plasma provides the substrate for conveying the sperm to the mare, and ensuring final maturation. Some of its major functions are the provision of energy and protection of the sperm from changes in osmotic pressure, pH and from oxidization. It also contains a gel, which forms a partial clot in semen, the function of which is unclear (Davies Morel, 1999).

Males of most species have this series of accessory glands, the relative size of the glands reflecting the relative importance of their secretions within the seminal plasma.

2.3.2. The bulbourethral glands

The bulbourethral glands, or Cowpers glands, are the accessory glands situated nearest to the roots of the penis. They are paired and oval in structure, approximately 2 × 3 cm and lying on either side of the urethra. Their secretions are clear, thin and watery and form part of the main sperm-rich fraction; they may also contribute to the pre-sperm fraction. As part of the pre-sperm fraction, they aid in clearing urine and bacteria collected within the urethra prior to ejaculation. Their secretion may also act as a lubricant, easing the passage of sperm (Mann, 1975; Weber and Woods, 1993).

2.3.3. The prostate

The prostate gland is a bilobed structure with a single exit to the urethra, situated between the bulbourethral glands and the ampulla. Prostate secretions in the stallion are alkaline and high in proteins, citric acid and zinc. The significance of these is unclear. Secretions of the prostate gland make a significant contribution to the pre-sperm fraction (Little and Woods, 1987).

2.3.4. The vesicular glands

The vesicular glands, or seminal vesicles, are again paired in structure and lie on either side of the bladder, and are about 16–20 cm in length. They are lobed and can be compared with large walnuts in external appearance. They secrete a major amount of seminal plasma, with a high concentration of potassium, citric acid and gel. Their secretions form part of both the sperm-rich and gel-like post-sperm fractions. Their function, and therefore the volume of secretion, is dependent upon circulating testosterone concentrations. As such, their contribution to seminal plasma significantly declines during the non-breeding season (Thompson *et al.*, 1980; Weber *et al.*, 1990).

2.3.5. The ampulla

The ampulla glands are paired outfoldings of the vas deferens where it meets the urethra. The outfolding nature of the ampulla, as opposed to a discrete structure with connecting duct, is the reason why these glands are excluded as accessory glands by some authors. However, they make a considerable contribution to both the pre-sperm and sperm-rich fractions. Their secretions are high in ergothionine, an antioxidizing agent, which acts to 'mop up' by-products of sperm metabolism.

2.3.6. The vas deferens

The vas deferens connects the epididymis of the testis to the urethra before passing the accessory glands and on into the penis. It has a diameter of 0.5–1 cm with a thick muscular wall, made up of three layers of muscle: inner oblique, middle circular and outer longitudinal (Fig. 2.5). These muscle layers actively propel the sperm plus surrounding fluid from the testis to the penis. The lumen of the duct is small and folded, especially near the epididymis, maximizing the surface area and so aiding sperm storage and the reabsorption of testicular fluids.

The vas deferens, the testicular nerve supply, arterial and venous blood vessels and the cremaster muscles pass out of the body cavity through the inguinal canal (Fig. 2.6). The cremaster muscle, which is divided into internal and external sections, is responsible for drawing the testis up towards the body in response to cold, fear, etc. (see Section 2.5).

2.4. The Epididymis

The epididymis in the stallion lies over the top of the testis (Figs 2.3 and 2.7). It consists of long convoluted tubules and is subdivided into three sections: the head (caput), the body (corpus) and the tail (cauda). The head of the epididymis is connected by several ducts to the rete testis within the body of the testis; as it continues towards the tail

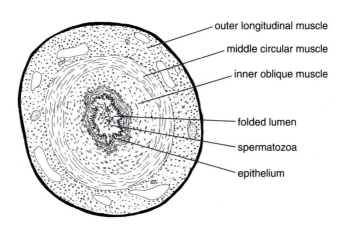

Fig. 2.5. Cross section through the vas deferens of the stallion.

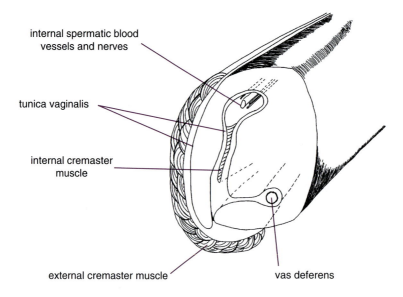

internal spermatic blood
vessels and nerves

tunica vaginalis

internal cremaster
muscle

external cremaster muscle

vas deferens

Fig. 2.6. Cross section through the inguinal canal connection between the testes and the stallion's body.

end these ducts merge and form a single duct, the vas deferens. The lining of these tubules is highly folded, very similar to that of the epididymal end of the vas deferens. These folds have additional microvilli, further increasing the surface area and so facilitating the reabsorption of testicular secretions in order to concentrate the sperm and increase its storage capacity (Thompson, 1992).

It is essential that all sperm spend a period of time, at least 48 h, within the epididymis in order to undergo maturation. Such maturation is essential so that released sperm are capable of movement via the beating of their tails and of further development (capacitation) within the female tract, enabling them to fertilize the awaiting ova. If they are not passed up to the vas deferens as a result of ejaculation, they degenerate and are reabsorbed over time, allowing a continual supply of fresh sperm to be available (Thompson *et al.*, 1979). The tail end of the vas deferens, therefore, acts as a storage site for sperm and is also a minor contributor to seminal plasma by secreting glycerylphosphorylcholine (GPC; Samper, 1995a).

2.5. The Testis

The testes are gametogenic (the site of sperm production) and endocrine in function. Figures 2.7 and 2.8 illustrate the structure of the testis.

The testes hang outside the body of the stallion in order to maintain a temperature of approximately 3°C below that of body temperature (i.e. 35–36°C rather than 39°C). Sperm production is maximized at this lower temperature. Increases in testicular temperature due to disease or inflammation of the scrotum, testis or epididymis result in a significant decrease in spermatogenesis. This is transitory, as the duration of testicular dysfunction is related to the duration of temperature elevation. Testicular temperature is normally controlled by means of the cremaster muscles, an abundance of scrotal sweat glands and the arteriovenous countercurrent heat exchange mechanism provided by the pampiniform plexus (Roberts, 1986b; Friedman *et al.*, 1991). Contraction of the cremaster muscle draws the testes up closer to the body, so increasing testicular temperature, and relaxation allows them to drop lower and so cool down. The pampiniform plexus (Figs 2.3 and 2.7) marks the dense capillary network where the testicular arterial and venous supplies come into close contact. In so doing, warm blood entering the testes via the artery loses heat to the cooler venous return. Such an arrangement ensures that the testicular artery cools down prior to entry into the testes and the testicular vein warms up prior to its re-entry into the main body.

The testes lie within a skin covering, termed the scrotum, under which lies the tunica vaginalis,

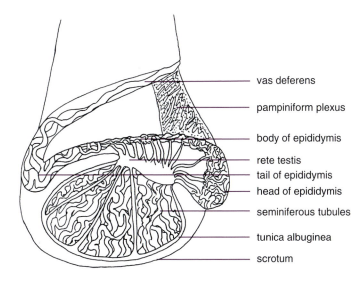

Fig. 2.7. Diagrammatic illustration of a vertical cross section through the testes of the stallion.

(A)

(B)

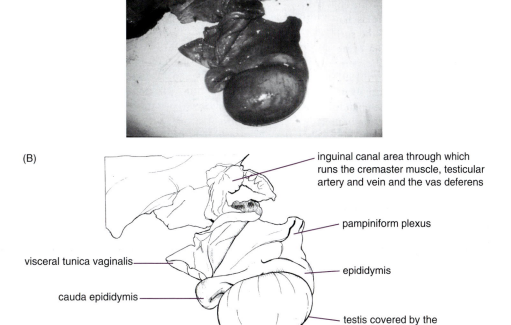

Fig. 2.8. The dissected stallion testis illustrating testicular tissue, the epididymis, vas deferens and pampiniform plexus.

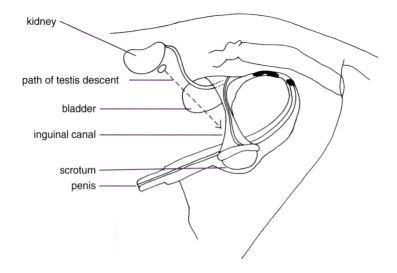

kidney

path of testis descent

bladder

inguinal canal

scrotum

penis

Fig. 2.9. The normal passage of descent of the testes in the stallion.

which is continuous with the peritoneal lining of the body cavity up through the inguinal canal. In the fetus, the testes descend from a position near the kidney, through the inguinal canal and into the scrotum at, or soon after, birth (Fig. 2.9). The failure of one or both testes to descend fully results in a condition termed cryptorchidism, with the stallion often being referred to as a rig (Searle *et al.*, 1999). In such stallions the retention of the testes within the body cavity results in elevated testicular temperature, which causes a reduction in sperm production, but has much less of an adverse effect on endocrine function. Such stallions may, therefore, still be fertile, though sperm count is normally much reduced, and may exhibit near normal stallion-like behaviour due to the continuing endocrine function of the retained testes. The condition may be further defined as unilateral, bilateral, inguinal or abdominal, depending on whether one or both testes have failed to descend and how far descent has progressed (see Section 19.3.2.4.1; Figs 19.7 and 19.8).

The testes of the stallion normally lie with their long axis horizontally, unless drawn up towards the body, when they may turn slightly. The long axis is normally 6–12 cm with the height and width being 4–7 cm and 5–6 cm, respectively. Testes may weigh 300–350 g per pair. Their size increases allometrically with (at the same rate as) general body growth until final body size has been reached, at approximately 5 years of age.

Under the tunica vaginalis, a fibrous capsule, the tunica albuginea, surrounds each separate testis. Sheets of this fibrous tissue invade the body of the testis and divide it up into lobes. Each lobe is a mass of convoluted seminiferous tubules with intertubular areas (Figs 2.7 and 2.10). Each area is largely responsible for one of two functions: gametogenic (seminiferous tubules and Sertoli cells) or endocrine (intertubular tissue and Leydig cells). The Sertoli cells lining the seminiferous tubules act as nurse cells, nourishing and aiding the developing spermatozoa in the lumen of the tubules. In addition, they are also phagocytic, digesting degenerating germinal cells and residual bodies; they secrete luminal fluid and proteins, and also form a blood-testis barrier, providing protection to the sperm from immunological rejection and provide cell-to-cell communication. The number of Sertoli cells varies with season, being significantly greater during the breeding season, when they are actively involved in sperm production. As such, Sertoli cells have a dominant control over sperm production (Johnson and Tatum, 1989). This increase in Sertoli cell number is accompanied by a corresponding increase in seminiferous tubule length (Johnson and Nguyen, 1986).

The Leydig cells, found in the intertubular spaces around the seminiferous tubules, secrete hormones responsible for sperm production and the development of general male bodily characteristics and

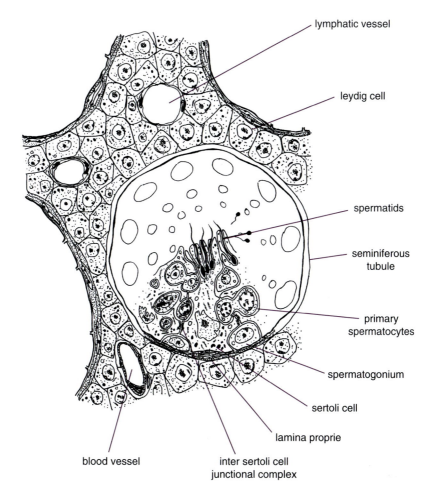

lymphatic vessel

leydig cell

spermatids

seminiferous tubule

primary spermatocytes

spermatogonium

sertoli cell

lamina proprie

inter sertoli cell junctional complex

blood vessel

Fig. 2.10. A cross-sectional view through a seminiferous tubule within the stallion's testis, illustrating the gradual meiotic division of spermatogonia to spermatozoa.

behaviour (see Section 4.2.1; Setchell, 1991; Amann, 1993a,b).

2.6. Sperm

Structurally sperm consist of three areas: the head, the mid-piece and the tail (Fig. 2.11), with three distinct functions.

The head is mainly made up of nuclear material, containing the haploid number of chromosomes (half the normal number, 32, to allow fusion with the ova to give the normal diploid complement of 64). The head of the sperm has a double membrane – the outer cell membrane and the inner nuclear membrane – except in the acrosome region at the top of

the head, where there is an additional acrosome membrane. The importance of this membrane will become apparent when fertilization is considered, as it is responsible for the breakdown of the cell membrane and the nuclear membrane at fertilization, allowing the fusion of the male and female nuclei. The mid-piece of the sperm contains a high proportion of mitochondria, organelles within the cell that produce energy. The mid-piece is, therefore, often termed the power plant of the sperm, providing the energy for metabolism and to drive the tail. The tail is made up of a series of muscle fibrils, equivalent to those found in the major muscle blocks of the body. Using the energy provided by the mid-piece, the tail is whipped from side to side, driving the sperm

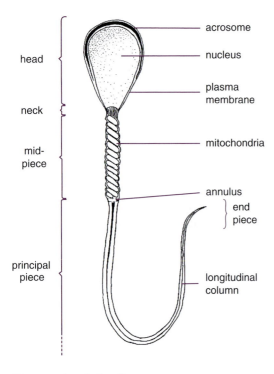

head {

acrosome

nucleus

plasma
membrane

neck {

mid-
piece {

mitochondria

annulus

end
piece

principal
piece {

longitudinal
column

Fig. 2.11. A typical stallion sperm.

movement in a wave-like motion (Amann and Graham, 1993; Davies Morel, 1999).

2.6.1. Spermatogenesis

As discussed, sperm are produced within the seminiferous tubules and are supported, or nursed, by Sertoli cells. They start as underdeveloped germinal cells or spermatogonia attached to the wall of the seminiferous tubules (Fig. 2.10), and then by the process of spermatogenesis progressively develop into mature sperm (Figs 2.11 and 2.12; Amann, 1981b; Amann and Graham, 1993). Spermatogenesis can be divided into spermatocytogenesis (19.4 days), meiosis (19.4 days) and spermiogenesis (18.6 days). The total time for spermatogenesis in the stallion is 57 days.

2.6.1.1. Spermatocytogenesis

Spermatocytogenesis is the first stage of spermatogenesis and in the stallion takes 19.4 days (Fig. 2.12). Spermatocytogenesis is the development of spermatogonia by spermatozoal division from the underdeveloped germinal cells or gonocytes in the

base of the seminiferous tubules. The exact number of spermatozoal divisions for the horse is unclear but the number ranges, in other mammals, from 1 to 14. At the end of the spermatozoal divisions A_1 spermatogonia are produced, which then enter the spermatocytogenic phase and multiply by mitosis. In the horse five different types of spermatogonia are evident through the spermatocytogenic phase: A_1, A_2, A_3, B_1 and B_2 (Johnson, 1991a).

The division of the A_1 spermatogonia has two main functions: first, to produce more stem cell spermatogonia (uncommitted A_1 spermatogonia) by mitosis which continue the supply of spermatogonia for future spermatozoa production; and second, to produce committed A_1 spermatogonia which go on to produce and multiply into A_2, A_3, B_1 and B_2 spermatogonia, and then primary spermatocytes and eventually spermatozoa (Johnson, 1991a; Johnson et al., 1997).

Those committed A_1 spermatogonia destined to produce primary spermatocytes once committed to this line of development divide by mitosis in four stages to give initially a pair ($A_{1.2}$), then four ($A_{1.4}$), then eight ($A_{1.8}$) and finally potential 16 A_2 spermatogonia; throughout this division the groups of spermatogonia originating from a single A_1 stay together connected by intercellular bridges (Johnson, 1991a). These A_2 spermatogonia then, again by mitosis, divide to give two A_3 then again to give two B_1 and finally to give two B_2 differentiated spermatogonia. Finally the B_2 spermatogonia divide to form two primary spermatocytes which enter the first division of meiosis. In theory, one A_1 spermatogonia can give rise to 128 B_2 spermatogonia ready to become primary spermatocytes and enter the next phase of spermatogenesis; however, this is not the case in practice as many spermatogonia degenerate as they progress through spermatocytogenesis. This rate of degeneration is particularly evident in seasonal breeders, such as the stallion, where it significantly increases in the non-breeding season, resulting in the characteristically lower sperm count (Johnson and Tatum, 1989; Hochereau-de Riviers et al., 1990; Johnson, 1991b). Spermatocytogenesis does not cause any reduction in chromosome number and hence the diploid (64 chromosome) A_1 spermatogonia give rise to diploid primary spermatocytes.

2.6.1.2. Meiosis

Meiosis is the means by which a single diploid cell divides to produce two haploid cells. This process

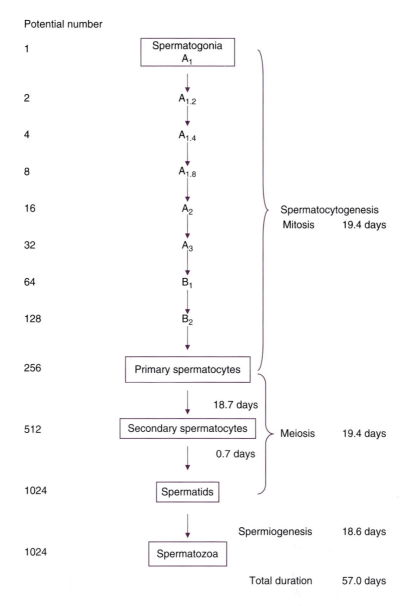

Potential number

1   Spermatogonia A₁

2   A₁.₂

4   A₁.₄

8   A₁.₈

16   A₂   Spermatocytogenesis
                 Mitosis   19.4 days

32   A₃

64   B₁

128 B₂

256 Primary spermatocytes

          18.7 days

512 Secondary spermatocytes   Meiosis   19.4 days

          0.7 days

1024 Spermatids

                 Spermiogenesis   18.6 days

1024 Spermatozoa

                 Total duration   57.0 days

Fig. 2.12. The cell divisions within spermatogenesis in the stallion.

only occurs in the gonads of the male and the female, allowing the production of spermatids and ova that are haploid in nature. This process also allows for the exchange of genetic material between chromosomes in the dividing cell (Johnson *et al.*, 1997). In the stallion meiosis follows spermatocytogenesis and takes 19.4 days; it starts with the primary spermatocytes (diploid), each of which results in the development of four spermatids (hap-

loid). Meiosis may be divided into two stages: the first and second meiotic division. The first meiotic division involves the multiplication and exchange of genetic material and results in two diploid secondary spermatocytes; it is by far the longest phase, taking 18.7 days. The second division results in the halving of the genetic material and the production of two haploid spermatids per single secondary spermatocyte and takes only 0.7 days.

2.6.1.3. Spermiogenesis

Spermiogenesis is the final stage of spermatogenesis, lasting 18.6 days, and is the process by which spermatids are differentiated into spermatozoa (Johnson *et al.*, 1997). Spermiogenesis is divided into four phases: the Golgi, the cap, the acrosome and maturation phases; the division of the phases is largely based on the development of the acrosome region.

2.6.1.4. Spermiation

Although not strictly a stage within spermatogenesis, spermiation is important as the stage at which the fully formed spermatids, now called spermatozoa, are released into the lumen of the seminiferous tubules. From here they pass to the rete testes and on to the epididymis for final maturation.

As the sperm develop through spermatogenesis they migrate, attached to their Sertoli cells, away from the wall of the seminiferous tubules towards the open lumen. Once mature, they are then freed by their Sertoli cells, released into the lumen and pushed along the seminiferous tubules by rhythmic contractions of the tubules and the surrounding secretory fluid. By this stage they have lost a considerable amount of cytoplasm but have developed tails. The tails are not functional until epididymal maturation has occurred. The whole cycle of development takes 57 days and occurs in waves, ensuring a continual supply of mature sperm for ejaculation. The mean daily sperm production of a mature stallion is in the order of $7–8 \times 10^9$ sperm and it may take up to 8–11 days for sperm to pass from the testes to the exterior (Dinger and Noiles, 1986b; Johnson *et al.*, 1997; Davies Morel, 1999). Total sperm production is related to testis size or volume, which can be assessed by using calipers or ultrasonically. The use of calipers in the stallion is not as easy as in other livestock that have more pendulous testis, so ultrasonic assessment is preferred and gives a more accurate result (Love *et al.*, 1991).

2.7. Semen

Semen is the term applied to seminal plasma plus sperm, and in the stallion is a milky white gelatinous fluid. The sperm concentration of semen varies with the fraction examined. As previously mentioned, there are three identifiable fractions: pre-sperm, sperm-rich and the post-sperm fraction. The pre-sperm fraction is the initial fraction and contains no sperm. Its function is to clean and lubricate the urethra of stale urine and bacteria prior to ejaculation. The high concentration of bacteria in this fraction means that its collection should ideally be avoided when collecting semen for artificial insemination.

The sperm-rich fraction is the major deposit by the stallion and commences as soon as the glans penis swells to force entry into the cervix. This fraction is normally 40–80 ml in volume and contains 80–90% of the sperm and the biochemical components of semen.

Table 2.1. The effect of age on the seminal characteristics of stallions. Means in the same row with different subscripts differ significantly. (From Squires *et al.*, 1979b.)

| | Age (years) | | | |
Characteristics	2–3 ($n = 7$)	4–6 ($n = 16$)	9–16 ($n = 21$)	P
Seminal volume (ml)				
Gel	2.1	5.1	13.3	NS
Gel-free	14.2$_a$	26.2$_b$	29.8$_b$	0.05
Total	16.2$_a$	31.4$_{ab}$	43.2	<0.01
Spermatozoa				
Concentration (10^6)	120.4	160.9	161.3	NS
Total (10^9)	1.8$_a$	3.6$_{ab}$	4.5$_b$	<0.05
Motility (%)	55.0	63.1	59.9	NS
pH	7.68$_a$	7.64$_{ab}$	7.59$_b$	<0.05

NS = not significant.

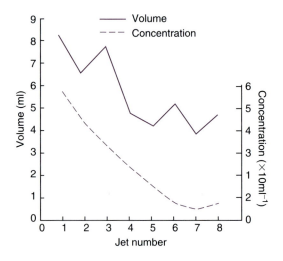

Fig. 2.13. Volume and sperm concentration of successive jets. (From Kosiniak, 1975.)

The third fraction is the post-sperm or gel fraction. Its volume varies enormously from none at all to 80 ml and is dependent on libido: the higher the libido, the greater is the gel fraction. Season also affects the gel fraction volume, this being lower in the non-breeding season. Breed and previous use also have an effect. If a stallion is used more than once per day, the second ejaculate frequently has half the gel fraction of the first. The volume of gel-free fraction is also affected by age, as are sperm numbers, and concentration as illustrated in Table 2.1 (Squires *et al.*, 1979b).

At ejaculation the stallion secretes semen in a series of up to nine jets, the average volume of semen and the sperm concentration decreasing with successive jets (Fig. 2.13; Tischner *et al.*, 1974; Kosiniak, 1975; Mann, 1975).

2.8. Conclusion

It can be concluded that the male reproductive tract is specifically designed for the efficient production, storage and subsequent deposition of sperm within the female tract. It also ensures that the sperm are deposited in a medium (seminal plasma) which is able to provide all the elements for their survival, so maximizing the chances of fertilization.

3 Control of Reproduction in the Mare

3.1. Introduction

The mare is naturally a seasonal breeder, showing sexual activity only during the spring, summer and autumn months. This is termed the breeding season, and the non-breeding season is termed anoestrus. On average, the breeding season lasts from April until November in the northern hemisphere and from October to May in the southern hemisphere (8–12 cycles) though there is significant variation between mares. Breed of the mare also has an effect on breeding season. It is not unknown for well-fed stabled horses to be still showing regular oestrous cycles as late as the January prior to going into a short seasonal anoestrus. During the breeding season, the mare shows a series of spontaneous oestrous cycles at regular 21-day intervals during which she ovulates regardless of being mated; she is, therefore, termed a seasonal polyoestrous spontaneous ovulator. Figure 3.1 illustrates the major milestones in the mare's reproductive life.

The mare's oestrous cycles commence at puberty (10 and 24 months of age). Each cycle lasts on average 21 days (range 20–22 days). Each 21-day cycle is a pattern of physiological and behavioural events under hormonal control and can be divided into two periods according to the mare's behaviour: oestrus, when the mare is sexually receptive, normally 4–5 days, and dioestrus, when she will reject sexual advances, normally 16 days. On either side of truly receptive oestrus two other phases have been suggested: pro-oestrus, as the mare comes into oestrus; and metaoestrus, as the mare goes out. These periods are more evident in mares than in other farm livestock, as oestrus is longer and less distinct (Gordon, 1997). The exact times of these periods vary considerably between individuals and with season and age, tending to be longer in the transition periods into and out of the breeding season and in older mares. In general, any variation in cycle length is due to a variation in the pro-oestrous, oestrous and metaoestrous phases, rather than dioestrus. For example, a cycle length of 20 days is likely to be due to 15 days dioestrus and 5 days pro-oestrus, oestrus and metaoestrus; for a mare showing a 26-day cycle the respective times would be 15 and 11 days (J. Newcombe, Wales, 2001, personal communication). Ovulation normally occurs 24–36 h before the end of oestrus and is denoted by day 0. Days 1–21 then denote the remainder of the cycle until ovulation recurs (Ginther, 1992). The cycle may also be divided according to ovarian physiology and function into luteal (corpus luteum (CL) is dominant and equivalent to dioestrus) and follicular (follicle development is dominant and equivalent to oestrus) phases.

Oestrous cycles continue throughout the mare's lifetime and only cease during the non-breeding season. The mare is an efficient breeder, showing oestrous cycles during lactation, unlike some other seasonal breeders such as the ewe; she is, therefore, capable of being pregnant and lactating at the same time (Fig. 3.1). The mare shows her first oestrus after foaling often within 4–10 days; this oestrus is termed her foal heat. After the foal heat, the mare may start to show her regular 21-day cycles, but in many cases, due to the effects of lactation, it takes a while for the system to settle down to a regular pattern again (Mathews et al., 1967; Allen, 1978; Ginther, 1992; Watson et al., 1994b).

For ease of understanding, the following discussion on the oestrous cycle of the mare has been divided into two sections: the associated physiological changes and the behavioural changes.

3.2. Physiological Changes

The major physiological events associated with reproductive activity in the mare are endocrine changes, which in turn govern and drive the other physiological changes as well as her behavioural activity.

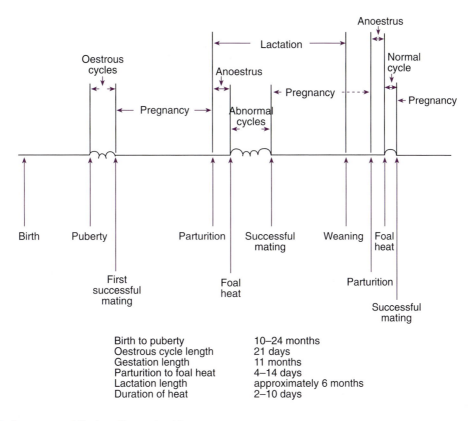

Birth to puberty   10–24 months
Oestrous cycle length   21 days
Gestation length   11 months
Parturition to foal heat 4–14 days
Lactation length   approximately 6 months
Duration of heat   2–10 days

Fig. 3.1. A summary of the breeding cycle of the mare.

3.2.1. Seasonality

The mare is a seasonal breeder and is primarily driven by photoperiod, though nutrition and body condition, environmental temperature, mare age, breed and close association of stallions and mares also play a minor role. Although photoperiod is the main controller of seasonality, it is evident that even in constant photoperiod the mare displays a circannual endogenous rhythm of reproductive activity and non-reproductive activity (Palmer *et al.*, 1982; Nagy *et al.*, 2000). Photoperiod and the other environmental cues act to entrain this natural circannual pattern so that foals are born at the most opportune time – spring. In latitudes where photoperiod does not change significantly throughout the year, other environmental cues, in particular nutrition, play a greater role in entraining the natural circannual rhythm; hence, mares in the tropics tend to enter the breeding season in the rainy season when food is plentiful (Dowsett *et al.*, 1993).

3.2.1.1. Photoperiod

Day length (photoperiod) is perceived by the pineal gland in the base of the brain, which, by means of the hormone melatonin, controls the activity of the hypothalamic–pituitary–ovarian axis, which in turn is responsible for controlling reproduction (see Section 3.2.2). Melatonin is produced nocturnally by the pineal gland, and under the influence of short day lengths dominates the reproductive system, inhibiting the activity of the axis. As day length increases, inhibition of the axis is removed, allowing gonadotrophin-releasing hormone (GnRH) to be produced by the hypothalamus, so driving luteinizing hormone (LH) and to a lesser extent follicle-stimulating hormone (FSH) production by the anterior pituitary (Fitzgerald *et al.*, 1987). Melatonin is secreted by the pineal gland in two phases: photophase (daytime) and scotophase (night time). It therefore demonstrates a circadian secretion, with the highest levels of secretion being evident during the scotophase. The presence or absence of daylight is perceived by the pineal gland via neural messages from the retina of the eye. In the absence of light the conversion of tryptophan to melatonin in the pineal gland is driven (Grubaugh, 1982; Kilmer *et al.*, 1982; Cleaver *et al.*, 1991; Sharp and Clever, 1993). The exact means by which melatonin then controls the hypothalamus is unclear, but seems likely to involve dopamine and/or endogenous opioids including β endorphin (Kilmer *et al.*, 1982; Aurich *et al.*, 1994, 1995; Guerin and Wang, 1994; Besognek *et al.*, 1995; Daels *et al.*, 2000). Dopamine appears to be the main contender and is reported to be positively correlated with melatonin and negatively correlated to LH, and hence evident only in low concentrations in cerebrospinal fluid during the breeding season (Melrose *et al.*, 1990; Nagy *et al.*, 2000).

Prolactin, another major seasonally affected hormone, is suggested by some to be responsible in the horse for non-reproductive seasonal changes such as changes in metabolic rate and increase in the food conversion efficiency during the winter months, a time of food deprivation. Especially evident in the more native breeds, this demonstrates an innate ability of the equine body to anticipate environmental conditions and respond accordingly. As might be expected, therefore, exposing mares in the non-breeding season to 16h light per day causes an increase in prolactin concentrations (Evans *et al.*, 1991). Prolactin was previously thought, therefore, to translate primarily the changes in day length to seasonal changes in non-reproductive physiology, with only a limited effect on reproductive seasonality. However, other work suggests that prolactin may have a role to play in controlling breeding, as prolactin receptors have been identified in large follicles. During the non-breeding season, high dopamine concentrations appear to inhibit prolactin production, which in turn reduces stimulation of the follicle and hence follicle growth and ovulation (Bennett *et al.*, 1998).

The link or mechanism by which melatonin and prolactin secretions interact in the horse is unclear but is better understood in other seasonal breeders (Johnson, 1986a, 1987b; Thompson *et al.*, 1986; Roser *et al.*, 1987; Thompson and Johnson, 1987; Worthy *et al.*, 1987; Evans *et al.*, 1991; Nequin *et al.*, 1993; Besognek *et al.*, 1995).

3.2.1.2. Environmental conditions, temperature, nutrition and stallion proximity

The onset of the breeding season is reported to be closely correlated to environmental temperature, so an early warm spring would be associated with an earlier start to the breeding season (Allen, 1987; Guerin and Wang, 1994).

Nutritional intake also plays a role; a high energy intake shortens the interval to first oestrus of the season (Kubiak *et al.*, 1987). Increasing protein intake also appears to have a similar effect (Van Niekerk and Van Heerden, 1997). This can be seen in practice where oestrous activity of anoestrous mares is advanced, and somewhat synchronized, by turn out on to lush pasture (Carnevele and Ginther, 1997). A nutritional effect may also account for the later onset of breeding in lactating mares and old mares, both of which have a higher nutritional requirement than non-lactating mature mares.

Housing mares in close proximity to stallions has also been associated with advancement in the start of the breeding season, though whether this is due to an auditory, olfactory or visual cue is unclear (Guerin and Wang, 1994).

3.2.1.3. Mare body condition

Not only nutrition but also body condition is reported to affect the timing of the onset of breeding. Work by Gentry *et al.* (2002a,b) demonstrates that mares in high body condition demonstrated more ovarian activity (indicated by follicle size), when monitored in January, than those in a poor condition. Similarly when challenged with GnRH those mares in high body condition reacted immediately with a significant release of LH, whereas those in low body condition hardly reacted at all. It is not surprising, therefore, that mares in better body condition (3, on a scale of 0–5) are reported to ovulate earlier than mares in poor body condition (Henneke *et al.*, 1984; Van Niekerk and Van Heerden, 1997). The perception of body condition by the hypothalamus is suggested to be via circulating concentrations of free fatty acids, glucose and possibly leptin (Fitzgerald *et al.*, 2002). Kubiak *et al.* (1987) even suggest that 15% body fat content is the important figure and that below this oestrus onset is delayed.

3.2.1.4. Mare age

Age also appears to moderate the commencement of breeding; young mares, up to about 5 years of age, are reported to start breeding at a similar time to mature mares but to cease breeding on average 2 months earlier. At the other end of the age range, mares 15 years or older appear to commence breeding later but cease at the same time as younger mature mares (Wesson and Ginther, 1981; Ginther *et al.*, 2004).

3.2.1.5. Mare breed

Finally, mare breed also has an effect on the timing of the breeding season. The more native-type, cold-blooded horses (mainly ponies) tend to have shorter more distinct breeding seasons that commence later in the year than the more hot-blooded horses (Ginther, 1992).

3.2.2. Endocrinological control of the oestrous cycle

The endocrinological control of the oestrous cycle is governed by the hypothalamic–pituitary–gonadal axis (Fig. 3.2), a similar axis to that which controls

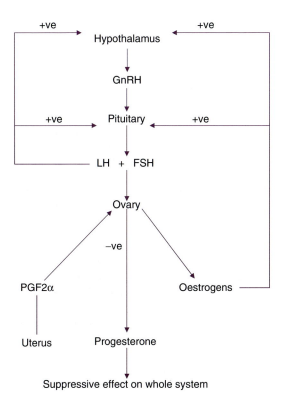

Fig. 3.2. The hypothalamic–pituitary–ovarian axis that governs reproduction in the mare. GnRH, gonadotrophin-releasing hormone; LH, luteinizing hormone; FSH, follicle-stimulating hormone; PGF2α, prostaglandin F2α.

stallion reproduction; the gonads in the case of the mare are the ovaries.

When environmental cues allow, inhibition of the hypothalamus, evident in the non-breeding season, is lifted and GnRH, the first hormone in the cascade of hormones through the hypothalamic–pituitary–ovarian axis, is produced.

3.2.2.1. Gonadotrophin-releasing hormone

GnRH release, in common with other reproductive hormones, is tonic and pulsatile in manner. Tonic secretion relates to the background continual level of secretion, whereas pulsatile secretion is the secretion superimposed upon this as a series of pulses or episodes of higher levels. Both the level of tonic secretion and the amplitude and frequency of episodes can vary throughout the cycle. An increase in episode amplitude, frequency or tonic secretions causes an increase in average hormone concentrations. Eighty per cent of GnRH released is passed directly down a specialized portal system, the hypothalamic–pituitary portal vessels, to have a direct effect on the anterior pituitary (adenohypophysis), with 20% passing back to the central nervous system (CNS) to affect behaviour. The level of GnRH in the mare's circulatory system is, therefore, relatively low, as its passage to the anterior pituitary is directed along these specialized portal vessels. In response to GnRH the anterior pituitary produces the gonadotrophins (hormones that affect growth and development (troph) of the ovaries (gonad)) FSH and LH, the target organs for which are the ovaries (Alexander and Irvine, 1993; Irvine and Alexander, 1993a,b).

3.2.2.2. Follicle-stimulating hormone

FSH, as its name suggests, is responsible for the stimulation of follicle development. Along with LH it is one of the two major gonadotrophins or gonadotrophic hormones that drive the development of the gonads in the mare. It is passed into the general circulatory system and its concentration suggests a biphasic release, with elevated levels at days 9–12 of the cycle and at ovulation. The largest rise starts at day 15 with levels of 4 ng ml⁻¹, increasing to reach concentrations of 9 ng ml⁻¹ during oestrus (Fig. 3.3). This biphasic release adds weight to the theory that follicular development in the mare occurs over a 21-day period, in contrast to the sheep, cow and pig, where it is much shorter (3–6

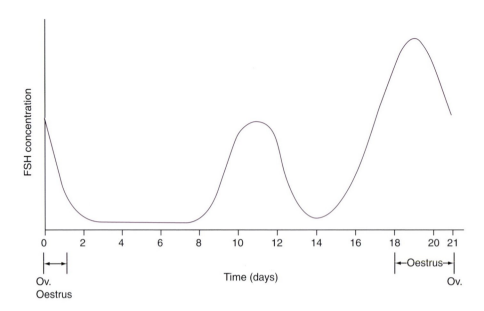

Fig. 3.3. Variations in the relative plasma concentrations of FSH in the non-pregnant mare.

days). It is likely that the peak at ovulation may serve two purposes: (i) to complete final follicle development prior to ovulation; and (ii) to start the development of a new cohort of follicles in readiness for the next ovulation 21 days later (Fay and Douglas, 1987; Alexander and Irvine, 1993; Bergfeldt and Ginther, 1993; Ginther and Bergfeldt, 1993; Irvine and Alexander, 1993a, 1994). Up to ten follicles may be initially affected by the rise in FSH but only a select one or two develop to a stage that can react to the final message to ovulate (Section 1.9.1).

Figure 3.3 along with the subsequent graphs illustrating plasma hormone concentrations are drawn to give an appreciation of the relative, rather than absolute, hormone concentrations. As with GnRH, all the hormones discussed here in relation to reproduction are secreted in a tonic and episodic fashion. The following series of graphs only indicate the average hormone concentrations. Absolute levels reported vary considerably between different scientific reports. Where known, concentrations are discussed within the text.

3.2.2.3. Inhibin and activin

It is suggested that the decline in FSH after its peak, especially near ovulation, is brought about,

at least in part, by the secretion of inhibin by large follicles as they near ovulation (Tanaka *et al.*, 2000). Inhibin acts as a negative feedback on FSH production by modulating the anterior pituitary response to GnRH, in the form of reducing FSH secretion (Watson *et al.*, 2002). Activin has also been isolated in follicular fluid and is reported to have a similar, but positive, feedback effect again specifically on FSH secretion (Piquette *et al.*, 1990; Nett, 1993b). Inhibin and activin appear, therefore, to be very much involved in the development of a dominant pre-ovulatory follicle which then suppresses other follicles, so enhancing its/their own dominance and chances of ovulation (see also Section 1.9.1).

3.2.2.4. Oestrogen

As the follicles develop, they secrete oestrogens which are responsible for the behavioural changes in the mare associated with oestrus and sexual receptivity. The major oestrogen is oestradiol-17β, an ovarian steroidal oestrogen produced from cholesterol by an interrelationship between the theca and the granulosa cells within the developing follicle (Fig. 3.4). The theca cells convert androgen precursors, such as cholesterol, to progesterone, which diffuses across to the neighbouring granu-

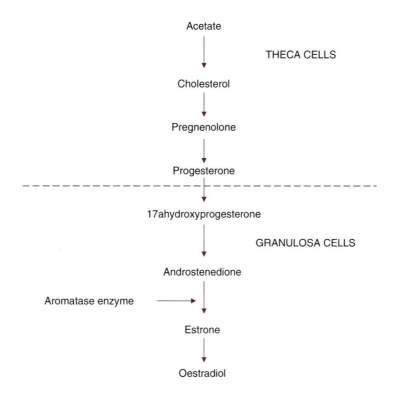

Fig. 3.4. The conversion of androgen precursors such as cholesterol to progesterone and oestrogens in the theca and granulosa cells of the equine follicle.

losa cells where it is converted to oestradiol-17β. This final conversion within the granulosa cells depends upon the enzyme aromatase, whose activity is FSH-dependent. Oestradiol-17β is secreted into the main circulatory system and 24–48 h prior to ovulation reaches a peak of 10–15 pg ml⁻¹, dropping to basal levels immediately post oestrus. This decline in oestrogen secretion is associated with the release of the granulosa cells into the follicular fluid as part of the ovulation process (see Section 1.9.1), leaving the theca cells to produce progesterone but no granulosa cells for its conversion to oestradiol-17β (Tucker *et al.*, 1991).

As FSH levels rise, follicle size increases; as follicle size increases, theca and granulosa cell populations increase; and so oestradiol levels also increase. Hence, FSH and oestradiol levels reach a peak within oestrus, thus ensuring that maximum follicular development, in readiness for ovulation, and oestrus are synchronized (Knudsen and Velle, 1961; Garcia *et al.*, 1979; Nett, 1993a; Weedman *et al.*, 1993; Fig. 3.5).

3.2.2.5. Luteinizing hormone

LH causes ovulation of the dominant follicle(s). LH, like FSH, is secreted by the anterior pituitary. At oestrus, both tonic and episodic concentrations of LH rise to a peak. However, it is the increase in episodic pulse frequency and amplitude that is largely responsible for peak LH concentrations. Receptors for LH on the follicular theca cells increase in number as LH concentrations rise. Increasing LH thus drives additional androgen precursor production, providing more progesterone to diffuse across to the granulosa cells for conversion to oestradiol-17β, which in turn further drives oestrous behaviour. Hence, rising LH levels induce increasing oestradiol-17β secretion, further ensuring the synchronization of ovulation and oestrous behaviour. LH levels begin to rise from their basal levels of less than 1 ng ml⁻¹, with a pulse frequency of 1.4 pulses 24 h⁻¹, several days before the onset of oestrus. They then reportedly reach a peak of 10–16 ng ml⁻¹ just after ovulation (Whitmore *et al.*, 1973). It has been suggested by some that LH not

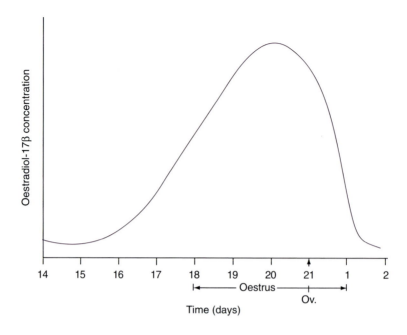

Fig. 3.5. Variations in the relative plasma concentrations of oestradiol in the non-pregnant mare.

only drives final follicular development and induces ovulation but is also involved in the formation and establishment of the CL, possibly explaining why peak concentrations are not reached until just after ovulation. LH declines from peak concentrations to low dioestrous levels within a few days of ovulation (Pattison *et al.*, 1972; Evans and Irvine, 1979; Alexander and Irvine; 1982, 1993; Pantke *et al.*, 1991; Irvine and Alexander, 1993a, 1994; Aurich *et al.*, 1994; Fig. 3.6).

3.2.2.6. Progesterone

Ovulation of a follicle results in the formation of a CL within the collapsed follicle lumen left after the ovum (or ova) and follicular fluid have been released (see Section 1.9.1). The luteal tissue contained within the CL is largely derived from the old theca cells and as there are no longer any granulosa cells, progesterone is released. Progesterone levels, therefore, rise post ovulation, commencing within 24–48 h. Maximum concentrations (10 ng ml^{-1}) are reached 5–6 days post ovulation and are maintained until days 15–16 of the oestrous cycle. If the mare has not conceived, progesterone levels drop dramatically 4–5 days prior to the next ovulation to give basal levels again during oestrus (Fig. 3.7).

Progesterone has an inhibitory effect on the release of gonadotrophins (FSH and LH) in most farm livestock. Oestrus cannot, therefore, begin until progesterone levels have fallen to below 1 ng ml^{-1}. However, the block to gonadotrophin release in the mare is not so complete. Elevated progesterone levels appear to have an inhibitory effect on the release of LH, preventing any rise in LH until progesterone levels decline. However, progesterone does not seem to have such an inhibitory effect on FSH. Indeed, unique to the mare, a second rise of FSH is apparent 10–12 days after ovulation, despite elevated progesterone concentrations. However, if the mare fails to conceive, progesterone levels must still decline in order to allow the mare to return to oestrus and LH levels to rise and cause ovulation on day 21. The decline in progesterone occurs in the absence of a message of pregnancy: in the presence of a conceptus, progesterone secretion is maintained; if no conceptus is detected, the mare's system automatically assumes there is no pregnancy. In response to this the uterus produces prostaglandin F2α (PGF2α).

3.2.2.7. Prostaglandin F2α

PGF2α is responsible for the luteolysis (destruction) of the CL in order to allow oestrus and

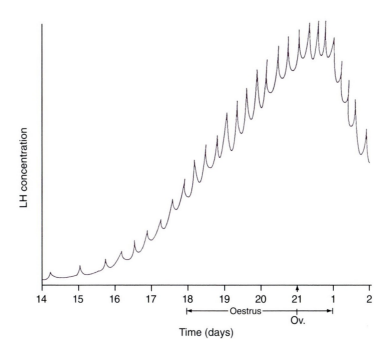

Fig. 3.6. Variations in the relative plasma concentration of LH in the non-pregnant mare.

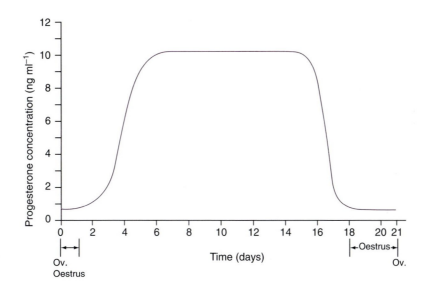

Fig. 3.7. Variations in the relative plasma concentrations of progesterone in the non-pregnant mare.

ovulation to recur. It is difficult to measure in the peripheral circulatory system because of its short half-life and pulsatile manner of release. However, PGF2α has a metabolic breakdown product pros-

taglandin F metabolite (PGFM), which has a longer half-life and so is easier to measure in blood serum and plasma. As a metabolite of PGF2α, it closely mimics changes in PGF2α concentration.

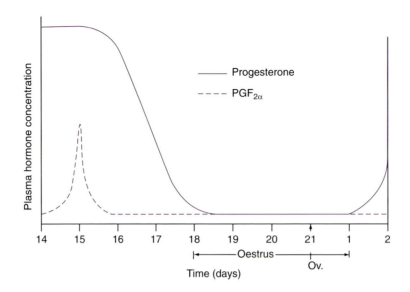

Fig. 3.8. Variations in the relative plasma concentrations of prostaglandin F2α (PGF2α) and progesterone in the non-pregnant mare.

Using levels of PGFM as a guide it can be seen that PGF2α levels rise between days 14 and 17 post ovulation, immediately before progesterone levels start to decline (Fig. 3.8). In mares suffering from retained CL or those that are pregnant, no such rise is detected. PGF2α is secreted by the uterine endometrium and reaches the ovary where it causes luteolysis of the CL, via the main circulatory system, not by a local countercurrent transport system as seen in the ewe and cow (Ginther and First, 1971).

The decline in progesterone levels, in response to PGF2α secretion, removes any inhibition of gonadotrophin release, allowing the hormone changes associated with oestrus and ovulation to commence.

3.2.2.8. Oxytocin

The message of pregnancy or non-pregnancy is also thought to involve the reaction of the uterus to circulating oxytocin. At 14–17 days after ovulation, oxytocin receptors on the uterine endometrium reach a peak, allowing oxytocin, which in the mare appears to be primarily secreted by the endometrium, to bind to the endometrium and so drive production of PGF2α, again produced by the endometrium (Starbuck *et al.*, 1998; Stout *et al.*, 2000). Oxytocin seems to be produced by the CL in other livestock, e.g. the ewe, but there is some doubt as to whether this also occurs in

the mare (Stevenson *et al.*, 1991). If a pregnancy is present in the uterus, the development of oxytocin receptors is inhibited and so oxytocin binding is prevented and PGF2α release significantly reduced (Tetzke *et al.*, 1987; Nett, 1993b; Lamming and Mann, 1995; Hansen *et al.*, 1999).

Figure 3.9 summarizes simply the major fluctuations in hormone concentrations during a single oestrous cycle of a non-pregnant mare.

The following is a summary of the major events that occur in the mare's oestrous cycle.

Day 0	Ovulation
	LH rising
	FSH falling
	Oestradiol falling
	Oestrus
Day 1	LH peak
	Metaoestrus
Day 2	Oestrus ends
	Dioestrus begins
	LH declining
	FSH approaching basal levels
	Oestradiol approaching basal levels
	Progesterone rising
Day 5	Progesterone at maximum
Day 9	FSH rising
Day 11	FSH peak
Day 13	FSH at basal levels
Day 15	PGF2α peak
	Progesterone begins to fall

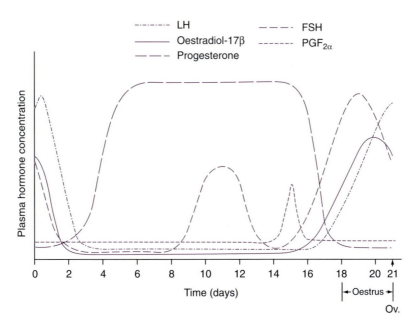

Fig. 3.9. A simplified summary of the major plasma hormone concentration changes during the oestrous cycle of the non-pregnant mare. LH, luteinizing hormone; FSH, follicle-stimulating hormone; PGF2α, prostaglandin F2α.

Day 16	FSH rising
	Progesterone falling
Day 18	FSH rising
	Progesterone basal
	Oestradiol rising
	Pro-oestrus
	LH rising
Day 20	Progesterone basal
	FSH peak
	LH rising
	Oestradiol reaching a peak
	Oestrus
Day 21/0	Ovulation
	LH rising
	FSH falling
	Oestradiol falling
	Oestrus

3.2.3. Physiological changes of the genital tract

In addition to the cyclical changes in hormone concentration, changes in the mare's reproductive tract may also be observed; these are driven by the fluctuations in hormone levels. In the uterus the epithelium proliferates during early dioestrus in preparation for embryo implantation. The epithelial cells are activated and appear tall and columnar during dioestrus, becoming deactivated and cuboidal in nature as the next early oestrus approaches (Ginther, 1992). The epithelial glands also change configuration with the cycle, becoming more active and secretory and appear vacuolated (with an obvious lumen) during dioestrus. Characteristic changes in uterine secretions can be detected by ultrasonic scanning typically producing 'cartwheel'-like images caused by oedema within the endometrial folds at the beginning of oestrus as progesterone levels decline and oestradiol levels increase (Pycock, 2000; J. Newcombe, Wales, 2000, personal communication; Fig. 13.13). Leucocyte concentrations within the uterus also vary, increasing during oestrus and so helping to combat infection at a vulnerable time. This increase is thought to be associated with elevated circulating oestradiol concentrations at this time.

Uterine myometrial (muscle cell) contractility also varies with the cycle, being more active during oestrus. This activity encourages the expulsion of uterine exudates, as well as excess sperm and seminal plasma if the mare is mated, which is particularly important at a time when the tract is most vulnerable to uterine infection. Failure of uterine myometrial contraction leads to post-coital endometritis (Section 19.2.2.5.3.2.2).

In general, the changes within the uterus result in an increase in uterine wall thickness and turgidity as the mare goes from oestrus into dioestrus in preparation for the imminent implantation of an embryo. If pregnancy does not occur, luteolysis results in a reduction of the thickness of the uterine wall and a reversal of these changes as oestrus approaches.

Cervical changes also occur within the cycle. Cervical appearance, as viewed by a vaginascope, can be used as a diagnostic aid in the detection of reproductive activity. During dioestrus, the cervix is tightly closed, forming a tight seal against entry into the uterus. Its appearance is white, firm and dry. During oestrus, the cervix relaxes, opening the cervical seal to allow entry of the penis at mating. During oestrus, the cervix appears moist, red and dilated as the secretions of the uterine epithelial cells and cervical cells increase (Warszawsky *et al.*, 1972). The presence or absence of these secretions within the vagina is also indicative of the stage of the oestrous cycle. For example, it is often very hard to insert a vaginascope into the vagina of a dioestrous mare, due to the thick, sticky nature of the secretions.

3.2.4. Variations in cyclic changes

The mare is notorious for variations or abnormalities in her reproductive cycle. This is in contrast to other farm livestock which have been specifically bred over time for their ability to reproduce rather than perform athletically.

A wide variation in the length of oestrus is evident among mares, the extremes being between 1 and 50 days. In general a variation can be seen with the time of year, longer and less-distinct oestrous periods being evident during the beginning and end of the breeding season. Nutritional intake also causes variation in oestrous length. When nutrition is limited oestrus tends to be longer and less distinct, making it less likely that the mare will conceive during such a non-ideal time. This effect of poor nutrition may be an additional signal to the mare, indirectly indicating seasonal and, therefore, day length changes (Sharp, 1980; Daels and Hughes, 1993).

The length of dioestrus also varies between mares, with the extremes being 10 days to several months. This delay is normally due to one of the three reasons: first, a silent ovulation – ovulation occurred but it was not accompanied by oestrus, giving the impression that the mare has been in dioestrus for a prolonged period of time; second, the existence of a persistent CL – a CL that has not reacted to PGF2α or has not received enough PGF2α to elicit a response; or third, inactive ovaries – usually associated with the transition into or out of the non-seasonal state or true anoestrus. Other variations with the cycle do occur – for example, ovulation in dioestrus. LH is normally released in a low episodic fashion (1–4 ng ml^{-1}) during dioestrus; occasionally these episodes are large enough to cause mid-cycle ovulation, despite high dioestrous progesterone levels (Vandeplassche *et al.*, 1979b). This evidence of dioestrus rises in LH and the previously discussed second FSH mid-cycle peak indicates that, unlike many other species, progesterone does not serve to completely block gonadotrophin release in the mare. The converse, oestrus with no ovulation, has also been reported, normally in mares out of the breeding season (Hughes and Stabenfeldt, 1977; Daels and Hughes, 1993). As mentioned previously foal heat may occur as early as 4 days post partum and mating on the foal heat is often unsuccessful, as fertility rates are normally low. Additionally, the oestrous cycles following this foal heat have an increased chance of being irregular, often showing prolonged oestrus and dioestrus, until steady cyclicity is achieved (Loy, 1980; Blanchard and Varner, 1993a; Camillo *et al.*, 1997).

The causes of many of these variations can be attributed to managerial or environment influences, e.g. nutrition, environmental temperature and day length. Occasionally they are due to genetic faults, lactational effects or embryonic death.

3.2.5. Multiple ovulation

Multiple ovulations – the release of more than one ovum per oestrus – are increasingly common in mares. Release of the ova may occur within oestrus (synchronous) or occur over time, including early dioestrus (asynchronous). All such ovulations may be considered as multiple ovulations as they have the potential to be fertilized and yield viable embryos (Ginther and Bergfelt, 1988). However, the more distant over time the ovulations occur, the less likely the chance of fertilization, and those occurring more than 3 days apart rarely result in multiple conceptuses. The reported incidence of multiple ovulations in mares is very variable at 0.83% to 42.8% (Arthur and Allen, 1972; Warszawsky *et al.*, 1972; Wesson and Ginther, 1981; Ginther *et al.*, 1982; Newcombe, 1995; Davies Morel and O'Sullivan, 2001). The issue of multiple ovulations and multiple pregnancies presents many dilemmas and is of significant economic importance to horse breeders; this issue will be further considered in Section 5.4.3.

3.3. Behavioural Changes

Cyclical, hormonal changes govern the mare's behavioural patterns in association with oestrus and dioestrus, elevated oestradiol concentrations and the absence or presence of progesterone being major factors, stimulating behavioural centres of the brain. As considered previously, GnRH is also known to play a minor role in oestrous behaviour (Irvine and Alexander, 1993b). There are many variations between individuals in the extent and strength of behavioural changes (Munro *et al.*, 1979). Details of the signs of oestrus and their interpretation are given in Chapter 13 (this volume). A summary of the major behavioural changes is given below.

Oestrus initiated by elevated oestradiol and low progesterone concentrations
Signs of oestrus in the presence of a stallion:

- Docility;
- Urination stance;
- Lengthening and eversion of the vulva;
- Exposure of the clitoris (winking);
- Tail raised;
- Urine bright yellow with a characteristic odour;
- Acceptance of the stallion's advances.

Dioestrus evident in the absence of oestradiol and presence of progesterone
Signs of dioestrus in the presence of a stallion:

- Hostility;
- Rejection of stallion's advances.

3.4. Conclusion

The prime aim of all the control mechanisms for the female reproductive system is to synchronize the physiological and behavioural events associated with oestrus, in order to synchronize mating with ovulation and so achieve fertilization, and subsequently synchronize embryo and uterine development.

4 Control of Reproduction in the Stallion

4.1. Introduction

The stallion, like the mare, is a seasonal breeder but tends to show a less-distinct season, and, unlike her, if given enough encouragement, is capable of breeding all year round. However, season does have an effect upon the efficiency of reproduction, semen volume, sperm concentration, total sperm per ejaculate, the number of mounts per ejaculate and reaction time to the mare all being poorer during the non-breeding season (Pickett *et al.*, 1970, 1975a; Pickett and Voss, 1972; Johnson, 1991b). Figures 4.1–4.5 demonstrate the effect of season on reproductive parameters.

Sperm production, unlike ova production, is a continual process and is not governed by cyclical hormonal changes and as with ovulation in the mare commences at puberty; sperm production then continues for the rest of a stallion's lifetime, though there is suggestion by some researchers that semen quality declines after 20 years of age (Johnson and Thompson, 1983; Fukuda *et al.*, 2001; Madill, 2002). The exact timing of puberty is unclear and varies with breed and stallion development. Various researchers have used histological changes within the testis, especially in association with the Leydig cells, to indicate the timing of puberty. Ages of 1–2.2 years have been suggested using such parameters (Cornwall, 1972; Naden *et al.*, 1990; Clay and Clay, 1992). However, other works using testicular weights, daily sperm production, testosterone concentrations and Leydig and Sertoli cell number and volume (Johnson and Neaves, 1981; Thompson and Honey, 1984; Berndston and Jones, 1989) suggest that true puberty may occur nearer 3 years of age. It is generally accepted that stallions of 3 years of age are spermatogenically active and so perfectly capable of fertilizing mares covered; however, they have a limited sperm-producing capacity. By 5 years of age they are capable of producing ade-quate numbers of spermatozoa to cover a full complement of mares (Johnson *et al.*, 1991) and at 5–6 years of age (when they reach mature body weight) will have attained full adult reproductive ability.

As with the mare, the reproductive activity of the stallion can be divided into physiological and behavioural changes and will be detailed in turn.

4.2. Physiological Changes

Hormone patterns are the major physiological events associated with stallion reproductive activity and indeed govern the remaining physiological and behavioural characteristics.

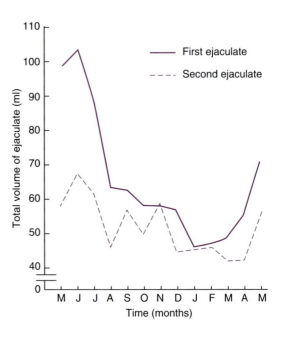

Fig. 4.1. Total semen volume produced throughout the year. (From Pickett and Voss, 1972.)

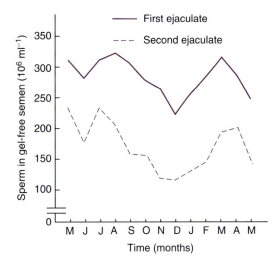

Fig. 4.2. Number of sperm in the gel-free fraction of semen throughout the year. (From Pickett and Voss, 1972.)

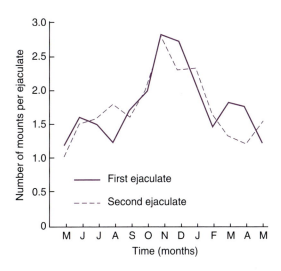

Fig. 4.4. Mean number of mounts required per ejaculate throughout the year. (From Pickett and Voss, 1972.)

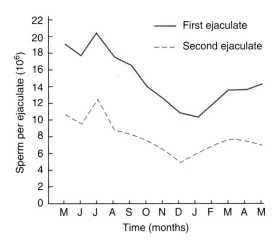

Fig. 4.3. Number of sperm per ejaculate throughout the year. (From Pickett and Voss, 1972.)

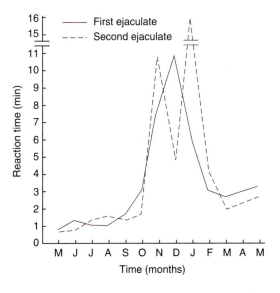

Fig. 4.5. The effect of season on sexual behaviour as measured by reaction time. (From Pickett and Voss, 1972.)

4.2.1. The endocrinological control of stallion reproduction

Control of stallion reproduction is governed by the hypothalamic–pituitary–gonad axis as seen in the mare, except that in the stallion the testes are the gonads (Fig. 4.6; Amann, 1981a).

Environmental stimuli, in the form of day length, temperature and nutrition, have an over-riding effect on the hypothalamic–pituitary–testis axis. As discussed for the mare, season is governed by the secretion of melatonin from the pineal gland in response to day length (see Section 3.2.1; Thompson *et al.*, 1977; Burns *et al.*, 1982;

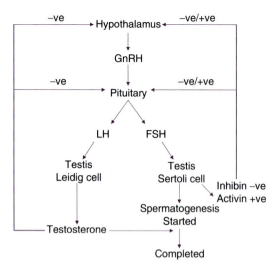

Fig. 4.6. The hypothalamic–pituitary–testis axis that governs reproduction in the stallion.

Clay *et al.*, 1987). Both melatonin and photoperiod can be manipulated, as in the mare, to alter the timing of the breeding season, but overstimulation with artificial long days for a prolonged period of time results in refractoriness to the photo stimulation (Clay *et al.*, 1987; Cox *et al.*, 1988; Argo *et al.*, 1991).

Again, as discussed in detail for the mare, prolactin concentrations are also affected by day length, increasing as day length increases (4–5 μg ml^{-1} in summer, 1.7 μg ml^{-1} in winter). Again it was thought that prolactin is the means by which photoperiod controls non-reproductive activity such as coat growth, weight gain, food conversion efficiency, etc. (Tucker and Wetterman, 1976). Unlike the mare there is as yet no work that indicates a role for prolactin in reproductive seasonal activity in the stallion.

4.2.1.1. Luteinizing hormone and follicle-stimulating hormone

The anterior pituitary is stimulated, via gonadotrophin-releasing hormone (GnRH), to produce follicle-stimulating hormone (FSH) and luteinizing hormone (LH; Irvine, 1984a). The plasma concentrations of LH and FSH in the sexually active stallion are in the order of 3–4 ng ml^{-1} and 7–7.5 ng ml^{-1}, respectively (Cupps, 1991; Seamens *et al.*, 1991; Amann, 1993b; Kainer, 1993).

The testes, which are the target organs for LH and FSH, consist of two major cell types, Leydig and Sertoli cells, plus associated structural tissue (Fig. 2.10). Leydig cells are found within the intertubular spaces or interstitial tissue of the testes and are responsible for the production of testosterone. These cells and, therefore, testosterone secretion are controlled by LH (Amann, 1981a). Sertoli cells are found lining the seminiferous tubules and act as nurse cells for developing spermatids. These cells, and therefore sperm production, are controlled by FSH and testosterone.

4.2.1.2. Testosterone

Testosterone, produced by the Leidig cells, is driven by the pulsatile release of LH resulting in a pulsatile release of testosterone. The pulsatile nature of testosterone release means that analysis of a single blood sample for the hormone can give erroneous results; a hormone profile taken over a period of time and averaged is a much more accurate indication of true testosterone levels (Amann, 1993b).

Testosterone, produced by the Leydig cells, passes via attachment to androgen-binding protein to the neighbouring Sertoli cells where it takes its effect. Sertoli cells are controlled by both FSH and testosterone. FSH is known to start the process of spermatogenesis, developing spermatogonia to secondary spermatocytes. Testosterone then completes their development from secondary spermatocytes to spermatozoa ready for passage to the epididymis for maturation (Davies Morel, 1999).

Additionally, testosterone controls the development of male genitalia, testes descent in the fetus or neonate, pubertal changes and accelerated growth, plus the maintenance and function of the accessory glands. It is also responsible for male libido and sexual behaviour by stimulation of the central nervous system, plus development of personality, stallion behaviour and muscular development. Testosterone also feeds back negatively on pituitary function to reduce the release of LH and FSH, and hence acts as a brake on its own production (Irvine *et al.*, 1986; Flink, 1988). A derivative of testosterone, dihydrotestosterone, also feeds back negatively on the pituitary as well as having limited additional effect on the other testosterone-driven stallion characteristics. This negative feedback ensures that the system does not overrun itself; for every driver there needs to be a brake. Testosterone

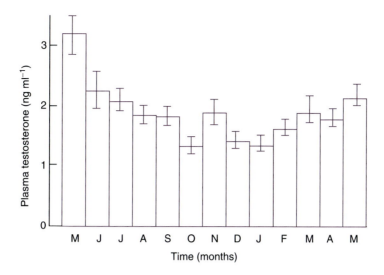

Fig. 4.7. Concentration (mean + standard error) of testosterone in the peripheral plasma of mature stallions over a 13-month period. (From Berndston *et al.*, 1974.)

may also be produced by a limited population of Leydig cells occasionally found in the wall of the vas deferens; this is evident in some geldings that have been successfully gelded but continue to demonstrate stallion-like characteristics. Testosterone is also produced by the adrenal glands in both the stallion and the gelding and it is this testosterone that is responsible for the continued, but reduced, male characteristics of the gelding.

The production of testosterone is not only dependent upon season (Fig. 4.7), but also a diurnal rhythm is suggested. Testosterone concentrations have been reported to be elevated at 06.00 and 18.00 h. It has been postulated that in the wild this ensures that mating activity is greatest at dawn and dusk, times of least risk to stallions and mares from predators (Pickett *et al.*, 1989).

4.2.1.3. Inhibin and activin

Two other hormones are also involved in the control of male reproduction: inhibin and activin. Both are produced by the Sertoli cells in response to total sperm production and have additional feedback effects on hypothalamic and pituitary function, specifically on FSH production. Inhibin acts as a negative and activin as a positive feedback (Roser, 1997). The precise mode of action of inhibin and activin in the horse is as yet largely unclear.

4.2.1.4. Prolactin

The central role that prolactin may play in translating day length into non-reproductive seasonal physiological changes has already been discussed in this chapter and in Chapter 3 (this volume). In addition, based upon work in other farm livestock, prolactin may also have a role in enhancing the effect of LH on the Leydig cells' activity and on the functioning of the accessory glands (Thomson *et al.*, 1996). The importance of prolactin specifically in the horse is as yet unclear.

4.2.1.5. Oestrogens

The stallion's testes are interesting in that they contain a higher concentration of oestrogens and oestrones (150–200 pg ml^{-1}) than the testes of other mammals. The control and significance of such testicular oestrogen is unclear (Raeside, 1969; Seamens *et al.*, 1991; Amann, 1993b). However, the elevated levels of oestrogen in the plasma of entire stallions serve as a useful way of determining if an apparent gelding is in fact a cryptorchid stallion (Sections 2.5 and 19.2.2.4.1).

4.2.1.6. Control of semen deposition

There are three stages to semen deposition: erection, emission and ejaculation (see Section 2.2.1).

Erection is initiated by a sexual stimulatory (erotogenic) signal, which may be visual, site of a mare or facilities associated with mating or semen collection; olfactory, the presence of smells or pheromones associated with covering; or auditory, the sounds of a mare or other sounds associated with mating. This stimulatory signal is transmitted to the behavioural centres of the brain within the hypothalamus where neurons synapse with the parasympathetic and sympathetic efferent neurons which control vasoconstriction and vasodilation in the penis. Sexual excitation causes parasympathetic dominance which, via the splanchnic nerves, causes relaxation of the penis retractor muscle allowing the penis to extrude from its sheath. Additionally, the parasympathetic stimulation overrides the normal sympathetic stimulation and the parasympathetic nerve endings fire releasing nitrous oxide (NO) from their terminals within the penis. NO activates the enzyme guanylate cyclase that is responsible for the conversion of guanylate triphosphate (GTP) to cyclic guanosine monophosphate (cGMP), which causes relaxation of the smooth muscle in the penis. This causes vasodilation and so increases blood flow, in particular to the erectile tissue of the corpus carvenosus penis, which results in engorgement, which in turn causes turgid pressure to be achieved within the penis.

However, in order for intromission (entry into the mare) to be successful the blood pressure within the stallion's penis must be further increased to intromission pressure. This is achieved by further engorgement of the erectile tissue via two means. First, as blood flow increases to the penis, penile blood pressure continues to increase; this compresses the veins exiting from the penis against the ischial arch of the pelvis, restricting venous return of blood from the penis and so further increasing blood pressure. Second, central nervous system stimulation causes contraction of the muscles associated with the penis, in particular the ischiocarvenosus muscle which draws the penis up against the ischial arch, further restricting venous blood flow out of the penis and so further increasing blood pressure. This, along with an increase in heart rate and a general increase in circulatory blood pressure, causes intromission pressure to be reached within the penis in readiness for mating (Beckett et al., 1972; Korenman, 1998).

The second and third stages of semen deposition are emission, the passage of seminal plasma from the accessory glands and sperm from the epididymis to the urethral area of the penis; and ejaculation, the passage of this seminal plasma and sperm (semen) along the penis and into the mare after intromission. The control of these two stages is very similar and they occur as a continuum. Sensory stimulation, such as temperature or pressure, of the glans penis activates sensory afferent nerves which send impulses to the lumbosacral region of the spine. This results in a direct efferent neural return which activates the muscles associated with the penis. This activation causes wave-like, or peristaltic, contraction of the myometrial cells within the wall of the epididymis, vas deferens, accessory glands and penis as well as contraction of the urethralis and bulbospongiosus muscles and further contraction of the ischiocarvenosus muscle. This results in semen deposition into the mare in a series of 6–9 jets (Tischner et al., 1974; Kosiniak, 1975; Weber and Woods, 1993).

Ejaculation follows shortly after entry into the mare and is signalled by the rhythmical flagging of the tail. Such mating behaviour is directly affected by circulating testosterone concentrations. At the beginning and the end of the season reaction time to a mare is longer and the number of mounts per ejaculate is greater, these being direct indications that when testosterone levels are declining sexual enthusiasm or libido is also waning (Figs 4.4 and 4.5; Weber and Woods, 1993).

4.3. Behavioural Changes

Testosterone is the prime driver of male behaviour, especially that associated with reproduction. There is much variation between individuals, but in summary the following generalized behaviour is controlled by testosterone (Pickett et al., 1975a). On sight of a mare the frequency and amplitude of GnRH and hence LH and FSH release increase, driving testosterone production and hence producing stallion behaviour (Irvine and Alexander, 1991). In addition, 20% of GnRH released acts directly on higher brain centres (Merchenthaler et al., 1989; Pozor et al., 1991). Behavioural patterns of a stallion include fixation of his eyes upon the mare, neck arching, stamping or pawing the ground and general elevated stallion stance. He will often show the characteristic flehman behaviour of drawing back his top lip as if tasting the air, accompanied by roaring. If the mare seems receptive, he will approach her from the front, muzzle to muzzle, and in the absence of hostility will work his way over her neck, back, rump towards her perineum and vulva. If the mare still stands with no objection he will turn and approach her from behind and to one side, nearly always the left-hand side. He will then mount her from that side, possibly after a few initial dummy mounts to test her reaction and confirm that she is willing to stand.

4.4. Conclusion

The control of reproduction in the stallion is designed to ensure continual reproductive activity rather than cyclic activity, as in the mare. The only constraint on the stallion's reproductive activity is season, a limitation that ensures that offspring are more likely to be born at a time of year most appropriate to their survival.

5

The Anatomy and Physiology of Pregnancy in the Mare

5.1. Introduction

The anatomy of pregnancy in the mare can be divided into four main sections for ease of consideration: fertilization, early embryo development, placentation and organ growth. More detailed accounts may be found in other texts such as Douglas and Ginther (1975), Betteridge *et al.* (1982), Flood *et al.* (1982), Ginther (1992), Asbury and Le Blanc (1993) and Flood (1993).

5.2. Fertilization

The ovum released by the follicle is directed down the Fallopian tube by the fimbriae lining, where it waits in the ampulla region, that nearest the infundibulum, for the arrival of the sperm. It is unable to pass through the utero-tubular junction until it has been fertilized.

The sperm, having been ejaculated into the top of the cervix/bottom of the uterus, make their way up through the uterus to the utero-tubular junction. They move by means of contractions of the female tract and the driving action of their own tails and many are lost along the way (Campbell and England, 2006). It is thought that they are attracted towards the ovum by chemical attractants produced by the ovum awaiting fertilization. On arrival at the utero-tubular junction, they pass through to the Fallopian tube and, if the timing is correct, meet a waiting ovum in the ampulla. As the sperm pass up through the female tract, they come in contact with uterine secretions, which induce a capacitation response in the acrosome region of the sperm heads, a response that is essential before they are capable of fertilizing an ovum. Capacitation activates the enzymes within the acrosome region in readiness for penetration. Once in the vicinity of the ovum, sperm stick to the outer gelatinous layer, this layer having replaced the corona radiata cells prior to ovulation (Fig. 5.1; Gadella *et al.*, 2001).

Sperm, by means of the whipping action of their tails, force themselves through the outer gelatinous layer to the zona pellucida. They then penetrate the zona pellucida by dissolving a pathway using the enzyme acrosin released from the sperm head of capacitated sperm.

As the sperm head meets the vitelline membrane of the ovum, the two fuse. This fusion initiates the final meiotic division, resulting in three polar bodies and the single ovum nucleus. The nuclei of the sperm and the egg (often termed the pronuclei) unite, their haploid compliment (32) of chromosomes joining together to give the full diploid (64) of the new individual. This newly combined genetic material now dictates all the characteristics of the new individual.

There is some variation in the reported length of time that the equine ovum remains viable; figures varying between 4 and 36h have been reported. Approximately 5 days after fertilization, the ovum is known to actively control, possibly via the localized secretion of prostaglandin E (PGE), its passage through the utero-tubular junction to the uterine horn, overtaking on its way any unfertilized ova from that or previous ovulations. Any ova not fertilized may take several months to degenerate

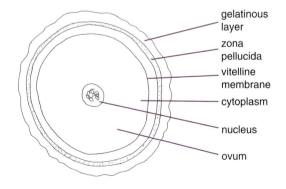

Fig. 5.1. The equine ovum prior to ovulation.

gelatinous layer

zona pellucida

vitelline membrane

cytoplasm

nucleus

ovum

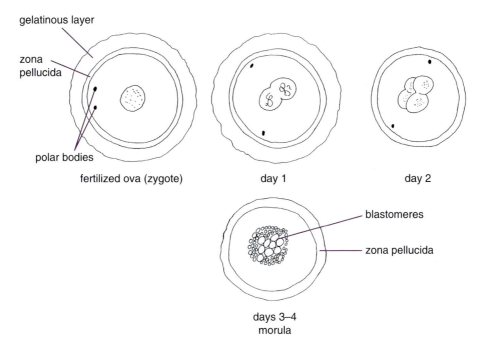

gelatinous layer

zona pellucida

polar bodies

fertilized ova (zygote)   day 1   day 2

blastomeres

zona pellucida

days 3–4
morula

Fig. 5.2. The developmental stages from fertilized ovum to morula in the equine conceptus, illustrating the loss of the gelatinous outer layer by day 2.

(Betteridge and Mitchell, 1975; Onuma and Ohnami, 1975; Flood *et al.*, 1979; Ball and Brinsko, 1992).

In order to ensure the successful fusion of one male pronucleus and one female pronucleus, it is essential to ensure that only one sperm penetrates the vitelline membrane of the ovum. Polyspermy (penetration by more than one sperm) is prevented by an instantaneous block, which occurs as soon as one sperm touches the vitelline membrane. This instantaneous response involves a chemical reaction within the vitelline membrane, forcing a gap between itself and the zona pellucida. No sperm can cross this gap and hence an instantaneous block to polyspermy is ensured.

5.3. Early Embryo Development

Twenty-four hours after mating, the fertilized ovum, now termed a zygote, has divided by mitosis (growth by cell division) into two cells. At this stage, the outer gelatinous layer is lost and the fertilized ovum, still within the zona pellucida, continues to divide into four, 16, 32 cells, etc. At 4 days old, it is a bundle of cells, again still contained within its zona pellucida. It is now termed a morula (Fig. 5.2).

At this stage the total volume and external size of the bundle of cells has not changed from the two-cell zygote stage. The cytoplasm of the original ovum has either been divided up between all the cells in the morula or used for energy. Nevertheless, the amount of genetic material has dramatically increased, giving a full identical complement to all cells of the morula.

As the cells continue to divide, the morula makes its way towards the utero-tubular junction by anti-clockwise rotational swimming. At this stage the morula begins to secrete very low levels of PGE. PGE causes relaxation of smooth muscle and the low localized nature of its secretion by the conceptus results in the relaxation of the utero-tubular junction sphincter, so allowing the conceptus to pass through and on into the uterus at days 5–6 (Allen *et al.*, 2006). At day 5, a thin acellular glycoprotein layer, termed the capsule (see Section 5.3.1), appears in the perivitilline space between the trophectoderm (outer layer of the morula) and the zona pellucida (Oriol, 1994). From day 6, the total size of the embryo starts to increase; this helps to force the thinning of the zona pellucida, which eventually breaks. The embryo then hatches through this break and is left surrounded by just its capsule. At this time, the conceptus starts to derive nutrients for its growth and cell division from the surrounding uterine secretions, as by this stage it has used up all its own reserves. The provision of such additional nutrients allows a further increase in size. The morula is now in its mobility phase, floating freely within the uterus (Section 5.4.1), deriving all its nutritional requirements from secretions of the endometrial glands which produce uterine histotroph (milk), a secretion designed to match exactly the requirements of the developing conceptus.

At day 8, the cells of the morula become differentiated (organized) and three distinct areas can be identified: the embryonic disc (shield or mass), the blastocoel and the trophoblast (Fig. 5.3). The morula is now termed a blastocyst.

These three areas go to form the embryo proper (embryonic disc), the yolk sack (blastocoel) and the placenta (trophoblast). This cell differentiation marks the beginning of the switching on and off of various genes, cells then becoming destined to pursue set lines of development. Prior to this differentiation, all cells in theory were capable, if extracted from the morula, of each developing into a new individual as none of its genes had been switched off. After differentiation, this is no longer possible, as certain cells have been given the message to only pursue set lines of development. The mechanism behind this switching on and off of genes and its trigger are unknown in the horse. It is important to note that, at this differentiation stage, the conceptus is very susceptible to external physical effects, e.g. drugs, other chemicals, disease, radiation, etc. These can disrupt the differentiation process, resulting in deformities, abnormalities and a high risk of abortion or reabsorption.

From this stage on further differentiation takes place. Day 9 marks the differentiation of two germ layers (cell layers): the ectoderm consisting of the outer blastocyst cell layers; and the endoderm consisting of the inner cell lining (Fig. 5.4).

The endoderm grows and develops, working its way around the inside of the trophoblast to give a complete inner layer. The endoderm and ectoderm together form the yolk sack wall and provide the means by which the embryonic disc receives its nourishment from the surrounding uterine secretions. The blastocoel, or fluid-filled centre, is now termed the yolk sack and acts as a temporary nutrient store (Fig. 5.5). This remains the major source of nutrients to the embryo until implantation or fixation occurs. The equine embryo is unique in being free-living within the uterus for up to 18 days prior to final implantation; this period of time is termed the mobility phase (see Section 5.4.1).

At day 14, when the conceptus has reached 1.3 cm in diameter, the mesoderm or third germ-cell layer begins to develop. It becomes progressively evident between the ectoderm and endoderm, in the centre of the yolk sack wall, again working its way down from the embryonic disc to enclose the whole blastocyst. These three germ-cell layers are the cell layers from which all placental and embryonic tissue development originates. In the case of the

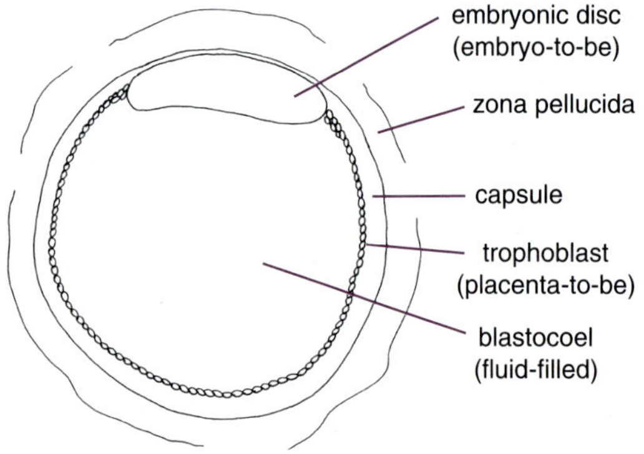

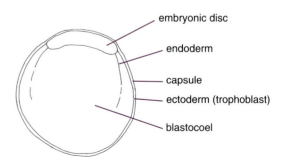

Fig. 5.3. The equine blastocyst at day 8 post fertilization, showing the differentiation of three areas.

Fig. 5.4. The equine conceptus at day 9 post fertilization, illustrating the differentiation of the ectoderm and endoderm layers.

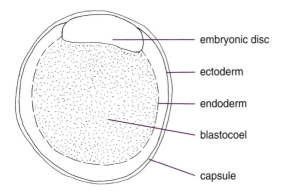

Fig. 5.5. The equine conceptus at day 12 post fertilization, illustrating the yolk sack, which at this stage provides a store for the nutrients required by the developing conceptus.

placenta, the ectoderm forms the outer cell layers nearest the uterine epithelium, the mesoderm forms the blood vessels and nutrient transport system within the placenta and the endoderm forms the inner cell lining that will become the allantoic sack (Fig. 5.6).

At day 16, folds appear in the outer cell layers, and the beginnings of the protective layers, which will surround the embryo, become evident. The ectoderm folds over the top of the embryonic disc, taking the mesoderm with it. The outer layer of these folds is now made up of the ectoderm plus a mesoderm layer and is termed the chorion. By day 18, these two folds fuse, producing a fluid-filled protective space for the embryonic disc; this is the amniotic sack containing the amniotic fluid (Fig. 5.7). At this stage, the first fixation of the embryo

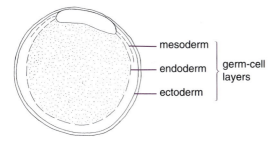

Fig. 5.6. The equine conceptus at day 14 post fertilization, illustrating the developing mesoderm, which forms the blood vessels and nutrient transport system of the conceptus.

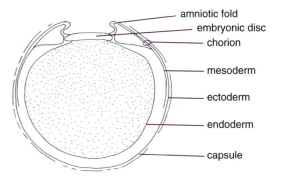

Fig. 5.7. The equine conceptus at day 16 post fertilization, illustrating the formation of the amniotic folds over the embryonic disc.

to the uterine epithelium is reported to occur, though this attachment may only be temporary (Waelchi et al., 1996).

The membrane encompassing the amnion and separating it from the surrounding allantoic fluid (discussed later) is termed the allantoamniotic membrane. Initially, the amnion is visible as a clear fluid-filled bubble surrounding the embryo. As pregnancy progresses, it tends to collapse and lie close to the fetus. Throughout its life *in utero* the amniotic sack provides a clean and protective environment in which the embryo can develop. The source of its surrounding amniotic fluid is not clear. However, its composition is very much like blood serum, and exchange of fluids between the amniotic sack and the kidneys, intestine and respiratory tract is known to occur. The fetus in later stages appears to breathe in and swallows its surrounding amniotic fluid. The volume of amniotic fluid surrounding the fetus is about 0.4 l at 100 days post fertilization and increases to 3.5 l at full term.

At day 16, when the amnion is first evident, the mesoderm has not yet spread to enclose the whole of the yolk sack. The area over which the mesoderm has spread and which, therefore, has the three layers ectoderm, mesoderm and endoderm and is nearest to the embryonic disc is called the trilaminar omphalopleur. The area into which the mesoderm has not yet spread and which has only ectoderm and endoderm is termed the bilaminar omphalopleur. The junction of these two areas, i.e. the line delineating the limit of mesoderm migration, is called the sinus terminalis (Fig. 5.8).

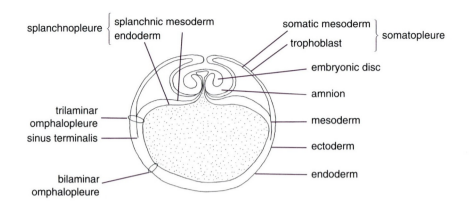

splanchnopleure ⎰ splanchnic mesoderm
             ⎱ endoderm

somatic mesoderm ⎱ somatopleure
trophoblast   ⎰

embryonic disc

amnion

trilaminar omphalopleure

mesoderm

sinus terminalis

ectoderm

endoderm

bilaminar omphalopleure

Fig. 5.8. The equine conceptus at day 18 post fertilization, illustrating the near completion of the amniotic sack surrounding the embryonic disc. The trilaminar omphalopleur is shown, nearest the embryo, consisting of the endoderm, mesoderm and ectoderm, and the bilaminar omphalopleur, into which the mesoderm has not yet spread.

5.3.1. The equine capsule

The equine conceptus is relatively unique in having a thin acellular glycoprotein/glycocalyx capsule which develops around the conceptus at day 5 and is certainly present until day 20 and even as late as day 35 of pregnancy (Fig. 5.9). This capsule appears in the perivitelline space between the trophectoderm and the zona pellucida (Oriol, 1994). The function of this capsule is unclear. It may have a protective role in that it is strong enough to retain the spherical shape of the conceptus up until implantation. It may also have a role in embryo mobility, preventing the adhesion of the embryo to the endometrium,

hence allowing the prolonged mobility phase characteristic of equine concepti. It may also have a role in driving embryo expansion, which occurs from day 5 (Crossett *et al.*, 1995). From day 6, the conceptus increases in size; this forces a thinning of the zona pellucida, which eventually breaks. The embryo then hatches and is left surrounded by just its capsule. At this time (day 6), the conceptus starts to derive nutrients for its growth and cell division from the surrounding uterine secretions, as by this stage it has used up all its own reserves. The capsule by nature of its negative electrostatic charge and unusual glycocalyx configuration (Oriol *et al.*, 1993) is very sticky to proteins within the surrounding uterine secretions. The capsule, therefore, attracts a whole host of proteins and other components as it moves through the uterus during the period of embryo mobility. These then diffuse, or are actively transferred by carrier proteins, across the capsule into the yolk sack to provide nutrients for the growing conceptus (Stewart *et al.*, 1995). The provision of such additional nutrients allows a further increase in size. The morula is now in its mobility phase, floating freely within the uterus (see Section 5.4.1), deriving all its nutritional requirements from uterine histotroph.

From day 16 onwards, it is increasingly evident that embryology can be dealt with in two sections: placentation and organ development.

5.4. Placentation

The placenta has two major functions: first, protection; and second, regulation of fetal environment,

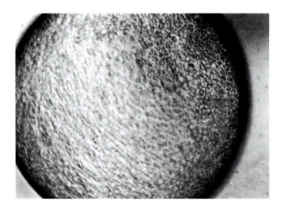

Fig. 5.9. A view of the surface of an equine embryo illustrating the outer trophoblastic cells. The capsule is evident as a clear area encircling the whole conceptus. (Photograph courtesy of Alison Crook.)

in the form of nutrient intake and waste output. The placenta develops from the extraembryonic membranes, the trophoblast of the blastocyst. The first source of nutrients, and, therefore, a form of primitive placenta, is the yolk sack or blastocoel. This provides both a temporary store and a transport system for nutrients derived from uterine secretions which have attached to, and then diffused across, the capsule.

Day 14 sees the first evidence of blood vessels developing in the centre of the yolk sack wall in the spreading mesoderm. These will become the blood system of the placenta. By day 18, the vitelline artery, carrying blood towards the mother, and the vitelline vein, carrying blood away from the mother, are identifiable.

On day 20, an outpushing of the embryonic hindgut can be seen immediately below the placenta. This is termed the allantois and continues to grow with the embryo. This sack is filled with allantoic fluid; it is encompassed by the allantochorionic membrane or placenta. The allantoic fluid consists of secretions of the allantochorion, along with urinary fluid, which is excreted from the fetal bladder via the urachus within the umbilical cord (Figs 5.10 and 5.11).

The volume of the allantois at day 45 is approximately 100 ml, increasing to 8.5 l by day 310, a considerably larger volume than seen in the amniotic sack. The allantoic fluid increases in volume as the fetus grows, producing more urinary fluid to be stored. During the first trimester (3–4 months), it is

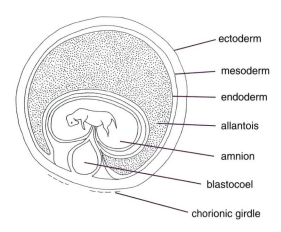

Fig. 5.11. The further development of the equine placenta at day 40 post fertilization. The allantois now dominates the conceptus.

clear yellow in colour, changing to brown/yellow as pregnancy progresses (Figs 5.12 and 5.13).

This developing allantoic sack moves over the top of the embryo as its contents increase, forcing the embryo down to the bottom of the blastocyst, reducing, as it goes, the extent of the yolk sack, until the yolk sack is hardly visible. As the allantoic sack increases in size the umbilical cord becomes evident. The attachment point of the umbilical cord normally corresponds to the position of initial implantation. It consists of two vitelline (umbilical) arteries, one vitelline vein and the urachus plus some supporting and connective tissue. The arteries and veins are responsible for blood transfer to and from the

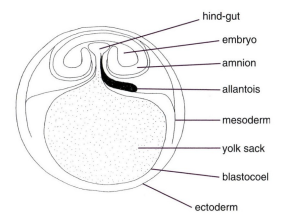

Fig. 5.10. The development of the equine placenta at day 20 post fertilization, illustrating the development of the allantois (allantoic sack).

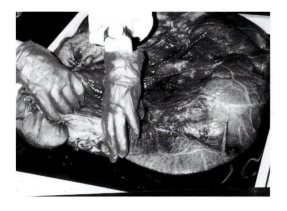

Fig. 5.12. The equine fetus at 200–220 days of gestation, illustrating the breaking and removal of the allantochorion (placenta) and the significant amount of allantoic fluid released as a result.

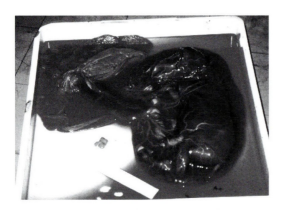

Fig. 5.13. The equine fetus at 200–220 days of gestation, illustrating the fetus within the amniochorion (amniotic sack) after removal of the allantochorion (placenta). The umbilical cord can be seen connecting the fetus and the allantochorion, which has been removed from around the fetus and is lying in the top left-hand corner. The, as yet small, hippomane can be seen just above the ruler.

placenta to the fetal system and the urachus transfers waste products from the bladder to the allantois; as such, it extends no further than the allantois and does not reach the placenta (Fig. 5.14).

As the fetus develops, its nutrient demand increases. The nutrients provided via the yolk sack are soon not enough to meet this demand; thus, a more intimate relationship needs to develop between the mother and the embryo, and hence its period of mobility ceases and it begins to implant. This occurs as a very gradual process from day 16 onwards, from which point the movement of the conceptus slows and it begins to derive increasingly more nutrition directly from the uterine endometrium. Initially

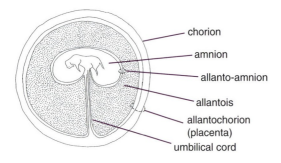

Fig. 5.14. The placental arrangement of the equine conceptus near term.

the amount of nutrition obtained in this way is very limited and the yolk sack continues to function as a nutrient store for a fair while. At this stage the capsule begins to degenerate, though remnants have been reported as late as day 35 (Enders *et al.*, 1993; Oriol, 1994).

5.4.1. Embryo mobility and fixation

The period of embryo mobility in the mare is relatively long and lasts for 16–18 days during which time the conceptus moves freely within and between the uterine horns and body (Gastal *et al.*, 1996; Stout and Allen, 2001). Mobility is due to the uterine myometrial contractions (uterine contractility), which is controlled by an interplay between localized secretion of PGE and prostaglandin F2α (PGF2α) by the conceptus and progesterone secreted by the corpus luteum (CL) of the mare's ovary. Provided the uterine environment is dominated by circulating progesterone, PGE and PGF2α secreted by the conceptus cause localized uterine contractility, driving movement of the conceptus (Kastelic *et al.*, 1987). Embryo movement slowly reduces as: the conceptus size increases, making movement harder; an increase in Na and Cl ion concentrations within the conceptus changes the conceptus osmolarity and the fluid and electrolyte exchange with the surrounding uterine fluid; uterine lumen reduces as the endometrial folds increase in size with increasing oestrogen levels around days 25–30 (see Section 6.3.3); and finally uterine contractility decreases as the blastocyst slowly reduces its prostaglandin (PG) production (Griffin and Ginther, 1990; Gastal *et al.*, 1996). The site of implantation is normally the junction of the uterine horn and body, and appears to be independent of the site of ovulation (Silva *et al.*, 2005).

5.4.1.1. Areolae temporary attachment

The first identifiable attachment between mother and fetus occurs around days 18–20 via temporary areolae (tufts of allantochorion which break through the now degenerating capsule and invade the mouths of the uterine endometrial glands) that encircle the conceptus at the limit of the extent of the spread of the mesoderm. These areolae provide not only an anchorage (though quite loose and temporary) for the conceptus but also more efficient absorption of secretions directly from the mouths of the endometrial glands (Allen, 2001).

5.4.1.2. Chorionic girdle

About 8 days (days 25–35 of pregnancy) after the embryo has become stationary in the uterus a thickening appears on the outer trophoblast (allantochorion) forming a band of shallow folds where the enlarging allantois butts up against the shrinking yolk sack. This is the chorionic girdle. Cells within this girdle elongate to form ridges and invade, through the now broken capsule, the uterine endometrium engulfing some of the epithelial cells. This girdle forms in an area of the conceptus where there is no mesoderm, i.e. in the bilaminar omphalopleur, and where the yolk sack is gradually being restricted by the developing allantois (Fig. 5.15). The growth of the chorionic girdle is likely to be governed in part by insulin-like growth factors (IGF) such as IGF II (Enders and Lui, 1991; Enders *et al.*, 1993). This attachment, though also only tenuous, does provide an increasingly significant exchange unit (Enders *et al.*, 1993). At day 38, the fetal cells migrate into the maternal endometrium and the girdle detaches from the allantochorion. Initially it was thought that this freed the conceptus to migrate again within the uterus. However, this is now thought to be unlikely and the next attachment structure, the endometrial cups, develop quite rapidly and take over as the attachment mechanism. The exact time of takeover by endometrial cups may vary, possibly occurring later in older mares (Carnevale and Ginther, 1992).

5.4.1.3. Endometrial cups

The invading fetal girdle cells stream down into the lumen of the endometrial glands and break through into the stroma of the endometrium. At day 40, they suddenly stop migrating, enlarge and tightly pack together within the endometrial stroma, forming a series of pale raised areas on the surface of the endometrium encircling the conceptus; these are termed endometrial cups fixing the conceptus at the junction of the uterine body and gravid (pregnant) horn, at the original position of the chorionic girdle (Fig. 5.15; Allen and Stewart, 1993). Their development is associated with increasing lymphatic activity (Enders and Lui, 1991). These endometrial cups secrete equine chorionic gonadotrophin (eCG), sometimes referred to as pregnant mare serum gonadotrophin (PMSG), which is essential for the maintenance of early pregnancy. eCG will be discussed in detail in Section 6.3.2.

Around day 90, the endometrial cups begin to degenerate and slough away from the uterine endometrium. The reason for this seeming rejection is not fully understood, but may be a maternal immunological rejection of the 'foreign' fetal tissue, which is specific to the endometrial cups and not the allantochorion (Asbury and Le Blanc, 1993). The duration of the endometrial cups is very variable, being longer in sibling matings, primiparous mares (mares not previously pregnant) and foal-heat matings (Spincemaille *et al.*, 1975; Bell and Bristol, 1991;

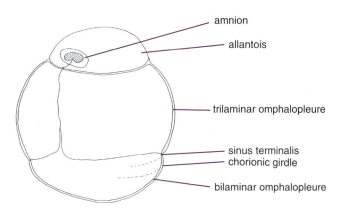

amnion

allantois

trilaminar omphalopleure

sinus terminalis
chorionic girdle

bilaminar omphalopleure

Fig. 5.15. The beginnings of the development of the placenta of the equine conceptus at day 25 post fertilization, illustrating the position of the chorionic girdle attachment.

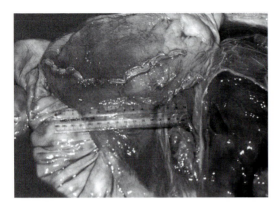

Fig. 5.16. The remains of the endometrial cups can be seen in a band running across the uterine endometrium.

Allen *et al.*, 1993). The remains of these sloughed-off endometrial cups may be reabsorbed by the fetus during the remainder of the pregnancy or they may be seen in the placenta at birth as invaginations or pouches in the allantochorion (Fig. 5.16).

5.4.1.4. Placenta

Gradually, over time as the endometrial cup attachment is lost, the rest of the fetal allantochorion begins to attach to the uterine epithelium. This attachment begins between days 45 and 70 and gradually becomes firmer over the next 100 days, being fully attached by day 150. At days 45–70, the allantochorion takes on a velvety appearance, created by fine microvilli over its entire surface; hence, the equine placenta is termed diffuse. These microvilli organize themselves into discrete microscopic bundles or tufts which invade into receiving invaginations in the uterine epithelium, These bundles of microvilli are termed microcotyledons, and their attachment develops over a period of time, being fully complete and functional by day 150 (Fig. 5.17).

A strong attachment is formed between the fetus and the mother. The equine placenta is relatively thick, with six cell layers and four basement membranes. The three cell layers on the fetal side are mesoderm (endothelium – blood vessel wall), endoderm (connective tissue) and ectoderm (allantochorion); and the three on the maternal side are epithelium, endometrium (connective tissue) and endothelial (blood vessel wall). The equine placenta is therefore termed epitheliochorial and covers the whole surface of the uterus, except the cervix (the cervical star) and the two utero-tubular junctions (Figs 5.12, 5.13, 5.18 and 5.19; MacDonald and Fowden, 1997).

The presence of the microcotyledons serves to increase the surface area of the placenta and, therefore, the area for nutrient and gas exchange. Within each microcotyledon, the maternal and fetal blood supply systems come in close proximity, allowing efficient diffusion.

However, the thickness of the placental attachment prevents the diffusion of large molecules such as immunoglobulins (large protein molecules). Hence, the attainment of passive immunity in the foal by diffusion across the placenta is very limited. Passage of immunoglobulins via colostrum is, therefore, of utmost importance in the mare, as will be discussed in further detail in Chapters 15 and 16

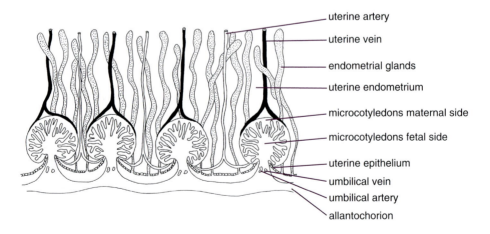

uterine artery
uterine vein
endometrial glands
uterine endometrium
microcotyledons maternal side
microcotyledons fetal side
uterine epithelium
umbilical vein
umbilical artery
allantochorion

Fig. 5.17. Equine placental microcotyledons in the fully developed placenta.

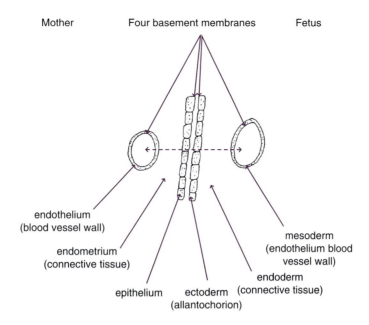

Mother   Four basement membranes   Fetus

endothelium
(blood vessel wall)

mesoderm
(endothelium blood
vessel wall)

endometrium
(connective tissue)

endoderm
(connective tissue)

epithelium   ectoderm
(allantochorion)

Fig. 5.18. The structure of the equine placenta.

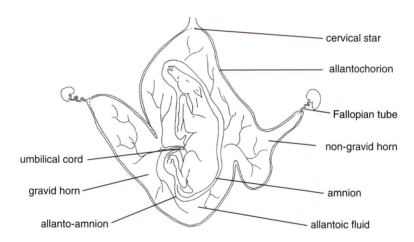

cervical star

allantochorion

Fallopian tube

non-gravid horn

umbilical cord

gravid horn

amnion

allanto-amnion

allantoic fluid

Fig. 5.19. The equine fetus and placenta near term.

(this volume). The thickness and arrangement of the placenta vary in different mammals, but in general, the thicker the placenta, the less efficient is the transfer of passive immunity *in utero*, and hence the greater the reliance on colostrum. However, a thicker placenta as seen in the mare has the advantage of providing extra protection to the fetus from harmful maternal blood-borne factors (Silver *et al.*, 1973).

5.4.2. Placental efficiency

Despite the thickness of the placenta, nutrient and gaseous exchange across the mare's placenta is relatively efficient when compared to other farm livestock. This is due to the diffuse (attachment to the uterus over the whole surface of the placenta) nature of the equine placenta compared to the cotyledonary (attachment to the uterus just at discrete areas) nature of the ewe and cow placentas. However, it must be remembered that measurements taken on placental efficiency involve the acute catheterization of the umbilical arteries, and hence the technique itself may affect the results obtained. Silver *et al.* (1973) demonstrated the relative efficiency of the mare's placenta, as changes in maternal blood oxygen and glucose concentrations were mimicked more closely by changes in the fetal blood concentration than in sheep. Free fatty acids and lactate also follow the same pattern. It may well be deduced, therefore, that factors affecting the mare will have a greater effect on the fetus than is evident in ruminants, though such an association has yet to be confirmed.

As pregnancy progresses, the maternal epithelium stretches as the uterus increases in size. As a result, the placenta also stretches and becomes thinner, and hence the resistance to gaseous and nutrient exchange decreases, the placenta becoming more efficient as the demands of the fetus increase. By full term, the placenta of a 15–16 hh horse weighs about 4 kg. Its surface area is approximately 14,000 cm^2 and it is about 1 mm thick. The foal's birth weight is directly proportional to the surface area of the placenta, as this is the limiting factor controlling nutrient and gas exchange and hence their availability to the developing fetus. The surface area of a placenta may be restricted for several reasons, including the presence of twins.

5.4.3. Twins

Twinning is an increasing problem in stud management, especially in intensively bred horses such as the Thoroughbred. The incidence of twin ovulations, which have the potential to result in twin conceptuses in the Thoroughbred, is 20–25% (Davies Morel and O'Sullivan, 2001). Of this potential number of twins, significant natural reduction to one conceptus does occur. Seventy per cent of twins are initially unilateral (both in the same uterine horn), of which 85% naturally reduce; 30% are bilateral (one conceptus in each horn), none of which naturally reduce (Ginther, 1989a,b; Ginther and Griffin, 1994). If twins do develop to the placentation stage, the area of the uterus available for each placenta is restricted by the presence of the other fetus (Fig. 5.21). If the division of uterine surface area available to each twin is equal, then both twins have an equal chance of survival but their birth weights will be reduced. If the division is unequal,

then the smaller one may cause the whole pregnancy to abort or, if the pregnancy is not well advanced, it may die and become mummified. If mummification occurs, the pregnancy may well continue; if this occurs after around day 100 the placenta of the larger surviving fetus cannot expand and attach to the uterine surface originally occupied by the now dead fetus as the extent of placental attachment has already been fixed (Fig. 5.20). At term, therefore, a single foal will be born, but with a reduced birth weight due to placental restriction (Fig. 5.21).

5.4.4. Placental blood supply

As mentioned previously, the mesoderm of the blastocyst surrounding the yolk sack forms the first placental blood supply to the fetus. Early in pregnancy, a clear network of blood vessels can be identified within the trilaminar omphalopleur, with two major vessels connecting the network to the rudimentary heart. Additional pathways develop to feed areas of considerable growth. Hence, when the yolk sack degenerates and the nutrient supply to the fetus is taken over by the allantochorionic placenta, a well-formed network of blood vessels lining the allantochorion already exists. This fine network enlarges and invaginates into the micro-

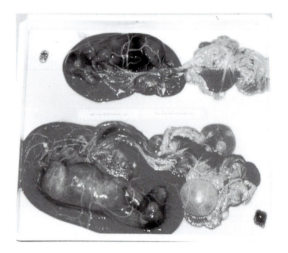

Fig. 5.20. The fetuses of a twin pregnancy dissected out at post-mortem. The different size of the twins is evident as a result of placental restriction of the smaller twin. If left to go to term, the smaller twin would eventually have died and probably have caused the abortion of the whole pregnancy.

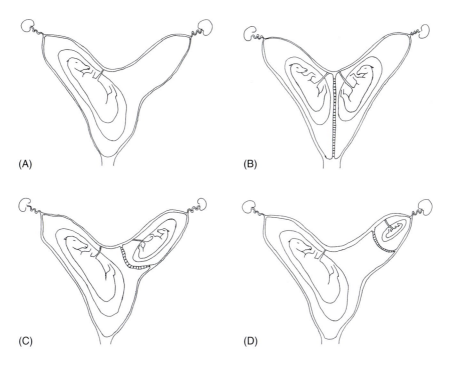

Fig. 5.21. Placental configurations in the equine singleton and twin pregnancies: (A) singleton; (B) equal split (50%:50%); (C) unequal split (60%:40%); and (D) unequal split (80%:20%).

cotyledons of the placenta. Each microcotyledon is supplied on the fetal side by several arteries, but exit back to the fetal heart is via a single vein. This arrangement slows down the flow of blood through the microcotyledons and encourages more efficient diffusion and gas exchange. The oxygenated and nutritionally replenished blood returns to the fetal heart by the umbilical vein (Fig. 5.22). The umbilical cord, therefore, contains two fetal arteries and one fetal vein plus the urachus from the bladder.

It should be remembered that in the fetus, because of the bypass of the non-functional lungs, deoxygenated blood is carried to the placenta via arteries and oxygenated blood passed back to the heart via the veins (Fig. 16.4).

On the maternal side a very similar arrangement exists. Oxygenated and nutritionally enriched blood approaches the microcotyledons in a fine network of arteries, but the drainage back to the maternal system is also via a single vein, again slowing down the passage of blood and increasing the efficiency of nutrient and gaseous exchange. This transfer across the utero-fetal placental barrier can be compared in many ways to the gaseous exchange within the mammalian lung.

5.5. Organ Development

Organ development arises from the reorganization of cell populations within the embryonic disc itself. This organization is related to that which occurs in placentation, previously discussed. This can be divided into two basic sections: gastrulation and neuralation. The former can be subdivided into segregation, delamination and involution. Further accounts of organ development can be found in Douglas and Ginther (1975), Van Niekerk and Allen (1975), Betteridge *et al.* (1982), Flood *et al.* (1982) and Enders *et al.* (1988).

5.5.1. Gastrulation

Gastrulation is defined as the organization of the embryo into three germ layers: ectoderm, mesoderm and endoderm. This involves primarily the cells of the embryonic disc but also those of the placental

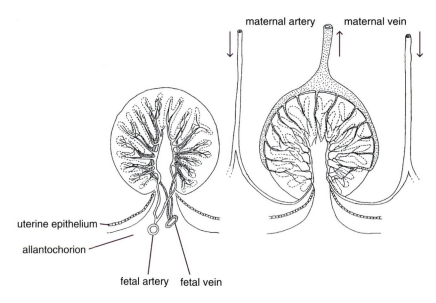

maternal artery   maternal vein

uterine epithelium

allantochorion

fetal artery   fetal vein

Fig. 5.22. The venous and arterial blood supply to the microcotyledons within the equine placenta, allowing the transfer of nutrients and waste products from the maternal (right) to the fetal (left) system and vice versa.

tissue. The first stage of gastrulation is segregation, during which the central blastomeres or cells of the embryonic disc organize themselves into smaller outer and larger inner blastomeres (Fig. 5.23).

The larger blastomeres collect underneath the disc and migrate in two directions. First, they migrate to line the remaining ectoderm of the blastocyst, forming the endoderm. Second, they migrate within the embryonic disc, creating at day 11 the first asymmetry, a thicker area at the caudal end (future tail end) and a thinner area at the cranial end (future head end; Fig. 5.24).

The second stage of gastrulation is termed delamination. This commences at day 12 and marks the first evidence of epiblast cells, hypoblast cells and the primitive gut (Fig. 5.25).

The epiblast cells are those of the embryonic disc. The hypoblast cells are the migrating endoderm and within this ring of hypoblast is the yolk

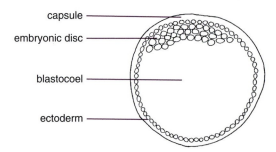

capsule

embryonic disc

blastocoel

ectoderm

Fig. 5.23. Day 9, segregation in the equine conceptus, illustrating the larger inner blastomeres and smaller outer blastomeres within the embryonic disc.

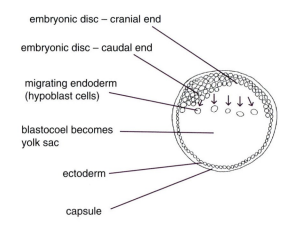

embryonic disc – cranial end

embryonic disc – caudal end

migrating endoderm (hypoblast cells)

blastocoel becomes yolk sac

ectoderm

capsule

Fig. 5.24. Day 11, segregation in the equine conceptus, indicating the migration of the large blastomeres from the lower part of the embryonic disc to line the remaining ectoderm.

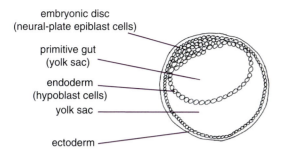

Fig. 5.25. Day 12, delamination in the equine conceptus, showing the future development of the endoderm layer.

Labels in figure:
embryonic disc (neural-plate epiblast cells)
primitive gut (yolk sac)
endoderm (hypoblast cells)
yolk sac
ectoderm

sack or primitive gut. At day 14, a change in this embryonic disc becomes evident. This change forms the beginning of the primitive streak identified within the epiblast cells. At this stage, it is about 1 cm in length.

The third and last stage of gastrulation is involution, when the epiblast cells move inwards to the centre of the caudal end of the disc (Fig. 5.26). At this stage, three types of cells – ectoderm (epiblast cells), mesoderm and endoderm (hypoblast cells) – are evident within the embryonic disc as seen in the extraembryonic tissue (Fig. 5.7). These three cell layers will go to form all the main body structures.

The moving ectoderm or epiblast cells reappear as mesoderm between the ectoderm and the hypoblast cells or endoderm. As the cells move through to the lower level they leave a depression in the upper surface of the epiblast. These migrating epiblast cells tend to move in greater concentrations at the caudal end of the primitive streak, making it wider. The primitive streak so formed makes the future longitudinal axis of the embryo.

At day 15, epiblast cell movement tends to slow down; the slight indentation along the longitudinal axis of the primitive streak becomes deeper, as cells continue to move out from underneath to form the mesoderm and are not replaced by migrating epiblast cells above. This deep groove is now termed the primitive groove. The cells associated with the primitive groove are termed node cells, to differentiate them from the cells of the remainder of the embryo. At day 15, these node cells can be identified as precursors of future body organs. The ectoderm node cells form the neural plate, running the length of the top of the primitive groove, the cranial end of which goes to form the head. The spreading mesoderm in the immediate vicinity of the neural plate goes to form the somites, or body trunk, and the mesoderm immediately below the primitive groove goes to form the notochord (spine and central nervous system (CNS)). Finally, the wide caudal end forms the tail end of the fetus (Fig. 5.27).

The process of gastrulation is now completed, the major cell blocks are identifiable and the longitudinal axis of the embryo is determined.

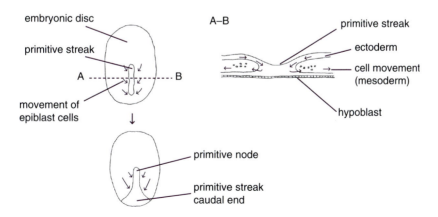

Labels in figure:
embryonic disc
primitive streak
A — B
movement of epiblast cells
A–B
primitive streak
ectoderm
cell movement (mesoderm)
hypoblast
primitive node
primitive streak caudal end

Fig. 5.26. Day 14, involution of the equine conceptus. A bird's-eye view of the embryonic disc, along with a cross-sectional view through A–B, illustrating the passage of ectoderm cells through the primitive streak to reappear between the ectoderm and endoderm, forming the mesoderm. Further cell movement results in the flattening of the caudal end of the primitive streak.

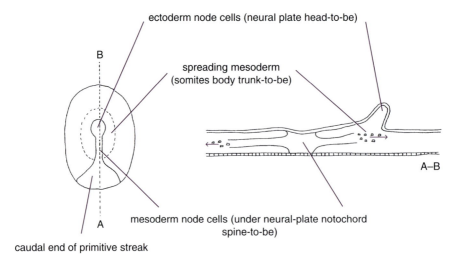

ectoderm node cells (neural plate head-to-be)

B

spreading mesoderm
(somites body trunk-to-be)

A–B

mesoderm node cells (under neural-plate notochord
spine-to-be)

A

caudal end of primitive streak

Fig. 5.27. Day 15, completed gastrulation in the equine conceptus. A bird's-eye view and cross-sectional view through A–B of the embryonic disc. The formation of the head from the cranial end of the primitive groove is illustrated, along with the somites, or body trunk, formation from the mesoderm in the immediate vicinity of the primitive groove and the spine and central nervous system formation from the mesoderm immediately below the primitive groove.

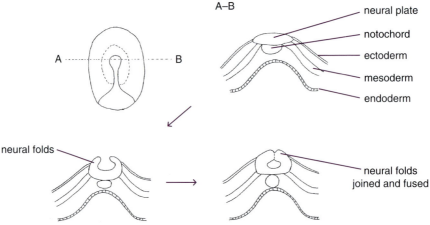

A–B

neural plate
notochord
ectoderm
mesoderm
endoderm

A .... B

neural folds

neural folds
joined and fused

Fig. 5.28. Days 16–17, neuralation of the equine conceptus. The ectoderm near the neural plate thickens and two neural folds develop on either side of the neural plate and join to enclose a hollow tube, the future spine and central nervous system.

5.5.2. Neuralation

The next stage, termed neuralation, involves the development of the CNS, gut and heart. Day 16 sees three major changes. First, the ectoderm near the neural plate thickens and two neural folds develop on either side of the neural plate. The neural plate becomes depressed and the neural folds fold over, join and then fuse to enclose a hollow tube, the spine-to-be and CNS-to-be (Fig. 5.28).

Second, the mesoderm on either side of the neural plate organizes itself into 14 somites (future muscle blocks). Third, at the cranial end of the neural plate an increase in cell growth above the surface becomes apparent, with an accompanying increase in the length of the neural plate. This cell growth folds over to form the head process, heart and pharynx.

By day 18, lateral folds are beginning to develop on either side of the head process. As cells move into this area and cell division increases, the cranial end of the neural plate lifts away from the underlying tissue (Figs 5.29 and 5.30).

These lateral folds move down from the cranial end to the caudal end, lifting the whole body away from the underlining tissue (Fig. 5.30).

This lifting away from the remaining tissue leaves just one attachment point in the centre, the first evidence of the umbilical cord. The embryo continues to lift off the underlying tissue and the head and tail processes fold back down to give the embryo its characteristic C-shape configuration. At this stage two more somites are evident, making 16 in total.

The gut tube also now begins to develop from the pharynx fold by closure of the endoderm folds, in a way similar to that by which the neural tube was formed from folds in the ectoderm of the neural plate. The hindgut of the fetus now extends out into the blastocoel to form the allantois, as illustrated in Figs 5.10 and 5.11 and blood is also now evident in the lumen of the tubular heart (Cottrill *et al.*, 1997), which can be seen to beat from day 21 (Van Niekerk and Allen, 1975).

The embryo now lies away from the underlying tissue, the placental tissue, and is connected directly to the mother only by the umbilical cord, which contains a blood system derived from the mesoderm, along with supporting connective tissue. The embryo now has an identifiable neural tube, the forerunner of the CNS, and a head process with enlarged neural tube, the brain-to-be. Its pharynx and gut tube are also present, as are the somites, or body muscle blocks. Therefore, by day 23, all the basic bodily structures are evident, though only in a rudimentary form.

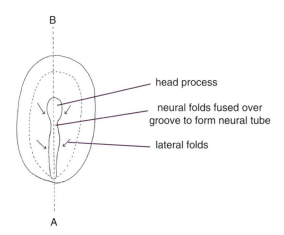

Fig. 5.29. Day 18, neuralation of the equine conceptus. A bird's-eye view of the embryonic shield, illustrating cell movement in towards the cranial end of the neural plate, which subsequently lifts away from the underlying tissue.

5.6. Organ Growth

From day 23 onwards, development is in the form of fine differentiation and organ growth. By day 40, all the main body features are evident, e.g. limbs, tail, nostrils, pigmented eyes, ears, elbows and stifle regions, eyelids, etc., and the embryo is now termed

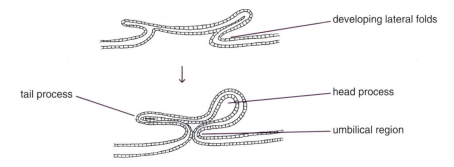

Fig. 5.30. A cross-sectional view (A–B) of Fig. 5.29, illustrating the gradual appearance of the head and tail processes during neuralation in the 19-day-old equine conceptus. The embryo now begins to take up the characteristic C-shape.

a fetus. Days 39–45 herald sexual differentiation and evidence of external genitalia. The weight of the fetal gonads reaches a maximum at days 180–200, the weight of fetal testes and ovaries being equivalent and develop to the following pattern (Douglas and Ginther, 1975).

Day 80 – 1.4 g
Day 140 – 18.7 g
Day 200 – 48.0 g
Day 320 – 31.4 g

The increase and decrease in size is due to a proliferation and degeneration of interstitial cells (Walt *et al.*, 1979) and appears to correspond to the period of masculinization or feminization of the fetus. The reason for the relatively large size of the fetal gonads in the horse is unclear, but may be related to the secretion of significant levels of oestrogen at this time.

At this stage, most of the development is complete and increase in growth now occurs (Fig. 5.31). At day 60, the eyelids close and finer eye development occurs, teats are present and the oral palate is fused. Day 160 sees the first evidence of hair around the eyes and muzzle, and, by day 180, hair has begun to develop at the tip of the tail and the beginnings of a mane are evident. By day 270, hair covers the whole of the body surface (Table 5.1).

From day 150 onwards, the hippomane, an accumulation of waste minerals within the allan-tois, becomes apparent. The hippomane increases in size with pregnancy (Fig. 5.32). From day 320 onwards, the testes in the male fetus may descend through the inguinal canal; however, this does not occur in all colt fetuses, as some drop neonatally.

The main milestones in equine fetal development are summarized in Table 5.1 (Ginther, 1995; Reef, 1998; Sertich, 1998).

Full term, normally at 320 days in ponies and up to 2 weeks later in Thoroughbred and riding-type horses, heralds the birth of a very well-developed fetus, typical of a preyed-upon, plain-dwelling animal. At birth foals are capable of all basic bodily functions, including walking within 30–60 min. Details of the foal's adaptation to the extrauterine environment are given in Chapter 16 (this volume).

5.7. Conclusion

Our understanding of embryo development specific to the equine is still incomplete especially in early pregnancy. Continuing development of our knowledge is essential if we are to understand and hence minimize embryo mortality, a significant cause of apparent infertility in the mare. When the factors affecting embryo survival are more fully understood, our management of the equine can be further directed towards minimizing losses.

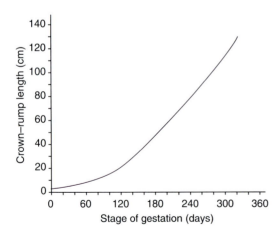

Fig. 5.31. The increase in fetal crown-rump length throughout gestation. (From Evans and Sack, 1973; Ginther, 1992.)

Fig. 5.32. The hippomane often found within the placenta of the mare. This example is 11 cm long.

Table 5.1. A summary of the major milestones for fetal development throughout pregnancy.

Day of gestation	Major milestones
1	Zygote, 2 cells
4	Morula, 16 cells plus
5	Capsule formation
6	Hatching of morula
8	Blastocyst, differentiated into embryonic mass, blastocoel and trophoblast
9	Ectoderm and endoderm germ layers evident, gastrulation begins
11	Segregation giving first embryonic asymmetry, caudal and cranial ends evident
12	Delamination, epiblast cells, hypoblast cells and primitive gut evident
14	Mesoderm evident, primitive streak appearing, involution commencing
15	Primitive streak now evident as a groove
16	Neuralation starts, folds leading to the formation of the amnion seen, first blood vessels evident in mesoderm, chorionic vesicle 2–4 cm diameter
18	Vitelline artery and vein identifiable, fetus begins to take on characteristic C-shape
20	Allantois forming from outpushing of the fetal hindgut, chorionic vesicle oval in shape (2.5–4.5 cm diameter), eye vesicle and ear present Capsule begins to degenerate
21	Amnion complete
23	All basic bodily structures evident, though in rudimentary state
25	Chorionic girdle, first evident attachment of fetus
26	Forelimb bud seen, three branchial arches present, eye visible
30	Genital tubercle present, eye lens seen
36	Rudimentary three digits seen on hoof, facial clefts closing, eyes pigmented and acoustic groove forming
40	Endometrial cups forming, ear forming, nostrils seen, eyelids seen, all limbs evident and elbow and stifle joint areas identifiable, chorionic vesicle 4.5–7.5 cm diameter
42	Ear triangle in shape, mammary buds seen along ridge
45	External genitalia evident, allantoic sac volume 110 ml
47	Palate fused
49	Mammary teats evident
55	Ear covers acoustic groove, eyelids closing
60	Chorionic vesicle 13.3 × 8.9 cm
63	Eyelids fused, fine eye development occurring, hoof, sole and frog areas of hoof evident
75	Female clitoris prominent
80	Scrotum clearly seen
90	Endometrial cups degenerate, chorionic vesicle 14 × 23 cm
95	Hoof appears yellow in colour
112	Tactile hairs on lips growing
120	Fine hair on muzzle, chin and eyelashes beginning to grow, eye prominent and ergot evident
150	Full attachment of placental microcotyledons, eyelashes seen, enlargement of mammary gland
180	Mane and tail evident
240	Hair of poll, ears, chin, muzzle and throat evident
270	Whole of body covered with fine hair, longer mane and tail hair clearly seen
310	Allantoic sack volume 8.5 l
320	Testes may drop from this time onwards
320–340	Birth of fully developed fetus

6 Endocrine Control of Pregnancy in the Mare

6.1. Introduction

When examining the endocrinological control of pregnancy in the mare, gestation can be divided into two stages: early (fertilization to day 150) and late (day 150 to full term).

6.2. Early Pregnancy

By day 6, the conceptus has migrated to the uterus and exists by deriving nutrients from the uterine hystotroph or secretions. No major changes from the non-pregnant cycle are evident as yet (Freeman et al., 1991). However, by day 15, a message has to be received by the reproductive system of the mare if it is to continue in a pregnancy mode: blocking the drop in progesterone, and hence allowing progesterone levels to remain elevated (Fig. 6.1). This is essential for the initial maintenance of pregnancy. If the mare is pregnant, then ovarian progesterone production, from one or a number of corpora lutea (CL) has to be maintained until at least day 75. If there is a failure of the functional CL during this time, the pregnancy will fail.

6.3. Maternal Recognition of Pregnancy

The importance of day 15, with regard to the recognition of pregnancy, has been demonstrated in experiments with early-pregnant mares. If the embryo is removed from a pregnant mare prior to day 15, she will return to oestrus at her normal time (21 days after the last). If, however, the embryo is removed at day 16 or later, the mare will not return to oestrus as expected and will show a prolonged dioestrus, due to the persistence of the CL. The length of the delay will depend to a certain extent on the age of the embryo at removal.

Experiments show clearly that D-Day as far as the hormonal control of pregnancy is concerned is day 15 (Hershman and Douglas, 1979); however, embryo transfer work suggests there may be some progressive recognition of pregnancy from day 8 onwards (Goff et al., 1987). The exact nature of the message informing the mare of the presence of a conceptus is unclear. However, there are several candidates, the most likely being oestrogens, which equine concepti are capable of synthesizing as early as day 12 (Flood et al., 1979a; Heap et al., 1982; Sharp, 1993; Choi et al., 1995). As such, oestrogens are the likely candidate acting locally on the uterine epithelium to inform the mother of the presence of the fetus (Flood et al., 1979a; Stewart et al., 1982a; Daels et al., 1990). However, it has been reported that the equine conceptus also produces three unknown proteins around days 12–14 (Sissener et al., 1996) and/or a single protein of molecular weight 6000 Da (Heap et al., 1982). Other species such as ruminants produce trophoblastic proteins (interferon tau), which act to inform the maternal system of a pregnancy; however, they are of a much higher molecular weight (17,000 Da; Hansen et al., 1999). Thus, in the mare it is likely that proteins produced by the conceptus along with oestrogens act as the messengers of pregnancy.

By whatever means the message is delivered, the result is the maintenance of the CL beyond day 15 (Ball et al., 1991). In the non-pregnant mare the CL is destroyed at about day 15 by prostaglandin F2α (PGF2α), allowing the cyclical changes associated with oestrus and ovulation to begin (Chapter 3, this volume). Therefore, in the pregnant mare this action of PGF2α must be blocked.

The exact mechanism for preventing the action of PGF2α is unknown, but there are several hypotheses. First, it has been suggested that PGF2α binding to the CL is reduced. However, doubt has been placed upon this hypothesis, as it appears that the CL concentration of PGF2α receptors is high during the period 16–18 days post ovulation in both the pregnant and non-pregnant mare. The second hypothesis is that an alternative component is produced by the uterus which competitively binds with the PGF2α

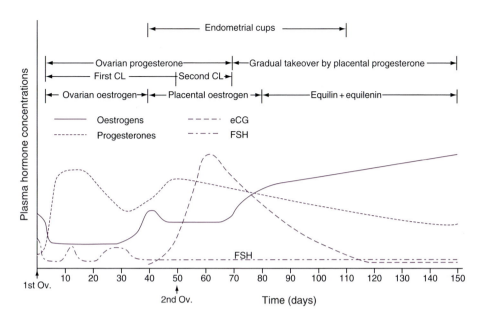

Fig. 6.1. A summary of the plasma hormone concentrations during early pregnancy in the mare (days 0–150). CL, corpus luteum; eCG, equine chorionic gonadotrophin; FSH, follicle-stimulating hormone.

receptors on the CL. A suitable candidate would be prostaglandin E (PGE), which is very similar in structure to PGF2α but biologically inactive with regard to CL regression (Allen, 1970; Heap, 1972; Vernon *et al.*, 1979). Finally the third, and currently favoured, option is that the secretion of PGF2α is reduced. This third hypothesis is supported by the reported reduction in the concentration of PGF2α in pregnant mare uterine washings, the reduction in PGF2α metabolite (PGFM) in the mare's circulation (Kindahl *et al.*, 1982; Zavy *et al.*, 1984) and the ability of the conceptus at days 12–14 to suppress the endometrial production of PGF2α *in vitro* (Bazer *et al.*, 1994; Sissener *et al.*, 1996). The production of PGF2α in the non-pregnant mare is driven by oxytocin (Section 3.2.2.8). In the pregnant mare the number of oxytocin receptors on the uterine endometrium is significantly reduced, presumably due to the influence of oestrogens and proteins produced by the conceptus. As a result, the endometrium response to oxytocin, in the form of PGF2α production, is significantly reduced and so the CL and pregnancy are maintained (Starbuck *et al.*, 1998; Stout *et al.*, 2000). The equine conceptus is unique in having a relatively long period of mobility (up to 18 days) and it is apparent that contact with much of the uterus by the moving of the conceptus is required to maintain pregnancy. If

movement is restricted by ligation, then the greater the restriction, the greater is the chance of CL regression and pregnancy failure (McDowell *et al.*, 1988).

From day 15 onwards, maternal progesterone and fetal oestrogens are the dominant hormones within the uterus and are important in the production and composition of uterine hystotroph and pregnancy-specific proteins, collectively termed uterine milk. The composition of uterine milk is particularly important in the mare as the conceptus survives in a free-living form for a relatively long period of time, no form of implantation occurring in the mare until day 18.

By considering the concentration of individual hormones a picture of how the mare maintains a developing conceptus can be developed.

6.3.1. Progesterone

Between days 6 and 14, the plasma concentration of progesterone is 8–15 ng ml⁻¹, similar to that seen during dioestrus of the non-pregnant oestrous cycle. Day 15 – decision time – heralds the divergence of progesterone concentrations between the pregnant and non-pregnant mare. Progesterone levels in pregnant mares decline slightly after day 16 (but not to the extent seen in non-pregnant mares) to reach concentrations of approximately 6 ng ml⁻¹ by day 30. Levels subsequently rise again to reach 8–10 ng ml⁻¹ by days 45–55, and remain at this level, or possibly falling slightly, until day 150 (Holtan *et al.*, 1975b; Schwab, 1990). Experiments in the mare indicate that ovarian progesterone, i.e. that produced by CL, is essential for the maintenance of all pregnancies, at least until day 75 and in some cases up to day 150. After day 150, placental progesterone is adequate to take over and maintain a pregnancy. All pregnant mares ovariectomized (the ovaries and hence the functional CL removed) before day 75 will abort. If mares are ovariectomized in the period of days 75–150, differing reports indicate differing abortion rates. After day 150, ovariectomy has no effect, and all mares successfully carry fetuses to full term (Holtan *et al.*, 1979). This demonstrates that ovarian progesterone is essential prior to day 75 in all mares; in the period of days 75–150, placental progesterone gradually takes over and by day 150, ovarian progesterone is not required. Indeed at this time the CL on the ovary can be seen to have regressed.

Ovarian progesterone is not secreted continually by a single CL. In the mare, an increase in ovarian activity is evident between days 20 and 30 post coitum: follicles develop, driven by follicle-stimulating hormone (FSH) surges similar to those seen during dioestrus (Bergfeldt and Ginther, 1992). Dominant follicles become apparent and luteinize between days 40 and 60 forming secondary CL (Chavatte *et al.*, 1997a). These secondary CL are unusual and relatively unique to the mare. During the period of days 40–70 the secondary CL gradually take over the production of progesterone, though the primary CL do not necessarily regress. From day 70 onwards, the placenta begins to produce progesterone, gradually taking over the function of all the CL by day 150 (Allen and Hadley, 1974; Evans and Irvine, 1975; Squires and Ginther, 1975; Squires, 1993a).

6.3.2. Equine chorionic gonadotrophin

As discussed in Chapter 5 (this volume), days 35–40 mark the appearance of endometrial cups. These are responsible for the secretion of equine chorionic gonadotrophin (eCG), also known as pregnant mare serum gonadotrophin (PMSG). eCG is secreted into the mare's circulatory system from around day 40 and reaches a maximum concentration between days 50 and 70 post coitum. It is secreted by the fetal tissue within the endometrial cups and maximum concentrations achieved vary considerably, different reports giving levels of between 10 and 100 IU ml⁻¹ (Allen and Moor, 1972). Levels are known to be affected by genotype, maximum levels reached and the duration of these levels being greater in mares mated to close relatives, e.g. brother-to-sister matings (Stewart *et al.*, 1977; Allen and Stewart, 1993). The parous state of the mare also has an effect on eCG levels, multiparous (having had previous pregnancies) having lower levels than primiparous (first-time pregnant) mares. Concentrations of eCG always decline and normally reach basal levels by days 100–120 (Allen, 1982a; Allen and Stewart, 1993). The importance of eCG and why it is only secreted for a short period of pregnancy are unclear. However, several hypotheses suggest it may have a role in the prevention of fetal immunological rejection by the mother and in the formation and maintenance of the secondary CL.

The uterus is a privileged site and is the only place in the body that, under the influence of the hormones of pregnancy, will tolerate a foreign body, the conceptus. The genetic make-up of the conceptus is only half that of its mother and, therefore, is foreign and, under normal conditions, such an invasion would set up an immunological response and the foreign body (the conceptus) would be rejected. For some reason, this rejection is prevented in the pregnant mare and all other mammals that carry fetuses *in utero*. It appears that an immunological barrier is set up between the fetus and its mother, blocking the expected immunological response. It has been suggested that eCG may be involved in this prevention of rejection from day 35 onwards. However, as it is not secreted prior to day 35, some other component must be responsible for preventing rejection during earlier pregnancy, but what this is remains unclear (Squires, 1993c; Koets, 1995).

An immunological involvement is also implicated in the means by which eCG secretion declines. It appears that the eCG-secreting fetal side of the endometrial cups is slowly rejected by the maternal system and, as a result, eCG secretion declines until day 100, at which stage all the cups have regressed. This hypothesis of an immunological rejection of eCG-secreting fetal cells is supported in work done on sister-to-brother matings; the fetus from such a mating has a relatively similar genetic make-up to its mother, and in such pregnancies eCG is secreted much longer than is seen in non-related matings and endometrial cups regress much later. Presumably the relatively similar genetic make-up of the fetus does not elicit the same rejection response as in non-related matings.

A hormonal role for eCG also appears likely, as eCG is known to have follicle stimulating harmone (FSH) like and luteinizing hormone (LH)-like biological properties in the mare and is used in other farm livestock as a super ovulation agent. It has been suggested, therefore, that eCG may be involved in follicular development in readiness for the formation of the secondary CL (Koets, 1995). However, eCG is not secreted until around days 35–40, though follicular growth, in readiness for secondary CL, may start as early as day 20. It is, therefore, unlikely to have a major role in follicle development prior to secondary CL formation. This pre day 35 follicle development may be driven by the continuing FSH pulses synonymous with those observed during dioestrus of the non-pregnant mare. The LH-like properties of eCG, however, may suggest a role in luteinizing the developed follicles to form the secondary CL; additionally, a role in the maintenance of the primary CL has also been suggested (Allen, 1975; Nett et al., 1975; Daels et al., 1991; Allen and Stewart, 1993; Koets, 1995).

6.3.3. Maternal oestrogens

The plasma concentration of maternal oestrogens also varies within pregnancy. Between days 0 and 35, levels remain very similar to those seen during the non-pregnant dioestrous period, but then rise sharply to reach 3–5 ng ml^{-1} by day 40. In the period of days 40–45, they decline slightly and subsequently remain at this constant level until days 60–70, after which they slowly rise again (Darenius et al., 1988; Le Blanc, 1991; Stabenfeldt et al., 1991).

These rising levels of maternal oestrogens between days 35 and 40 are thought to be secreted by the follicles developing prior to the formation of the secondary CL, in much the same way as oestrogens are produced by developing follicles prior to ovulation and oestrus in the normal oestrous cycle (Daels et al., 1991). These rising levels of oestrogens, however, do not result in normal oestrous behaviour. Evidence for the ovarian origin of these oestrogens is their absence in pregnant mares, which have undergone ovariectomy prior to day 40 and the delayed decline in concentrations seen after fetal death not accompanied by immediate CL regression (Daels et al., 1990, 1991, 1995;

Stabenfeldt et al., 1991). However, the second rise in oestrogen at days 60–70 is unaffected by ovariectomy but is affected by induced or spontaneous fetal death (Darenius et al., 1988) and can, therefore, be assumed to originate from the feto-placental unit. The precursors for such production originate in the fetal gonads (Pashan and Allen, 1979a). By day 85, oestrogens in the mare's peripheral blood system are higher than those detected in non-pregnant mares, and are diagnostic of pregnancy. The continuing rise in oestrogens after day 80 is due to increased feto-placental production of equilin and equilenin, two oestrogens unique to the pregnant mare (Raeside et al., 1979).

6.4. Late Pregnancy

As far as the discussion on hormone control is concerned late pregnancy can be classified as day 150 onwards (Fig. 6.2).

6.4.1. Progesterone

Progesterone levels, which, prior to day 150, were elevated and possibly slowly declining, remain at a steady 1–3 ng ml^{-1} until days 240–300, after which they steadily decline to basal levels pre partum. It has previously been reported that progesterone concentrations increase during the last 30–50 days of gestation. However, it is now known that this is not the case, erroneous results arising from the cross reactivity in the assays used between progesterone and its metabolites, in particular 5αpregnane, which do increase in the last 30–50 days of gestation (Hamon et al., 1991; Houghton et al., 1991; Squires, 1993a). These high concentrations of progesterone metabolites (progestagens) may themselves have a role pre partum. They have a similar structure to progesterone but are biologically inactive. As such, by binding to progesterone receptors high levels of progestagen cause the reproductive system to perceive even lower levels of circulating progesterone (Barnes et al., 1975; Moss et al., 1979; Holtan et al., 1991).

6.4.2. Oestrogens

Oestrogen levels within the maternal system continue to rise in late pregnancy, reaching a peak, between days 210 and 280, of approximately 8 ng ml^{-1}, the two main oestrogens being the equine-specific equilin and equilenin. Oestrogen levels

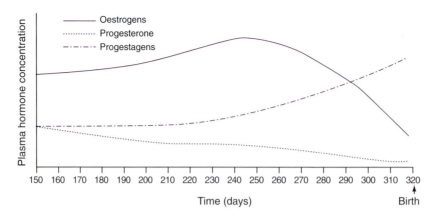

Fig. 6.2. A summary of the plasma hormone concentrations in the mare during late pregnancy (days 150–parturition).

then decline as parturition approaches, reaching levels in the order of 2 ng ml^{-1} at parturition (Raeside *et al.*, 1979; Holtan and Silver, 1992; Le Blanc, 1997).

6.4.3. Prostaglandin F2α

During the major part of late pregnancy, PGF2α remains at low levels, equivalent to those seen during early pregnancy. However, near to term, levels do increase slightly, in the form of short pulses, but significant elevations in concentration are not detected until parturition has started, when they play a major role in uterine myometrial contraction (see Chapter 8, this volume; Barnes *et al.*, 1978).

6.5. Conclusion

It is evident that the control of pregnancy in the mare follows a similar but not identical pattern to that seen in other mammals, with a dominant role played by progesterone and oestrogen throughout. The period of mobility, the production of eCG and the reliance on two successive populations of CL are particularly unique features of pregnancy in the mare.

7 The Physical Process of Parturition

7.1. Introduction

Parturition is the active expulsion of the fetus, along with its associated fluid and placental membranes. Gestation length varies significantly in the mare but the average gestation length is 320–335 days (Davies Morel *et al.*, 2002). It tends to be up to 2 weeks shorter in ponies than in Thoroughbreds and horses of the larger riding type (Rophia *et al.*, 1969; Rossdale *et al.*, 1984).

There are several signs that indicate that parturition is approaching. These may become evident at any time in the last 3 weeks of pregnancy. It must be remembered that these signs, detailed below, should not be used in isolation and that there is much variation between individuals and between successive pregnancies. Therefore, a combination of the following signs should be looked for, when watching for imminent parturition. It is also useful to have information on a mare's previous pregnancies, as general behavioural patterns may be characteristic to a particular mare.

7.2. Signs of Imminent Parturition

Changes in the appearance of the udder are one of the first signs of imminent parturition. During the last month of gestation, as lactogenesis (milk production) commences, the udder increases in size as colostrum is produced and stored (Forsyth *et al.*, 1975; Peaker *et al.*, 1979). The udder may feel relatively warm to the touch, as a result of the increased metabolic activity associated with milk production. At this time the udder may seem to increase in size at night, especially if the mare is kept in, and decrease during the day when she is let out and able to exercise; exercise increases circulation and reduces udder oedema (fluid accumulation). When there is no such apparent change in udder size between exercise (day) and standing in (night), parturition is imminent. At this stage, the udder is so full of milk that exercise no longer affects its size.

The extent to which udder size increases is dependent upon the size of the mare and her parity (number of previous foals; Rossdale and Ricketts, 1980). Figure 7.1 shows the udder of a Welsh Cob Section D mare 5 days prior to parturition.

The teats also change, initially becoming shorter and fatter as the udder fills and the bases of the teats are stretched. As the time for parturition approaches, the teats get filled as milk production increases; they elongate and become tender to touch. Some mares may even start to loose milk as production by the udder gets too great for its storage capacity and the sphincter at the end of the teat is breached. If a mare does start to loose milk, it is very important to minimize the loss. Milk at this stage is in fact colostrum, with a high concentration of immunoglobulins, and is vital for the transfer of passive immunity to the foal. As there is a finite amount of colostrum produced, if a mare is seen to loose milk or habitually does so, it is a good idea to milk her out a bit and store the collected colostrum for feeding to the foal immediately post partum. Colostrum can be successfully frozen for more than a year for use at a later date (Peaker *et al.*, 1979).

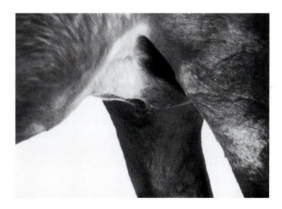

Fig. 7.1. The udder of a Section D mare 5 days prior to parturition.

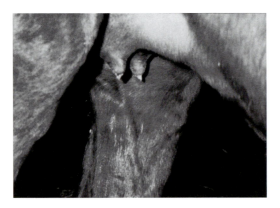

Fig. 7.2. One of the signs of imminent parturition is the accumulation of dried colostrum on the teats of the mare, termed waxing up; dried colostrum may also be seen on the insides of the hind legs.

Fig. 7.3. A further sign of imminent parturition is a hollowing of the hindquarters above the pelvis, as a result of a relaxation of the birth canal.

Many mares 'wax up', a term given to the clotting of colostrum at the end of the teat (Fig. 7.2). This is a good sign of imminent parturition. However, the lack of wax is not indicative that parturition is not imminent as these plugs of colostrum can easily be dislodged, especially in active mares.

The concentration of several minerals – sodium (Na), phosphorus (P), calcium (Ca) and potassium (K) – within the mammary gland secretions as parturition approaches is also indicative of the imminence of parturition. These parameters are advocated for use when attempting to assess fetal maturity prior to the artificial induction of parturition (see Section 15.3.1; Ousey *et al.*, 1984). In particular, Ca concentrations can be assessed via water hardener strips and used to indicate the closeness of parturition (Ley *et al.*, 1993). Ca concentrations in excess of $10 \, \mathrm{mmol} \, \mathrm{l}^{-1}$ are reported to be indicative of parturition (Ousey *et al.*, 1984; Cash *et al.*, 1985; Brook, 1987).

Changes in the birth canal also become apparent as parturition approaches. Approximately 3 weeks prior to parturition, hollowness may appear on either side of the tail root as the muscles and ligaments within the pelvic area relax. The whole area may appear to sink with this relaxation and so allow expansion of the birth canal during the passage of the foal. If the area on either side of the tail root is felt daily in the last 3–4 weeks of pregnancy, it may be possible to detect a change as the muscle tone relaxes (Fig. 7.3).

Changes in the mare's abdomen may also be evident in late pregnancy. As the fetus increases in size, the abdomen expands correspondingly, becoming characteristically large and pendulous. However, in the final stages of pregnancy, the abdomen appears to shrink as the foal moves up out of the lower abdomen and into the birth canal ready for delivery (Fig. 7.4).

As parturition approaches closer still, the mare becomes restless and agitated, especially as she enters first-stage labour. Some restlessness may also be apparent in late gestation; in feral herds, at this stage, the mare would move away to the periphery of the group in readiness to move away completely once labour started. As the mare moves into first-stage labour, her body temperature increases and she may sweat profusely (Haluska and Wilkins, 1989). Internally, her cervix will dilate and the vulva may appear to gape and secretions may be seen (Volkmann *et al.*, 1995). During first-stage labour she may show signs very similar to those indicative of colic, e.g. walking in circles, swishing tail, looking around at her sides, kicking her abdomen, etc. If a mare does show signs of colic in late gestation, it is pertinent to consider that it may in fact be first-stage labour, and so, her eating, drinking and defecating should be monitored.

As discussed, not all mares show all these symptoms, but a combination of one or two will give an accurate prediction that foaling is imminent. Some commercial products have been produced in an attempt to aid in the diagnosis. These make use of

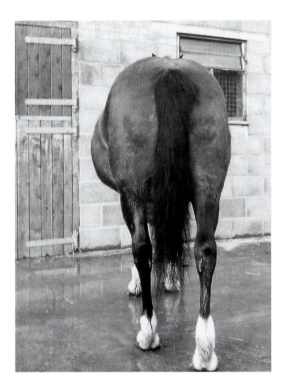

Fig. 7.4. One of the obvious signs of pregnancy is a large pendulous abdomen. However, immediately prior to parturition, the size of the abdomen appears to shrink as the foal moves up into the birth canal.

some of the mare's natural signals, such as reacting to the mare's movements, increase in body temperature or sweating, stretching of the vulval lips, etc.

(Shaw *et al.*, 1988; Amann *et al.*, 1989). Once triggered, they normally produce a signal transmitted to an audio and/or visual receiver. Close circuit television is also popular, allowing discrete observation.

7.3. The Process of Parturition

Parturition as in most mammals involves three distinct stages: stage 1, positioning of the foal and preparation of the internal structures for delivery; stage 2, the actual birth of the foal; and stage 3, the expulsion of the allantochorion (placental membranes). All three stages involve considerable myometrial activity, mainly within the uterus itself, but with some involvement of the abdominal muscles (Card and Hillman, 1993).

7.3.1. First stage of labour

Stage 1 involves uterine myometrial contractions and positions the foal ready for birth; it lasts between 1 and 4 h though exact timing is unclear as the mare may not show obvious signs of stage 1 labour immediately. Figure 7.5 illustrates the forces involved in this stage (Ginther, 1993).

The uterine muscles contract in mild waves from the tip of the uterine horn towards the cervix. These contractions, helped by the movement of the mare and, to a certain extent, by those of the foal, result in the repositioning of the foal and its passage into the birth canal, the area of least resistance (Haluska *et al.*, 1987a,b). Throughout late pregnancy the foal lies in a ventral-flexed position (its

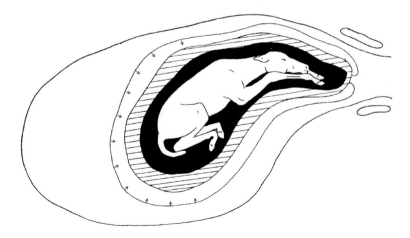

Fig. 7.5. The forces involved in the first stage of labour are provided by contractions of the uterine myometrium, as indicated by the arrows.

vertebrae lying along the line of the mother's abdomen) with its forelimbs flexed. As first-stage labour approaches and during first stage it rotates into an extended dorsal position, with its forelimbs, head and neck fully extended, and engaged in the birth canal (Fig. 7.6). This movement occurs normally 3–4 h prior to the commencement of second-stage labour. In some cases, a mare may show signs of first-stage labour and then cool off, only to show further signs several hours later; this is particularly evident in Thoroughbred mares (Jeffcote and Rossdale, 1979; Rossdale and Ricketts, 1980).

If parturition is to continue, other changes, in addition to the engaging of the foal, must occur. The cervix gradually dilates; this is encouraged during the later part of first-stage labour by the pressure of the allantochorionic membrane and the foal's forelimbs against the uterine side of the cervix. During the birth of a dead fetus, dilation of the cervix is less complete and slower, presumably as cervical dilation is actively

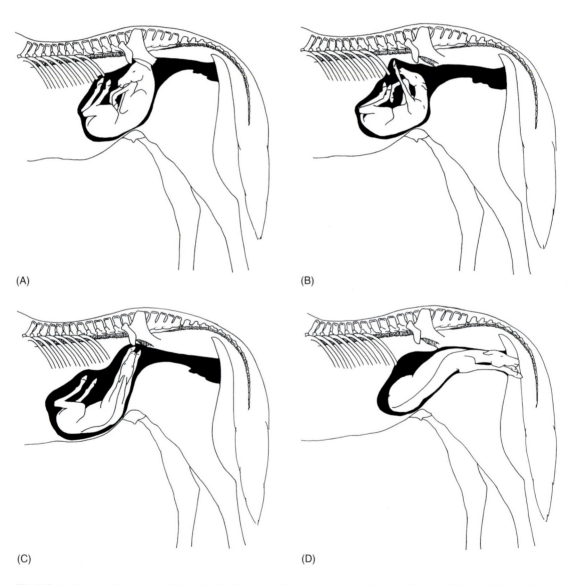

(A)

(B)

(C)

(D)

Fig. 7.6. During the first stage of labour the foal is gradually rotated and positioned within the birth canal in readiness for expulsion during the contractions of second-stage labour.

encouraged by the movements of the foal (Volkmann *et al.*, 1995). The exact mechanism by which parturition is induced and controlled is unclear (see Chapter 8, this volume). However, active movement of the foal against the cervix and within the birth canal has been shown to increase prostaglandin F metabolite (PGFM) levels, and hence, by inference, prostaglandin F2α (PGF2α) levels. Such an increase in PGF2α is not seen during the birth of a dead foal. The cervix has a high concentration of collagen, the ratio of collagen to muscle fibres progressively increasing from the uterine horn through the uterus to the cervix. It is thought that PGF2α affects this collagen, causing it to change its configuration and so allow the cervix to relax. A similar effect has been shown in the ewe (Volkmann *et al.*, 1995). The hormone relaxin may also have a role (Bryant-Greenwood, 1982).

During this time the vulva continues to relax and secretions increasingly collect within the vagina. At the end of first-stage labour, the foal's forelegs and muzzle push their way through the dilating cervix, taking with them the allantochorion. At the cervix, this membrane is termed the cervical star and is one of the three sites devoid of microcotyledons and, therefore, there is no attachment to the maternal endometrium. The other two are at entry to each Fallopian tube. The cervical star is the thinnest area of the placenta and hence it is this area that ruptures as the pressure of the myometrial contractions against the placental fluids increases, forcing them and the fetus through the cervix. The subsequent release of the allantoic fluid (breaking of the waters) is the trigger for the beginning of the second stage of labour.

7.3.2. Second stage of labour

The release of allantoic fluid at the start of stage 2 lubricates the vagina and is thought to trigger the stronger uterine contractions of the second stage. These strong contractions continue until the birth of the foal, normally within 20 min (acceptable range 5–60 min; Rossdale and Ricketts, 1980). At the start of stage 2 the amniotic sack is often visible bulging through the vulva (Fig. 7.7), within which the foal's forelegs and muzzle can be felt (Figs 7.8 and 7.9). Second-stage labour involves stronger contractions of the uterine myometrium, supplemented by abdominal muscle contractions.

The supplementary force provided by the abdominal muscles is termed voluntary straining. During voluntary straining, the mare inspires deeply, holding the rib cage and diaphragm at maximum inspiration

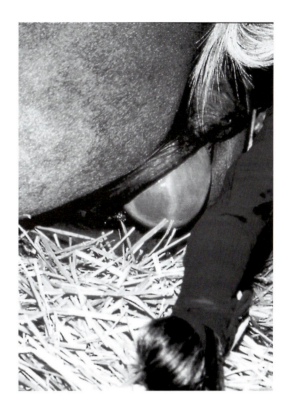

Fig. 7.7. The amniotic sack should be seen as a white membrane protruding from the mare's vulva. (Photograph courtesy of Steve Rufus.)

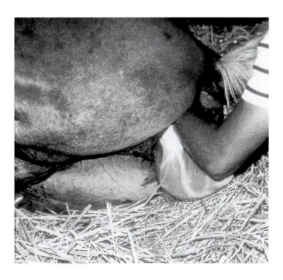

Fig. 7.8. At the start of second-stage labour, a brief internal examination may be made to ascertain whether or not the foal is correctly presented. (Photograph courtesy of Steve Rufus.)

Fig. 7.9. The foal's forelegs should be seen within the amniotic sack. (Photograph courtesy of Steve Rufus.)

At the end of first-stage labour, the foal lies in a dorsal-extended position. The soft tissue and surrounding bones govern the shape of the birth canal. The pelvis delineates the sides and ventral (bottom) part of the canal, and the sacral and coccygeal vertebrae delineate the dorsal (top, near the mare's vertebra) part. The diameter of the entry into the birth canal (20–24 cm) is slightly larger than the exit diameter (15–20 cm), and slightly more dorsal (nearer the mare's vertebra) than the exit. Hence, the foal is funnelled through the birth canal in a curved manner, being expelled ventrally (down towards the mare's hind legs; Figs 7.11 and 7.12; Rossdale and Ricketts, 1980).

The foal is delivered forelegs first, followed by the head lying extended between the forelegs, parallel with the knees. The two forelegs are not delivered aligned: normally one leg is delivered slightly in advance of the other, the fetlock of one being in line with the hoof of the other (Fig. 7.13). This misalignment of the forelegs reduces the cross-sectional diameter of the foal's thorax, which is the widest part of the foal, so reducing trauma to both mare and foal at birth and easing foaling. Once the thorax and shoulders have passed through the birth canal, the remainder of the birth process is relatively easy (Rossdale and Ricketts, 1980). At the end of stage 2 the foal lies with its head near the mare's hind legs, with its own hind limbs still within the mare (Fig. 7.12). The presence of the foal's legs within the mare's vagina has an apparent tranquillizing effect, most mares being reluctant to rise immediately post stage 2 (Rossdale and Ricketts, 1980).

and increasing pressure on the abdomen. The rib cage and abdominal muscles react by contracting against the pressure; this additional contraction force is transferred to the uterine contents, adding extra impetus to the expulsion of the uterine contents (Fig. 7.10). The area of least resistance is the cervix, and so the foal gets pushed forcefully further into the birth canal and out through the cervix and vagina.

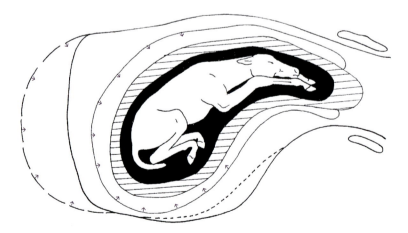

Fig. 7.10. Second-stage labour involves stronger contractions of the uterine myometrium, supplemented by contractions of the abdominal muscles, as indicated by the arrows.

Chapter 7

Fig. 7.11. The angle of the birth canal dictates that the foal is delivered in a curved manner, being expelled down towards the mare's hind legs. (Photograph courtesy of Steve Rufus.)

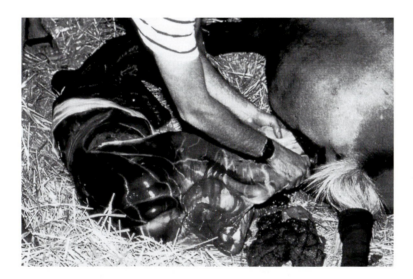

Fig. 7.12. The end of the second stage of labour is marked by the foal lying with its hind legs still within the vulva of the mare. (Photograph courtesy of Steve Rufus.)

7.3.3. Third stage of labour

Stage 3 of labour is normally completed within 3 h of the end of stage 2. Uterine contractions continue at a level similar to that evident during stage 1 labour, again originating at the uterine horns and passing down in waves to the cervix. At the same time, the allantochorion begins to shrink, as blood is drawn away from it towards the foal's pulmonary system (see Section 16.2.2.1). The blood vessels constrict and draw the allantochorion away from the uterine endometrium. This releases the remaining attachments between the allantochorion and the uterine epithelium and forces the placenta to be expelled inside out. The placenta is delivered with its red, velvety outer allantochorion innermost and the white smooth, inner allantochorion outermost (Figs

Fig. 7.14. The amniotic sack, the white membrane in the background and the allantochorion (placenta), in the foreground, after expulsion during the third stage of labour. As can be seen, the placenta is expelled inside out, the red, velvety, outer allantochorion being inside and the white smooth inner allantochorion being outermost.

Fig. 7.13. In order to reduce the diameter of the foal's thorax passing through the birth canal the foal is delivered with one foot slightly in advance of the other. (Photograph courtesy of Steve Rufus.)

7.14 and 7.15). The contractions of third-stage labour also help to expel any remaining fluids and assist in uterine involution (recovery).

7.4. Hippomane

Passed out either during second-stage labour or along with the placenta during third-stage labour is a small, brown, leathery, structure termed the hippomane (Fig. 5.32). It is found within the allantoic fluid and is an accumulation of waste salts and minerals collected throughout pregnancy. It has a high concentration of calcium, magnesium, nitrogen, phosphorous and potassium, and is first evident at around day 85 of pregnancy. It has been associated with much folklore, including aphrodisiac properties and responsibility for keeping the foal's mouth open. There is, however, no evidence to support these claims.

7.5. Conclusion

The diagnosis of immanent parturition is challenging in the mare, which may demonstrate a range of

Fig. 7.15. A close-up of the placenta in its inside out state, illustrating the innermost red allantochorion and the outermost smooth white allantochorion.

signs over a period of time. However, once started, parturition itself occurs rapidly over a matter of an hour or so and without problems requires minimal interference from man.

8   The Endocrine Control of Parturition

8.1. Introduction

Parturition in the mare occurs at approximately 11 months (320–335 days), though ranges as large as 315–388 days have been reported to result in viable full-term foals (Davies Morel *et al.*, 2002; Perez *et al.*, 2003). The exact length of gestation, however, varies with the type of horse, for example, in general, ponies tend to foal on average 2 weeks before Thoroughbred mares (Rophia *et al.*, 1969; Rossdale *et al.*, 1984).

Within these averages, many factors may influence the exact timing of parturition including environmental, fetal and maternal factors. Environmental factors include the season of mating. Mares mated early in the season tend to have longer gestations than those mated later on. This is presumably nature's way of compensating for early and late matings, trying to ensure all mares foal at the most optimum time of the year for foal survival – i.e. during spring (Howell and Rollins, 1951; Hodge *et al.*, 1982). Climate (Astudillo *et al.*, 1960), year of breeding or foaling (Howell and Rollins, 1951; Rophia *et al.*, 1969) and nutrition are also reported to have an effect. Lastly, nutritional deprivation in the final trimester is reported to cause early parturition.

Fetal factors include the genotype of the offspring. This can be demonstrated by comparing the gestation lengths of various crosses within the equine species. A stallion-cross-mare fetus has an average gestation of 340 days, a stallion-cross-jennet fetus 350 days, a jack-cross-mare fetus 355 days and a jack-cross-jennet 365 days (Rollins and Howell, 1951; Ginther, 1992). Breed of foal is also a reported factor (Jochle, 1957), as is foal gender; colt foals have pregnancies on average 2.5 days longer than filly foals (Rophia *et al.*, 1969; Hevia *et al.*, 1994; Panchal *et al.*, 1995; Davies Morel *et al.*, 2002). Multiple births also have shorter gestations than singles (Jeffcote and Whitwell, 1973).

Finally, maternal factors, the mare herself has some control over the exact time of delivery. The majority of mares foal at night undisturbed (Fig. 8.1; Rossdale and Short, 1967; Goater *et al.*, 1981; Roberts, 1986c; Hines *et al.*, 1987). There is evidence that this is linked to a circadian rhythm in oxytocin secretion around parturition (Nathanielsz *et al.*, 1997). In addition, the mare's age (Bos and Van der May, 1980), parity (Britton and Howell, 1943; Panchal *et al.*, 1995), foaling-to-conception interval (Britton and Howell, 1943), genotype (Rophia *et al.*, 1969) and breeding-to-ovulation interval (Ganowiczow and Ganowicz, 1966) have all been reported to affect gestation length.

8.2. Initiation of Parturition

Birth involves the expulsion of the fetus plus all associated placental membranes and fluid and is achieved by myometrial activity within the uterus and surrounding structures (see Chapter 7, this volume). Myometrial activity is, therefore, central to parturition and is inhibited by elevated progesterone and depressed oestrogen concentrations, characteristics of pregnancy in most mammals. At full term, the ratio of progesterone to oestrogen reverses, removing any inhibition and allowing myometrial activity to be driven by elevated prostaglandin F2α (PGF2α) and oxytocin concentrations, both of which can be seen to rise at parturition. Efficient expulsion of the fetus and placental tissue is dependent upon sequential contraction of the whole uterine myometrium. There must, therefore, be immediate activation of muscle cells and efficient cell-to-cell excitation. This message transfer is affected by circulating hormone concentrations; elevated progesterone concentrations reduce the spread of muscle cell excitation and contraction (Liggins, 1979; Holtan *et al.*, 1991; Rossdale *et al.*, 1997). The exact mechanism for the initiation of this myometrial contraction for parturition in the horse or any other *Equidae* is as yet unclear. However, in other mammals two alternative mechanisms are apparent.

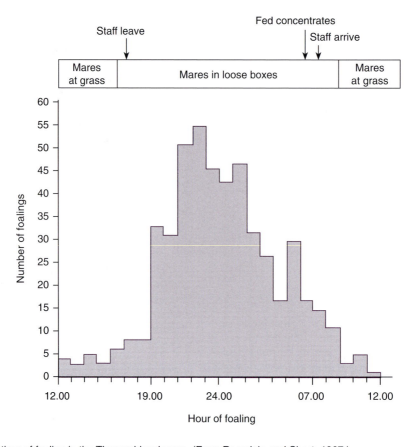

Fig. 8.1. The time of foaling in the Thoroughbred mare. (From Rossdale and Short, 1967.)

First, as seen in the ewe, nanny goat, sow and cow, the fetus itself actively controls the initiation of its own parturition. Towards the end of gestation, the fetus comes under increasing stress, due to hypoxia (a shortage of oxygen), physical restriction within the uterus and an inability of the placenta to provide enough nutrients for growth and adequately remove waste products. These increasing stress levels activate the fetal hypothalamic–pituitary–adrenal axis causing the production of adrenocorticotrophic hormone (ACTH) by the fetal anterior pituitary. ACTH activates the fetal adrenals to produce corticoids, so fetal cortisol levels increase. Cortisol passes to the placenta where it affects the metabolic pathways involved in conversion of progesterone to oestradiol. Under the influence of fetal cortisol, three enzymes in this pathway are activated, and so increasing the conversion of progesterone to oestrogen (Flint *et al.*,

1979; Power and Challis, 1987). As a result, the characteristic rise in oestrogens and fall in progesterone as required for myometrial activity is achieved.

The second apparent method by which parturition is initiated is seen in primates. In such mammals, the start of parturition is determined by a genetically controlled maturation signal linked closely to the time of gestation. It appears to be this maturation signal, and not fetal stress, that activates the fetal hypothalamic–pituitary–adrenal axis. In response, the adrenals produce increased levels of androgens, the precursor for placental oestrogens. Hence, elevated oestrogens are observed. In primates there is no involvement of the fetal adrenocorticoids. However, the result is the same, increasing oestrogen concentrations (Nathanielsz *et al.*, 1997).

Regardless of the exact means of initiating parturition, the end result is an increase in the ratio of oestrogen to progesterone. Such a ratio, particularly elevated oestrogen levels, is known to be linked to increasing oxytocin production by, and oxytocin receptors on, the endometrium (Fuchs *et al.*, 1983; Wu and Nathanielsz, 1994). Rising oestrogen levels result in rising oxytocin levels and as oxytocin is one of the two major inducers of uterine muscle contractions it plays a central role in parturition. The second inducer of myometrial contractions is PGF2α, which, as seen previously (see Section 3.2.2.7), is produced by the uterine endometrium in response to rising oxytocin levels. So elevated oestrogen levels, characteristic of the end of pregnancy, drive the production of both oxytocin and PGF2α; both these are major activators of uterine myometrial activity required during parturition. Initially it was thought that PGF2α played the major role, being involved in all three stages of parturition, whereas oxytocin was thought to be primarily involved in stages 2 and 3. However, the importance of oxytocin is becoming increasingly evident, and it may indeed play more of a central role in driving the contractions of labour than previously thought (Nathanielsz *et al.*, 1997).

In addition to a purely hormonal drive to myometrial contraction there is a neural involvement. The increasing pressure from the allantoic fluid and fetus, on the inside of the cervix, as it is pushed up into the birth canal sends a neural message to the hypothalamus, via the spinal chord. The hypothalamus activates the posterior pituitary, which in response produces oxytocin, which significantly elevates circulating oxytocin concentrations, which in turn further drives the major uterine contractions, particularly those in stage 2 (Wu and Nathanielsz, 1994; Nathanielsz *et al.*, 1997).

There is also evidence of a circadian rhythm in oxytocin production. This rhythm is inherent but may be modulated by daylight, night-time being associated with increased oxytocin production (Nathanielsz *et al.*, 1997), so in part explaining the increased incidence of parturition during the hours of darkness. The mare herself also appears to have some fine control over the exact timing of oxytocin release.

In summary, elevated oestrogen levels drive the production of oxytocin and PGF2α, which in turn, provided progesterone levels are low, drive the uterine myometrial activity associated with contractions of labour. Additional posterior pituitary oxytocin provides the extra impetus required for second-stage labour.

Although there is no conclusive evidence as to which mechanism occurs in the mare, most of the evidence available indicates a system similar to that seen in the ewe (Silver, 1990; Silver and Fowden, 1994; Ousey *et al.*, 2004). Prolonged elevated corticosteroid levels in the later part of gestation have not been reported in the equine fetus, though this may be due to difficulties encountered in catheterizing such fetuses. However, significantly elevated corticosteroid levels have been reported in the equine fetus 72–96 h prior to parturition, reaching a peak 30–60 min post partum (Silver and Fowden, 1994). It has been demonstrated that cortisol is essential in foals for the final maturation of several internal organs, including the respiratory and digestive tract. Pashan and Allen (1979a,b) presented evidence that parturition in the equine is influenced by fetal stress, via an interaction between fetal and placental size, and that fetal constriction may be a trigger. This hypothesis is further supported by Rossdale *et al.* (1992), who demonstrated that treatment of fetuses *in utero* with ACTH resulted in an increase in corticosteroid production by the fetal adrenals, which caused premature parturition, as observed in ewes. It appears that in foals cortisol levels immediately prior to parturition are elevated more dramatically but over a shorter period of time than is evident in the ewe and cow. The relatively short period of elevated cortisol levels apparently reflects the rapid maturation of the equine fetal adrenals, which is a necessity for post-partum survival and which occurs in the last 3–5 days of gestation (Chavatte *et al.*, 1997b; Le Blanc, 1997; Rossdale *et al.*, 1997).

8.3. Endocrine Concentrations

The hormones involved in parturition will be considered in turn (Fig. 8.2).

8.3.1. Oestrogen

Oestrogen concentrations in the maternal blood system fall over the last 30 days of gestation, and, as they originate from the fetal-placental unit, reach basal levels within hours of parturition. Concentrations are approximately 6 ng ml^{-1} at 30 days prior to parturition and fall to less than 2 ng ml^{-1} after parturition (Nett *et al.*, 1973; Pashan, 1984).

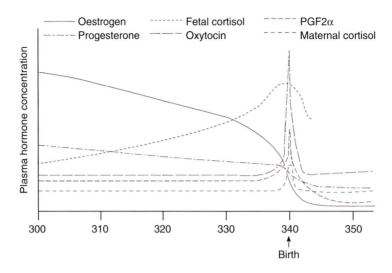

Fig. 8.2. A summary of the main changes in hormone concentration evident at parturition.

8.3.2. Progesterone

Progesterone concentrations decline to basal levels at parturition, though concentrations of progesterone metabolites such as 5αpregnanes peak at 10–15 ng ml⁻¹ 10 days pre partum (Hamon *et al.*, 1991; Holtan *et al.*, 1991; Schutzer and Holton, 1995). These metabolite concentrations reflect an alteration in placental metabolism towards oestrogen production during late pregnancy and parturition. These progestagens also occupy the progesterone-binding sites on the uterine endometrium and, as such, their increase in concentration is perceived by the uterus as a further decrease in circulating progesterone. It is likely that it is the ratio of oestrogen to progesterone, rather than the absolute levels, which is important. Therefore, even though oestrogen levels decline, it is possible that their decline is significantly less than that perceived for progesterone, with, in relative terms, the ratio of oestrogen to progesterone increasing, allowing myometrial activity to commence (Purvis, 1972; Moss *et al.*, 1979; Rossdale *et al.*, 1991).

8.3.3. Prostaglandin F2α

As discussed previously (Chapter 3, this volume) prostaglandin F2α metabolite (PGFM) is used as an indicator of PGF2α concentrations. From measurements of PGFM, it can be deduced that PGF2α concentrations rise sharply in the peripheral plasma of the mare at term, mainly during the second stage of labour. Fetal PGF2α levels, as determined by catheterization, increase more gradually over the final weeks (40 days) of pregnancy (Barnes *et al.*, 1978; Cooper, 1979; Silver *et al.*, 1979). PGF2α may also be detected in the allantoic fluid near to parturition. The major functions of PGF2α are to induce cervical dilation and act as a strong inducer of uterine myometrial activity. PGF2α is primarily associated with first- and second-stage labour.

8.3.4. Oxytocin

The actions of PGF2α and oxytocin are closely linked and they tend to show the same pattern of release. Oxytocin concentrations in the maternal system rise sharply at parturition, especially during the second stage. The central role of oxytocin is increasingly becoming evident and, like PGF2α, its primary role is in the induction of strong myometrial contractions, in particular those of second-stage labour, when it works in concert with PGF2α (Fuchs *et al.*, 1983; Haluska and Currie, 1988). Oxytocin is also involved in the myometrial contractions of third-stage labour, the expulsion of the placenta; hence its use in cases of retained placenta (Hillman and Ganjam, 1979). The reported circadian rhythm to oxytocin release has been discussed previously (see Section 8.2).

8.3.5. Cortisol

Cortisol concentrations do not change significantly in the maternal system during pregnancy, though they rise, due to stress, at parturition. Changes within the fetal system occur in late pregnancy with a sudden rise in cortisol evident 3–5 days prior to parturition (Nathanielsz *et al.*, 1975; Liggins *et al.*, 1979). As previously discussed (Section 8.2), this increase is shorter and sharper than the gradual increase observed in the sheep and goat and is known to be associated with the maturation of the fetal adrenal cortex and its ability to react to circulating ACTH (Silver and Fowden, 1994; Nathanielsz *et al.*, 1997). It may also be involved in final organ maturation – for example, the respiratory system – in the equine fetus (Rossdale *et al.*, 1973; Alm *et al.*, 1975).

8.3.6. Prolactin

Prolactin concentrations are reported to increase in the last 7–10 days. It is not apparent that prolactin has a direct role in parturition, but its increase at this time may indicate a role in equine lactation, as seen in other mammals (see Chapter 10, this volume; Forsyth *et al.*, 1975; Worthy *et al.*, 1986; Nett, 1993b).

8.3.7. Relaxin

Plasma concentrations of relaxin are reported to be elevated in late pregnancy/parturition, its production site being the placenta. In late pregnancy, it is thought to maintain the quiescent nature of the uterine myometrium, but during parturition it is overcome by the stronger activation of oxytocin and PGF2α. Relaxin may also have a role in the relaxation and softening of pelvic ligaments and cervix as parturition approaches (Bryant-Greenwood, 1982, Stewart *et al.*, 1982b).

8.4. Conclusion

It is evident that the mechanism for the initiation of parturition in the mare is unclear, but is likely, at least in part, to be due to fetal stress. Clarification of the hormonal control of equine parturition would be very beneficial as it would enable more accurate prediction of parturition and the successful artificial induction of parturition in the case of emergencies.

9 The Anatomy and Physiology of Lactation

9.1. Introduction

The mammary glands are situated along the ventral midline in all mammals in a varying number of pairs. The mare normally has four glands – two pairs. In most mammals, each gland exits via its own teat; however, in the mare each pair of glands on either side of the midline exits via a single teat. The anatomy and physiology of lactation specific to the equine have not been widely studied. Other reviews on the subject include Jacobson and McGillard (1984), Mepham (1987), Kainer (1993) and McCue (1993).

9.2. Anatomy

The mammary glands of the mare are situated in the inguinal region between the hind legs, covered and protected by skin and hair, except for the teats, which are devoid of hair. The whole of the skin surface is supplied with nerve endings, the concentrations of which are increased in the teat area, enhancing the response to touch. The mare normally has four glands, two larger cranial ones and two smaller caudal ones, though six glands have been reported in the occasional mare. Each of the four mammary glands is completely independent, with no passage of milk from one quarter to another. They are separated and contained within a fibroelastic capsule and supported by the medial suspensory ligament, running along the mare's midline. Further support is provided by the lateral suspensory ligaments, running over the surface of the mammary gland itself under the skin, and by laminae, developing off the suspensory ligament and penetrating the mammary tissue in sheets (Fig. 9.1; Sisson, 1975; Kainer, 1993).

Each udder half, on either side of the midline, is made up of two quarters and the openings from these two quarters exteriorize via a single teat (Fig. 9.2).

The mammary tissue itself is made up of millions of alveoli and connecting ducts. This arrangement can be compared to a bunch of grapes, each alveolus being equated to a grape and the ducts to the branches (Fig. 9.3).

The alveoli are grouped together in lobules and then the lobes into lobes. These lobes join together via a branching duct system, which eventually leads to the gland cistern. Each quarter has its own gland cistern draining into a teat cistern and on to the streak canal, one from each quarter on that side. At the end of each teat is the Rosette of Fustenburg, a tight sphincter that prevents the leakage of milk between sucklings. This sphincter can withstand a considerable build-up of milk pressure, though occasionally it may be breached, as in the case of mares that lose milk when parturition is immanent (Kainer, 1993).

The alveoli, which are the milk-secreting structures, are lined by a single layer of lactating epithelial cells, surrounding a central cavity or lumen. This alveolar lumen is continuous with the mammary duct system. Milk is secreted by the lactating cells into the alveolar lumen across the luminal or apical membrane. Surrounding each alveolus is a basket network of myoepithelial cells. These muscle cells also surround the smaller ducts, and their contraction is activated as part of the milk ejection reflex. Surrounding these myoepithelial cells is a capillary network supplying the alveoli with milk precursors and providing hormonal control; there are also a series of lymph vessels. In addition, the alveoli have a nerve supply, which is responsible for the activation of the myoepithelial basket cells and also vasodilation and constriction of the capillary supply network (Fig. 9.4; Mepham, 1987).

The mammary gland as a whole is supplied with blood via two mammary or external pudendal arteries, one on each side of the midline and entering the caudal end of the gland. Venous return from the mammary gland is via the venous plexus at the base of the gland and then on to the superficial vein of the thoracic wall (the subcutaneous abdominal milk vein) or via the external pudendal vein. Both the external pudendal artery and vein enter and leave the body in the inguinal region. The subcutaneous

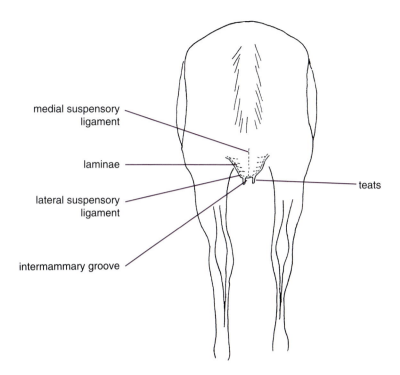

Fig. 9.1. A caudal (tail end) view of the mare's udder illustrating the suspensory apparatus of the mammary gland.

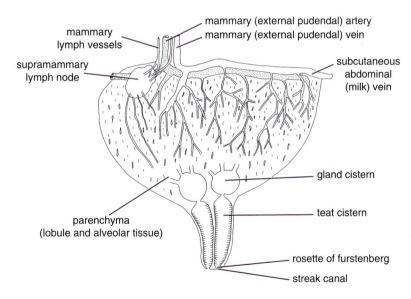

Fig. 9.2. A cross section through the mammary gland of the mare illustrating the exit of two quarters via a single teat.

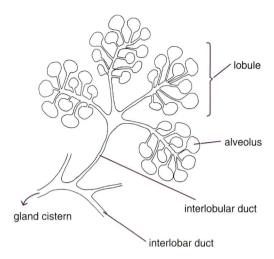

Fig. 9.3. A single lobe of the mammary gland made up of several lobules which in turn are made up of numerous alveoli.

abdominal vein, which runs along the abdomen of the mare, can be seen more clearly in lactating mares and is hence sometimes referred to as the milk vein. The mammary gland also has two supramammary lymph nodes, one on either side of the midline, at the base of the udder, and connecting the main circulatory lymph system to that of the mammary gland itself (Saar and Getty, 1975).

9.3. Mammogenesis

Mammogenesis, or mammary development, is first evident in the embryo. Glands develop along either side of the midline in the inguinal region. Cells in this region proliferate to form nodules that develop to form mammary buds, evident from day 50 of gestation. At birth, teats are present, along with a few short branching ducts within the connective tissue associated with each teat.

From birth to puberty, mammary gland growth is isometric with (at the same rate as) body growth. Most of this prepubertal growth is an increase in fat and connective tissue, rather than duct development. Puberty marks a change, as mammary development becomes allometric with (at a greater rate than) body growth. Beyond puberty, mammary growth increases and decreases with the oestrous cycle. The amount of mammary development within these cycles depends on the length of the dioestrous phase of the oestrous cycle, as elevated progesterone levels are responsible for mammary lobular-alveolar development. In the mare, the duration of dioestrous is such that just limited lobular-alveolar development takes place.

During pregnancy, elevated progesterone levels again cause significant lobular-alveolar development, especially in the last trimester. In the last 2–4 weeks of pregnancy, lactogenesis (milk production) also predominates (Leadon *et al.*, 1984; Ousey *et al.*,

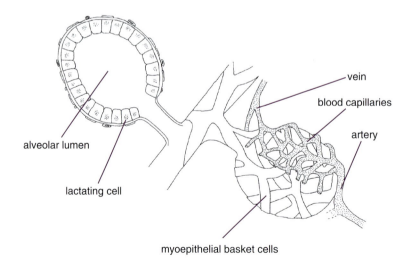

Fig. 9.4. The mammalian alveolus. On the left, a cross-sectional view illustrating the lactating cells surrounding the alveolar lumen, which is continuous with the mammary duct system. On the right, an alveolus illustrating the myoepithelial basket cells and alveolar blood supply.

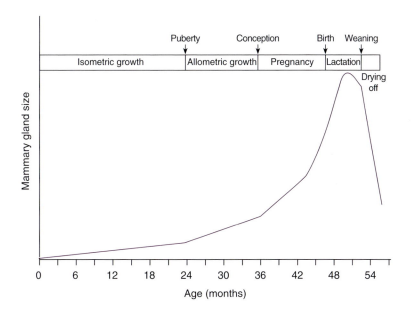

Fig. 9.5. Equine mammary development from birth to 54 months, including development during a mare's first pregnancy.

1984; Mepham, 1987). During lactation mammogenesis continues, as cell division increases in line with milk production to satisfy the increasing demands of the foal. Cell division then decreases after the maximum yield has been reached. At the same time, the size of the mammary gland slowly decreases until it returns to its normal non-lactating size post weaning (Fig. 9.5).

9.4. Lactation Curve and Milk Quality

There has been significantly less research conducted into the lactation of the mare compared with other livestock, especially the cow. Except in a very few cultures, mare's milk is of indirect rather than direct commercial importance, its value being assessed via the standard of foal reared, rather than directly by milk yield. As such, it is often not given the attention it warrants. The following discussion will concentrate on conclusions drawn either directly from experiments using mares or extrapolated from other species studied.

9.4.1. Lactation curve

There is much variation in the lactation curve demonstrated by different mares, largely due to man's interference and early weaning. As a general trend,

milk yields in mares tend to increase during the first 2–3 months post partum. Initial levels in the first 2 weeks are in the order of 4–8 l day^{-1} for Thoroughbreds and 2–4 l day^{-1} for ponies (Fig. 9.6).

Milk production reflects demand, which in turn reflects the size of the foal, and production therefore continues to rise as the foal grows until 2–3 months post partum, when maximum levels of 10–18 l day^{-1} in Thoroughbreds and 8–12 l day^{-1} for ponies are reached (Tyznik, 1972; Oftedal *et al.*, 1983; Doreau *et al.*, 1990; McCue, 1993). After 3 months, the foal's demand for nourishment from its mother decreases, as it starts to increasingly investigate grass or hay and its mother's hard feed. As the weaning process progresses towards full weaning, the lactational yield drops off further with decreasing demand (Jacobson and McGillard, 1984; Doreau and Boulet, 1989; Smolders *et al.*, 1990). As will be discussed later, milk quality also declines at this time, further encouraging the foal to seek nourishment elsewhere and hasten the weaning process.

Lactation naturally lasts nearly a full year, the mare drying up completely a few weeks before she is due to deliver the following year's foal. However, in today's managed systems, humans normally dictate the length of lactation by weaning foals at about 6 months, at which time milk yield is less than that immediately post partum. At this stage

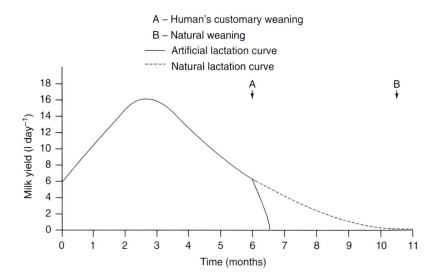

A – Human's customary weaning
B – Natural weaning
—— Artificial lactation curve
----- Natural lactation curve

Fig. 9.6. The average lactation curve for a mare illustrating the natural extent of lactation along with that customarily imposed by man.

the foal is obtaining little of its nourishment from its mother, deriving most from roughage and/or concentrate feeds provided the foal is doing well. Weaning at 6 months, therefore, has little long-term effect on the foal's development (Fig. 9.6).

The total milk yield of a Thoroughbred or one of the larger riding-type mares is 2000–3000 kg of milk per lactation. As a rough guide, in these larger horses, the natural daily milk yield averaged out over the whole lactation is 2–3 kg 100 kg^{-1} body weight. The corresponding equation for ponies is 5 kg 100 kg^{-1} body weight (Oftedal *et al.*, 1983). The foal normally suckles up to four times per hour during the first week, reducing to once an hour by week 10. Initial suckling ensures an intake of little but often; however, with age, the frequency of suckling declines and the intake per suckle increases up to 250 g or so at each suckling for larger riding-type horses (Frape, 1998). The number of suckles per day and the amount of milk taken per suckle reduce from peak lactation towards weaning.

9.4.2. Milk quality and composition

The composition of milk reflects the requirement of the young of that particular species and provides the energy and precursors needed for growth throughout lactation. In the case of some mammals, including the foal, milk additionally provides

immunoglobulins during the initial stages of lactation (Table 9.1; Ullrey *et al.*, 1966; Peaker *et al.*, 1979; Mepham, 1987; Smolders *et al.*, 1990; Malacarne *et al.*, 2002).

9.4.2.1. Colostrum

Colostrum, the first milk produced ready for the foal immediately after parturition, contains a relatively high concentration of proteins, immunoglobulins. Protein concentration in colostrum is in the order of 13.5% compared to 2–4% in the main lactational milk. The main protein immunoglobulin in mare's colostrum is Immunoglobulin G (IgG; 8911 mg dl^{-1}); IgA (957 mg dl^{-1}) and IgM (122 mg dl^{-1}) are of less

Table 9.1. Comparative milk compositions of several species, expressed as percentages. (From Jennes and Sloane, 1970.)

Species	Total solids	Fat	Casein protein	Whey protein	Lactose
Human	12.4	3.8	0.4	0.6	7.0
Cow	12.7	3.7	2.8	0.6	4.8
Goat	13.2	4.5	2.5	0.4	4.1
Sheep	19.3	7.4	4.6	0.9	4.8
Horse	11.2	1.9	1.3	1.2	6.2

Chapter 9

importance (Kohn *et al.*, 1989; McCue, 1993). This high protein concentration is at the expense of fats, which are present in relatively low concentrations. However, within 12–24 h protein levels fall dramatically and fat levels rise. The relative concentrations within milk now stabilize, though both protein and lipid concentrations tend to decline gradually over time. Lactose remains largely unchanged throughout the remainder of lactation (Table 9.2; Ullrey *et al.*, 1966; Forsyth *et al.*, 1975; Gibbs *et al.*, 1982; Smolders *et al.*, 1990). The digestive system of the foal is 'permeable' to the complete protein molecules, such as immunoglobulins, for the first 24 h of life. This 'permeability' is due to enterocytes within the wall of the small intestine which absorb whole proteins via pinocytosis. After 24 h, this ability is irreversibly lost, as the enterocytes are replaced (McCue, 1993). It is essential, therefore, that a newborn foal receives its colostrum well within 24 h of birth, as after this time it cannot take advantage of the immunoglobulins carried by colostrum and they will be broken down by proteolytic enzymes within the intestine into their component amino acids and absorbed as such. The average composition of milk during the main part of lactation in the mare is given in Table 9.2.

Table 9.2. The average composition of the milk during the main part of lactation in the mare. (From Ullrey *et al.*, 1966; Oftedal *et al.*, 1983; Schryver *et al.*, 1986; Frape, 1989; Doreau *et al.*, 1990; Saastamoinen *et al.*, 1990; Martin *et al.*, 1991; Caspo *et al.*, 1995; Malacarne *et al.*, 2002; National Research Council, 2007.)

Component	
Water (%)	89.0
Protein (g kg^{-1})	19–40
Lactose (g kg^{-1})	51–69
Fat (g kg^{-1})	6–20
Energy (kcal 100 g^{-1})	46–60
Ash (minerals, vitamins, etc.; g kg^{-1})	0.6–3
Ca (mg kg^{-1})	600–1200
P (mg kg^{-1})	230–800
Mg (mg kg^{-1})	30–100
K (mg kg^{-1})	400–700
Na (mg kg^{-1})	160–246
Cu (µg kg^{-1})	200–450
Zn (µg kg^{-1})	1800–2500

9.4.2.2. Fat

The concentration of fats or lipids in mare's milk is reported to be relatively low when compared with other species. However, there is some suggestion that this may be due to sampling error, as the highest concentration of fat is evident in the last milk milked out, and which is not easily obtained. Fat is present in milk in the form of globules of saturated fat, cholesterol and unsaturated fats, as free fatty acids, phospholipids and triglycerides. The 8% concentration of triglycerides as a proportion of total fats is much lower than the 79% in cows. These fat globules exist as an emulsion within the milk and contain a high concentration of short-chain fatty acids, less than 16 carbons (C) in length.

9.4.2.3. Proteins

Proteins during the main lactation are present in the form of near equal proportions, approximately 1.3% caseins and 1.2% whey. Caseins are unique to milk and have several functions. Under the influence of the stomach's acid pH, they form a clot with the enzyme rennin. This clot facilitates the digestion of proteins by the proteolytic enzymes of the digestive system. Caseins also contain essential amino acids and aid in the transport of minerals from the mare to the foal via milk. Caseins associate with calcium (Ca), phosphate (P) and magnesium (Mg) ions to form micelles, thus allowing a higher concentration of these minerals to be transported in milk than would be possible in a simple aqueous solution.

Two types of whey proteins are found in mare's milk and, unlike caseins, do not precipitate in acid pH. The whey proteins are divided into those that are specific to milk and those that can be found in both milk and blood. Those specific to milk can be further subdivided into β lactoglobulin (28–60% of whey proteins) and α lactalbumin (26–50% of whey proteins; Gibbs *et al.*, 1982). α lactalbumin is a good source of amino acids and is rich in essential amino acids such as tryptophan. It is also the B component which, along with an A component, makes up the two halves of the enzyme lactase synthetase. Lactase synthetase is the terminal enzyme in the synthesis of lactose, the major sugar component of mare's milk. The second type of whey proteins found in mare's milk constitutes ones also found in blood: serum albumin (2–15% of whey proteins) and serum globulin (11–21% of whey proteins;

Gibbs *et al.*, 1982). Serum albumin is identical to blood serum albumin and is directly transferred unchanged from the blood through the lactating cell to the alveolar lumen. It is, therefore, only found in small concentrations, unless there has been cellular damage or haemorrhage within the mammary tissue. Serum globulin, on the other hand, is the immunological fraction of milk and, therefore, its concentration is very high in colostrum. Antibodies attach themselves to these globulins and it is via these that the foal attains its passive immunity.

9.4.2.4. *Lactose*

Lactose is the energy component of mare's milk (5.9–6.9%). Unique to mammals, each lactose molecule consists of a molecule of galactose and one of glucose. In the foal's intestine lactose is split into its two component parts, galactose then being easily converted into glucose. Lactose is therefore, in essence, two molecules of glucose. The question then arises as to why lactose, not glucose, is present in milk, especially as there is an energy cost in converting glucose to lactose and vice versa. The answer lies in the effect of glucose on the osmotic pressure of milk relative to blood. The osmotic pressure of the two must be the same, and the component of milk that has the largest effect on osmotic pressure is the small molecule of lactose. However, if glucose were present, it would have an even greater effect on the difference in osmotic pressure. Additionally, one molecule of lactose gives rise to two molecules of glucose; that is, one molecule of lactose has twice the calorific value per molecule than glucose, and hence also per unit of osmotic pressure. It has also been suggested that lactose provides a more beneficial medium for intestinal activity, regulates bacterial flora and stabilizes pH, so aiding the absorption of minerals (Mepham, 1987; Smolders *et al.*, 1990; McCue, 1993).

9.4.2.5. *Minerals*

Mineral concentrations also vary with the stage of lactation. Potassium (K) and sodium (Na) concentrations in colostrum tend to be high, up to 1200 and 500 mg kg^{-1}, respectively, dropping to 700 and 225 mg kg^{-1} within 1 week and further dropping to 500 and 150 mg kg^{-1} in weeks 9–21. Ca tends to be slightly raised in colostrum, but then falls slightly within hours only to rise again to a peak of 1200 mg kg^{-1} at 3 weeks post partum, but then

drops as lactation progresses to 800 mg kg^{-1} at weeks 9–21. Mg levels are also elevated, at 500 mg kg^{-1}, in colostrum but fall off rapidly in the first 12 h and then continue to decline slowly throughout the rest of lactation to 45 mg kg^{-1} in weeks 9–21. Phosphorous (P) remains relatively steady at 400–700 mg kg^{-1}, for the first 8 weeks of lactation and then concentration declines slightly (Ullrey *et al.*, 1966; Oftedal *et al.*, 1983; Schryver *et al.*, 1986; Saastamoinen *et al.*, 1990; Martin *et al.*, 1991; National Research Council, 2007).

It is evident that there is a trend for the concentration of all the nutrient components of milk to decline as lactation proceeds. This is nature's way of encouraging the foal to obtain its nourishment elsewhere (Smolders *et al.*, 1990).

9.5. Milk Synthesis

Milk is synthesized in the epithelial or lactating cells lining each alveolus. The precursors of, and components for, milk are obtained from the blood system supplying the udder. These components cross the basal membrane into the lactating cells. There is little information on how they pass across this membrane, but as the molecules are small it seems likely that the majority pass by diffusion. The protein, fat and lactose components of milk are then built up within the lactating cells and pass across the cell membrane to the lumen of the alveolus (Mepham, 1987). Each of the major components of milk is discussed below.

9.5.1. Proteins

Proteins are built up from amino acids within the lactating cells. The total amount of nitrogen that crosses the basal membrane is equal to that within milk; however, there is a change in amino acids. Non-essential amino acids are synthesized within the cell and are built up into proteins along the mRNA within the ribosomes of the rough endoplasmic reticulum (RER). These proteins are then secreted into milk. Essential amino acids are passed unchanged across the basal membrane of the lactating cell and are incorporated into proteins, along with the non-essential amino acids synthesized within the cell.

9.5.2. Lactose

Blood glucose is the primary precursor of lactose. However, glycerol, acetate and amino acids are also thought to contribute. The amount of glucose absorbed by the gland is much more than is needed solely for conversion to lactose. The difference is used as energy for general cell metabolism. The conversion of glucose to lactose involves five enzymes, the fifth being lactose synthetase, which is made up of two components A and B. As discussed previously, component B is the major milk protein α lactalbumin. The biochemical pathways involved in the conversion of glucose to lactose are summarized in Fig. 9.7.

9.5.3. Fat

The fat globules within milk are made up of esterified glycerol and free fatty acids, which aggregate to form a fat droplet emulsion within milk. There is much variation in the length of free fatty acids making up the fat globules in the milk from females of varying species. The horse tends to have a higher concentration of short-chain fatty acids (less than 16 C atoms in length).

Fatty acids are derived mainly from three sources: glucose, triglycerides and free fatty acids. Glucose C is a significant precursor of free fatty acids in the non-ruminant – for example – the horse. Glucose is absorbed across the basal membrane and converted to acetyl coenzyme A (CoA) and on to malonyl CoA within the cytosol of the cell. Malonyl CoA is then built up, using a multienzyme complex, to free fatty acids, which tend to be short chain (<16 C).

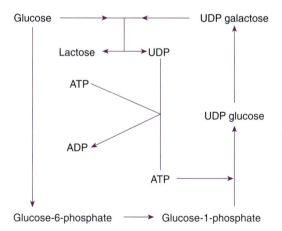

Fig. 9.7. A summary of the conversion of glucose to lactose within the lactating cell. UDP, uridine phosphate; ATP, adenosine triphosphate; ADP, adenosine diphosphate.

Blood triglycerides provide an alternative source of free fatty acids for the lactating cell; these are broken down into glycerol plus free fatty acids within the cell. The free fatty acids obtained from triglycerides are longer-chain fatty acids, typically 16–18 C in length. Triglycerides, therefore, are not a very important source of fatty acids in the mare. The triglycerides are either broken down into amino acids and glycerol in the blood – similarly to the way that proteins are broken down into amino acids and then absorbed into the cell – or they are absorbed directly.

Glycerol that combines with the free fatty acids is derived again by three different methods: from the breakdown of triglycerides within the cell; by the absorption of free glycerol in the blood; or, finally, from the breakdown of glucose within the cell.

The free fatty acids and glycerol within the cell combine by esterification within the endoplasmic reticulum. These molecules then aggregate together to form the fat droplets within milk.

9.6. Milk Secretion

All the components of milk produced by the lactating cells have to pass across the apical membrane of the lactating cell into the alveolar lumen (Fig. 9.8). The different components of milk pass by different mechanisms.

9.6.1. Fat

The fat droplet size increases as the free fatty acids and glycerol continue to combine by esterification and the resulting molecules aggregate into increasingly larger droplets as they migrate towards the apical membrane. In the vicinity of this membrane, strong London-Van der Waals forces attract these fat droplets and envelop them in the membrane, forming a bulge in the apical membrane surrounding the droplet (Fig. 9.9).

The droplet and surrounding plasmalemma move away from the apical membrane into the lumen, forming a narrow bridge. This bulge then pinches off as the bridge gets narrower and releases the fat droplet plus surrounding plasmalemma into the alveolar lumen. The process is termed pinocytosis. Occasionally, part of the cell cytoplasm, sometimes including cell organelles, is enclosed in the bulge of the apical membrane along with the fat droplets, and then gets secreted into the alveolar lumen along with the milk fat. The formation of these

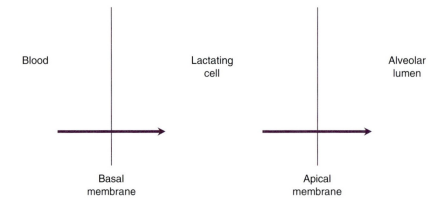

|Blood|Lactating cell|Alveolar lumen|
|Basal membrane| |Apical membrane|

Fig. 9.8. The route of passage for all milk components from the mare's blood supply on the left through the lactating cell to the lumen of the alveolus.

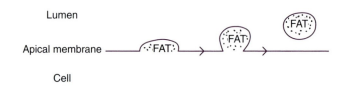

Lumen

Apical membrane

Cell

Fig. 9.9. The secretion of milk fat from the lactating cell (below) into the alveolar lumen (above) by pinocytosis.

structures, termed signets, occurs more in the lower order of mammals, but they are evident occasionally in mare's milk (Mepham, 1987).

9.6.2. Protein

Proteins are built up from their constituent amino acids along the RER within the cell and then pass on to the Golgi apparatus. They accumulate as granules of proteins within the Golgi; this Golgi apparatus then migrates towards the apical membrane. The membrane of the Golgi apparatus fuses with the apical membrane and this releases the proteins into the alveolar lumen by reverse pinocytosis (Fig. 9.10).

By this reverse pinocytosis, plasmalemma lost during the secretion of milk fat is replaced during the secretion of milk proteins.

9.6.3. Lactose

The secretion of molecules of lactose, unlike that of milk fat and protein, is not visible using electron microscopy, and so the method of secretion is less clear. As discussed previously, one of the mare's milk proteins is α lactalbumin, and this protein is the B component of the fifth and last enzyme involved in lactose synthesis. It, therefore, seems likely that lactose secretion is closely linked to that of milk protein. The A protein of the enzyme lactase synthetase is known to be closely associated with the membrane of the Golgi apparatus. The B component (α lactalbumin) is synthesized, as are all other milk proteins, on the RER and then passed on to the Golgi apparatus. While the B component is in the Golgi apparatus, it becomes associated with the A component already there, and together they form active lactase synthetase. This enzyme catalyses the conversion of uridine diphosphate galactose and glucose to lactose. The lactose is then presumed to be secreted along with the milk proteins by reverse pinocytosis (Figs 9.10 and 9.11).

9.6.4. Minerals

Milk has a relatively high concentration of K when compared to Na, and this is similar to the relative

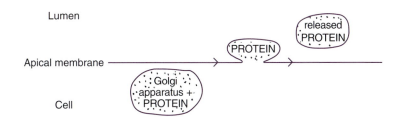

Fig. 9.10. The secretion of milk protein from the lactating cell (below) into the alveolar lumen (above) by reverse pinocytosis.

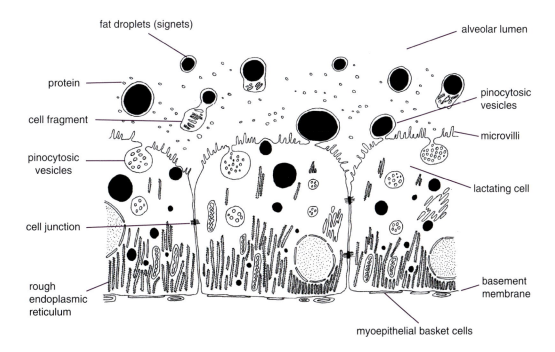

Fig. 9.11. Diagrammatic representation of a mammary secretary cell illustrating the build-up of protein and fat for release into the alveolar lumen.

concentrations within the cytoplasm of the cell. Na and K in milk are derived from the intercellular fluid. The components within the lactating cell are derived from the blood system but the K/Na ratio in blood is the reverse of that within the cell and milk. Therefore, there must be an active transfer system across the basal membrane. This is via an Na pump, which pumps Na away from the cell cytoplasm and into the blood system and pumps K the opposite way towards the cell. This maintains the high K/Na ratio within the cell. As the K/Na ratio in the milk is the same as that in the cell, Na and K are presumed to pass across the apical membrane to the alveolar lumen by simple diffusion (Mepham, 1987).

The concentrations of Ca, Mg and P are higher in milk than in the cell cytoplasm. Therefore, their passage must also be via an active transport system. The exact mechanism is unclear but all three ions are known to be closely associated with the milk protein casein. It is, therefore, assumed that this association occurs within the Golgi where the casein proteins are synthesized. These ions are then passed into the milk, along with proteins, via reverse pinocytosis (Mepham, 1987).

Iron is also secreted in association with proteins by reverse pinocytosis, as it is specifically bound to a minor milk protein, lactoferrin (Mepham, 1987).

9.6.5. Water

Water passes to the alveolar lumen from the cell cytoplasm by osmotic pressure. Fat and protein molecules in milk are in the form of large droplets, and so their effect on osmotic pressure is minimal. However, lactose and free ions are much smaller and it is these that affect osmotic pressure and hence drive water diffusion from the cell into milk (Mepham, 1987).

9.6.6. Immunoglobulins

Colostrum, as discussed previously, has a high concentration of proteins; these proteins are immunoglobulins and their associated antibodies. These immunoglobulins are combined into large corpuscles termed bodies of Donne. The mechanism by which these are secreted is unclear. It is possible that engorgement of the lactating cell in late pregnancy results in the breakage of some of the junctions within the cell membranes, especially the basal cell membranes of the lactating cells. This allows the serum proteins to pass into milk unchanged. There is also evidence suggesting an active transport system for these serum proteins, but the exact mechanism is as yet unclear.

9.7. Conclusion

Our present knowledge specifically regarding equine lactation is still limited; however, by extrapolation from other species, a reasonable understanding can be achieved. Caution must be practised, however, in making definitive statements until these assumptions and extrapolations have been confirmed or refuted by more detailed research, specifically on equine lactation.

10 Control of Lactation in the Mare

10.1. Introduction

The control of lactation in most mammals, including the mare, is via both nervous and hormonal pathways. Unfortunately, information specific to the mare is very limited, though it is assumed that the control of lactation is very similar to that evident in other mammals. The information discussed in this chapter is gleaned from the limited experiments carried out on horses and extrapolation from other mammals where appropriate.

Lactation may be divided into three stages as far as its control is concerned: lactogenesis, galactopoiesis and milk ejection.

10.2. Lactogenesis

Lactogenesis refers to the initial milk secretion that occurs in late pregnancy prior to parturition, resulting in a build-up of colostrum within the mammary glands. In the mare, lactogenesis is evident well before parturition occurs, as demonstrated by the presence of lactose, proteins and fat within the mammary secretions (Peaker et al., 1979).

Control of lactogenesis is hormonal. High progesterone concentrations characteristic of pregnancy drive lobular-alveolar development, but inhibit milk secretion. Hence, with the decline of progesterone in late pregnancy, inhibition of milk production is removed (Mepham, 1987). In addition, elevated prolactin, growth hormone, cortisol and insulin-like growth factor (IGF) seem to be involved in driving milk production (Heidler et al., 2003). It appears that in the mare prolactin may play a significant role as concentrations are observed to increase in the last 2 weeks of pregnancy (Neuschaefer et al., 1991). Lactogenesis, therefore, increases in late pregnancy and reaches a maximum immediately prior to parturition (Forsyth et al., 1975; Peaker et al., 1979; Worthy et al., 1986).

In other mammals – for example, ruminants and humans – a placental lactogen has been identified and found to have an additional effect on lactogenesis. No such placental lactogen has been identified in equines (McCue, 1993).

10.3. Galactopoiesis

Galactopoiesis is the term given to the maintenance of milk production. Again, little information specific to the horse is available. However, it is assumed that control is similar to that in the sheep and the cow, where it is under the control of prolactin, growth hormone and cortisol, increasing concentrations of which act to drive galactopoiesis (Neuschaefer et al., 1991). Galactopoiesis is driven by, and mimics, the foal's demand for milk, and this, in turn, dictates and governs the shape of the lactation curve and the quantity of milk produced.

10.4. Milk Ejection

Milk ejection, also termed milk let-down, differs from the other stages of lactation in that its control is both neural and hormonal. A nervous reflex acts as the stimulus or afferent pathway and hormones form the efferent path. Nerve receptors within the teats are stimulated by the action of suckling, and the nervous afferent pathway is activated, resulting first in a localized effect causing localized myometrial cell contraction. Second, this afferent nervous pathway acts via the central nervous system (CNS) to stimulate the paraventricular nucleus within the mare's hypothalamus. The hypothalamus then activates the posterior pituitary, which in response produces the hormone oxytocin. The efferent pathway of the milk ejection reflex is formed by the oxytocin, which passes into the systemic blood system and hence to the mammary gland. The effectors that react to oxytocin are the myoepithelial basket cells surrounding each alveolus and the small ducts, causing them to contract further and forcing milk out of the alveolus, along the ducts, to the gland cistern and on to the teat cistern ready to

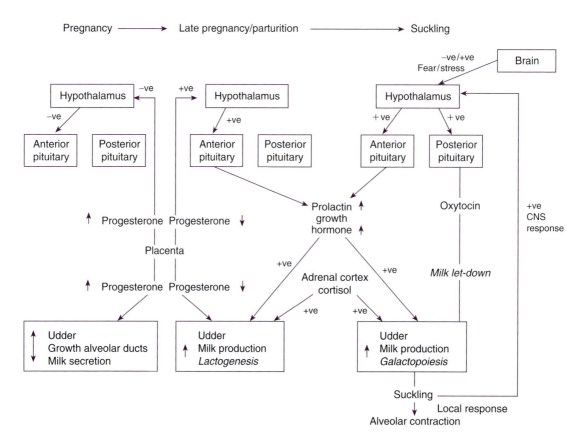

Fig. 10.1. Schematic representation of the control of lactation in the mare.

be removed by the suckling action of the foal (Ellendorff and Schams, 1988; Nett, 1993b). Hence, at suckling, the milk initially available is that within the gland and teat cisterns, which is removed by the negative pressure exerted by the suckling action of the foal. This is then closely followed by the milk ejection reflex, which replenishes the milk within the gland and teat cisterns, making more available to the foal.

In addition to the above control mechanisms, the CNS has an overriding effect. For example, stress, especially as a result of fear or shock, reduces the effectiveness of the milk ejection reflex by increas-ing the levels of circulating adrenalin. Adrenalin causes vasoconstriction, so reducing the amount of oxytocin reaching the alveoli and hence the effectiveness of the reflex (Fig. 10.1).

10.5. Conclusion

It is evident that there is much left to be learnt about the control of lactation in the mare. The lack of direct commercial value has led to little research into equine milk production and, as such, many extrapolations and assumptions from the cow have been applied to the horse.

11 Selection of the Mare and Stallion for Breeding

11.1. Introduction

The choice of both mare and stallion for breeding can be a very complicated and time-consuming process. Often not enough importance is placed upon this selection, the result of which is an over-supply of mediocre or poor stock and unnecessary difficulties with mares at covering, during pregnancy, at foaling and during lactation.

One of the most obvious selection criteria is that of performance or athletic ability. If you are looking to produce a quality show horse, you will obviously be selecting for conformation and show success on both the mare and stallion's side and in their breeding. If you are looking to produce a racehorse or competition horse, you will be interested primarily in the proven performance of the dam and sire and select accordingly. Finally, if you are looking for a leisure horse, you will be interested primarily in temperament and possibly hardiness.

It is beyond the scope of this book to discuss in detail the wide range of performance criteria that a breeder may look for in order to achieve that specific, ideal horse. These performance criteria vary considerably, with the individual breeder and type of horse required. Many books and articles have been written on the subject, but ultimately it is a personal decision.

Regardless of the performance criteria used for selection, stock should also be selected on reproductive competence. However, all too often, such criteria are not considered, with potentially serious consequences for the individual breeder and the equine breed as a whole. Regardless of the type of horse you intend to breed, reproductive competence and the ability to produce healthy offspring with minimal danger to the life and well-being of the dam should also be of prime importance. Today's horse, unlike other farm livestock, has been selected primarily for performance ability, often at the expense of reproductive competence.

As a result, there are many potential reproductive problems that the breeder should be aware of in selecting both the mare and stallion.

The following sections will concentrate solely upon the criteria and techniques that can be used in the selection for reproductive competence and will assume that the selection criteria for performance and conformation have already been met. All the possible techniques will be included in the following sections, many of which are costly in time and money. The extent to which these techniques are used depends on personal choice and the value of the breeding stock concerned and potential offspring. Further information specific to infertility and hence an expansion of some of the issues raised here is included in Chapter 19 (this volume).

11.2. The Mare

The selection criteria for reproductive competence in the mare can be listed as follows:

- History;
- Temperament;
- Age;
- General conformation and condition;
- Reproductive tract examination;
- Infections;
- Blood sampling;
- Chromosomal abnormalities.

11.2.1. History

The history of the mare includes both her specific breeding and general history (Shideler, 1993a). Many mares these days, especially those of any value, come with historical documentation. Such records are invaluable in assessing her ability to produce a foal, as well as easing management. If there are no records available, contact should be made with her previous owners to find out as much information as possible.

11.2.1.1. Reproductive history

Details of her past breeding performance should ideally include answers to the following questions: Does she show regular oestrous cycles? What is the length of her normal oestrous cycle? When does her breeding season normally start? How long does her season normally last? Does she show oestrous well and are there any characteristic signs? How long is her typical oestrous? Does she demonstrate oestrous better under certain circumstances? etc.

This information will indicate, among other things, whether or not she will be easy to detect in oestrous and to cover. Mares that do not demonstrate oestrous well tend to be hard to cover. This leads to frustration, due to missed oestrouses and wasted journeys to the stallion, all increasing costs. In addition, mares that are habitually hard, or even dangerous, to cover may be refused by some studs, reducing the pool of potential stallions or necessitating the use of artificial insemination (AI).

Other questions asked should include whether she has bred before. If so, has she had problems holding to the stallion? This will necessitate return journeys to the stud or a prolonged period of time away. Has she had problems during any of her pregnancies? Has she ever reabsorbed or aborted? Does she or her family throw twins? If a mare has a history of habitual reabsorption or abortion or the need for holding injections (artificial progesterone supplementation) post service, then she is really not a good candidate for a brood mare. Such problems may indicate an inherent inability to carry a pregnancy to term, due to hormonal imbalance, uterine incompetence and/or genetic abnormality.

Repeated incidences of twins are a problem, in that the likelihood of one or both of the fetuses aborting is very high. If only one of the twins aborts, the remaining one is often born weaker and smaller than would normally be expected. Spontaneous abortion of twins is not due to an inherent inability to carry a foal to term, and such mares are perfectly capable of producing a foal provided a single pregnancy is conceived. They may, however, need to be checked routinely for twin pregnancies and managed appropriately to eliminate one twin at an early stage or to induce abortion, and the mare returned to the stallion (Ginther, 1982; Chapter 14, this volume).

It is important to find out whether there may have been any incidences of dystocia. It is rare for abnormal fetal positions to occur repeatedly, but past difficulties may have caused internal lacera-

tions or damage. It is particularly important to know if the cervix has been affected. Limited damage to the uterus, vagina and vulva will repair quite effectively, though they may leave areas of weakness or adhesions. Damage to the cervix may cause cervical incompetence, allowing bacteria to enter the higher reproductive tract and thus causing infection. Conversely, adhesions may hinder the passage of sperm at covering, make it less able to dilate during parturition, necessitating a Caesarean delivery, and prevent natural drainage after covering, so increasing the incidence of post-coital endometritis.

Previous post-foaling problems may also be indicated, including rejection or even habitual attacking of foals. This could be a sign of a general temperament fault, or may not show up in subsequent pregnancies, particularly if such behaviour was associated with a first foaling. Records may show that the mare is not a good mother, producing ill-thrifty foals. It is, therefore, helpful to look at foal birth weight and subsequent development and growth rates. Such effects may be due to poor milk yield, which in turn may reflect faults in nutritional management rather than a specific mare problem, but if she has consistently produced foals that did not do well, it may be worth investigating further.

Records should also show any incidences of infections and the treatments given. Minor infections may show no long-term effects. However, infections of the reproductive tract are largely responsible for the relatively high infertility rates in horses and are the single major cause of fertilization failure and abortion in mares (McKinnon and Voss, 1993). Previous mastitic infections, though rare, should be considered and such mares examined to make sure that her udder has not suffered permanent damage.

Detailed mare records not only prove invaluable as an aid to selection; they also provide useful information for the stud to which you intend to send your mare. This is especially important if she is to stay at stud for a prolonged period of time and oestrous detection is to be carried out by them.

11.2.1.2. General history

In addition to specific breeding records, information on the mare's general history is also very useful. These should indicate her vaccination and worming status as well as accidents, especially if these involved injury to the pelvic area, abdominal muscles, internal injuries and damage to limbs or muscles. These may be exacerbated by pregnancy, especially in the later stages, and predispose the mare to problems and may prevent natural parturition.

Ongoing or past conditions should also be noted. Mares with a history of respiratory or circulatory disorders, recurrent airway obstruction (RAO), exercise-induced pulmonary haemorrhage (EIPH), umbilical hernias or vaginal/uterine prolapse would possibly not stand up to the strain of pregnancy. Such conditions may also be exacerbated by pregnancy itself (Turner and McIlwraith, 1982; Shideler, 1993a,b). Severe laminitis, navicular disease, tendonitis, etc. may also have a bearing on the ability of a mare to carry a pregnancy to term (Woods, 1989). Mares with disorders that are potentially heritable may be able to carry a pregnancy to term, but it is debatable whether such mares should be bred and hence perpetuate a problem within the population.

11.2.2. Temperament

The temperament of the mare is obviously important for ease of management, especially at birth and with a young foal, when many mares tend to be antisocial. Her temperament also has a significant effect on the temperament of her foal. The temperament of any animal has both a genotypic (inherited) and phenotypic (environmental) component. Both parents have an equal genotypic effect on the temperament of the foal. However, the phenotypic component of temperament is affected much more by the mare, in whose company the foal spends much of its early life until weaning. The temperament of the mare could, therefore, be argued to be more important than that of the stallion in determining the temperament of the foal.

A mare with a quiet and gentle disposition is much easier to handle and manage. Such mares tend to show oestrous more readily and are, therefore, easier to cover. Anxious and highly strung mares with poor temperaments may show signs of aggression towards a teaser or stallion even though she is in oestrous; stress may often mask the signs of oestrous in such mares. Such aggressive mares are obviously more difficult and dangerous to cover. Many need to be twitched and/or hobbled, which further increases stress and is not conducive to optimum fertilization rates or easy management.

Such mares can, of course, be covered using AI but the problem of detecting oestrous remains. Highly strung and nervous mares may also suffer from higher abortion rates. Finally, a masculinized temperament may be indicative of hormonal abnormalities, for example granulosa cell tumours (Hinrichs and Hunt, 1990).

11.2.3. Age

The age of a mare at her first breeding influences the ease of her pregnancy and may have potential long-term effects, especially in mares under 5 or over 12 years of age (Day, 1939). As mares reach puberty between 18 and 24 months of age, it is theoretically possible to breed a mare at 18 months to produce a foal at 29–30 months of age. However, evidence, especially from research into sheep, demonstrates that embryo mortality rates are significantly higher in maiden pubertal animals, and return rates in young maiden mares do tend to be higher than in older multiparous mares.

Young mares may not only be hard to get in foal but they may suffer detrimental long-term effects if nutrition is inadequate. A horse does not attain its mature body size until, on average, 5 years of age, so for a mare under 5 years, there are the additional demands of growth on top of those of pregnancy and maintenance. This should be reflected in her nutritional intake. To breed a mare early in life is a decision not to be taken lightly. She must be well grown for her age and in a good, but not over-fat, physical condition. There must also be the wherewithal to feed her additional good-quality food, especially in late pregnancy, and to provide good accommodation for her in order to minimize her body's maintenance requirement.

At the other end of the spectrum, maiden mares over the age of 12 may also find it difficult to carry a pregnancy and bring up a foal. There is evidence to suggest that embryo survival rates, as well as fertilization rates, are reduced in old mares (Ball, 1988, 1993a; Ball and Brinsko, 1992). This is due to an increase in the incidence of embryonic defects, a reduction in ova viability and an increase in age-related degenerative endometritis (Chapter 19, this volume; Ball, 1988, 1993a; Ricketts and Alonso, 1991). In addition, the mere fact that she has been alive longer also means that she has had a greater chance of exposure to reproductive tract infections, which, if severe, may have permanently affected her ability to conceive. Older mares may have also spent the majority of their lives as barren mares without the attention of the stallion; they tend, therefore, to reject or be antagonistic towards a stallion's approaches.

Older maiden mares may also have problems associated with their previous life. Many such mares are ex-performance horses and, as such, they have been trained and kept in top athletic condition, which often disrupts reproductive activity. Prime athletic condition is associated with a variety of reproductive malfunctions – for example, delayed oestrous, prolonged dioestrous, complete reproductive failure, etc. These mares need a readjustment period before they are physically and psychologically capable of conceiving, bearing and rearing a foal. A prolonged period of rest, at least 6 months, allows the mare's system to settle down into a non-athletic state; some mares may take as long as 18 months to adjust to their new way of life. If a mare has been treated with drugs, such as corticosteroids and/or anabolic steroids, illegally or legally, her system will require time to eliminate them, during which time reproductive function may continue to be disrupted (Shoemaker *et al.*, 1989). A performance horse, in top athletic condition, develops musculature, especially in the abdominal and pelvic region, not necessarily conducive with an easy pregnancy and parturition. Time should, therefore, be allowed for muscle tone to relax.

The ideal age to breed a maiden mare is at 5–6 years of age, at which time she will have reached her mature size. She will also not be old enough to have become set in her ways and will be less likely to have developed aggressive tendencies towards the stallion and to have contracted uterine infections. However, breeding at this age will not suit all systems. Most mares will not have proved their worth so the decision to breed may not yet have been made. One increasingly popular way of overcoming this is the use of embryo transfer, discussed in detail in Chapter 21 (this volume), which allows performance mares to breed but not at the expense of their performance careers.

Once a mare has borne one foal, she is more able to cope with the demands of subsequent pregnancies and, as such, is much more likely to breed successfully as an older mare well into her teens (Allen, 1992). If a mare has been a brood mare all her life, then age is less important. She may naturally be barren during the occasional year, and this has been suggested as nature's way of allowing recovery and regeneration. More care and attention is needed with older age, and it is not normally advisable to breed a mare over 20 years old, though this largely depends on the breed, type and condition of the individual mare. Mares of the native type tend to breed more successfully into later life than the hot blood/warm blood-type animals, though of course there are exceptions to every rule.

11.2.4. General conformation and condition

A mare's general conformation is of importance, not only to ensure that her offspring are well conformed, but also to ease her pregnancy. She needs to have a strong back and legs to enable her to carry the considerable extra weight of the fetus during late pregnancy. She should have correct pelvic conformation, with a pelvic opening adequate for a safe delivery. Ideally, she should also possess good heart and lung room across the chest and have plenty of abdominal space. In general, rather fine, tucked-up mares tend to be poorer breeders or have more problems during pregnancy.

A mare's general body condition is considered to have implications on reproductive activity, though evidence is conflicting (Gentry *et al.*, 2002a,b; Godoi *et al.*, 2002). Recent work, however, suggests that there may be a link between leptin levels, which are related to body condition, and reproductive ability, especially in relation to seasonality (Fitzgerald and McManus, 2000; Fitzgerald *et al.*, 2002). The body condition of horses can be classified on a scale of 0–5, 0 being emaciated and 5 obese (Figs 11.1–11.3). The ideal condition score for a mare at mating is 3. Such mares have a good covering of flesh; the ribs may be felt with some pressure, as can the vertebrae of the backbone. It is widely believed that mares in condition score 3 have the highest fertilization rate and subsequent reproductive success (Fig. 11.2; Van Niekerk and Van Heerden, 1972; Kubiak *et al.*, 1987; Morris *et al.*, 1987).

Thin mares may show prolonged anoestrous or very long/delayed oestrous cycles, along with suspension of all reproductive activity in cases of emaciation. At the other end of the spectrum, over-fat mares (Fig. 11.3) may suffer reproductive failure, due to excess fat deposition on the reproductive tract limiting its ability to move and expand with a developing pregnancy. Fat may also be deposited on the ovaries and around the Fallopian tubes, interfering with the process of ovulation.

It has been demonstrated that, in barren mares, an increasing plane of nutrition 5–6 weeks prior to

CONDITION SCORE					
5	4	3	2	1	0
General					
Obese	Fat	Good	Fair	Thin	Emaciated
Back/Pelvis/Croup					
Ribs cannot be felt, buried in fat. Flat backbone with deep 'gutter' continuing over croup to base of dock. Pelvis buried in very firm fatty tissue and cannot be felt	Ribs well covered, only felt with firm pressure. Backbone well covered with slight 'gutter' continuing over croup. Pelvis buried in fatty tissue and only felt with firm pressure	Ribs, backbone and pelvis rounded and covered in fat but felt with light pressure. No 'gutter'	Ribs just visible. Backbone just covered in fat, spinous processes easily felt but not visible. Croup well defined with some fat under skin. Pelvis easily felt, slight depression under tail	Ribs and backbone easily seen covered by slack skin. Pelvis and croup well defined with no fat under skin. Deep depression under tail	Ribs, backbone and pelvis very easily seen, prominent and sharp, covered by slack skin. Deep cavity under tail
Neck					
Very wide and firm with excessive fatty tissue. Marked crest in mares and stallions	Wide and firm with folds of fat. Slight crest in mares and stallion.	Firm with some fat. Crest only in stallions	Narrow but firm. No crest	Ewe neck, narrow and slack at base	Ewe neck, very narrow and slack at base

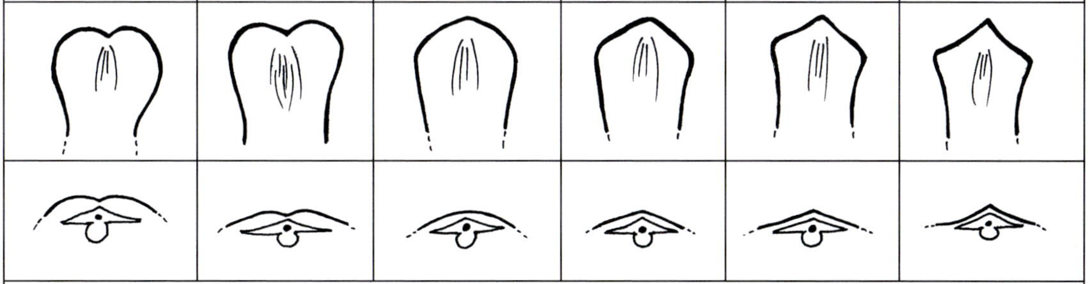

Fig. 11.1. Body condition scoring in the horse.

mating, with an accompanying gradual improvement in body condition up to a score of 3, results in the best ovulation rates and reproductive success (Van Niekerk and Van Heerden, 1972). This process is termed flushing and is a common practice in sheep and cattle management. It is discussed further in Chapter 12 (this volume). Finally, work done by Swinker *et al.* (1993) indicates that body condition can also affect the mare's response to hormonal treatment used in the manipulation of reproduction.

11.2.5. External examination of the reproductive tract

Poor external conformation of the mare's reproductive tract can have severe implications on reproductive performance (see Chapter 1, this volume).

The perineal area forms the outer vulval seal of the tract (see Chapter 1, this volume) and so should be examined. First, the presence of lacerations, damage, puckers, scars, etc. should be noted, as they may be indicative of more extensive internal damage. Second, the general conformation of the perineal area should be assessed. If the vulval seal is incompetent, this allows bacteria and airborne pathogens to enter the vagina and challenge the upper reproductive system (Chapter 1, this volume; Figs 1.4, 1.6 and 1.7). This predisposes the mare to pneumovagina and urovagina. Both of these are common causes of reproductive tract infection and hence infertility. Such conditions are prevalent in lean athletic mares such as Thoroughbreds, mares in poor condition and mares with reduced vulval tone due to injury, damage, age, oestrous or foaling (Pascoe, 1979a).

As discussed in Chapter 1 (this volume), the height of the pelvic floor also has a bearing on bacterial contamination and so should also be considered when selecting mares. It may be assessed by inserting a sterile probe into the vagina and resting it on the pelvic floor (Fig. 11.4).

In a normal conformation, at least 80% of the vulva should lie below the pelvic floor in order for the vaginal seal to be fully competent. In a mare

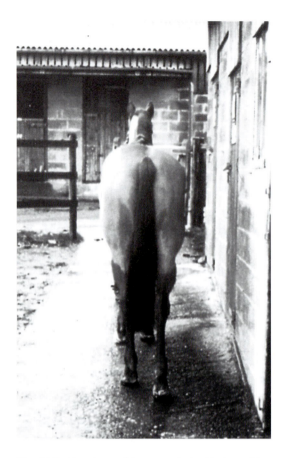

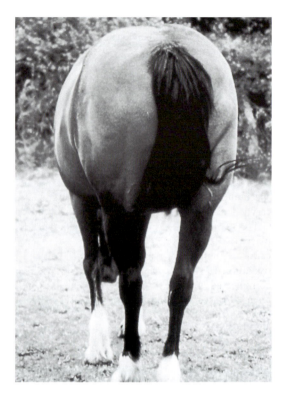

Fig. 11.3. Mare in over-fat condition, condition score 5.

Fig. 11.2. Mare in condition score 3, the ideal condition for covering.

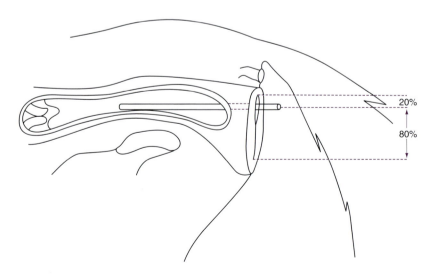

20%

80%

Fig. 11.4. The assessment of vaginal and vulval competence and hence the likelihood of pneumovagina.

with a low pelvic floor, the risk of contamination of the reproductive tract can be reduced if perineal conformation is correct and hence the vulval seal is competent.

The external perineal conformation, and hence vulval seal competence, can be assessed by eye. A simple ruler and protractor can be used to allocate a Caslick index (Chapter 1, this volume) and so indicate the likelihood of infection and the need for a Caslick vulvoplasty (Pascoe, 1979a). The competence of the vaginal seal can be assessed via the use of a sterile probe as indicated or may require a speculum. Speculum examination of the vagina and cervix is discussed in detail in the following section.

The selection of a mare with a Caslick vulvoplasty for breeding must be made with considerable caution. There are a finite number of times that the operation can be performed and, unless the more complicated Pouret operation (Pouret, 1982) is resorted to, there is, by implication, a limit to the mare's breeding career. An additional consideration is one of equine welfare and whether it is appropriate that a mare should go through such a procedure repeatedly. It may also be questioned as to whether such poorly conformed mares should be bred from at all as, in so doing, the trait is perpetuated within the equine population. However, a Caslick vulvoplasty operation certainly does reduce the incidence of repeated uterine infection, which in itself may be considered a welfare issue (Hemberg *et al.*, 2005).

11.2.6. Internal examination of the reproductive tract

Examination of the internal reproductive tract of the mare is a skilled veterinary surgeon's job, necessitating the use of several examination techniques. Information given by these assessments can be indispensable in assessing the reproductive potential of a mare.

11.2.6.1. Vulva and vagina

Vaginal internal assessment can be carried out by means of a speculum (Fig. 11.5). The most commonly used is a Caslick's speculum (Fig. 11.6).

The speculum consists of either an expandable, or non-expandable, hollow metal tube, with a light source attached. The sterilized and well-lubricated speculum is inserted into the mare's vagina. The expandable speculum is inserted in the closed position, and is expanded slowly opening up the vagina.

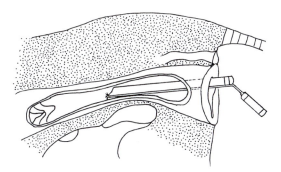

Fig. 11.5. Vaginal speculum examination of the mare's vagina. The attached light source allows the inside of the vagina and cervix to be illuminated and hence easily viewed.

Fig. 11.6. A non-expandable Caslick vaginal speculum.

The light source then illuminates the mucous membrane lining and the cervix, allowing examination in detail. Disposable speculums of plastic or coated cardboard are also available, but these need an independent light source, and so the use of two hands, making techniques such as swabbing very difficult. The use of a vaginal speculum must be accompanied by several precautions as, when in position, it opens up both the vulval and vaginal seals. The procedure should be carried out under conditions as sterile as possible, in a dust-free environment. The mare's tail should be bandaged and the perineal area thoroughly washed (Fig. 11.7).

A more sophisticated technique, endoscopy, is often used today for viewing the mare's vagina and cervix before it is then passed up into, and used to view, the rest of the internal reproductive tract. Old endoscopes were rigid and, therefore, had limitations to their use; the newer flexible fibre-optic endoscopes are much more versatile and can be used to view internal structures either through the body wall and into the body cavity; via the oral/nasal cavity into the digestive/respiratory system; via the rectum into the large intestine; or, finally, via the vagina into the reproductive tract.

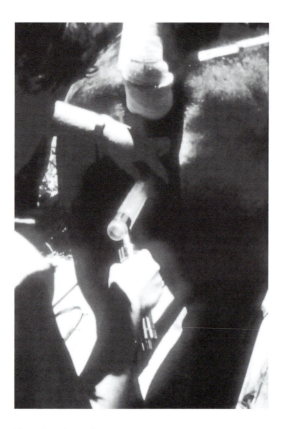

Fig. 11.7. Use of a vaginal speculum in the mare.

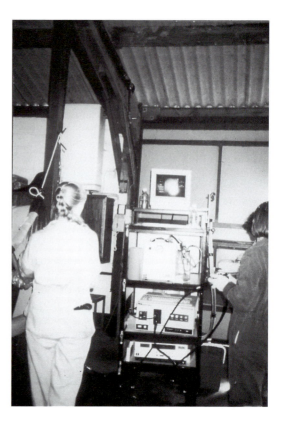

Fig. 11.8. Endoscopic examination of the reproductive tract of the mare. The attendant on the right controls the fibre-optic camera, while the attendant on the left guides the endoscope within the mare's reproductive tract. The image produced is viewed on the TV monitor. (From Thoroughbred Breeders Equine Fertility Unit, Newmarket, UK.)

The endoscope consists of a series of flexible carbon-fibre filaments with a light source and camera attached. The flexible carbon-fibre rod is passed through the vagina and up into the inner reproductive tract. One set of the carbon-fibre filaments allows the passage of light from the external light source down the endoscope, illuminating the internal structures. The other set allows the transmission of the image back to the camera and on to the television monitor. The positioning and angle of view of the endoscope can be controlled remotely from the exterior end of the endoscope (Fig. 11.8). The use of the endoscope is expensive and is usually confined to examination of the upper, less-accessible parts of the reproductive tract (Fig. 11.8; Le Blanc, 1993a; Threlfall and Carleton, 1996).

In the healthy mare, whether viewed via a speculum or endoscope, the mucous membranes lining the vulva and vagina should appear a healthy pink colour. Any mucus present should be clear and not cloudy or yellow-white in colour. An infected vagina appears red and inflamed and possibly covered in a cloudy mucus. Infections of the vagina are not uncommon and will be discussed in the following section on infections and also in Chapter 19 (this volume). The vulva and inner vagina should also be free from bruising, irritation, scars and tears. They are susceptible to damage and injury at parturition and so they should be carefully checked. Adhesions, if present, are caused by scar tissue and may render the mare unserviceable or cause her pain at mating and hence rejection of the stallion. AI is a possible alternative but she is likely to still have problems at foaling.

Neoplasms of the vulva and vagina are relatively rare, though if present, may cause problems at covering. Three kinds of neoplasms can be seen in horses: first, melanomas, malignant growths of the melanin-containing pigment cells, particularly prevalent in grey mares and not necessarily confined to the vulva and vagina; second, carcinomas or malignant neoplasms of the epithelial cells and finally papillomas or benign neoplasms. If a mare does show evidence of neoplasms, it is debatable whether she should be bred, as there is evidence to suggest that the tendency to develop some neoplasms is heritable.

Finally urinovagina (pooling of urine within the floor of the vagina) may be observed. Stagnant urine held within the vagina will increase the chance of infection. Urine pooling is evident in mares with poor perineal conformation and in older multiparous mares in which vaginal tone has been lost. It may also result from past vaginal damage or injury.

11.2.6.2. The cervix

The cervix may be considered as the connection and final seal between the outer reproductive tract and the inner, more susceptible tract. As such, its competence as a seal is very important. This may be assessed by means of a speculum or endoscope as described above for vaginal examination. The cervix varies greatly with the stage of the oestrous cycle and pregnancy, and, as such, can aid diagnosis. During oestrous, the cervix is fairly relaxed, pink in colour, its tone flaccid and any secretions quite thin, and it appears to 'flower' (relax) into the vagina. This is to ease the passage of the penis at copulation. During dioestrous, and to a greater extent, during anoestrous, the cervical tone increases, it appears whiter in colour, the seal becomes tighter and secretions thicker. During pregnancy, the cervix is again tightly sealed and white in colour, with a mucous plug acting to enhance the effectiveness of the seal. The tone of the cervix may also be assessed, via rectal palpation. In general, the state of the cervix should correspond to ovarian activity and that of the rest of the tract. If not, infections or abnormalities should be suspected. Infection of the cervical mucosa, or cervicitis, will be discussed in detail later (see Sections 19.2.2.4.4 and 19.2.2.5.4). Cervicitis is characterized by a red/purple swollen cervix, often protruding into the vagina and covered with mucus, clearly seen at speculum examination. In severe cases, cysts may be evident and may result in adhesions and long-term permanent damage, pre-

venting correct closure of the cervical seal, with resultant infection. Adhesions as a result of cervical damage may not only close it completely, preventing entry of the penis, and so the passage of sperm to the upper tract, but also prevent the drainage of fluid from the uterus, resulting in serious uterine infections. Minor cervical damage and/or adhesions can be helped by surgery.

11.2.6.3. Uterus

Examination of the mare's uterus is a more complicated procedure, as the uterus forms part of the inner and less-accessible reproductive tract. An appreciation of its gross structure can be obtained by rectal palpation and/or ultrasonic scanning. The colour and condition of the mucous membranes of the uterus may be inspected by means of an endoscope and further examined by uterine biopsy.

Rectal palpation is a commonly used and relatively effective way of obtaining a tactile impression of the inner reproductive tract and so identifying structural abnormalities (Figs 11.9 and 11.10). The procedure involves restraining the mare in stocks and inserting a well-lubricated, gloved hand and arm through the anus and into the rectum of the mare. The wall of the rectum is fairly thin, so the reproductive tract that lies immediately below it can be palpated through the rectum wall. The procedure, however, must only be carried out by an experienced operator, as rupturing the rectum is a risk. Rectal palpation can be used to assess the tone, size and texture of the uterus, uterine horns, Fallopian tubes and ovaries. In addition, cysts, tumours, neoplasms, stretched broad ligaments, uterine endometritis, sacculations, adhesions, lacerations, scars and delayed involution can

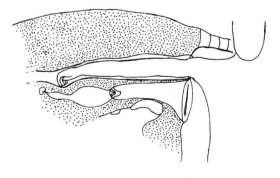

Fig. 11.9. Rectal palpation in the mare.

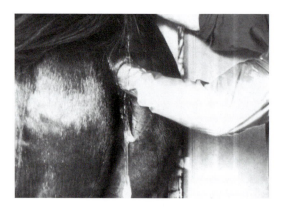

Fig. 11.10. Rectal palpation in the mare allows the practitioner to feel the reproductive tract through the relatively thin wall of the rectum.

be identified. As such, it is a useful, cheap and immediate aid to selection of mares (Greenhof and Kenney, 1975; Shideler, 1993c).

Ultrasonic scanning is a newer and more expensive method of assessment, but gives an immediate, visual impression of the upper reproductive tract, rather than the tactile interpretation obtained with rectal palpation (Jones, 1995; Reef, 1998; Sertich, 1998). The ultrasonic scanner is based on the Doppler principle. When high-frequency sound waves hit an object, they are absorbed or deflected back to a varying extent depending on the density of that object. The reflected sound waves are transduced to appear as an image on a visual display unit (VDU). Solid objects appear white and fluid appears black, with many variations of grey in between (Fig. 11.11; Toal, 1996). In the case of ultrasonic scanners used in the assessment of a mare's reproductive status the transducer or ultrasonic emitting and receiving device can be inserted into the rectum or less commonly placed on the mare's flank, and directed towards the reproductive tract (Fig. 11.12; Ginther and Pierson, 1983). The resultant image can be used to identify abnormalities similar to those assessed via rectal palpation but is particularly useful in detecting cysts, intrauterine fluid, air, debris, neoplasms, etc. (Ginther, 1984; McKinnon, 1987a–c, 1998). The technique can also be used to assess the size and shape of the uterus and ovaries and so can be used to determine ovarian activity (Adams *et al.*, 1987; Carnevale, 1998; McKinnon, 1998).

When assessed via rectal palpation or ultrasonic scanning the healthy uterus should appear either flaccid with little tone if the mare is in oestrous, or

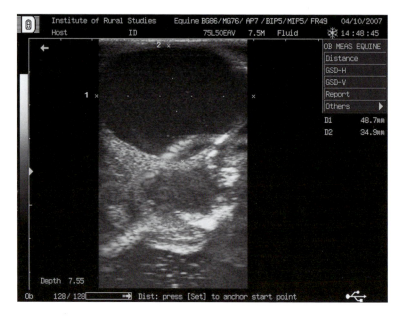

Fig. 11.11. The image produced by an ultrasonic scanner; fluid areas show up as black and solid structures as white, with variations in between.

Fig. 11.12. The use of ultrasonic scanning machine in the mare.

turgid with plenty of tone if she is in dioestrous. If the palpater works his/her way gradually from either uterine horn and across the uterine body, local thickenings, cysts or fibrotic areas can be detected and the diameter of each horn estimated and compared. The two horns should match. Mares that have had a history of infections may show ventral outpushings of the uterus, especially at the junction between the uterine body and horns. These mares may well have difficulty in carrying another pregnancy to term as they often lose the fetus at the time of, or soon after, implantation. Incomplete involution (recovery) of the uterus post partum can also be detected, indicating infection, weakening or rupture of the uterine wall, and, in severe cases, the uterus may never completely involute. This leaves a flaccid area of uterus with inadequate tone compared to the remainder. Again, such mares may well have difficulty carrying a pregnancy to term. Haemorrhages within the broad ligaments may be identified as local hard swellings. Stretched broad ligaments may also be detected, often due to excessive strain, accidents or damage. Again, such mares may be incapable of carrying a pregnancy to term, the broad ligaments being unable to support the weight of a full-term fetus.

An endoscope, as described earlier in this chapter, may be used to examine the internal surfaces of the reproductive tract and to give a real-time image.

Once the endoscope is inserted, sterile air or oxygen is passed into the uterus, easing the viewing of the mucous membranes and uterine epithelium (Fig. 11.8; Mather *et al.*, 1979; Le Blanc, 1993b). The normal healthy uterine epithelium is pale pink in colour and any mucus present is clear. Any signs of redness or discoloration within the uterus, especially if it is associated with cloudy or creamy mucus, are indicative of uterine infection – endometritis. Evidence of cysts, blood clots, thickening of the endometrium or scar tissue may indicate possible future problems in the mare's reproductive life.

The use of the endoscope is as yet limited in routine mare selection, due to the cost of the instrument and its delicate nature. Endoscopes at present are normally restricted to large specialist veterinary practices and research.

If any areas of concern within the mare's uterus are detected, a uterine biopsy may be performed to aid further investigation. Uterine biopsies are taken during endoscopy and involve the removal of a small section of uterine endometrium by means of a small pair of forceps (Figs 11.13 and 11.14). The biopsy forceps are passed into the vagina and guided up through the cervix by a well-lubricated gloved hand. The index finger guides the forceps through the cervix and into the uterus. The gloved hand is then removed and inserted into the rectum to guide the forceps to the section of the uterus that

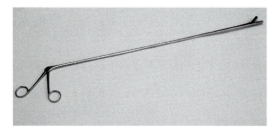

Fig. 11.13. Uterine biopsy forceps.

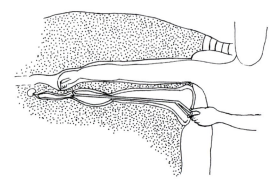

Fig. 11.14. The use of uterine biopsy forceps in the mare. The arm inserted via the rectum is used to ease the wall of the uterus into the jaws of the biopsy forceps.

is to be biopsied. The jaws of the forceps are opened and the section of endometrium to be sampled is pushed between the jaws by this hand. The jaws are then shut and the forceps drawn back slowly until the pressure of the endometrium is felt. A sharp tug will then release the sample, which is immediately removed and fixed in readiness for histological examination.

Samples taken from the mid-uterine horn have been shown to be representative of the uterine endometrium as a whole (Bergman and Kenney, 1975), so they are suitable for routine biopsies where no specific abnormality is being investigated. In biopsies to further investigate suspected abnormalities, samples should be taken from each suspected area and another sample from a seemingly normal area of endometrium. Biopsies can be taken at any time of the oestrous cycle, but normally standard biopsies are taken in mid-dioestrous in order to standardize results (Kenney, 1978; Doig and Waelchi, 1993).

Such samples can allow identification of abnormalities, plus indicate evidence of uterine degeneration. The examiner looks for changes in cell structure, especially luminal cells, which may be diagnostic of inflammation, fibrosis or necrosis (Doig and Waelchi, 1993). The process is reported not to disrupt the mare's reproductive cycle, though some people report a delay in the next oestrous period (Kenney, 1977).

Not all these methods discussed are necessarily used when selecting a mare, it largely depends on the value of the mare. Some techniques may be more appropriate to investigate specific abnormalities which have been indicated by the simpler more routine procedures of rectal palpation and ultrasonic scanning.

11.2.6.4. Fallopian tubes

The competence of the Fallopian tubes (oviducts) is more difficult to assess. As mentioned previously, uniformity in size and shape can be assessed via rectal palpation. In such an examination, the Fallopian tubes should feel wiry and uniform in consistency when rolled between the palpater's fingers. Severe abnormalities may be detected this way, especially adhesions connecting the infundibulum to the uterus or ovaries and scar tissue. However, salpingitis (inflammation of the Fallopian tubes) is hard to detect. Although it has been reported to be relatively uncommon in mares, any such inflammation is of importance, as it may hinder the passage of ova and/or sperm along the Fallopian tube and may even result in complete blockage of the oviduct.

Assessment of Fallopian tube blockage is very difficult. The traditional starch-grain test is relatively successful but complicated to carry out and, therefore, of limited practical use. A starch-grain solution is injected through the mare's back at the sub-luminal fossa, using a long needle. The ovary is manipulated, via rectal palpation, to lie immediately beneath the needle. Starch-grain solution (approximately 5 ml) is then injected on to the surface of the ovary. Washings are collected from the anterior vagina and cervix 24 h later. The presence of starch grains in the flushings indicates that the Fallopian tube associated with the ovary on to which the starch grains were injected is patent.

A more invasive form of investigation is the use of laparotomy under general anaesthesia. This technique involves the exteriorization of the uterine horns and Fallopian tubes through an incision

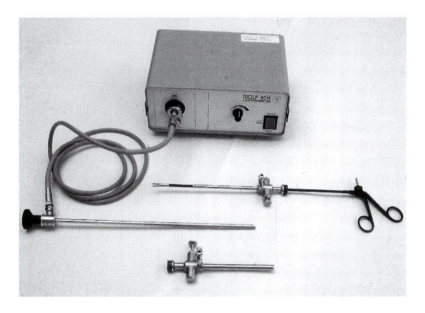

Fig. 11.15. A laparoscope, which may be used to view the internal structures of the mare's reproductive tract via the abdominal cavity. The trocars pictured in the foreground are inserted through puncture wounds in the mare's abdomen. The laparoscope is then passed through the larger one for viewing the internal organs, and the manipulating forceps through the smaller one for moving internal organs.

made either in the mare's flank or in the abdomen. The Fallopian tubes and ovaries can then be examined in detail. Such techniques are expensive and highly invasive; they also run the normal risks associated with general anaesthesia and are, therefore, rarely used.

Laparoscopy, involving the insertion of a rigid endoscope with a light source (Fig. 11.15) into the body cavity via the abdominal wall, is a further possibility, allowing the upper tract to be directly viewed *in situ*. This technique is slightly less invasive but does not allow such detailed examination. Its expense and the risks of general anaesthesia again deem it a rare specialist technique.

11.2.6.5. Ovaries

Ovarian activity can be assessed most simply by rectal palpation and more extensively by ultrasonic scanning. The appearance of the mare's ovaries varies considerably with season and reproductive activity. These changes are detailed and discussed in Section 1.9. Rectal palpation has traditionally been used to assess the stage of the mare's oestrous cycle and, therefore, help in the timing of mating. It can be used in this context of selection for breeding to

ensure that the mare is reproductively active and that follicles and corpora lutea (CL) are being produced. It can also be used to ensure that the reproductive stage indicated by the ovaries is synchronized with developmental changes in the remainder of the mare's tract. Such techniques can also indicate the presence of adhesions and neoplasms, as well as cystic follicles, ova fossa cysts and other ovarian abnormalities, which may disrupt the mare's reproductive activity. Further detail on the assessment of the reproductive stage of the mare by ovarian examination is given in Section 13.4.1.

Laparoscopy has been used experimentally to elucidate ovarian problems. It is not used in practical stud farm management due to cost and the need for general anaesthesia.

11.2.7. Infections

Mares are notorious for being susceptible to reproductive tract infections, in particular endometritis (Ricketts and Mackintosh, 1987). Not only can these cause temporary and possibly permanent infertility, but they can also be transferred relatively easily to the stallion and hence on to other mares. It is imperative, therefore, that infected mares are identified so that they can be discarded in the selection process and/or treated. One of the standard procedures used to detect such infections is swabbing. Swabs may be taken from the uterus, cervix, urethral opening and/or clitoral area and assessed for pathogenic organisms – for example, venereal disease (VD) bacteria such as *Klebsialla pneumoniae*, *Pseudomonas aeroginosa* and *Taylorella equigenitalis* which are true sexually transmitted bacteria, or *Streptococcus zooepidemicus*, *Esherichia coli* and *Staphlococcus* which, though not strictly VD bacteria, may also be transferred at covering and cause uterine infections.

Uterine swabs should only be taken during oestrous when the cervix is relaxed and moist, easing the swab's passage, and at the time when the mare's natural immunological response to accidentally introduced bacteria is heightened due to the dominance of oestrogen. The swabs may be guided in through the cervix using a speculum or guided via a gloved, lubricated arm, as in the case of uterine biopsies. Once the swab is in the lumen of the uterus it is rotated against the endometrium to absorb uterine secretions and any bacteria. The risk of accidental contamination of the swab by either airborne pathogens or bacteria present in the cervix and/or vagina is minimized by careful washing of the mare and the use of a guarded swab. Such swabs are contained within two sterile tubes. The first tube is passed through the cervix and the second tube is telescoped into the uterus. The swab can then be pushed out through the second tube far enough to reach the endometrium. A reversal of this process will reduce the contamination on retraction of the swab (Figs 11.16 and 11.17; Greenhof and Kenney, 1975).

Cervical and urethral-opening swabs are easier to obtain and may be collected at any time of the cycle.

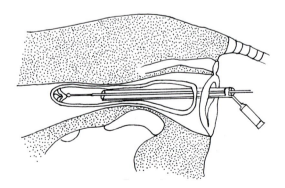

Fig. 11.16. Cervical swabbing in the mare.

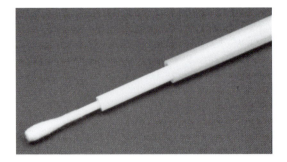

Fig. 11.17. A guarded swab used to collect samples from the mare's reproductive tract for bacterial assessment.

The samples may be taken using a vaginascope by a similar process to that described for the uterus.

Clitoral swabs are taken from the clitoral sinuses around the clitoral fossa. Gentle squeezing of the clitoris may produce smegma secretions for swabbing; care, however, should be taken as some mares object. The clitoral fossa is notorious for harbouring, among other pathogens, *Taylorella equigenitalis*, the causal agent for contagious equine metritis (CEM).

Once the sample has been taken, the swab can be applied to a variety of growth mediums and incubated under various conditions (temperature, humidity, atmospheric pressure and oxygen content) in order to identify infective organisms. Alternatively, swabs can be smeared on to glass slides for cytological examination (Ricketts *et al.*, 1993).

Swabbing is a very effective method of assessing bacterial contamination of the mare's reproductive tract, but results must be treated with some caution. The presence of bacteria within the vagina and cervix does not necessarily indicate uterine infection, especially if the cervical seal is fully competent. The natural microflora of many mares contains pathogenic organisms. The actual process of taking a sample also increases the chance of reproductive tract contamination by airborne bacteria; hence the technique must be carried out under as clean and sterile conditions as possible. In general, therefore, as indicated above swabbing is best carried out during oestrous.

Swabbing may also be used to collect samples of cells that line the reproductive tract which can then be assessed under the microscope (cytological examination) to identify any abnormal cell populations.

Providing the possible drawbacks are borne in mind, swabbing is a simple procedure to carry out in the routine selection of mares for breeding. Indeed as a result of the Thoroughbred Breeders Association Annual Code of Practice (1978 onwards) it is advised that all Thoroughbred mares, and many of those in other breed societies, should be swabbed prior to arrival at the stud and again at the oestrous of service, to ensure the absence of VD bacteria. Only mares with a negative certificate will be accepted at the stud and only those with a second negative certificate will be covered.

11.2.8. Blood sampling

If there are reasons to believe that a mare may have problems in carrying a foal to term, and no anatomical abnormalities have been detected using the previously discussed techniques, then the problem may lie in hormonal inadequacies. The endocrine profiles of mares can be determined by sequential blood sampling. Any deviations from the normal profile can be identified and the specific area of failure, i.e. follicle development, ovulation signal, oestrous behaviour or CL regression, can be identified. In the light of these results, appropriate hormone therapy may be possible to compensate for the natural deficiencies, or it can be decided that such a mare is not worth the risk or cost of such therapy. Details on the normal endocrine profiles for mares are given in Chapter 3 (this volume).

Blood samples can also be used to indicate the general health status of a mare, bringing to light specific deficiencies related to diet, low-grade infections, blood loss, cancer and parasite burdens. Low red cell counts, i.e. below 10×10^6 ml^{-1}, indicate anaemia. Pack cell volume is a quick and easy assessment of red blood cell/fluid balance, and normal levels are 40–50%. Assessment of the colour of the supernatant in pack cell volume test is also a useful problem indicator. It is normally straw-like in colour, and discoloration can indicate problems; for example, red/pink indicates a breakdown of red cells and release of haemoglobin. Haemoglobin levels themselves are a well-established indicator of anaemia: levels <12–17 gl^{-1} indicating problems.

A high white blood cell count above 12,000–14,000 ml^{-1} indicates the presence of disease – in particular infection or cancer. A total protein index (protein content of serum after clotting) of <15 gl^{-1} indicates blood loss, starvation or liver, kidney or gastrointestinal disease. Fibrinogen levels may also be indicative of abnormalities; high levels >10 gl^{-1} suggest inflammatory, neoplastic or traumatic disease (Table 11.1; Varner *et al.*, 1991; Pickett, 1993c).

Table 11.1. Blood parameters indicative of disease.

Parameter	Value indicative of disease
Red blood cell count	$<10 \times 10^6$ ml^{-1}
Pack cell volume	<40–50%
White blood cell count	>12,000–14,000 ml^{-1}
Total protein index	<15 g l^{-1}
Haemoglobin	<12–17 g l^{-1}
Fibrinogen	>10 g l^{-1}

11.2.9. Chromosomal abnormalities

Blood samples may also be used for chromosomal analysis, if genetic abnormalities are suspected (Bowling *et al.*, 1987). Genetic abnormalities reported include a missing X chromosome, Turner's syndrome (63XO), chimerism or mosaic (63XO: 64XX), sex reversal (64XY) and extra X chromosome (65XXX; Hughes *et al.*, 1975; Halnan, 1985; Bowling, 1996).

11.3. The Stallion

The selection criteria for reproductive competence in the stallion are similar to those of the mare and can be listed as follows:

- History;
- Temperament and libido;
- Age;
- General conformation;
- Reproductive tract examination;
- Semen evaluation;
- Chromosomal abnormalities;
- Blood sampling;
- Infections;
- General stud management.

Reproductive evaluations are necessary prior to purchase but may also be used routinely prior to each breeding season or if a problem is suspected (Thompson, 1994).

11.3.1. History

Records of a stallion's history are invaluable in aiding selection and, as with the mare, can be divided into his breeding and general history. Records for stallions do not tend to be as detailed or as readily available as those for mares.

11.3.1.1. Reproductive history

Records of his past breeding performance, if available, should answer questions such as: When does his season normally start and end? How many mares is he used to covering in a season? What are his return rates like? What is his semen quality like?

The answers to these questions will indicate his reproductive ability (Van Buiten *et al.*, 1999). Stallions with short seasons will be less able to cover as many mares and may suffer from low libido. The number of mares he has served per season in the past and the return rates, along with semen analysis, will give an indication of what workload he will be capable of. If his return rates are high, especially if a significant decrease is seen with an increase in workload, this may indicate the natural limit of the number of mares he is able to cover. The routine of covering may affect performance and can be tailored to suit the stallion. Routines may involve one or two covers per day for 6 days with a day's rest, or two covers per day for 8 days followed by 2 days' rest or numerous variations on these themes. Most stallions do need a rest day but should be able to cover mares at the rough frequency of the systems given above (Pickett *et al.*, 1985). If there are indications that a stallion is not capable of such workloads and requires more rest days to maintain his fertility rates, then his selection should be queried, especially if you are looking for a stallion to purchase. Return rates are a good guide to a stallion's ability (Van Buiten *et al.*, 1999) but it must be remembered that the fertility of a stallion is only as good as the fertility of the mares he is presented with.

Any previous semen analysis should also be detailed in his records. Many valuable stallions have a semen analysis carried out routinely at the beginning of each season. This, along with a blood sample, normally taken at the same time, allows any potential problems to be identified in time for remedial action to be taken before the breeding season starts. Any past reproductive tract infections should also be detailed in a stallion's records, along with any treatment given and the outcome. Any long-term effects of infection should be evident in the stallion's workload and return rates for the rest of that season and any subsequent seasons.

11.3.1.2. General history

The stallion's general history should indicate his vaccination and worming status, along with the incidence of injuries and accidents. Damage to his hindquarters or limbs may restrict his ability to mount a mare, as may laminitis and neurological disorders (Griffin, 2000). AI may be an alternative (Davies Morel, 1999); even so, he is likely to need the occasional mount for the collection of semen samples, though the number of mounts per mare fertilized will be significantly reduced and the unpredictability of mounting a mare avoided. Such stallions are not advised for purchase. As with the mare, if there were any suggestions that any damage or weaknesses may be heritable, selection would not be advised. Injuries to a stallion's genitalia, usually as a result of a kick from a mare, will cause degenerative and scar tissue within the penis and/or testes, which will reduce his ability to mate a mare and reduce the number of testicular germinal and Sertoli cells and hence fertility rates of the stallion. Severe damage resulting in the removal of a testicle should also be noted in a stallion's records to reassure potential purchasers that he is not a rig. Such stallions are capable of fertilizing a mare, but the workload may have to be reduced. Severe injuries to a stallion during mating often have psychological effects, drastically reducing his libido, possibly to such an extent that he is unwilling to cover naturally.

Past illnesses should also be indicated in his records. Illnesses associated with the respiratory or circulatory systems may indicate that the stallion will not be capable of working a full season, limiting the numbers of nominations available or that can be sold. Again, if there is a possibility that such weaknesses could be heritable, the stallion should be avoided. Any illnesses resulting in a fever can disrupt spermatogenesis, due to the elevated temperature (Johnson *et al.*, 1997). This may result in temporary infertility, though this may not be evident for several months as the spermatogenic cycle takes 57 days (Davies Morel, 1999). Systemic infections, for example, strangles or flu, can cause inflammation within the testes and, if this results in a significant amount of tissue degeneration, permanent sub-fertility or even infertility may result.

The sorts of stallion records that should be available to a potential purchaser are illustrated in Fig. 11.18.

11.3.2. Temperament and libido

The temperament of the stallion is very important for ease of management and as a heritable trait.

Date of mare's arrival	Mare/owner	First service date	Second service date	Return date	Return date	Positive pregnancy diagnosis	Live foal date of birth (dob)	Comments
16.1.06	Trefaes Fancy Mr T. James	20.3.06	22.3.06	–	–	4.4.06	22.2.01 filly	Foal at foot d.o.b. 12.2.06
18.1.00	Katie Jane M. Davies	15.3.06	16.3.06	7.4.06	–	21.4.06	Abortion 10.12.06	Maiden mare
30.3.00	Penpontpren Mina Mrs M. Morel	30.3.06	1.5.06	21.4.06	22.4.06	15.5.06	15.3.07 colt	Barren mare
1.2.00	Lluest Megan Lluest Stud	30.4.06	1.5.06	21.5.06	–	–	–	Foal at foot d.o.b. 30.3.06 delayed uterine involution
12.2.00	Rose Mr A. Jones	18.3.06 10.4.06	–	–	–	8.4.06 twins aborted 3.5.06	– 30.3.01	Foal at foot, d.o.b. 12.3.06 Twins detected 8.4.06 Aboted and recovered
12.2.00	Lady Jane Barnswood stud	19.3.06	20.3.06 21.3.06	–	–	5.4.06 twins, 1 pinched	2.3.07	Barren, twins detected 5.4.06, smallest pinched

Fig 11.18. An example of the type of records that should be kept for each stallion during his breeding season, indicating the mares he has been put to and the result, along with any noteworthy comments.

A stallion of a quiet and kind disposition is a great asset and will be much easier and safer to handle (Fig. 11.19). A stallion that is rough to his mares will not only run the risk of inflicting permanent damage to them but may also be hurt himself if they retaliate. A rough stallion will prove unpopular and it may be difficult to get him enough mares to make his use economic. Some protection, in the form of neck guards, can be given to mares that are mated to stallions that tend to bite during covering, but no protection can be given against stallions that are downright vicious, and as such they should be avoided at all costs. There is some evidence to suggest that stallions brought up in an intensive/isolated environment with little social interaction are more likely to show such traits as well as poorer libido than stallions brought up in a more natural environment (Christensen et al., 2002).

Fig. 11.19. A well-behaved stallion is an asset to any stud, easing his management and reducing the danger to his handlers.

Ideally, records should indicate the stallion's temperament and any specific characteristics that he might have. It is to be hoped that his bad habits, especially those that might prove dangerous, will also be indicated. To be forewarned is to be forearmed and might lead you to reject an unsatisfactory stallion.

Bad behaviour in many stallions is a direct result of the conditions and management under which they are kept (Chapter 18, this volume). Therefore, especially in the case of a stallion that seems to have developed bad habits later in life or after a change of owner or management, the conditions under which he is kept should be assessed before he is rejected for covering a mare. However, as a potential purchase he is not a good choice, as such habits are difficult to break. Bad behaviour tends to perpetuate itself, as, due to the potential danger, such stallions are kept confined for longer periods of time and hence away from other companions. Their boredom is exacerbated and their bad habits develop further. Stereotypies (habits) to be aware of include: weaving, crib-biting, and wind-sucking, all signs of boredom. Additionally, there is a commonly held belief, though not supported by scientific research, that other horses may copy stereotypies.

Stereotypies such as self-masturbation were once frowned upon but are now considered natural behaviour, of no consequence except the potential embarrassment to owners. Some stallions also indulge in self-mutilation, especially after mating, biting themselves in areas where the smell of the mare lingers. Though thorough washing post mating can reduce the incidence, the potential for self-harm, and the added management time and expense may preclude their selection.

A stallion's reproductive temperament and willingness to cover is termed his libido, which partly determines his reproductive potential. Libido is governed, like all other sexual activity, by season (see Section 4.1). Hence, those stallions with longer seasons tend to show a higher libido and, therefore, willingness to mate early on in the season, extending the time in which he can be worked. Ideally, if selecting a stallion to purchase, he should be seen teasing and covering a mare. A stallion with a low libido will need to mount a mare several times before ejaculation, taking up to 20 min to cover a mare, or he may fail completely; he may also show initial interest very reluctantly. The number of mounts per ejaculation and the time between actual intromission and ejaculation are good indications of libido. The number of mounts per ejaculation should be as near to one as possible and the time between intromission and ejaculation a matter of seconds (Thompson, 1994).

11.3.3. Age

The age of the stallion is less important than that of the mare, as far as reproductive ability is concerned. The significance of age in the selection of the stallion depends on what that stallion is required for, i.e. for a single mating to a selected mare or as a potential purchase for long-term future use. If you are selecting him for service of a single mare, then as far as you are concerned, he will be required to perform on just a couple of occasions; his age is of limited importance provided he is capable of covering. However, if you are looking to select a stallion for purchase and, therefore, long-term future use, you have to ensure that he is young and fit enough to give you plenty of seasons but old enough to have proved his worth.

As far as a lower limit is concerned, most colts reach puberty at 18–24 months of age (see Section 4.1; Clay and Clay, 1992). A colt can, in theory, be used as soon as he reaches puberty, but care must be taken to introduce him to the job gradually and not to overwork him too soon or give him awkward mares, which may affect his as yet delicate ego and reproductive confidence (Johnson *et al.*, 1991). Further details on early stallion management are given in Section 18.4. The purchase and use of such young stallions is risky, as they have no proven performance record.

As far as an upper age limit is concerned, this really depends on the stallion's general health and condition. If he has no problems with lameness, stamina, wind, injury, etc., he may well be capable of working well into his teens and even twenties, though in the latter years his workload may have to be reduced. There is reported evidence that reproductive capability is inherently reduced with age (Johnson and Thompson, 1983; Amann, 1993a,b; Naden *et al.*, 1990; Fukuda *et al.*, 2001; Madill, 2002). However, other work disputes this, suggesting that any decline in reproductive performance with old age is indirect due to reducing libido from injury, arthritic conditions, etc., and not a decline in spermatogenesis per se (Johnson, 1991a).

As discussed in the case of mare selection, if an older ex-performance horse is being considered, it must be borne in mind that he will require a prolonged period of time to adjust physically and psychologically to his new role in life. Details of the problems associated with using performance horses as stallions are given in Section 12.4.

11.3.4. General conformation and condition

A stallion's general conformation is of importance, not only as it will be passed on to his offspring, but also to ensure that he is capable of withstanding a full breeding season. A stallion with poor limb conformation, especially the hindquarters, will also be weak in this area and may, therefore, be unable to withstand the heavy workload of a full breeding season, limiting his economic viability.

Particular note should be made of his physical ability to cover mares. He should be free of all signs of lameness, especially in the hind limbs. His legs should be checked before and after exercise and a comparison made, to ensure that there is no sign of swelling, a sign of possible weakness. He should be free of all conditions such as arthritis, spinal or limb injury, wobbler syndrome, laminitis or any neurological disorder, all of which could cause pain (Griffin, 2000). A stallion's feet should also be in excellent condition, regularly trimmed to ensure they stay that way. Adequate heart room in a broad chest is also desirable and, if doubt is placed on the stallion's cardiovascular system, electrocardiography may be conducted.

Good general condition and physical fitness are very important for the breeding stallion. The condition of a stallion, like that of the mare, can be classified on a scale of 0–5 (0 emaciated, 5 obese; Fig. 11.1). The optimum body condition for a stallion in work is 3, i.e. he is well muscled up and in fit working condition (Fig. 11.20). Stallions in condition score less than 3 tend to have lower libido, and are physically less able to stand a heavy workload (Jainudeen and Hafez, 1993). If the stallion's condition is very poor, spermatogenesis may also suffer. At the other extreme obese stallions also tend to have low libido; they tend to be lazy and may be incapable of mounting a mare. In addition, the extra weight puts additional strain on the mare at mating and may cause her damage. It is to be remembered that the nutritional demands during the breeding season are similar to those of a performance horse, the workload of the two being approximately equivalent (Thompson, 1994; Griffin, 2000). Further details on stallion nutritional management are given in Section 18.5.3.

11.3.5. External examination of the reproductive tract

An external examination of the stallion's reproductive genitalia is an essential selection procedure, as

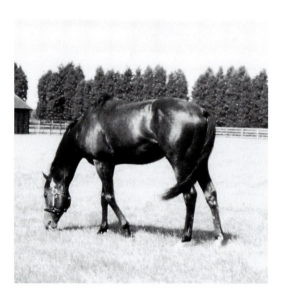

Fig. 11.20. A stallion in good, fit, well-muscled working condition (body condition score) ready for the breeding season.

Malignant or benign growths within the testes are rare but may be evident (Caron *et al.*, 1985; Schumacher and Varner, 1993). The skin of the scrotum should be checked for dermatitis, which can cause an increase in testicular temperature. The position of the epididymis should also be felt. Their normal position in the non-retracted relaxed testes is on the cranial (abdominal) side of the scrotum. Positioning elsewhere may indicate testes torsion or twist (Hurtgen, 1987; Threlfall *et al.*, 1990).

The vas deferens leaving the testes, plus testicular blood and nerve supply, passes up into the body of the stallion through the inguinal canal, which should be free from adhesions and hernias.

The penis and prepuce of the stallion should also be examined for injury, haematomas, squamous cell carcinoma, summer sores, sarcoids and general infections or injury. Examination can be carried out at washing prior to semen collection and should be a routine selection procedure. Details on VD infections and penile conditions are given in Sections 19.3.2.4.7 and 19.3.2.5.

11.3.6. Internal examination of the reproductive tract

As with the mare, examination of the internal reproductive tract of the stallion is a skilled veterinary surgeon's job. Information given by internal examination can be very useful in assessing the reproductive potential of a stallion, although internal examination is less useful than in the mare and may be limited by financial implications and the need for experienced personnel.

Access to the internal parts of the stallion's reproductive tract is very difficult. Some appreciation may be gained by rectal palpation and ultrasound (Little, 1998). Via rectal palpation, the vas deferens can be felt entering the body cavity at the inguinal canal and both, one on either side, should feel smooth and of uniform diameter. Alongside the vas deferens as they enter the body cavity lies the spermatic artery, the pulse of which should also be checked. An appreciation of testicular blood flow may also be gained by colour Doppler ultrasound (Pozor and McDonnell, 2004). Very low blood pressure, or a drop between successive examinations, may be indicative of a haemorrhage, blood clot or tumour, or the release of body fluids into a localized infection site. The accessory glands may also be palpated individually and their texture, size and shape assessed. Paired glands, such as the seminal vesicles,

his ability to perform is naturally a function of the condition of his reproductive organs (Griffin, 2000). He should have two normally functioning testes, which may be felt through the scrotum and palpated to ensure they are of a similar size and consistency, move easily within their tunicae and are not warm to the touch. Occasionally, the left testis is slightly larger than the right, but the difference is slight and should not be accompanied by an increase in heat. The surface of the testis should feel smooth, with the occasional blood vessel being felt running under the skin. Any adhesions preventing the testes moving up and down easily within the tunica are likely to indicate scar or fibrous tissue due to past injuries. This not only reduces the volume of functioning testicular tissue but may also interfere with spermatogenesis within the remaining tissue. Indeed testicular size is a good indicator of the spermatozoa-producing capacity of the stallion and hence his potential workload. As such, testicular volume has been advocated as an assessment criterion when selecting for reproductive potential (Love *et al.*, 1991; Pickett and Shiner, 1994; Parlevilet, 2000). Excessive fat within the scrotum as a result of excessive body condition will increase the insulation of the testes, with the danger of increasing testicular temperature and, therefore, decreasing sperm production.

should be checked for symmetry. Ultrasound may be used to give an indication of physical abnormalities and accessory-gland secretary function (Weber and Woods, 1993; Pozor and McDonnell, 2002). An indication of the function of the accessory glands may also be gained by semen evaluation as will be discussed in the following section.

11.3.7. Semen evaluation

Semen evaluation is a routine selection procedure. If a stallion is to cover mares throughout the breeding season with consistent success, his semen has to meet various minimum parameters (Colenbrander *et al.*, 2003). In many studs, all stallions routinely have their semen evaluated at the beginning of each season and if a problem is suspected. The quality of his semen has a direct effect on the stallion's ability to consistently and successfully cover a number of mares throughout the season (Jasko *et al.*, 1990a,b, 1991; Gastal *et al.*, 1991; Pickett, 1993a; Parlevliet and Colenbrander, 1999). Ideally, for an accurate evaluation, samples should be taken: (i) one after 3 days' sexual rest preceded by a double collection taken 2 and 1 h prior to test collection; (ii) as the last collection of a series of seven daily collections, preceded by a double collection taken 2 and 1 h prior to test collection; or (iii) both collections taken 1 h apart after 1 month's sexual rest (Pickett and Voss, 1972; Kenney, 1975a; Sullivan and Pickett, 1975; Swierstra *et al.*, 1975). In most commercial enterprises/AI programmes such regimes are not economically viable, and single sampling, interpreted with caution, can provide adequate information for most routine practices.

Collection of semen is normally by means of an artificial vagina (AV). Details of the collection and evaluation procedure are given in Sections 20.3 and 20.4 and elsewhere (Davies Morel, 1999). The normal parameters for semen are given in Table 20.3.

11.3.8. Infections

Like the mare, the stallion is susceptible to sexually transmitted diseases and, as such, all stallions should be tested for infections prior to purchase, either to eliminate them or to allow treatment to commence prior to their use.

As with the mare, swabs can identify infections of the genital tract; these are taken from the urethra, the urethral fossa and prepuce of the stallion's penis. Swabs should be taken from the erect penis, erection

being encouraged by an oestrous mare or tranquillizers. Three different swabs must be used and it is best to take the urethral fossa sample last, as this one can cause considerable discomfort and hence objection. Swabs or cultures of semen samples can also be assessed. The stallion's semen and penis have a natural microflora of bacteria and fungi and these should be distinguished from VD pathogens. The most noteworthy bacteria, classified as VD causers of acute endometritis, are *Klebsialla pneumoniae*, *Pseudomonas aeroginosa* and *Taylorella equigenitalis* (Couto and Hughes, 1993; Parlevliet *et al.*, 1997).

Swabbing is routinely carried out in many studs on all their stallions well before the season starts. This allows time, if infections are identified, for treatment to begin and take effect before the breeding season. Further details of infection of the stallion's reproductive tract and the effect upon reproduction are given in Section 19.3.2.5.

11.3.9. Blood sampling

Blood sampling of stallions can be used to assess their general health and can indicate low-grade infection, blood loss, cancer, nutritional deficiencies or parasite burdens (Pickett, 1993c). Details on the information, which can be gathered from blood sampling, have been given in the previous section on the selection of the mare (Section 11.2.8 and Table 11.1.). Any stallion showing these characteristics should not be considered for use until the problem has been identified and appropriate treatment commenced.

Blood samples are rarely used for hormone analysis because the considerable inter-stallion variation reduces the accuracy of such testing to assess potential reproductive performance (Roser, 1995). The episodic nature of testosterone release also necessitates a period of sequential blood sampling from which an average can be taken rather than a single representative sample. Low plasma testosterone concentrations have been associated with low libido and poor semen quality (Watson, 1997).

11.3.10. Chromosomal abnormalities

Chromosomal abnormalities are well documented in the mare but less so in the stallion (Long, 1988). However, conditions such as XX male syndrome (64XX), chimerism or mosaic (64XX: 64XY), Klinefelter's syndrome (65XXY) and 13 quarter/ deletion (64XY) are associated with an inability to impregnate mares or very low fertility rates, despite apparently normal genitalia (Halnan and Watson, 1982; Halnan *et al.*, 1982; Bowling *et al.*, 1987; Bowling, 1996; Makinen *et al.*, 2000).

11.3.11. General stud management

If your selection of a stallion is not for purchase but rather for use on one of your mares, you will also be interested in the management at the stud at which he stands. There are several things that will concern most owners selecting a stud to send their mare to, and these will be discussed in turn.

The system of breeding used is of prime importance. Is the stud appropriately equipped for visiting mares or are you expected to "walk in" your mare, i.e. bring her in for the day, having detected yourself that she is in oestrous, and take her away the same day after covering? Some studs allow mares to stay a few nights but have only limited facilities and may well expect mares to live out. This obviously has a bearing on the distance it is possible to travel. You should also consider the method of covering, varying from pasture breeding to intensive in-hand breeding. The various methods used are discussed in detail in Chapter 13 (this volume). Some studs will expect the mare to be taken home as soon as she has been covered; others will allow her to stay for re-covering if necessary and will only allow her home after a positive pregnancy diagnosis at scanning and/or rectal palpation, usually 2–4 weeks post mating.

Many of the smaller native pony-type studs do not have the facilities to foal down visiting mares, necessitating mares to be brought to stud very soon after foaling. This can be traumatic and dangerous for the foal, and precludes using a stud that is too far away. Larger studs tend to have the mares brought in to foal, normally 4–6 weeks prior to foaling. This allows the mare to be covered on her foal heat without the danger of travelling with a young foal.

The daily management at the stud should also be investigated and matched as closely as possible to the mare's normal routine. If not, her routine at home should be slowly altered to that at the stud to minimize the stress of change. All animals on the stud should be wormed regularly and vaccinated, and documented proof of adequate protection is usually required of all visiting mares. At the more intensive studs, mares standing to valuable stallions will also require negative certificates to a variety of VD bacteria (Horse Race Betting Levy Board, 2008).

A good impression of the standard of management of a yard can be gained by a general visit. The yard, whatever system in use, should be clean and tidy, all the mares and stallions should be in good condition, the pasture well tendered and the animals contented. If the mare is to foal there, the foaling facilities should be clean, safe and roomy with a good system for 24h monitoring by skilled staff. The facilities of the yard and the equipment and expertise available will reflect the type of stallion and his nomination fee.

The system that you choose is ultimately a personal choice depending on your priorities and the finances available. The more intensive systems tend to be associated with the Thoroughbred industry, where expense is of less concern but hygiene and protection of valuable stock are of paramount importance. In such intensive systems, mares are often taken to the stud to foal, are subsequently covered and possibly re-covered and remain at stud until pregnancy is confirmed, often at days 12 and 25. In such systems, the service fees are high and the costs of keep and veterinary attention are great, but this is offset by the value of the offspring and the risks are lower. At the other end of the spectrum, native studs will serve a mare that arrives in their yard and within half an hour she can be on her way home. In such systems, stallion fees are low, as are costs, but the offspring is often of low value and the risks are higher.

Further details on the management systems and principles for both the mare and stallion at breeding are given in Chapters 12–19 (this volume). When examining potential studs it is as well to bear in mind that the ideal is not normally achieved. It is unrealistic to expect a yard standing a cheaper stallion, with stud fees of £50–100, to have the facilities found in a Thoroughbred stud standing stallions with nomination fees up £50,000.

11.4. Conclusion

If more attention was paid to assessment and selection of breeding stock on the basis of reproductive competence significant amounts of money, time and effort would be saved in trying to breed from sub-fertile or infertile stock. Failure to select fertile stock leads not only to suffering for the mare and stallion, but also to frustration for all. Additionally, many problems are inherited and so will be perpetuated in subsequent generations, to the detriment of the equine population as a whole.

12 Preparation of the Mare and Stallion for Covering

12.1. Introduction

It is essential that preparation of the mare and stallion start in plenty of time prior to covering. It is no use making a last-minute decision that you wish to put a mare in foal or stand a stallion at stud and then wondering why she does not concieve or your return rates are high. A preparation time of at least 6 months is required in order to maximize the chance of conception. This chapter will concentrate mainly on this period, with some reference to any earlier preparation that may be required.

12.2. Preparation of the Barren and Maiden Mare

If a mare is destined to be a brood mare, then she must be brought up with this aim in mind, with close monitoring of her general condition and growth. Horses normally reach puberty at 1.5–3 years of age, depending on the breed and their nutritional status in early life. If you are considering breeding a mare at, or near, puberty, before she has reached her full mature size, then her stage of development and her general condition are of utmost importance. Breeding before attainment of mature body size places on the mare the additional burden of pregnancy as well as that of her own continued growth and development. Provided she is well grown and in good body condition, she should be able to cope with such additional demands, but this must be borne in mind when considering her management through pregnancy, especially that of nutrition.

Today, many horses, both mares and stallions, are bred for the first time relatively late in life, having already had a successful performance career, on the basis of which they have been chosen for breeding. In such cases, both mares and stallions that are older are more set in their ways and have been managed to date as athletes, not as breeding stock (Fig. 12.1).

Performance horses must be allowed plenty of time to unwind both physically and psychologically. This should start during the autumn prior to the planned covering, during which time workloads should be slowly reduced to a maintenance level. This is usually adequate for most horses, though much variation is evident between individuals, and some may take as long as 18 months to adjust. Intensive training is detrimental to reproduction in all mammals; this can be clearly demonstrated in women athletes who regularly fail to ovulate. Mares in peak athletic condition will characteristically demonstrate abnormal oestrous cycles, often showing delayed oestrus, silent heats and oestrous behaviour not accompanied by ovulation. However, given a long enough adjustment period, a mare should start showing regular oestrous cycles and can be successfully mated.

Careful attention should also be given to the mare's nutrition and, related to that, her exercise. Mares should be in good, not fat, body condition at covering. The body condition score to aim for at mating is 3 (see Section 11.2.4). In addition, it is generally believed that in maiden and barren mares better conception rates are obtained if they are on a rising plane of nutrition, in particular increasing energy, in the last 4–6 weeks prior to covering; this is termed flushing (Van Niekerk and Van Heerden, 1972; Kubiak *et al.*, 1987; Morris *et al.*, 1987). Flushing of barren mares has also been reported to advance the start of the breeding season by as many as 30 days (Fig. 12.2; J. Newcombe, Wales, 2008, personal communication). The best regime is to ensure that the mare is in condition score 2–3 in the autumn prior to covering. As the season approaches, her energy intake can be gradually increased by replacing some of her roughage intake with concentrates in the last 4–6 weeks before planned covering (Hintz, 1993a; Guerin and Wang, 1994; Frape, 1998).

If the mare is young, it may also be pertinent to supplement her diet during this period with protein,

Fig. 12.1. Many successful stallions are currently, or have been, performance horses. As such, they need careful management in order to perform well at both jobs.

Fig. 12.2. Mares being flushed on lush pasture prior to mating in order to encourage optimum fertilization rates.

calcium, phosphorous and vitamin A. Requirements for these elements are higher in young maiden mares than in mature mares.

Exercise is also important, helping to maintain body condition and prevent obesity. Gentle riding or hacking out provides a good form of exercise for barren mares during the preparation period (Fig. 12.3). All mares should be turned out daily, and ideally barren mares, especially those not broken, should live out except in the most inclement weather, thus providing them with *ad libitum* exercise (Fig. 12.4).

Chapter 12

Fig. 12.3. Gentle riding provides ideal exercise for barren mares, in order to keep them fit for covering and any resultant pregnancy.

Fig. 12.4. If mares are not broken to ride, they must be turned out to provide enough exercise in order to keep fit.

A period of 6 months' preparation also allows time for a mare to be tested for infection, and the appropriate treatment given and full recovery to occur. Any damage due to old infections can be investigated and either assisted in its repair or the mare discarded from the breeding scheme. If she is of sufficient genetic merit, the possibility of embryo transfer can be investigated and necessary arrangements made. All barren mares, especially maidens, should be introduced to new handling systems, buildings and surroundings associated with breeding during this period. This is of particular importance with ex-performance mares that are usually moved or sold in readiness for their new career. Familiarization with management practices associated with breeding, such as restraint in stocks, rectal palpation, ultrasonic scanning, teasing facilities, etc. should also be ensured prior to covering. Old-timers need only to be reintroduced a few weeks before the season starts. Any changes in diet must be introduced gradually and also the introduction of new companions should be done early enough to allow a settling-down period. Provided all changes are made gradually and in good time stress at covering can be minimized, so maximizing reproductive success.

12.3. Preparation of the Pregnant Mare

Care must also be taken in the preparation of the pregnant mare for re-covering the following spring. However, the presence of a pregnancy limits to a large extent how she can be managed. One advantage with a pregnant mare is that she has seen it all before, at least in the previous year, and no psychological and physiological adjustment is required. However, a careful eye should be kept on her condition in order to prevent obesity. Once a pregnant mare becomes overweight, it is very difficult to rectify, especially in late pregnancy, without endangering the fetus. Prevention is, therefore, much better than cure and it is essential that the mare is in a fit condition prior to initial covering and that condition score of 3 is maintained throughout pregnancy and into her next covering. Flushing of pregnant mares is not advised and has no effect or possibly a detrimental effect on conception rates to the foal heat (Frape, 1998). If there is a reasonable period of time between foaling and re-covering, this can be used to try and adjust body condition, but care must be taken as alteration in nutrition will affect milk yield and hence the foal at foot. There is conflicting evidence with regard to the effect of nutrition during late pregnancy and between parturition and foal heat on subsequent conception rates (Jordon, 1982; Henneke et al., 1984). However, it is generally accepted that a condition score of 3 is ideal during this period and that mares should be fed well in lactation to maximise fertility (Frape, 1998).

Exercise is an important aid in maintaining a body condition score of 3 in the pregnant mare. Pregnant mares can normally be safely ridden up to the sixth month of pregnancy; but this depends on the individual. By the sixth month, all strenuous work must be excluded. Mares that are not broken and those in late pregnancy should be turned out every day to help maintain fitness, blood circulation and prevent boredom. In an ideal world, mares would live out with plenty of opportunity to exercise: such systems are popular in temperate latitudes without the risk of adverse weather. Exercise not only helps to prevent obesity but also maintains the mare's fitness and muscle tone, both of which will be needed at and after parturition.

12.4. Preparation of the Stallion

If a colt is destined to become a working stallion, he must be brought up during his early life with this aim in mind, especially with regard to discipline. Many stallions become hard to handle, and in some cases downright dangerous, because discipline and respect for authority have not been established in early life.

Most of the problems encountered in stallions previously used as performance horses are behavioural abnormalities. They will have had several years during which they will have been actively discouraged from displaying any sexual behaviour. As a result, they may be severely inhibited at their first sight of an oestrous mare, anticipating punishment. They will often find it hard to revert to natural stallion behaviour and need varying amounts of time to adjust to their new career (Van Dierendonck and Goodwin, 2005). Many stallions take a few seasons to completely adjust and some never really do achieve complete adjustment. A stallion's libido may also be affected, and such stallions may, as a result, always prove to be slow to react to an oestrous mare and show clumsy mounting behaviour.

As well as their psychological adjustment, attention should be paid to a stallion's nutrition and exercise during this preparation period. They must be fit, not fat. A heavy covering season places significant demands on the stallion, especially in terms of energy, and he will often lose condition over the season. This loss in condition is minimized if the stallion's energy intake is increased by increasing the concentrate proportion of his diet and if he is fit and in a body condition score 3 as the season commences (Fig. 12.5). Both excess and low body weight reduce a stallion's libido. Exercise helps a stallion maintain good condition, preventing obesity and maintaining muscle tone and stamina. Stallions have a tendency to become obese, as they are regularly kept individually in stables or paddocks away from each other and mares. In the natural conditions they of course would be free to exercise at will.

If stallions are badly behaved, it is tempting to keep them confined, with only limited turnout.

Fig. 12.5. A stallion should be in a fit, not fat condition, that is, condition score 3 at the beginning of the breeding season. (Photograph courtesy of Derwen International Welsh Cob Stud.)

This perpetuates the problem and accentuates any misbehaviour due to boredom. Some stallions can be safely ridden or driven, which provides an excellent form of exercise as well as good discipline (Fig. 12.6). In the less-intensive studs, usually where stock is less valuable, some quiet stallions are turned out in July at the end of the season with either their mothers, an old mare or other quiet mares. They can then be brought back into riding work over the winter. This system allows a rest period after the season, followed by a fitness regime prior to the start of the next season. It also provides them with another purpose in life, which greatly helps discipline.

At least 4 weeks prior to his first mare of the season, the stallion should be brought into the stud environment. He should then be introduced or

Fig. 12.6. Riding or driving provides an excellent form of exercise as well as good discipline.

Preparation of the Mare and Stallion for Covering

reintroduced to the yard, handling systems, buildings, surroundings and specially the covering area with plenty of time to allow familiarization prior to the first covering. Any changes in diet should be introduced slowly, before the season starts, along with any new companions. Further details regarding stallion management are given in Chapter 18 (this volume).

12.5. General Aspects of Preparation for Breeding

12.5.1. Drugs

Many performance horses have been on various drug regimes during their performance careers. Plenty of time must be allowed for the body to eliminate these drugs from the system. Corticosteroids, used as anti-inflammatory drugs to treat various injuries, can have serious detrimental effects on reproductive performance in both stallions and mares. They can also impair the body's ability to fight infection, which is of particular significance in the mare. The use of anabolic steroids to boost muscle development and hence performance also has severe detrimental effects on reproductive performance and should be eliminated from the animal's system well in advance of the breeding season (Shoemaker *et al.*, 1989).

12.5.2. Testing for infections

Prior to covering, all mares, whether barren or pregnant, should be tested for reproductive tract infections. It is good practice for barren and maiden mares to have clitoral fossa swabs taken during their preparation and if done early enough and the result is positive, then there should be enough time for treatment, recovery and retesting before mating. It is the normal practice for many studs to require mares to be swabbed and proven to be clear of infection before being accepted for covering; indeed, the practice is contained within the Horse Race Betting Levy Board (HRBLB), Codes of Practice for Thoroughbred breeders. There are different recommended swabbing practices for high-risk (mares that have had previous contact with contagious equine metritis (CEM)) and low-risk (all those not classified as high-risk) mares and for those boarding at studs and walking in. It is very important that you make sure that you know and carry out the swabbing requirements before your mare goes to stud; they may change

but they are published annually in the HRBLB Codes of Practice. Normally, low-risk mares must have two clean swabs, one from the clitoral fossa before the oestrus of covering, and the second from the endometrium at the oestrus of covering. High-risk mares normally need three swabs, the two that low-risk mares have with an additional clitoral one taken before the mare arrives on the stud. All swabs must be sent to approved laboratories; the results are normally available within 48 h and so allow enough time for mares to be covered on the oestrus of testing. Swabs are tested primarily for *Taylorella equigenitalis*, the bacteria which causes CEM, but they are also tested for *Klebsiella pneumoniae* and *Pseudomonas aeroginosa*, both venereal disease (VD) bacteria; other bacteria such as *Escherichia coli*, *Streptococcus zooepidemicus* and *Staphylococcus aureus*, all present in the environment but which may still cause problems, may be also tested for. It must be remembered that in the UK, CEM is a notifiable disease and if isolated must be reported by the testing laboratory to the Divisional Veterinary Manager of the Department for Environment, Food and Rural Affairs (DEFRA). Current codes of practice have largely eliminated the disease in the UK but sporadic incidences do occur (Jackson *et al.*, 2002). Although the other bacterial diseases are not notifiable, if they are identified, covering should immediately stop and advice/treatment sort. Most Thoroughbred studs require the completion of a mare certificate to indicate her previous coverings and any positive swab results. Whatever stud you are taking your mare to (except many native pony studs), it is likely that they will require paperwork to confirm that your mare is free of infection. Native pony studs do not as a rule require swabbing to take place but this must be checked during your stud selection.

Apart from bacterial infection mares may also need to be tested for viral infections. Since 1995, equine viral arteritis has also been a notifiable disease. It is advised that all mares are blood-sampled twice, at least 14 days apart, at the beginning of the season to test for antibodies. If no antibodies are present or the antibody levels are stable or declining, then the mare is free of active infection and is safe to be covered. However, if increasing antibody levels are identified, this means that the mare has an active infection and should not be covered. Equine herpes virus (EHV) and strangles (caused by *S. equi*) can also have disastrous consequences on breeding mares, being highly contagious, and in the case of EHV causing high abortion rates. Neither is notifiable but any mares suspected of having contact with EHV or strangles must be isolated and certainly not sent to stud, and veterinary advice sort as to whether they can be covered at all that season. The HRBLB Codes of Practice also provide advice on these conditions.

Similarly, all stallions should be swabbed to test for infection. It is advised that they have two sets of swabs taken at an interval of no less than 7 days soon after 1 January in each covering season. Swabs should be taken from the urethra, urethral fossa and the sheath of the penis. For high-risk stallions, clitoral swabs must be taken from their first four mares of the season at 2 days after covering. It is good practice to also swab stallions in the middle of the season or as soon as any problem is suspected. As with the mare, CEM is a notifiable disease in the UK and so must be reported to DEFRA; if other bacteria are isolated, then covering should be immediately stopped and advice/treatment sort. Once swabs have been taken and the stallion is declared clean, the laboratory certificate confirming the stallion's disease-free status should be available to all mare owners.

Again, as with mares, equine viral arteritis is a notifiable VD in the stallion. It is advised that all stallions are blood-sampled at the beginning of the season, at least 28 days before their first mare, and the blood tested for antibodies. If no antibodies are present, then the stallion is free of infection and can be safely used for covering. However, if antibodies are identified this may not necessarily mean that the stallion has an active infection, veterinary advice should, therefore, be sought as to whether the stallion can be used.

Again in common with the mare, EHV and strangles are potential infections. Neither is notifiable but any animals suspected of having contact with EHV or strangles must not be allowed on to a stud and if either is confirmed in a stallion, the stud must be closed and veterinary advice sort as to whether covering can recommence that season. The HRBLB Codes of Practice again provide advice on these conditions (Crowhurst et al., 1979; Ricketts et al., 1993; Horse Race Betting Levy Board, 2008). Further details on VDs are given in Chapter 19 (this volume).

12.5.3. Nomination forms

Once a stud has been chosen and during the preparation period for your mare a nomination form (covering agreement) must be completed. This is a legal agreement between you and the stud owner for a nomination to a specific stallion. The exact information and agreement made varies with the stud but it will lay out the basic conditions.

Some studs will require a fee or deposit to be paid at the time of submitting the nomination form, the stud should then return acceptance as soon as possible. Payment of the nomination fee depends upon the stud. A straight fee may be payable regardless of whether the mare is in foal or not; this is often the case in native pony studs and those charging a lower fee. Alternatively 'no foal, no fee October 1st' terms may apply under which agreement fees are paid on covering but if the mare is proven not to be pregnant on 1 October the stud fee (excluding any keep fees) is returned. A similar arrangement is termed 'no foal, free return October 1st' in which instead of the fee being returned the mare has a free cover to the same stallion or a replacement the following year. Finally, for the most expensive stallions, a part payment arrangement may be made whereby 50% of the fee is due on covering and the balance paid if the mare is pregnant on 1 October, or alternatively a 'live foal' arrangement may be made whereby the stud fee is returned if the mare does not have a live foal, or one that survives for 48 h. Occasionally concessions can be given to certain mares in the form of reduced fee to encourage good mares whose offspring will be a good advertisement for the stallion. This is most often seen with young stallions. Again it is important to check such agreements when selecting a stud.

12.5.4. General preparation

Immediately prior to sending your mare to stud she should at least have her hind shoes removed, be in good physical condition and have up-to-date tetanus and flu vaccination certificates. She should also be wormed. Most studs will not require you to bring any tack or equipment with your mare, preferring to use their own so as to minimize the risk of losing other people's equipment. If your mare arrives at stud in good condition, then it is very likely that she will be returned to you in similar condition.

In order to have some control over the time of ovulation and hence the time to cover a mare, many mares have their reproductive activity controlled with hormones (Section 12.6). This is particularly the case with Thoroughbred mares where the start of the breeding season is advanced and in mares covered by artificial insemination (AI). Problem breeders may also be put on hormone programmes under veterinary advice.

Immediately prior to the season the stallion should ideally have all shoes removed to minimize damage to mares at mounting. He should be up to date with his vaccinations, including influenza and tetanus and any other necessary vaccinations, as well as recently wormed. Especially in studs where the mares are walked in by their owners for covering, the stallion should be well turned out and in good condition; he should be a good advertisement for himself.

12.6. Manipulation of the Oestrous Cycle in the Mare

As the majority of equine matings are today influenced by humans we have, therefore, attempted to control the timing of covering to our advantage. There are several methods by which the timing of mating can be manipulated: first, by altering the beginning of the season; and, second, by manipulating the timing of the mare's oestrus and ovulation within that season.

12.6.1. Advancing the breeding season

The horse is classified as a long-day breeder with a breeding season extending on average from April to November in the northern hemisphere and October to May in the southern hemisphere (see Section 3.2.1). The Thoroughbred industry, and increasingly other breed societies, registers the birth of all foals on 1 January (northern hemisphere) or 1 July (southern hemisphere), regardless of their actual birth date. It is, therefore, desirable, in order to achieve perceived maximum advantage in the racing industry, to ensure that mares foal as soon after 1 January as possible. As mares have an 11-month gestation, they need to be covered ideally in February. Hence, the arbitrary covering season for Thoroughbreds in the northern hemisphere runs from 15 February to 1 July and in the southern hemisphere from 15 August to 1 January. This arbitrary breeding season, therefore, starts well before the natural season. Other breed societies also have similar arbitrary breeding seasons. Even in societies that do not stick to set breeding seasons, there is a desire to have foals born as early as

possible in the year in order to maximize their size during the showing and event season and enhance their chances of success.

The existence of an arbitrary breeding season that does not correspond to the natural one is a major limiting factor in breeding mares. Manipulation of the mare's breeding cycle is required to advance the timing of oestrus and ovulation in order to accommodate these artificial limits. There are several means by which the breeding season in the mare may be advanced. These include the use of light and hormone therapy, or a combination of the two.

12.6.1.1. Light treatment

In a population of intensively managed mares only 10% will voluntarily show oestrus and ovulation during the non-breeding season. This figure is much lower for extensively managed feral populations. Artificial manipulation of light, along with nutrition, temperature and close association of mares and stallions will significantly increase this percentage. Of these environmental factors, manipulation of light is the most successful (Meyers, 1997).

Light treatment of mares to advance the breeding season was pioneered by Burkhardt (1947) with later work by Kooistra and Ginther (1975) among others. Mares can be introduced to a 16h light/8h dark regime either suddenly or gradually. Light can be delivered by means of a 200W incandescent source per 4 × 4m loose box, or equivalent. A slightly lower-watt light source may be used if the stables are lined by reflective material. Light treatment can be started any time from early November onwards, but it is essential that the mare experiences an initial autumnal reduction in day length prior to light manipulation. Commencement of light treatment early in December will result in coat loss within 4 weeks, followed by ovarian activity normally 2–4 weeks later. The season may be advanced by up to 3 months, but there is considerable individual variation (Kooistra and Loy, 1968; Kooistra and Ginther, 1975; Oxender et al., 1977). In general, the earlier in the year light treatment begins, the longer the time interval to oestrus and ovulation, though it is very difficult to induce mares to ovulate before February. Light treatment from early December is often chosen, as it results in ovarian activity during early February. More recent work suggests that a photosensitive period exists around 1.00–3.00 a.m. and that a window of 1h light 8h after dusk would suffice, resulting in the same effect as 16 h light (Palmer *et al.*, 1982; Scraba and Ginther, 1985). However, in practice, as the timing of dusk varies it is easier to administer a 16 h block of light.

The effects of light manipulation may be enhanced by increases in ambient temperature and nutritional levels, both of which are also associated with spring. However, the relatively minor effect that temperature and nutrition have on advancing the season means that their use beyond that of rugging up mares and increasing their energy intake is not justified (Allen, 1978; Meyers, 1997). Stallion effect, via pheromone and vocal stimuli, is also reported to have an additional effect on advancing the season (Allen, 1978). It is, therefore, advocated by some that stallions should be housed on the same yard as, or near to, mares during January/February in order to enhance the effect of light treatment.

Although light treatment is very successful in advancing the season, the timing of response within a group of mares is very variable. In an attempt to reduce this variation the use of hormone treatment was investigated.

12.6.1.2. *Exogenous hormonal treatment*

The use of exogenous hormones to induce out-of-season breeding is very successful in anoestrous ewes and, as such, is a useful management tool. Unfortunately, the anoestrous mare's ovary appears relatively insensitive to exogenous hormone therapy. Treatment with equine chorionic gonadotrophin (eCG), human chorionic gonadotrophin (hCG) or follicle-stimulating hormone (FSH) by single or multiple injection, as used in other livestock, results in little if any response. However, natural crude equine pituitary extract, if injected daily over a period of 2 weeks, does induce oestrus in the non-breeding season (Douglas *et al.*, 1974; Lapin and Ginther, 1977). Such a long period of treatment is required as the mare's ovary apparently requires a prolonged period of gonadotrophin stimulation in order to adequately develop follicles so that they can react to an ovulating agent (see Chapter 3, this volume). Short-term treatments will not develop follicles to an adequate stage to allow them to react to any ovulation stimulus.

12.6.1.2.1. GONADOTROPHIN-RELEASING HORMONE
As some success had been achieved using natural pituitary extract, the use of gonadotrophin-releasing hormone (GnRH), naturally responsible for induc-

ing FSH and luteinizing hormone (LH) production by the pituitary, was investigated. Both GnRH infusion, via mini pumps or subcutaneous implants, and injection over a period of time have been reported to give some success (Hyland and Jeffcote, 1988; Palmer and Quelier, 1988; Ainsworth and Hyland, 1991; Harrison *et al.*, 1991; Mumford *et al.*, 1994). GnRH is now commercially available (Deslorelin, Buserelin, etc.) and used with some success in commercial practice (Farquhar *et al.*, 2000; Morehead *et al.*, 2001), especially in mares in the transition period from anoestrus to the breeding season (Gentry *et al.*, 2002b).

Further development of the use of GnRH involved its inclusion in a hormone regime mimicking the natural concentrations of FSH, LH and progesterone in the normal oestrous cycle. A single injection of GnRH is followed by daily injections of progesterone for 8 days. This regime is repeated twice more to give three mini artificial oestrous cycles. This treatment, though not effective in mares in deep anoestrus, is reasonably successful in the transitional period (Evans and Irvine, 1976, 1977, 1979).

12.6.1.2.2. PROGESTERONE
Progesterone has been used to induce oestrus and ovulation in anoestrous mares, but also seems only to be effective in the transition period (Alexander and Irvine, 1991). It is known that it is the decline in progesterone prior to ovulation that encourages LH and FSH release, which drives final follicle maturation and ovulation. Without this progesterone decline, the reaction, in terms of concentration of LH and FSH release, is limited and, therefore, ovulation does not occur, or occurs in the absence of oestrus. The resulting corpus luteum (CL) of any such ovulations is often incompetent, and only at the second ovulation of the season are acceptable conception rates achieved. Therefore, if progesterone is artificially administered and then withdrawn, it mimics the natural decline in progesterone and helps induce the normal increase in LH and FSH required for ovulation and results in a competent CL (Freedman *et al.*, 1979). Ovulation accompanied by oestrus is then seen to occur within 7–10 days of cessation of a 10–12 days' progesterone treatment period; again the best response is obtained in the transition period.

Progesterone is commonly used in the management of breeding mares. Traditionally, it was administered as an artificial progestagen (altrenogest) either orally (Regumate) or via intramuscular injection. More recently, it has been available for administration via a progesterone-releasing intervaginal device (PRID; Rutten *et al.*, 1986), controlled internal drug-releasing device (CIDR; Lubbeke *et al.*, 1994), sponges (Dinger *et al.*, 1981) or Cu-mate, all of which are impregnated with natural progesterone, rather than progestagen, and placed within the mare's vagina for the treatment period. Oestrus and ovulation post progesterone treatment (PRID, CIDR and Cu-mate) are reported to be quicker than post progestagen treatment (Regumate or injection), by virtue of the fact that natural progesterone allows folliculogenesis to continue during treatment. Additionally, the elevated systemic progesterone levels resulting from treatment with PRID, CIDR or Cu-mate decline over time, as the finite amount of progesterone within the devices is absorbed. This further ensures that significant follicle development can occur, especially towards the end of the treatment period, prior to their removal. The interval from the end of treatment to oestrus and ovulation is reported to be 2–4 days for PRID and CIDR compared with 8–10 days for Regumate (Squires, 1993a,b; Arbeiter *et al.*, 1994; Newcombe and Wilson, 1997). In contrast, progestagen appears to depress follicular activity, which can only resume after the end of treatment, hence a longer interval to oestrus and ovulation. Intervaginal sponges, PRID and CIDR, have the advantage of being labour-saving, and require only to be inserted and removed. They also ensure that a standard dose is administered. However, they can be lost and can cause vaginitis, which, though of no real consequence to conception, looks unsightly. In common with these, the advantage of injection is that you can be assured that each mare has received her allotted dose. However, it is more expensive, especially when you consider the costs of the veterinary attendance, now required by law in the UK, for the administration of injections to horses. Oral administration (Regumate) has the advantage that no vet is required, but some mares may refuse to take the feed in which it is mixed. If only part of the food is eaten, it is impossible to know how much progestagen has been ingested. Furthermore, all mares have to be fed individually. While considering the use of progesterone, it is worth noting that its use has been associated with a decrease in neutrophil production in response to a bacterial challenge. This may well be significant in mares with poor perineal conformation or a history of uterine infections and is a disadvantage of long-term use; hence, the recommended length of

progesterone treatment is now much reduced. Additionally, this presents a potential problem to animals such as in performance mares where it is used to suppress reproductive activity during competition periods.

12.6.1.2.3. PROSTAGLANDINS A further alternative for use in late anoestrous mares is prostaglandin F2α (PGF2α), administered in a series of two or three injections at 48 h intervals. PGF2α is luteolytic in nature – that is, it destroys the CL and terminates the luteal phase of the cycle. In the natural cycle, in the absence of a pregnancy, prostaglandin is produced at a specific time (12–14 days) post ovulation. As such, it both marks and causes the termination of the luteal phase and the commencement of the endogenous hormone changes associated with oestrus and ovulation (see Section 3.2.2.7; Allen and Rowson, 1973; Loy *et al.*, 1979; Neely *et al.*, 1979; Cooper, 1981; Savage and Liptrap, 1987). It is this association with the termination of one cycle and commencement of the next which is exploited in this treatment. Ovulation has been reported in 73% of mares treated in this way in the transition period (Jochle *et al.*, 1987).

Prostaglandin use does have some side effects, including smooth muscle activation. Its use may, therefore, be linked to increased gastrointestinal activity, manifested as diarrhoea, sweating and possibly slight caudal ataxia (Cooper, 1981; Le Blanc, 1995). These side effects vary with the analogue used and the individual mare; however, provided the recommended dose rate is not exceeded, they are not serious.

12.6.1.2.4. PROLACTIN AND DOPAMINE ANTAGONISTS A link between prolactin and seasonal breeding activity is likely but the exact nature of this association is unclear. It is thought by some to involve dopamine (a neurotransmitter produced in the brain) which actively inhibits the secretion of prolactin and GnRH, both of which are involved in driving follicle development (see Section 3.2.1.1; Melrose *et al.*, 1990; Bennett *et al.*, 1998). Elevated prolactin and suppressed dopamine concentrations are characteristic of the breeding season. Treating anoestrous mares with prolactin results in rapid follicle development. As prolactin receptors have been identified on large follicles it is likely that prolactin has an effect at the level of the ovary (Bennett *et al.*, 1998). If mares in the transition period (January time) are treated with a dopamine antagonist, such as sulfide or domperidone, many show advancement in their first ovulation of the year (Brendemuehl and Cross, 2000; Daels et al., 2000). However, not all mares respond and there is a significant variation in the timing of response. If light treatment (16 h day^{-1}) is introduced 2 weeks prior to dopamine antagonist treatment, a better and tighter response is obtained (Daels et al., 2000). Neither of these treatment protocols is currently used in stud practice but may hold the key to future hormonal treatments.

12.6.1.3. Combination treatments

The use of light to advance the season is very successful but gives variability in timing of response. On the other hand, hormone therapy appears to work effectively and time oestrus and ovulation quite well but will only work in the transition period. A combination of the two may, therefore, be successful. Indeed light therapy can be used to advance mares into the transition period and then exogenous hormones used to time oestrus and ovulation more precisely. Progesterone alone plus light treatment can be successful. Light treatment (16 h light and 8 h dark) should be introduced during November/December (or May/June in the southern hemisphere), followed by progesterone for a period of 10–15 days from early to mid-January (July) onwards (Palmer, 1979; Squires et al., 1979a; Heeseman et al., 1980; Scheffrahn et al., 1980). Table 12.1 gives an example of such a regime. The

Table 12.1. The advancement of oestrus and ovulation in the mare, using light treatment and progesterone supplementation (considerable variation in individual mare's response may be observed).

Time	Drug to be administered/event
Day 0 (15 December)	Light treatment commenced 16 h light/8 h dark
Day 28 (12 January)	Coat loss in mare may be apparent
Day 42 (26 January)	Ovarian activity may be apparent
Day 43 (27 January)	Progesterone treatment started
Day 55 (8 February)	Progesterone treatment stopped
Day 60 + (13 February)	Oestrus commences
Day 62 + (15 February)	Ovulation may occur – covering/AI

exact timing of the progesterone treatment depends on when covering is planned and when light treatment is commenced.

Some of the best and most consistent results have been obtained using a similar regime but with the addition of PGF2α: 16 h light and 8 h dark for 6–8 weeks, followed by 10 days progesterone treatment; PGF2α is then administered on day 10 to induce luteolysis of any naturally occurring CL. It is reported that up to 82% of mares ovulate within 9–16 days of treatment with a conception rate of 65% (Bristol, 1986).

Although there are very many ways, in theory, of advancing the breeding season in the mare, most are too expensive and time-consuming to make them commercially viable and often achieve unreliable results. In practice, most commercial studs just use light treatment, possibly supplemented by rugging up mares, increasing energy intake and Regumate (progesterone) treatment in the transition period.

12.6.2. Synchronization and timing of oestrus

In addition to advancing the breeding season, there are many other reasons for manipulating the timing of oestrus – mainly related to easing mare management. In many countries of the world, e.g. South America and the USA, mares are run in large herds that roam over vast tracts of land. In such systems, handling needs to be kept to a minimum. It would, therefore, be ideal if mares could all be treated in batches right through from conception to birth and foal rearing. In order for this to be successful a reliable and exact method of timing ovulation and oestrus is required. If this could be achieved it would also alleviate the need for teasing, rectal palpation, scanning, etc., and would be most useful in conjunction with artificial insemination. Such treatment would also allow the ovulation and oestrus of a single mare to be timed precisely, again to ease her management, but specifically useful for AI and embryo transplant (ET).

The methods used to time or synchronize oestrus and ovulation work on the principle of either: artificially prolonging and then terminating the luteal phase of the oestrous cycle (progesterone); or prematurely terminating the natural luteal phase (PGF2α) or variations on these two themes (Fig. 12.7).

12.6.2.1. Progesterone

Progesterone supplementation and subsequent withdrawal may be used to time oestrus and ovulation.

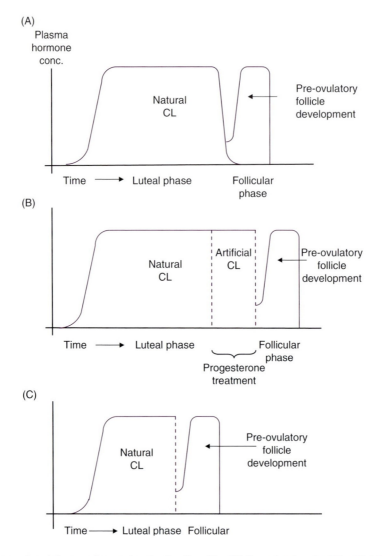

Fig. 12.7. Oestrus and ovulation can be synchronized or timed by: (A) the natural cycle; (B) artificially prolonging and then terminating the luteal phase of the cycle (progesterone treatment); or (C) prematurely terminating the natural luteal phase (PGF2α).

The use of progesterone or one of its analogues works on the principle of imitating the mare's natural dioestrus or luteal phase. This is achieved by mimicking the natural progesterone production through the administration of exogenous progestagens. Termination of this artificial luteal state, achieved by the cessation of treatment, acts like the end of the natural luteal phase and so induces the changes in the mare's endogenous hormones responsible for oestrus and ovulation (Bristol, 1993).

Within 2–3 days of progesterone supplementation, a mare will normally cease all oestrous activity, which will remain suppressed until treatment is terminated (Loy and Swann, 1966). After 15 days' treatment, oestrous behaviour is apparent at 3–7 days and ovulation at 5–15 days post progesterone withdrawal (Table 12.2; Van Niekerk et al., 1973; Squires et al., 1983a). Conception rates are comparable to those associated with naturally occurring oestrus (Van Niekerk et al., 1973; Squires et al.,

Table 12.2. The timing of oestrus and ovulation in the mare, using progesterone supplementation (considerable variation in individual mare's response may be observed).

Time	Drug to be administered/event
Day 0–14	Progesterone supplementation (intervaginal sponges or PRID)
Day 17 onwards	Oestrus
Day 19 onwards	Ovulation may occur – covering/AI

PRID = progesterone intravaginal device.

1979a, 1983a). As discussed in Section 12.6.1.2.2, there are various methods of administering progesterone. Generally treatment is either with artificial progestagen, which tends to inhibit reproductive activity completely, or natural progesterone, which, as in normal dioestrus, allows some follicular development to occur. Hence, the mare returns to oestrus and ovulation sooner after progesterone rather than after progestagen treatment.

Traditionally, long periods of progesterone supplementation were used up to 20 days, and although oestrus was suppressed, ovulation occasionally did occur during treatment. Hence, the timing of ovulation was not that successful (Loy and Swann, 1966). Additionally, it is known that prolonged progesterone treatment is associated with an increase in endometritis due to suppression in the neutrophil defence mechanism. Shorter periods of progesterone supplementation are now being advocated. As such, the period of progesterone supplementation may not be long enough to ensure that the natural CL has regressed in all mares, a combination of progesterone supplementation and prostaglandin treatment is, therefore, used. This is discussed in the following section.

12.6.2.2. Prostaglandins

PGF2α (Prostin F2 Alpha and Dinoprostromethamine 5), or one of its analogues (alfa prostol, fluprostenol, prostalene or frenprostalene), provides a successful means of timing oestrus and ovulation in the mare (Bristol, 1987; Le Blanc, 1995). As discussed previously, PGF2α both marks and causes the termination of the luteal phase and the commencement of the endogenous hormone changes associated with oestrus and ovulation (see Section 3.2.2.7; Allen and Rowson, 1973; Loy et al., 1979; Neely et al., 1979; Cooper, 1981; Savage and Liptrap, 1987;

Bristol, 1993). As such, administration of exogenous prostaglandins, provided it is within certain time limits within the luteal phase, allows its termination to be controlled, and with it the timing of oestrus and ovulation.

The success of prostaglandin in timing oestrus in the mare is variable and depends upon the stage of the cycle. The CL of most mares is refractory to prostaglandin treatment prior to day 5 (Douglas and Ginther, 1972; Loy et al., 1979). A good response is normally obtained when treatment is given between days 6 and 9 (Loy et al., 1979). To be successful, the treatment must not only terminate the luteal phase but also induce ovulation. Considerable variation exists between the time of prostaglandin treatment and ovulation, a range of 24 h to 10 days is reported (Loy et al., 1979). The time interval is determined by the stage of follicular development at treatment. Follicles at 3–4 cm in diameter or greater ovulate on average within 4 days, though again considerable variation is reported. If the follicle ovulates within 72 h, it is often accompanied by an abbreviated oestrus or no oestrus at all. Occasionally, when a large follicle is present, prostaglandin treatment results in the regression of that follicle and the development and subsequent ovulation of another follicle; hence, there is a longer time interval between treatment and ovulation (longer than 8 days; Loy et al., 1979). The most consistent results are obtained when treating mares earlier in the cycle with small follicles (less than 3–4 cm in diameter) as less variation exists and the interval to ovulation is on average 6 days (Table 12.3).

The previously described use of prostaglandin relies upon a single injection, the major disadvantage of which is that the stage of the mare's oestrous cycle must be known. In smaller intensive studs, where individual mares are monitored, this may present no problems. However, in large groups of mares, kept in herd situations, or in mares whose

Table 12.3. The timing of oestrus and ovulation in the mare, using a single injection of Prostaglandin (considerable variation in individual mare's response may be observed).

Time	Drug to be administered/event
Day 0	Oestrus
Day 7	Prostaglandin
Day 9	Oestrus commences
Day 13	Ovulation may occur – covering/AI

AI = artificial insemination.

Table 12.4. The timing of oestrus and ovulation in the mare, using two injections of Prostaglandin (considerable variation in individual mare's response may be observed).

Time	Drug to be administered/event
Day 0	Prostaglandin
Day 14	Prostaglandin
Day 20	Oestrus commences
Day 22	Ovulation may occur – covering/AI

AI = artificial insemination.

Table 12.5. The timing of oestrus and ovulation in the mare, using two injections of prostaglandin and a single injection of hCG (considerable variation in individual mare's response may be observed).

Time	Drug to be administered/event
Day 0	Prostaglandin
Day 14	Prostaglandin
Day 18	Oestrus commences
Day 19	hCG
Day 20	Ovulation may occur – covering/AI

AI = artificial insemination.

stage of the oestrous cycle is unknown, a double injection of prostaglandin is required (Hyland and Bristol, 1979). These two prostaglandin injections need to be administered 12–16 days apart (Table 12.4). In this regime all mares between days 5 and 14 of the oestrous cycle will react to the first prostaglandin injection and will ovulate within the next 5–9 days. All those mares between day 14 and ovulation will naturally ovulate within the next 5 days. This leaves the remaining mares, those between days 0 and 5, which will not react to the prostaglandin injection as they do not have a fully functional CL. However, in 12–16 days' time when the second prostaglandin injection is given, all mares will be between days 5 and 16 of the cycle and so the vast majority will react and ovulate within 5 days or so, giving acceptable synchronization/timing of ovulation.

The timing of the onset of oestrus with such treatment is quite successful: 60% of mares are reported to commence oestrus within 4 days of the second injection and 92% show oestrus within 6 days. Nevertheless, the synchrony and timing of ovulation is still very variable. Ovulation may occur anywhere between 2 and 12 days after the second injection (Hyland and Bristol, 1979; Voss *et al.*, 1979; Squires *et al.*, 1981a; Squires, 1993a,b).

12.6.2.3. Human chorionic gonadotrophin

Further refinement of these protocols includes the use of hCG, a human placental gonadotrophin with LH- and FSH-like properties. As such, it enhances and supplements the natural release of gonadotrophins, which drive follicular development and, more specifically, ovulation. Its additional use is advocated to hasten ovulation and reduce the duration of oestrus (Table 12.5; Voss *et al.*, 1975; Harrison *et al.*, 1991).

Several timings for the injection of hCG have been advocated, most of them between 4 and 6 days after the second prostaglandin injection (Douglas and Ginther, 1972; Palmer and Jousett, 1975; Hyland and Bristol, 1979; Voss *et al.*, 1979; Bristol, 1981; Squires *et al.*, 1981a). Palmer and Jousett (1975) reported that 75.8% of mares ovulated within 72 h of an hCG injection, which was given 6 days post second prostaglandin injection. Yurdaydin *et al.* (1993) achieved similar success using hCG on day 5 post prostaglandin. When used on day 8 post PGF2α injection oestrus synchronization rates of 90% have been reported (Holtan *et al.*, 1977). Other work, however, has demonstrated a more variable response or no significant improvement with the use of hCG (Holtan *et al.*, 1977; Squires *et al.*, 1981a). It has been advocated that hCG be used twice, on day 7 (7 days post the first PGFα) and on day 21 (7 days post the second PGF2α injection; Table 12.6). The aim of this is to encourage the development of competent CL from the first prostaglandin injection, which would then react with less variation to the second prostaglandin injection. This regime is reported to

Table 12.6. The timing of oestrus and ovulation in the mare, using two injections of prostaglandin and two injections of hCG (considerable variation in individual mare's response may be observed).

Time	Drug to be administered/event
Day 0	Prostaglandin
Day 7	hCG
Day 14	Prostaglandin
Day 18	Oestrus commences
Day 21	hCG
Day 22	Ovulation may occur – covering/AI

AI = artificial insemination.

result in up to 95% of mares ovulating on either day 22 or 23 (Allen *et al.*, 1974; Palmer and Jousett, 1975; Voss, 1993).

Not only is hCG used in combination with other hormones, it is increasingly used alone in commercial practice to control the precise timing of ovulation in mares cycling naturally but which are monitored closely using ultrasonic scanning. In such mares, 750–5000 IU hCG can be administered once a large pre-ovulatory follicle (a follicle beginning to soften and with a diameter of 3 cm or greater) has been identified. The mare will then normally ovulate within 2–4 days (Harrison *et al.*, 1991; Brugmans, 1997; Barbacini *et al.*, 2000; Samper *et al.*, 2002), allowing timing of mating/AI. A reported disadvantage of such use of hCG is an increase in multiple ovulation rates (Perkins and Grimmett, 2001; Veronesi *et al.*, 2003); however, this is not supported by other workers (Davies Morel and Newcombe, 2008). Certainly the use of hCG is associated with increased fertility rates (Vanderwall *et al.*, 2001), presumably largely due to the better synchronization of covering and ovulation, and, as such, is popular in studs with high-value mares being covered by high-value stallions.

Although reasonably successful in inducing ovulation, repeated administration is reported to cause the mare to develop antibodies to, and therefore become refractory to, hCG (Roger *et al.*, 1979; Wilson *et al.*, 1990). Although this effect is not reported to be a problem by others (Barbacini *et al.*, 2000; J. Newcombe, Wales, 2008, personal communication), GnRH and its analogues have been advocated for use in its place.

12.6.2.4. Gonadotrophin-releasing hormone

GnRH acts to stimulate the natural release of LH and FSH from the anterior pituitary. As such, its administration as a series of multiple injections (four at 12 h intervals) or via a subcutaneous implant has been demonstrated to significantly advance the onset of ovulation in mares with follicles greater than 3 cm in diameter (Table 12.7).

Success rates of 88–100% of mares ovulating within 48 h of treatment (with deslorelin) have been reported (Johnson, 1986a,b; Harrison *et al.*, 1991; Meinert *et al.*, 1993; Jochle and Trigg, 1994; Mumford *et al.*, 1995). It has been suggested that GnRH may be more successful than hCG in inducing ovulation in larger, thicker-walled follicles. GnRH also does not have the possible disadvantage of

Table 12.7. The timing of oestrus and ovulation in the mare, using two injections of prostaglandin and GnRH implant (considerable variation in individual mare's response may be observed).

Time	Drug to be administered/event
Day 0	Prostaglandin
Day 15	Prostaglandin
Day 19	Oestrus commences
Day 21	GnRH implant
Day 23	Ovulation may occur – covering/AI

AI = artificial insemination.

inducing refractoriness of response due to antibody formation (Mumford *et al.*, 1995). However, much of the work done to date on using GnRH has been in mares during their natural oestrous period, rather than within a synchronized oestrus regime. The limited work done on using GnRH with prostaglandin to time oestrus and ovulation has indicated variable results as to whether there is a significant change in the timing of ovulation as a result of treatment, compared with the use of prostaglandin alone (Voss *et al.*, 1979; Booth *et al.*, 1980; Squires *et al.*, 1981a; McCue *et al.*, 2002). Therefore, though the regime suggested in Table 12.7 would be feasible, it is yet to be proved that it is a significant improvement on other regimes. Additionally, when compared to hCG the interval to ovulation is reported to be longer (McCue *et al.*, 2002).

12.6.2.5. Combination treatments

Several combination treatments are used, some of which have already been mentioned. Two most commonly used and not previously discussed in detail are progesterone and prostaglandin, and progesterone and oestradiol.

12.6.2.5.1. PROGESTERONE AND PROSTAGLANDINS

Combination treatments of progesterone and prostaglandin are increasingly popular; as such, treatments improve the timing of ovulation and may reduce the length of progesterone supplementation. Administration of progesterone via intervaginal sponges for 20 days with a PGF2α injection on the day of sponge removal was reported to result in oestrus at 1.8–2.2 days and by ovulation at 3.0–5.4 days, respectively, post PGF2α injection (Palmer, 1979; Draincourt and Palmer, 1982). Today, administration of progesterone is normally only for 7–9

Table 12.8. The timing of oestrus and ovulation in the mare, using progesterone supplementation, plus prostaglandin (considerable variation in individual mare's response may be observed).

Time	Drug to be administered/event
Day 0–8	Progesterone supplementation (intervaginal sponges or PRID)
Day 8	Prostaglandin
Day 12	Oestrus
Day 16 onwards	Ovulation may occur – covering/AI

PRID = progesterone-releasing intravaginal device; AI = artificial insemination.

Table 12.9. The timing of oestrus and ovulation in the mare, using progesterone and oestradiol treatment followed by prostaglandin (considerable variation in individual mare's response may be observed).

Time	Drug to be administered/event
Day 0–10	Progesterone and oestradiol treatment
Day 10	Prostaglandin
Day 20	Ovulation may occur – AI
Day 22	AI – continue every 48 h until oestrus ceases or ovulation has occurred

AI = artificial insemination.

days, with prostaglandin administered on the day progesterone treatment ceases. Using this protocol and again with the intervaginal sponges, Palmer (1979) demonstrated that, on average, oestrus occurred earlier (3.8 days) than figures suggested for progesterone treatment alone. The timing of ovulation was, however, very variable at 8–15 days after prostaglandin injection. Later work by Palmer *et al.* (1985) using the same sponges but inserted for 7 days with prostaglandin at sponge withdrawal suggested better synchrony of ovulation, at 10.1–14.0 days post sponge removal (Table 12.8).

It is evident that yet again considerable variation in response is observed, this variation is greater with shorter-term progesterone supplementation, especially during the early breeding season (Hughes and Loy, 1978; Palmer, 1979).

12.6.2.5.2. PROGESTERONE AND OESTRADIOL This combination treatment is increasingly popular. Both hormones may be administered daily via intramuscular injection for 10 days or more commonly and easily by using PRIDs containing progesterone plus 10 mg oestradiol (held within a gelatin capsule) inserted for 10 days (Rutten *et al.*, 1986). This may be followed, as in the previous protocols, with an injection of PGF2α at the end of the treatment (Table 12.9), and has proved to be successful, with a reported 81.3% of treated mares ovulating 10–12 days post PGF2α injection (Loy *et al.*, 1981). Normal pregnancy rates have been reported to AI after such treatment (Jasko *et al.*, 1993b). Further refinement of PRIDs or the development of slow-release subcutaneous capsules may further enhance the use of such combination treatments, removing the need for time-consuming daily injections. Increasingly hCG is also used with combination treatments, such as discussed above, to further encourage and more precisely time ovulation.

12.6.3. Suppressing oestrous activity

Although not strictly part of stud management, oestrous suppression is important in many performing mares. Oestrous activity is often associated with poor concentration and attitude to work and, therefore, poor performance. The most drastic means of stopping reproductive activity is ovariectomy, which has the major disadvantage of being irreversible. Alternatively hormone manipulation can be used. While progesterone concentrations are high, oestrous activity is suppressed, this naturally occurs during dioestrus; therefore, as mentioned previously, performance mares can be treated with exogenous progesterone (often Regumate) to suppress oestrous activity. More recent work suggests that the insertion of a small glass ball or marble (25–35 mm in diameter) into the uterus of the mare prolongs luteal function (Nie *et al.*, 2001, 2003), and as a result also suppresses oestrous activity. The explanation for why such a reaction should occur is unclear, but it is possible that the marble simulates a pregnancy and by its movement within, and contact with, the uterine endometrium prevents the release of prostaglandin, as seen in early pregnancy, and so the CL is maintained. Further, more recent work by Stout and Colenbrander (2004) suggests that preventing the action of GnRH by blocking its action on the pituitary, via GnRH vaccines, antagonists and agonists, is a possible alternative for reversibly suspending a mare's reproductive activity.

It is evident that the use of oestrous manipulation does not allow precise and accurate timing of ovulation in the mare. However, there is no evidence to suggest that the use of such treatment significantly affects conception rates (Palmer *et al.*, 1985), and some techniques are more successful than others at allowing the time of ovulation to be predicted. Many of these hormone regimes really only reduce the random spread of ovulations within a population rather than allowing the exact timing to be predicted. Even so, such an effect allows ovarian examination or teasing to be concentrated into a shorter period of time, so reducing labour costs, etc. Manipulation of the oestrous cycle is, therefore, regularly practised and, with the additional use of ultrasonography and rectal palpation, can allow an accurate and precise prediction of ovulation to be made.

Many of these synchronization techniques can also be used to induce ovulation and oestrus in mares with some abnormal ovarian conditions. These cases will be discussed further in Chapter 19 (this volume).

12.7. Manipulation of the Breeding Activity in the Stallion

The breeding season in the stallion, as in the mare, is governed by photoperiod, and the stallion reacts to increasing day length in a similar manner to that seen in the mare. The breeding season of the stallion can, therefore, be advanced by the introduction of a 16h light/8h dark regime in November/December. Continual stimulation, however, produces refracto-riness and a return to normal seasonal changes, despite the altered photoperiod (Argo *et al.*, 1991). Manipulation of stallion reproduction is not as essential as manipulation in the mare, as it is the mare's reproductive cycle that is normally limiting to performance. Given enough encouragement a stallion will naturally breed during the non-breeding season, but less efficiently. Manipulation of reproduction in the stallion is normally limited to suppression of activity as a possible temporary alternative to gelding (Stout, 2005).

12.8. Conclusion

It is evident that the preparation of both the mare and stallion for covering needs considerable thought, especially if the animals concerned are maiden and/or the added complication of an athletic career behind them. However, provided adequate time and forethought are invested, most horses that are anatomically and physiologically sound are capable of breeding, regardless of their previous careers. The manipulation of reproduction, though widely practised, will not in itself allow the precise timing of oestrus and ovulation. However, if used in conjunction with rectal palpation or ultrasonic scanning, it will allow accurate determination of timing of ovulation and, as such, is a very useful stud management tool.

13 Mating Management

13.1. Introduction

Reproductive activity is governed by season and commences at puberty, which occurs between 10 and 36 months of age. The mare will only allow mating to occur when she is in her sexually receptive oestrous phase, which can be referred to as oestrus, season or heat. This period of sexual receptivity occurs on average every 21 days during the breeding season and lasts 2–10 days (average 5 days). Oestrus is synchronized with ovulation and so ensures that the mare is mated or covered at the optimum time for fertilization (Chapters 1 and 3, this volume).

Ovulation occurs normally about 24 h before the end of oestrus, so in an average mare it will be on day 4 of a 5-day oestrus (Ginther, 1992). There is, however, considerable variation between mares. Sperm are reported to survive for 24–72 h within the mare's reproductive tract, but the exact time remains in doubt, as most experiments on longevity are carried out *in vitro* (Watson and Nikolakopoulos, 1996). Some reports suggest that sperm longevity may be up to 7 days (Newcombe, 1994). The ovum, on the other hand, is thought to survive only 8–12 h, though longer times have been reported in *in vitro* work (Hunter, 1990). The timing of mating is, therefore, very important, and in practice most mares are covered every 48 h while oestrus lasts or until the mare has ovulated.

13.2. Mating Systems

The human desire to control the covering process has led to a wide variety of mating systems, varying from the two extremes of natural covering to intensive in-hand covering.

13.2.1. Natural mating

In the natural system, stallions run with their mares and detect those in oestrus at will, examining them often for signs of sexual receptivity. This process is leisurely and unrushed and the signals used by the stallion to detect oestrus are smell and taste rather than sight (Stone, 1994). Natural courtship may occur over several days, as the mare slowly progresses from dioestrus into full oestrus, and takes place between a mare and stallion that are well known to each other. When she is fully in oestrus and receptive, the courtship culminates in mating. Courtship of a mare in full oestrus follows a sequence of events. The stallion will normally stand and fix his eyes upon a mare that he suspects is receptive, arch his back and neck and draw himself up to his full height, he will become restless, pacing around, pawing the ground and stamping his feet. He will show the typical facial grimace, termed flehman, or tasting of the air (olfactory stimulation), often accompanied by roaring or vocalization (Fig. 13.1). He will approach the mare and gauge her response to his attentions. The whole process takes some time, during which the mare, if she is truly receptive, will stand quietly and possibly nicker in response if interested. Once the mare's interest has been ascertained, the stallion will have the confidence to approach her more closely, normally from the head, working his way slowly down her neck, nickering as he does so; he may nudge her slightly or lightly bite her neck. All the time, he is watching for her response and for signs of rejection. If he feels confident, he will then work his way down her flanks and on to her hind quarters and to her perineal area. At any time, he may pause to reassure himself that she is still interested. If all is still amicable, he will nudge her vulva and clitoral area. If the mare is in full oestrus she will stand still, relatively passive throughout the whole procedure, showing her interest by curling her tail to one side, urinating – often bright yellow urine with a characteristic odour – or she may just take up the urinating stance. She will expose her clitoris by inverting the lips of the labia around the ventral commissure, termed winking (Fig. 13.2; Ginther, 1992; McDonnell, 1992).

Fig. 13.1. The typical facial grimace, termed flehman, shown by a stallion when in the presence of an oestrous mare.

Fig. 13.2. A mare 'showing' to the stallion by lifting her tail to one side, 'winking' her clitoral area and displaying a passive, quiet reaction to his attentions.

If she is not in oestrus, she will show hostility to the stallion, which will be unable to get closer to her other than the initial advance. In nature, the stallion will then turn away, transferring his attentions elsewhere, and return to her later in that day or the next (Bristol, 1982; Klingel, 1982; Keiper and Houpt,

1984; Kolter and Zimmermann, 1988; McDonnell, 2000a).

Once the stallion is sure that the mare is truly receptive he will mount her. His penis normally becomes erect as he approaches the mare and she shows interest. Mounting will only occur when his penis is fully erect. If the stallion is very sure of himself and the mare, he will often mount her directly and ejaculate immediately. A stallion with less confidence or one not so sure of the mare may nudge or push her forward slightly prior to mounting or half mount her a couple of times first, in order to ascertain her reaction, before he commits himself and runs the risk of injury. The number of mounts per ejaculate tends to be higher at the beginning and end of the season, and in stallions with a low libido (Pickett and Voss, 1972). Ejaculation follows a varying number of pelvic oscillations, during which time the stallion may dance on his hind feet. Successful ejaculation is signalled by rhythmic flagging of the tail and can be confirmed by feeling for the urethral contractions along the ventral side of the penis. Ejaculation is followed by a terminal inactive phase when the stallion remains quiescent on the mare while penile erection subsides. The stallion will not naturally dismount until his penile erection and engorgement of the glans penis has completely subsided (detumescence; Fig. 13.3; see Section 2.2.1; Boyle, 1992; Davies Morel, 1999).

The time taken to achieve ejaculation varies with stallion and circumstances. On average, a stallion will achieve erection within 2 min of contact with the mare, be ready to mount within 5–10 s of full erection, and achieve ejaculation within 5 s of mounting with a final post-coital quiescent stage of up to 30 s (McDonnell, 2000b).

It has been demonstrated, by means of an open-ended artificial vagina, that ejaculation begins as the pelvic thrusts cease and consists of 6–9 jets of semen, deposited over 6–8 s, each a result of urethral contraction. The volume of each successive jet decreases, the first three jets making up 70% of the total seminal volume. The remaining jets mainly comprise secretions of the seminal vesicles, the gel fraction of stallion semen (Kosiniak, 1975; Weber and Woods, 1993; Davies Morel, 1999).

In the natural system a stallion will cover a mare in oestrus many times, up to eight to ten matings in 24 h. Nature's system works extremely well, with high pregnancy rates (Bristol, 1982, 1987). It is a system, however, that is rarely practised today, and is only occasionally seen in pony studs, or with horses run on large expanses of land with minimal managerial input. For such a system to run completely naturally, the stallion will only cover the mares within his herd and no outside mares may be introduced solely for covering. The introduction of foreign mares causes a disruption in the hierarchy and can result in jealousy and hostility from other mares, and uncertainty between the stallion and the introduced mares (Ginther, 1983c; Ginther et al.,

Fig. 13.3. The final terminal stage after ejaculation, where the stallion remains inactive as detumescence occurs. (From Derwen International Welsh Cob Stud, UK.)

1983). The need to introduce outside mares to a natural mating system is accommodated in some native pony studs by running the stallion out with specific groups of mares (Davies Morel and Gunnarsson, 2000). Visiting mares can then be put in their own group and home-bred mares in another, while the stallion is moved around between the groups and allowed to cover the mares at will when they come into oestrus.

Although conception rates are higher in natural breeding than in a controlled system (Bristol, 1987; Davies Morel and Gunnarsson, 2000), the natural breeding system has many disadvantages from the breeder's point of view, when maximum financial return from the stallion is required. The natural system limits the number of mares the stallion can cover in a season, as each mare gets covered eight to ten times per oestrus, whereas if mating is timed correctly, in theory only a single service is required. As a result, breeders feel the need to control events to protect the investment made, maximize the number of mares covered per season and minimize the risk of injury to stock. This interference is at the root of many of the problems associated with stud management today.

13.2.2. Mating in hand

Mating in hand is practised by the majority of studs today and involves complete control over the events surrounding covering. In general, humans now control the life of the horse to such an extent that there is very little similarity between the natural life and the one imposed.

It is common practice to segregate fillies and colts early on in their lives, sometimes from birth, and often at weaning. Naturally, colts and fillies would run together as part of a herd. Colts would be disciplined by other mares and stallions, socially interact with fillies and learn respect at a young age, when the chance of serious damage is reduced (Bristol, 1982). In the intensive systems of today, this social introduction of the young stallion to fillies and mares is removed. In addition, stallions are expected to cover mares that are unknown to them. The failure to develop social awareness and respect for mares results in stallions being inept at the interaction associated with covering (McDonnell, 2000b; Christensen et al., 2002). When this is coupled with the strangeness of any unknown mare and the sexual tension present, it is not surprising that the risk of mare rejection and injury to both

parties is high. An additional consideration is the significant value of some stallions and the need to control events so as to protect one's investment – the stallion – and to minimize the number of coverings per pregnancy so as to optimize his use and maximize financial return.

To overcome the potential danger to stallions and to optimize their use, most systems mate in hand. Humans, therefore, have total control over the number of mares covered by each stallion (Umphenor et al., 1993). However, one of the major drawbacks of this system is that there is the need to interpret the mare's oestrous signals before the stallion is allowed near her. As discussed previously the stallion uses the senses of smell and taste to detect mares in oestrus; the breeder, however, has to rely on the sense of sight only (Palmer, 1979; Stone, 1994). This method leads to inaccuracies and can prove unreliable. Hence, a teaser stallion is employed.

13.3. Teasing

Teasing is the use of a stallion, often not that chosen for mating, to encourage a mare to demonstrate oestrous behaviour under controlled conditions. The principle being that, as soon as the mare is thought to be in oestrus, she is brought in contact with the teaser/an entire male horse under controlled conditions in an attempt to enhance the signs of oestrus and confirm the initial diagnosis. Once the mare is confirmed as being in true oestrus, she can then be prepared for covering by the chosen stallion. This system allows the stallion destined to cover the mare access to her only when she is in true oestrus and at the most optimum time for fertilization, so minimizing the risk of injury and maximizing the chance of conception (Squires, 1993d). Apart from allowing oestrus to be detected, teasing is now known to play a role in enhancing reproductive activity. Prolonged teasing results in elevated gonadotrophin-releasing hormone (GnRH) concentrations, which in turn increase gonadotrophin release in both the mare and stallion, thus advancing ovulation in the mare and enhancing libido in the stallion (Irvine and Alexander, 1991; Lieberman and Bowman, 1994; McDonnell and Murray, 1995). In addition, it is becoming increasingly evident that teasing plays an important role in uterine clearance and reduction of post-coital endometritis. Teasing causes the release of oxytocin and prostaglandin which, due to their ability to activate the uterine myometrium, cause mild uterine contractions, which not only help sperm transport towards the waiting ova, but also expel exudate and dead sperm from the tract (Nikolakopoulos *et al.*, 2000a,b; Stecco *et al.*, 2003; Campbell and England, 2004). Various stimuli are known to affect uterine contractility; work by Madill *et al.* (2000) suggests that teasing with a stallion results in the greatest uterine contractility, followed by artificial insemination (AI), then the sight of a stallion and finally the sound or calling of a stallion. Such uterine contractility is particularly important in mares which habitually suffer from post-coital endometritis (see Section 19.2.2.5.3.2.2). Prolonged teasing of such mares may well prove to be good practice.

The teaser stallion is often kept purely to detect mares in oestrus. He is often of low value. If he is injured or damaged by an objecting mare, then there is no significant loss. Often a pony stallion is used. In many native pony-breeding studs, the stallion to be used for covering also acts as the teaser for his mares. This allows the reproductive state of the mare to be confirmed, but does not give the stallion the protection of using a separate teaser. However, such stallions tend to be less-valuable, native mares show oestrus more readily and the mare may well be teased initially over a teasing board or equivalent, providing some protection.

There are problems associated with in-hand covering, mainly because the courtship that naturally takes place over a prolonged period is concentrated into a short space of time and forces the attentions of the stallion upon the mare. Some mares object to such forced attentions, even if they are in full oestrus, and such objection may mask the signs of oestrus. Mares with foals at foot often object, occasionally violently, to the removal of the foal prior to teasing and covering, a practice often carried out to protect the foal. In the natural system the foal would still be in the close vicinity of the mare, but seems to know instinctively to keep its distance. Teasing some mares before feeding or turnout can also give erroneous results, and environmental conditions of extreme heat, cold, rain or wind may mask signs of oestrus. Some mares need a longer time of teasing before they can be coaxed into demonstrating oestrus and, in a busy stud working to a tight schedule, there is little time for extended teasing and hence she may never seem to be ready to cover. Again, some mares will only show oestrus under certain circumstances, i.e. only in the covering yard, when the perineum is being washed, the tail bandaged prior to service, or when a twitch is applied. This is where the mare's records are invaluable in identifying any such idiosyncrasy (Lieberman and Bowman, 1994). There is also evidence that mares may have a preference for certain stallions and this may be associated with vocalization. It has been suggested that the more vocal a stallion, the greater is his popularity (Pickerel *et al.*, 1993).

When a mare is in oestrus she will be docile; accept the attentions of the stallion, take up the urination stance and expose her clitoris (referred to as 'winking' or 'showing') and demonstrate a general lack of hostility towards, and signs of acceptance of, the stallion (Figs 13.2 and 13.4A–D). There are a wide range of methods used to tease mares, depending on the stud, the value of the stock and the facilities available.

13.3.1. Trying board

One of the most common methods of teasing is a trying or teasing board. The mare and stallion are introduced one on either side of the board and their reactions monitored. The board is designed so as to provide protection for both and should be high enough to allow just the horse's heads and necks to reach over. It is solid in construction, often made of wood and ideally twice the length of the horses. Its top should be covered by curved rubber or equivalent, to provide protection if the stallion or mare attempt to attack each other over the board (Fig. 13.4A–D).

The approach of the teaser to the mare over the board should mimic that of the natural approach. Initially muzzle to muzzle, the teaser is then allowed to stretch his muzzle along the mare's neck, possibly gently nipping her. The attitude of the mare to this attention is closely observed; signs of hostility include laid-back ears, squealing, biting and kicking out, indicating that the mare is still in dioestrus. In contrast, leaning towards the stallion, raising of the tail and the other typical signs of oestrus indicate that she is ready to be covered. If the mare is interested, then her flank can be turned towards the trying board and the teaser allowed to work his way further down her body. It is, however, very important that direct contact with the mare's genital area is avoided in order to prevent possible disease transfer to other mares teased. After a few minutes of such attention, most mares will show definite signs of oestrus; some mares, as discussed, may take longer.

Using the same principle, a stable door may act as an alternative to a purpose-built trying board, with the teaser in the stable and the mare intro-

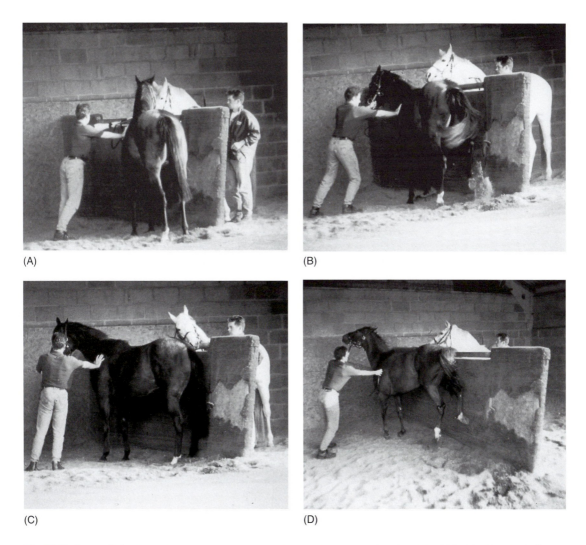

Fig. 13.4. A mare being teased over a trying or teasing board. The mare and stallion should be introduced initially muzzle to muzzle and the stallion allowed to work his way down to the mare's vulva (A, B, C). If she is in oestrus she will show little if any objection. A mare not in oestrus will usually object violently (D). (Photograph courtesy of Elizabeth Wood.)

duced to him outside the door. This is a popular practice in the smaller, more native-type studs, but can be dangerous unless the teaser is well known and is unlikely to be overzealous (Fig. 13.5).

13.3.2. Teasing over the paddock rail

Teasing over the paddock rail provides an alternative popular system, used in larger studs. The teaser is led in-hand to the paddock rail of fields containing mares, which are running loose, normally in small groups. A permanent trying board is often built into the paddock rail or a movable trying board is placed there. The reaction of the mares to the teaser is noted. Most mares in oestrus will approach the teaser by the fence and show definite signs of oestrus, others may show hostility, some may appear disinterested. Mares that show no reaction, often due to shyness or low social ranking, can be caught and brought up to the trying board to be tested individu-

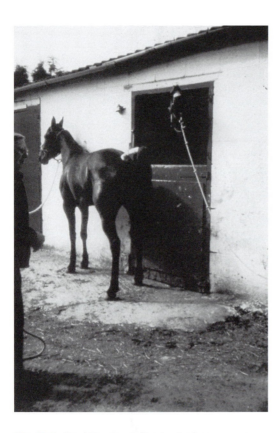

Fig. 13.5. Providing the stallion is of a known good temper, then he may tease the mares he is to cover over his stable door.

ally. Those that show interest can also be tried individually for confirmation or brought in for covering immediately. This is an efficient method for use with mares that are turned out, as it greatly reduces time and labour. It is best to avoid teasing mares by this method immediately after turnout or just before feeding, as this can give erroneous results.

Walking the teaser past the paddock fence on a regular basis, a version of the above, is of particular use if the teaser can be ridden or requires regular exercise in-hand. His daily route can then be organized to pass appropriate paddocks and the general reaction of the mares observed. Those that show interest can then be caught and taken to the covering yard.

This system does have potential problems when mares have foals at foot. Conflicting reports state either a considerable danger to the foal or that foals distance themselves, as they do naturally, and that mares, especially in small groups, are very careful to avoid damage to their foals.

Not all mares react to these two forms of teasing. Those that are less demonstrative may well be missed, especially if they are low within the hierarchy (Lofstedt, 1988). It is essential in such systems, therefore, that a detailed record of the mare's normal oestrous behaviour is available.

13.3.3. Teasing in chutes and crates

In the southern hemisphere, including South Africa and South America, mares tend to run in large herds, and are handled less frequently. In such enterprises, the mares may be run into chutes or crates and are held individually for a short period of time and teased from outside. This system reduces labour and enables large numbers of mares to be teased in as short a period of time as possible, with limited handling. The mares need to be accustomed to this system or errors may result (Fig. 13.6).

13.3.4. Teasing pen

A further alternative system is to confine the teaser in a railed or boarded area in the corner of a paddock. This system is particularly popular in Australia and South America. The area confining the teaser normally has high boards with a grill or meshed fencing, possibly with a hole through which the teaser can put his muzzle (Fig. 13.7). An alternative is the use of a small pony or miniature-breed stallion confined by a stout fence. If he escapes, it is not too disastrous, as his size limits his ability to cover the mares, though it is not impossible. These are again good systems for teasing a large number of mares, but they do require frequent observation of mares for signs of oestrus and, as with teasing over the paddock rail, it may be difficult to pick up shy mares. It must, of course, be remembered that the teaser can only be confined in the railed area for relatively short periods of time.

Along the same lines as the teasing pen is a central teasing pen surrounded by a series of individual pens in which mares can be held.

A further variation on this theme is the use of two adjacent boxes, divided by a grill, one for the stallion and another for the mare. Yet another involves the teaser being confined in a stable in the corner of a yard with the mare free within the yard to show to the stallion at will. Such an arrangement is used at the National Stud, Newmarket, UK, for maiden or shy mares that do not show oestrus well under more traditional methods (Fig. 13.8).

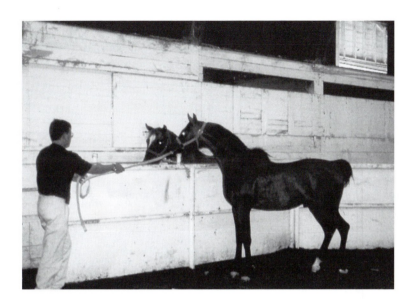

Fig. 13.6. Teasing of mares using a chute in the USA. Mares may be placed in the crate individually or the crate incorporated as part of a chute system. The stallion is restrained outside the crate and the mares stopped momentarily next to the stallion to gauge their response. (Photograph courtesy of Julie Baumber.)

Fig. 13.7. A teasing cage can be used to confine the stallion or teaser and allows any mares, either loose within the field or restrained, to approach at will. (Photograph courtesy of Julie Baumber.)

These last three systems are good for use with difficult mares, as they can be left alone to show in their own time without competition from other mares and can be observed from a discrete distance.

13.3.5. Vasectomized stallions

Rarely, vasectomized stallions may be used to run out with mares. This can be especially useful with maiden or difficult mares; those mounted can then

Fig. 13.8. An arrangement such as this can be used to tease shy or reluctant mares. The stable in the corner of a covering yard (open window in background) is used to confine the teaser, the yard (in the foreground) allows the mare to be free to exhibit oestrus at leisure. (From The National Stud, Newmarket, UK.)

be covered by the intended entire stallion. This system has the obvious danger of running unaccustomed mares and stallion out together and is, therefore, of limited use especially with valuable stock. However, it is an extremely reliable method of detecting oestrus and has the reported advantage of increasing the number of mares showing regular 21-day cycles (Barnisco and Potes, 1987). The other major disadvantage is that this system allows intromission to occur, and so enables the transfer of venereal disease. This can be averted by surgical retroversion of the penis, causing it to extend caudally (between the hind legs) at erection, making intromission impossible (Belonje, 1965). The risk of venereal-disease transmission via the stallion's muzzle is not, however, eliminated.

13.3.6. Hermaphrodite horses and androgenized mares

Hermaphrodite horses are very useful as teasers but are very rare and, therefore, not really a viable alternative. Androgenized mares (mares treated with testosterone) have been used successfully, but are not common practice (McDonnell *et al.*, 1986, 1988).

13.3.7. Teasing mares with a foal at foot

A mare with a foal at foot may present problems. The foal often becomes agitated with the unaccustomed attention to its mother and so distracts her. To overcome this, the foal may be penned or held within reach or sight of its mother, or removed completely from sight and sound while teasing occurs (Fig. 13.21). Prior knowledge of a mare's normal behaviour in such circumstances is very useful.

13.3.8. Conclusions on teasing

In any system of teasing, direct contact between the stallion's muzzle or penis and the mare's genitalia ideally must be avoided. Direct contact risks the transfer of disease to successive mares via the stallion. This is one of the major advantages of teasing over a trying board and a potential disadvantage of many of the other methods discussed.

Not all mares show under the above systems and some require specific management, prolonged individual teasing, etc. The key to success is careful observation, as it must be remembered that all mares react individually and no system is 100% reliable. The physiological value of teasing and its role in advancing ovulation and uterine clearance is increasingly becoming evident; however, further confirmation of a mare's reproductive activity is often required. This is achieved by veterinary examination.

13.4. Veterinary Examination

Routine veterinary examination to confirm a mare's reproductive state is used in many studs, especially those running valuable stallions with the aim of optimizing their use. Veterinary techniques may be used alone or to back up teasing and confirm diagnosis and optimize the timing of covering. There are three types of veterinary examination that may be used in this context: ultrasonic scanning; rectal palpation; and vaginal examination. All three techniques have been previously described (see Chapter 11, this volume). They are used to assess ovarian, uterine, cervical and vaginal activity. This assessment can be used to confirm a mare's sexual state, correlate coitus to ovulation and diagnose venereal infections. For all techniques, the mare should be restrained in stocks, with the perineal area thoroughly washed and her tail bandaged up out of the way (Fig. 13.9).

13.4.1. Ovarian assessment

The main activity assessed in the context of covering management is ovarian activity. The techniques used are ultrasonic scanning and rectal palpation (see Chapter 11, this volume). In particular, the presence of follicles and/or corpora lutea (CL), their consistency, appearance and position are noted so that the time of ovulation can be estimated and the most appropriate time for covering determined.

Follicles may develop on either ovary, often there are several follicles but many regress due to an inability to react to hormonal stimulation (see Section 1.9.1). As oestrus and ovulation approaches, normally one to three of these developing follicles can be identified as dominant. The diameter of these follicles used to be advocated as a good predictor of the imminence of ovulation (Greenhof and Kenney, 1975). However, though diameter may be used as a guide, pre-ovulatory follicular diameter varies con-

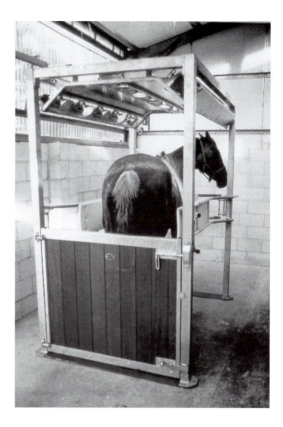

Fig. 13.9. Stocks for use in restraining mares for internal veterinary examination. (From Monarch Manufacturing, King Street, Willenhall, West Midlands, UK.)

siderably. Most mares ovulate follicles of 3.5–4.5 cm in diameter, others habitually ovulate follicles of up to 6.0–6.5 cm and still other mares as small as 2.0 cm (Ginther and Pierson, 1984a,b; Pierson and Ginther, 1985b). Multiple ovulations are seen; double ovulations have been reported to occur in 23% of ovulations in the Thoroughbred (Davies Morel and O'Sullivan, 2001) but are rarer in other breeds, such as draught and Arab mares and are very uncommon in native ponies (Arthur and Allen, 1972; Newcombe, 1995). The incidence of multiple ovulations also appears to increase with mare age (Davies Morel *et al.*, 2005). If multiple ovulations are present, they are likely to ovulate at a smaller size. The presence of multiple ovulations may well preclude a mare from covering. Despite the increasing success of managing twin conceptuses by pinching out (see Section 14.3), studs may still prefer not to cover multiple-ovulating mares but rather to leave them and advance the next oestrus, which will hopefully demonstrate only a single ovulation. Of further note with regard to multiple ovulations is that they may occur asynchronously but both may still be fertile. As such, scanning or rectal palpation may indicate a single dominant follicle but examination several days later may indicate multiple CL and a multiple pregnancy. Close regular monitoring, especially of mares with a high risk of multiple ovulations, e.g. older Thoroughbred mares, is, therefore, essential.

Ovulation occurs in two stages (see Section 1.9.1). These two stages can be used as an additional guide to the imminence of ovulation (J. Newcombe, Wales, 2008, personal communication). As ovulation becomes increasingly imminent, the follicular wall becomes thinner and the pressure of the follicular fluid contents decreases. On rectal palpation, this decline in follicular pressure can be felt. On scanning, the clear spherical shape of the follicle becomes less clear and the margins become thicker and more 'ragged' in appearance (Ginther and Pierson, 1984a,b; Pierson and Ginther, 1985b; Sertich, 1998). Such follicles would be expected to ovulate within 24h and most studies advise that the mare should be covered immediately. Further detailed scanning work suggests that not only the shape of the follicle can be used as a guide but also the echogenicity of the granulosa cells and the space between them and the theca cells lying below can be used to increase the accuracy of ovulation prediction (Chan et al., 2003). The best pregnancy rates are achieved by covering within the 48h period prior to ovulation (Woods et al., 1990; Katila et al., 1996). If at re-examination 24–48h later the mare has not ovulated, she should be re-covered immediately. In such a scenario human chorionic gonadotrophin (hCG) is often administered to advance ovulation (see Section 12.6.2.3). The presence of a CL within the ovary indicates that the mare has ovulated and that, unless the ovulation is very recent, within 12h, there is little point in covering the mare. Ova only remain viable for up to 12h post ovulation (Ginther, 1992), whereas sperm may remain viable for up to 7 days (Newcombe, 1994). Coverings later than 12h post ovulation result in higher rates of early embryonic death, with pregnancy rates for covering more than 24h after ovulation at zero (Woods et al., 1990; J. Newcombe, Wales, 2007, personal communication). CL appear as semi-solid structures within the cavity of the old follicle. At scanning, they are evident as a grey spherical shape (Figs 13.10–13.12; Ginther and Pierson, 1984a,b; Pierson and Ginther, 1985a).

At rectal palpation, new CL can be detected as a soft friable mass. At 24h post ovulation, they feel firmer and a pit may be felt. Later on, when assessing via rectal palpation, they may be hard to distinguish from follicles.

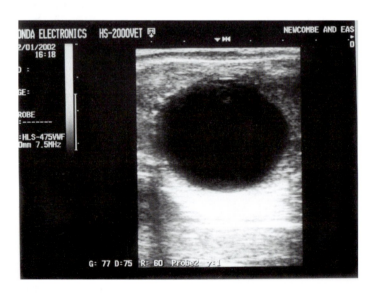

Fig. 13.10. An ultrasonic scanning photograph of a large 4 cm follicle approximately 48 h before ovulation. Note the clear spherical shape. (Photograph courtesy of Dr John Newcombe.)

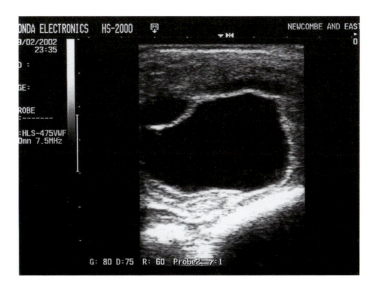

Fig. 13.11. An ultrasonic scanning photograph of a follicle immediately prior to ovulation; notice the thick wall and the loss of a clear spherical shape. (Photograph courtesy of Dr John Newcombe.)

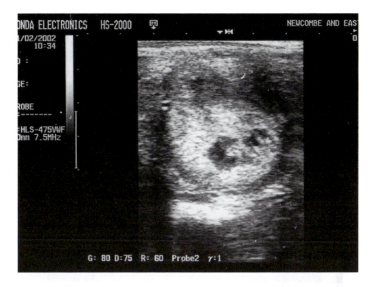

Fig. 13.12. An ultrasonic scanning photograph of a corpus luteum immediately post ovulation. (Photograph courtesy of Dr John Newcombe.)

13.4.2. Uterine assessment

Uterine activity and appearance can be used as a further guide to reproductive activity. As with ovarian activity, this is assessed via scanning and rectal palpation. Striking changes occur within the uterus between oestrus and dioestrus. This is thought to be due to the presence or absence of progesterone, rather than oestrogens. Once the dominance of progesterone has been removed, as the mare goes into oestrus, the uterine endometrial folds become oedematous and can clearly be visualized at scanning (Ginther and Pierson, 1984a,b; Hayes *et al.*, 1985; Ginther, 1992; Plata-Madrid *et al.*, 1994; Bergfelt, 2000). The dense

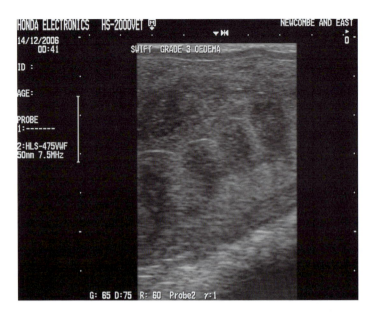

Fig. 13.13. The characteristic cartwheel appearance of the oedematous uterine endometrial folds seen at scanning in the uterus of a mare in oestrus particularly as oestrus commences. (Photograph courtesy of Dr John Newcombe.)

central portions of the folds appear echogenic (white/grey) and the oedematous portion non-echogenic (grey/black). Hence, when the uterine horn is viewed in cross section, it resembles a sliced orange or cartwheel (Fig. 13.13; J. Pycock, Wales, 2007, personal communication). This endometrial oedema can be scored and, as such, bears a close correlation to behaviour scores (Hayes *et al.*, 1985; Squires *et al.*, 1988). However, more recent work by Plata-Madrid *et al.* (1994) indicates that oedema may peak up to 6 days prior to ovulation. Despite this, uterine oedema is used in practice as an additional useful indicator of imminent ovulation and is a good indicator of basal progesterone levels and hence the likelihood of elevated oestrogens and oestrus.

Rectal palpation may also be used to indicate uterine changes associated with reproductive activity. The tone, size and thickness of the uterus are assessed, these tend to increase under the dominance of progesterone; so during dioestrus the uterus appears larger and more toned while during oestrus it appears smaller and more flaccid (Ginther, 1992; Bergfelt, 2000).

13.4.3. Cervical and vaginal examination

Cervical and vaginal examination is a less reliable determinant of reproductive activity; nevertheless, it can be a useful tool in confirming a mare's sexual state. The cervix and vagina are visualized via a vaginascope (see Section 11.2.6.1). The cervix during oestrus appears pink to glistening red in colour and is relaxed and appears to 'flower' into the vagina, and its lining is oedematous. Shortly after oestrus, the cervix begins to contract and becomes paler pink in colour, with a thick secretion. By dioestrus, the cervix is closed, pale in colour and dry. Pregnancy may be considered an extreme form of dioestrus; as such, the cervix is very tightly closed and white to pink in colour, with a central sticky mucous plug. Vaginal secretions originate from, and so mimic, cervical secretion. During oestrus the vagina appears red/pink in colour with fluid secretions; during dioestrus secretions appear more sticky and viscous, causing the vaginal walls to adhere together and make speculum examination more difficult. The vaginal secretion during pregnancy is thick, thickening even more as pregnancy progresses.

13.4.4. Hormone profiles

It has been suggested that declining oestradiol plasma concentrations can be used as an indicator of imminent ovulation (Allen *et al.*, 1995). However, due to individual variation, this is unlikely to be consistently diagnostic in itself but may, like cervical and vaginal examination, be an additional aid in diagnosis.

13.5. Preparation for Covering

Once it has been determined that the mare is in oestrus and ready for covering and that she has the appropriate negative swab certificates (see Section 12.5.2; Horse Race Betting Levy Board, 2008), attention must be turned to the preparation of the mare and the stallion for actual covering. Preparation depends entirely upon the system used for mating and varies from the strict codes of practice within the Thoroughbred industry to practically no preparation at all in the case of many native pony studs. The most cautionary preparation will be considered in the following account. Other studs dispense with some, if not all, of the preparation techniques. Exact management also depends on the size of the stud, labour available, etc. In larger studs there are often two to three breeding sessions per day, spaced at regular intervals, e.g. 9.00 a.m., 2.00 p.m. and 7.00 p.m. Most mares are covered twice within an oestrus, at 24–48 h intervals or until ovulation has been confirmed (Ginther, 1992; Newcombe, 1994).

13.5.1. The mare

The mare is prepared with all eventualities in mind. She is bridled and restrained, her tail bandaged and the perineal area washed thoroughly. When washing the mare, gloves should be used and there should be a different swab of cotton wool for each swipe. Each swipe should be taken from the buttocks towards the perineum and the cotton wool discarded immediately to prevent contamination of the washing solution. If soap is used it should be mild, non-detergent soap and the area thoroughly rinsed. Soap and disinfectants can act as spermicides and may also upset the natural microflora of the genital tract, opening up the opportunity for colonization by opportunistic bacteria and hence endometritis. A hose is an alternative to manual washing and is popular in the USA.

At this stage, the vulva of any mare with a Caslick vulvoplasty should be opened (episiotomy). If any sutures remain, these should also be removed to avoid damage to the stallion. Once the mare has been washed, she is led to the covering area, where felt covering boots may be fitted to her back feet, which should have had their shoes removed prior to arrival at the stud (Fig. 13.14).

She may also have a nose or ear twitch applied, depending on her temperament and past behaviour. Some studs twitch as standard, believing that prevention is better than cure as far as damage to

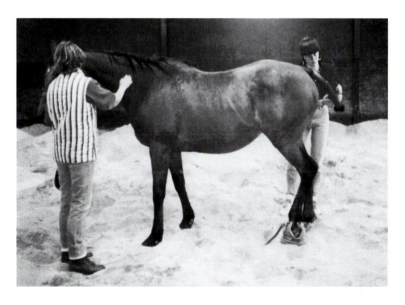

Fig. 13.14. Mare ready for covering, tail bandaged, perineum washed and wearing covering boots. (Photograph courtesy of Angela Stanfield.)

expensive stallions is concerned. If she has the reputation of being particularly bad-tempered she may need one of her forelegs held up in a carpal flexion, or hind leg hobbles may be fitted to prevent her lunging forward and objecting when the stallion mounts. Hobbles are particularly popular in the USA and also in Australia, but they are not commonly used in the UK. It is questionable whether a mare requiring such drastic restraint is truly in oestrus. In such cases, conception rates are likely to be adversely affected, due to both the incorrect timing of service and the stress of such treatment. Other articles used occasionally include blinkers, hood or blindfold, especially for highly strung maiden or difficult mares. A mare being covered by a stallion that tends to bite his mares can be protected by use of a shoulder or whither pad, with or without a biting roll (Fig. 13.15).

On very rare occasions, tranquillizers may be administered 15–30 min prior to covering; this can be of use in particularly nervous or vicious mares. If such extreme restraint is necessary, then the use of such a mare for breeding should be questioned, as there is a risk that such a temperament is heritable.

13.5.2. The stallion

A stallion's behaviour during the covering season can be unpredictable and, therefore, dangerous.

Management techniques can reduce this danger, such as by the regular tying up (racking up) of stallions in their stables for a period of time each day. This makes catching an enthusiastic stallion much safer when a mare arrives on the yard if he has been tied up and is used to the routine. The use of a specific bridle for exercise and covering is also advised so that a stallion is aware of what is required of him by the tack presented. For covering in intensive in-hand systems, a stallion is normally tacked up with his covering bridle and long rein; alternatively, a pole may be used giving a greater degree of control.

He is then led from his box to a washing-down area where his penis, genital area, belly and inside hind legs are washed with warm clean water. Ideally, the penis should be erect at washing. Erection may occur naturally, due to the anticipation of covering; others may require initial teasing. Antiseptic and/or soap solutions, once very popular, should be used with care, due to their spermicidal effects (Fig. 13.16; Betsch *et al.*, 1991). The stallion is then led to join the mare waiting in the covering area.

13.5.3. The mare and stallion

This preparation of the mare and stallion, as described, represents the extreme of full precaution

Fig. 13.15. Additional items used in covering, from the left: a twitch, separate biting roll, breeding roll, withers pad, covering boots and neck guard with biting roll. (From National Stud, Newmarket, UK.)

Fig. 13.16. Prior to covering, a stalllion's penis and genital area should be washed; erection can be encouraged by the close proximity of an oestrous mare. (Photograph courtesy of Julie Baumber).

to avoid all possible transfer of bacterial infection and is referred to as the minimal contamination technique (Kenney *et al.*, 1975a). In the Thoroughbred industry, where the value of stock is high, such precautions are economically justifiable. In other studs, with progressively lower turnovers and less-valuable stock, preparation becomes less cautionary in nature. It is interesting to note that several of these procedures are gradually being introduced into the less-intensive studs. For example, many breed societies now require swabbing prior to covering and some restrict the use of the yearly premium stallions to mares with negative swab certificates.

The other extreme to that practised in the Thoroughbred industry is seen in many native studs, where no preparation of the mare or stallion is practised, except the possible bridling of the pair for restraint and removal of the mare's hind shoes. Such extensive systems run a high risk of disease transfer. However, one of the saving graces is that native-type stock appear to suffer less from genital infections, though, of course, they are not completely immune.

13.6. Covering

The management of the actual covering procedure varies considerably. In-hand covering is the norm within the Thoroughbred industry in most countries and, with the increasing value of stock, in-hand breeding is now widely practised.

For each horse there should be at least one handler, ideally wearing stout footwear and a hard hat. The stallion handler should also carry a stick for reprimanding and, in many traditional systems, there is also an assistant to help the stallion gain intromission, if necessary, or hold the mare's tail. A fourth handler may also be present to hold the mare steady and prevent her tottering too far forward. It is essential that all handlers know their exact role and that an emergency procedure has been drawn up beforehand. Once prepared, the mare is normally taken into the covering area first to await the arrival of the stallion. The stallion is then brought in to cover the mare immediately or, in some systems, the stallion may initially be introduced to the mare over a trying board positioned within the covering area. This provides him with protection during initial contact, and is required by some stallions to avoid a prolonged period with the mare in the covering area before full erection and mounting occurs, a potentially dangerous situation. The mare is then led away from the trying board ready for covering. At this stage, she may be further restrained by means of a twitch or hobbles. Once the mare is ready, the stallion is allowed to approach her, normally at an angle several feet from her nearside to avoid startling her and causing her to kick out.

Once the stallion's penis is fully erect, he should be allowed to mount. Mounting before full erection should not be allowed. There are two blood pressures within the stallion's penis at erection: turgid and intromission pressure (see Section 2.2.1). Attainment of intromission pressure is essential to avoid damage to the stallion on entry into the mare (Fig. 13.17).

As the stallion is allowed to mount the mare, all handlers should stand to one side of the animals, usually the left. If problems do occur, then both the stallion and the mare can be pulled towards their handlers, turning their hindquarters away from anyone likely to be kicked.

As the stallion mounts the mare, she will probably totter forwards. This is to be allowed within reason. Excessive tottering will put a strain on the stallion, which will rest more of his weight on her back and also cause her possible strain or injury. Tottering too far forward can be prevented by placing the mare up against a protective barrier. This arrangement may also be used to protect handlers

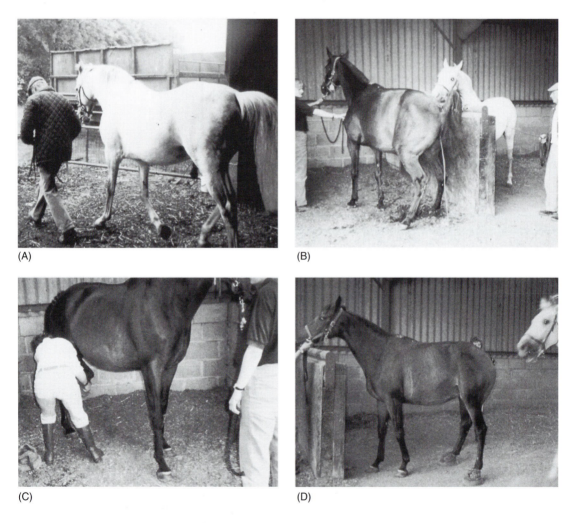

(A)

(B)

(C)

(D)

Fig. 13.17. The sequence of events associated with in-hand covering: (A) the stallion is bridled and brought into the covering area to meet the mare; (B) the mare may be teased by the stallion over a trying board immediately prior to covering; (C) covering boots are put on the mare's hind feet; (D) the stallion is allowed to cover the mare – ideally, the stallion should be calm and not over enthusiastic;

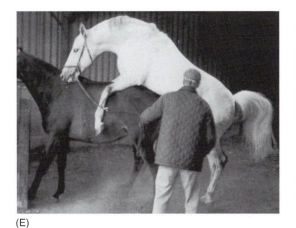

(E)

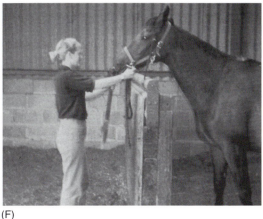

(F)

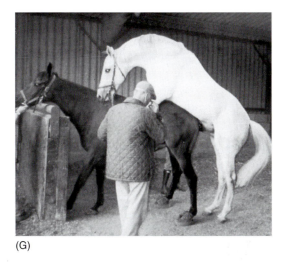

(G)

Fig. 13.17. *Continued* (E) ejaculation is marked by flagging of the stallion's tail; (F) the mare may be stood up against a trying board during covering to prevent excessive tottering forward; (G) ejaculation is followed by a quiescent period prior to dismount. (From The Elms Stud; photographs courtesy of Victoria Kingston.)

with mares that strike out (Fig. 13.17F). If a leg strap is being used to restrain the mare, it must be released at this stage. For this reason tight hobbles are not advocated, as they do not allow the mare to move in order to accommodate the weight of the mounting stallion.

Many traditional studs still advocate the use of an assistant stallion handler to pull the mare's tail to one side and guide the penis into the vagina. However, this is no longer considered good practice, as any unexpected external stimuli may disrupt the normal sequence of events leading up to ejaculation.

Successful ejaculation is signalled by the rhythmic flagging of the stallion's tail (Fig. 13.17E). In most stallions this is clearly evident, but if in doubt, the urethral contractions can be felt by a hand placed along the ventral side of the penis. However, possible disruption of the natural sequence of events must again be considered.

After ejaculation, the stallion should be allowed to relax on the mare (Figs 13.3 and 13.17G). During this period of relaxation, the glans penis, that has increased in size considerably at ejaculation, returns to normal, allowing the penis to be withdrawn. Early withdrawal can cause damage to the stallion and mare. He should then be allowed to dismount at will, which may take several minutes. The mare should then be turned towards her handler and walked slowly away, allowing the stallion to slide off and reducing the chance of him being hit should she lash out. The stallion should similarly be turned towards his handler after he has dismounted, again reducing the chance of injury to the mare or handlers. Some studs routinely collect a dismount semen sample from the drips on the penis at dismount. The presence of spermatozoa confirms that ejaculation has taken place. At this time a dismount swab may also be taken from the urethral area. Increasingly, in studs running high-value stallions, the mating process is videoed to provide additional evidence that covering has taken place.

After covering, the stallion should be allowed to wind down and walked to the washing area. There his penis should be washed and his genitalia examined for abrasions. Many practitioners advocate walking the mare around slowly after covering to prevent her straining and so losing semen. However, as the vast majority of the semen is deposited into the top of the cervix and uterus, beyond the cervical seal, it is very unlikely that there will be significant loss at post-coital straining. Any semen that is lost is likely to have been that deposited into the vagina and, as such, will have limited bearing on fertility.

Occasionally, the mare and stallion differ in size. If so, certain precautions should be taken. A very large stallion put on to a small mare should be considered carefully, as it may cause her problems in late pregnancy due to large fetal size near term. The reverse situation poses no such problems. In order to assist in the mating of unequal-size horses, some covering yards have a dip in the floor or use a breeding platform to equalize the horse's heights. In the USA hydraulic breeding platforms are becoming popular and can be adjusted to the exact height required.

A breeding roll may also be used if there is concern that the stallion's penis is too long or with maiden mares. The breeding roll is placed between the mare's perineum and the belly of the stallion, preventing him from penetrating too far and so causing damage. The breeding roll is usually a padded leather roll of 15–20 cm in diameter and approximately 50 cm long. In an attempt to maintain sterile conditions, it can be covered with a disposable examination glove (Fig. 13.15).

13.6.1. The covering yard

The covering yard can be any area that is quiet, dry and safe with a non-slip surface and away from the melee of the general yard. A paddock, open yard or specially designed covered area (Fig. 13.18) may be

Fig. 13.18. Covering in a field or paddock is common practice in smaller studs. (Photograph courtesy of Gillian Humphries.)

Fig. 13.19. The covering yard, if large enough, can double up as an exercise area. (From Derwen International Welsh Cob Stud, UK.)

used. If large enough, this covering yard can double up as an exercise area (Fig. 13.19).

If an area is to be specially designed, it should be at least 20 × 12 m, roofed, with two sets of wide doors to allow horses to enter and exit in different directions. It is very important that the floor of the covering yard is clean, non-slip and dust-free. Suitable surfaces include clay, chalk, peat moss, woodchip, or nowadays, rubber matting, which is increasingly popular (Fig. 13.20). The requirement for a non-slip surface makes

floorings such as concrete not advisable, though they are occasionally used.

13.6.2. Management of foal heat covering

A perceived major problem when covering mares is the danger to any foals at foot (Fig. 13.21). The majority of mares foal at the stud or arrive with

Fig. 13.20. Rubber matting may also be used as a suitable clean dust-free, non-slip surface in a covering yard.

Fig. 13.21. Mares with foals at foot may prove difficult to tease and cover successfully. In this case, the cage can be used to hold the foal, allowing reasonable access by the dam but ensuring the foal's safety during teasing and covering. (Photograph courtesy of Julie Baumber.)

foals of a very young age. In nature, this presents no significant problems, as foals naturally move away from the mating activity but stay within the vicinity of the mare. If teasing and/or mating is to occur in a large open area, the foal may be let loose and will normally keep well away from the proceedings, but within sight of its dam, reducing anxiety and stress. If mating is to occur in a small enclosed covering yard or if the mare is difficult, the foal should be restrained for its own safety. It may be removed and put in a loose box within sight of the mare and with a handler to watch it. However, some mares will become distracted by the antics of the foal. In such cases, more success may be achieved by removing the foal altogether, out of sight and earshot. The system used depends entirely on the character of the mare and foal, facilities available and personal preference. Further consideration of foal heat breeding is given in Section 16.5.2.

13.6.3. Management of mares susceptible to post-coital endometritis

All mares suffer from post-coital endometritis to some extent, but, some mares appear to be more susceptible than others and, in such a situation, conception rates can be significantly depressed (Troedsson, 1999; Knutti *et al.*, 2000; Watson, 2000). These mares can be managed in a number of ways, including prolonged teasing as mentioned previously, in order to reduce the post-coital inflammatory response. Details of the management techniques that can be used are given in Chapter 19 (this volume; Section 19.2.2.5.3.2.2).

13.7. Variations in Mating Management

As mentioned previously, there are many variations on the above theme, many being more cost-effective and used on the smaller native studs. The alternatives include covering in a small open paddock, yard or railed area with a non-slip surface, possibly earth or grass. Many stallions, especially in the Welsh cob industry, successfully cover mares in hand in a convenient field near to the main yard. This is how mares would have been covered historically, when stallions were walked around from farm to farm. If covering is to occur in a large field, then the stallion used must be well behaved and controlled, as his escape may incur considerable wasted time and disturbance to mares in adjacent fields.

The other extreme to in-hand breeding is pasture breeding, where the stallion is allowed to run free with his mares to cover them at will (Bristol, 1987). This system is the nearest to the natural situation but, as discussed at the beginning of this chapter, presents problems with visiting mares. Pasture breeding is popular in South Africa and South America, where mares are run in large herds over wide expanses of land with stallions running out freely (Ginther, 1983c; Ginther *et al.*, 1983). Fertility rates in such systems are very good (Bristol, 1982; Davies Morel and Gunnarsson, 2000).

A halfway house between in-hand breeding and pasture breeding is to allow the stallion to be loose in the field and introduce the mares to him individually, either restrained or free. This system allows visiting mares to be successfully covered in a system that is quite near to the natural one. Stallions used in this system must be well behaved, of an appropriate temperament and their characteristics well known by the mare handler, so that any necessary avoiding action may be taken.

13.8. Conclusion

It is evident that there is a vast array of management practices for covering mares. The system ultimately chosen is up to the individual stud, its facilities, normal practices and labour available. Financial considerations also have a considerable bearing on management choices. Intensive breeding systems are increasingly employed, providing more protection for both handlers and horses, but often at the expense of natural equine behaviour and reproductive performance.

14 Management of the Pregnant Mare

14.1. Introduction

Before the management of the pregnant mare is considered, it is essential to ascertain whether indeed she is pregnant. Once pregnancy has been confirmed the mare can be monitored and managed accordingly. The diagnosis of a twin pregnancy poses specific problems in the mare, these will also be considered.

14.2. Pregnancy Detection

Pregnancy detection may be carried out for a number of reasons. Ideally, it is conducted as soon as possible so that non-pregnant mares can be re-covered on their next oestrus and twin pregnancies identified and appropriate action taken. It may also be necessary for sale or for insurance purposes, particularly relevant if the mare is in the last two-thirds of her gestation. Finally, if the agreement at covering is no foal, no fee, it establishes whether a fee is due. The customary date on which the buyer of a stallion nomination has to pay in the northern hemisphere is 1 October; the corresponding date in the southern hemisphere is 1 April.

The mnemonic AEIOU describes the major aspects of an ideal pregnancy test:

A – Accurate
E – Early
I – Inexpensive
O – Once only
U – Uncomplicated

There are problems with many tests in that some fail to indicate embryo viability, the presence of twin pregnancies, anticipate fetal death or differentiate between embryonic vesicles and uterine cysts. Numerous methods of detection are available; however, none as yet meet all the above criteria (Collins and Buckley, 1993; Lofstedt and Newcombe, 1997; Henderson *et al.*, 1998).

14.2.1. Manual

Manual methods of pregnancy detection are the oldest and cheapest and still in common use. A major advantage is that they give immediate results, though they can only be accurately used after day 20 post coitum. Examination is carried out via rectal palpation or cervical/vaginal examination.

14.2.1.1. Rectal palpation

Rectal palpation, as described in Section 11.2.6, allows the uterus to be felt through the rectum wall. Initial work on this technique was carried out by Day (1940). This, and subsequent work, suggests that with experience, an accurate diagnosis can be made 20–30 days post coitum. Detection of pregnancy is reported to be possible at day 16, but poor accuracy precludes its use so early. Occasionally, with awkward mares, diagnosis is not possible until day 50. Normally, the accuracy of detection is very good and at its best at day 60, at which stage the age of the fetus can be estimated to within 1 week and the presence of twins detected (Van Niekerk, 1965; Bain, 1967; Roberts, 1986c). Table 14.1 indicates the size of the embryonic vesicle at various stages throughout early pregnancy.

Between days 16 and 20, a thickening of the uterine wall, rather than identification of the embryonic vesicle, may indicate pregnancy. Between days 20 and 60, the embryonic vesicle can be felt as a discrete swelling on either side of the midline, at the junction of the uterine horn and uterine body. Additionally, the uterus may feel turgid, rather than flaccid as in the non-pregnant state.

Post day 60, the fetus appears to move towards the uterine body and, due to its size, palpation of the margins of the fetal sack is difficult. In the case of twins, the fetal sacks merge and their diagnosis becomes less accurate. From this stage onwards, the whole uterus becomes progressively less turgid

Table 14.1. The size of the embryonic vesicle at various stages during early pregnancy in the mare.

Age of pregnancy (days post coitum)	Size of embryonic vesicle (diameter mm)	Comments
15	15–20	Feel for uterine tone
20	30–40	Embryonic vesicle first detected
30	40–50	
40	65	
50	80	
60	100–130	Most accurate diagnosis

and more distended, with no discrete swellings detected (Fig. 14.1). Hence, the accuracy of pregnancy detection at these later stages declines. However, as pregnancy progresses still further, it becomes possible to feel fetal structures, such as the head and ribs, through the uterine wall. After day 200, accuracy of detection returns to near 100%. At this stage, the fetus can be readily felt through the now much thinner uterine wall (McKinnon, 1993). Pregnancy may also be obvious from the mare's external appearance.

Pregnancy detection using rectal palpation is most commonly conducted between days 20 and 35 of gestation and is now considered accurate and safe, though some historical discussion with regard to a link with abortion has occurred (see Section 14.2.1.3).

The major disadvantage of rectal palpation is that it cannot be accurately used early enough to detect mares failing to conceive in time to allow them to be re-covered within one oestrous cycle. However, the technique is quick, simple and cheap and gives immediate results.

14.2.1.2. Cervical/vaginal examination

Vaginascopy – the examination of the mare's cervix and vagina via a speculum (see Section 11.2.6) – can be used as an aid to pregnancy detection, but is not accurate enough to be used in isolation (Roberts, 1986c). The pregnant cervix is white/light pink in colour, firm and with a sticky mucus. The vagina also contains sticky opaque mucus, making the insertion of the speculum difficult. Most cervical

and vaginal changes are evident from days 17–20 onwards. However, there is significant variation in cervical/vaginal characteristics between mares, making the technique inaccurate when used in isolation (Asbury, 1991). Used in conjunction with rectal palpation, however, it can be a useful aid to diagnosis in awkward mares.

14.2.1.3. Abortion risk

Historically, there are conflicting reports associating manual manipulation of the mare's reproductive tract with an increased risk of abortion. Allen (1974) and Voss and Pickett (1975) produced no evidence of such an association in pony mares. In fact, they suggest that abortion of a twin by squeezing (pinching out) of the fetal sack through the uterine wall per rectum at day 40 requires considerably more relative force than ordinary rectal palpation, but is reported to carry only a minimal risk to the remaining fetus. However, Osborne (1975) demonstrated that uterine myometrial activity does increase at palpation and that increased levels of such stress are associated with abortion. The current consensus of opinion is that any association between rectal palpation and abortion is due to the stress of unaccustomed handling, rather than the technique per se.

14.2.2. Blood tests

Blood tests are of use in small ponies and mares with injury or rectal/vaginal tears, which render rectal palpation and ultrasonic pregnancy detection not feasible. Blood plasma concentrations of one or several hormones may be used to indicate pregnancy. These tests can be very accurate but have the disadvantage of cost and a possible delay before the results are available. The tests used may be biological, immunological or chemical, and rely mainly upon the detection of equine chorionic gonadotrophin (eCG), progesterones and oestrogens.

14.2.2.1. Equine chorionic gonadotrophin

As discussed in Section 6.3.2, between days 35 and 100 of gestation, the endometrial cups produce eCG also known as pregnant mare serum gonadotrophin (PMSG), which can be detected in plasma samples between days 40 and 100 post coitum (Allen, 1969a,b). The presence of eCG

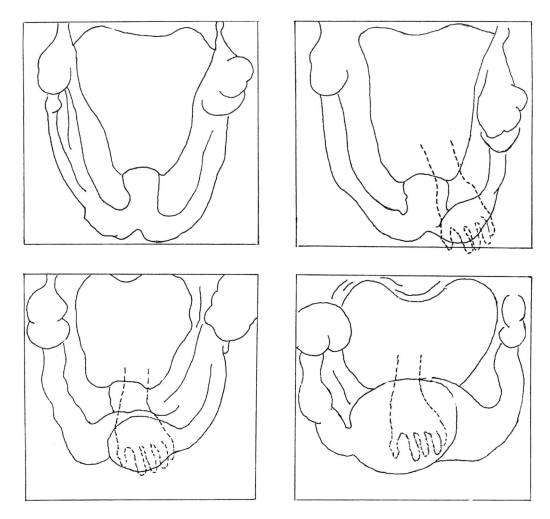

Fig. 14.1. The expected size of the fetal sack at various stages of pregnancy when undertaking rectal palpation. From the top-left to bottom-right: non-pregnant uterus, 25-, 45- and 60-day pregnant uterus.

was traditionally detected by biological tests, relying upon its effect on the reproductive tract of laboratory animals, such as the mouse (McCaughey *et al.*, 1973; Asbury, 1991). In the 1960s, a new immunological or antibody test was developed, called the haemagglutination inhibition test, now more commonly known as the mare immunological pregnancy (MIP) test (Wide and Wide, 1963; Allen, 1969a). This along with the more recent immunological tests, such as ELISA and radioimmunoassay (RIA), are the tests now widely used. During the period 45–100 days post coitum, all three tests have an accuracy of 60–100%. The

ELISA test, however, allows the earliest detection at day 35 with 43% accuracy and is more accurate (100%) at day 40 than the other two tests (31–37% accuracy; De Coster *et al.*, 1980; Squires *et al.*, 1983b). All three tests can be performed in less than 2 h.

Despite its accuracy, testing for eCG has the major disadvantage of being unable to determine whether the fetus is viable. After spontaneous or induced abortion, the endometrial cups continue to secrete eCG for several days and, therefore, an erroneous positive result may be obtained (Mitchell and Betteridge, 1973; McKinnon, 1993).

14.2.2.2. Progesterone

Progesterone is the hormone responsible for the maintenance of pregnancy (see Section 6.3.1) and, as might be expected, elevated levels are indicative of the presence of a fetus. When testing very early in pregnancy at the time of the mare's possible return to oestrus, elevated levels (those greater than 3–4 ng ml^{-1}) on days 16–17 post coitum are 71% accurate in indicating pregnancy. In non-pregnant mares progesterone levels should be declining at this time. However, prolonged dioestrus, which is quite common in mares, may be erroneously diagnosed as a pregnancy (Holtan et al., 1975a,b; Hunt et al., 1978; Villani et al., 2000). Commercial kits are now available for the rapid testing of plasma progesterone concentrations.

14.2.2.3. Oestrogen

Oestrone sulfate is one of the major oestrogens identified in the mare's plasma during pregnancy (Cox and Galina, 1970; Raeside et al., 1979). Elevated levels are observed at days 35–60; however, at this stage they are of limited diagnostic value but may be a preliminary indicator of fetal viability (Hyland and Langsford, 1990). However, elevated plasma oestrone sulfate concentrations from day 85 onwards, peaking around day 210 (Terqui and Palmer, 1979), provide an accurate, but late, diagnosis of a viable pregnancy. Schuler (1998) reported that plasma concentrations higher than 1.6 ng ml^{-1} are diagnostic of pregnancy and concentrations less than 0.8 ng ml^{-1} confirm a negative result. Concentrations between these two levels are inconclusive. Biological tests for oestrogens have largely been replaced by the chemical (Cuboni) test and now immunological (RIA and ELISA) tests (Cuboni, 1934; Terqui and Palmer, 1979; Evans et al., 1984). Most recently, the use of ELISA dipstick tests have been investigated (Henderson and Stewart, 2000, 2002). Oestrone sulfate analysis has the advantage of indicating fetal viability, as the fetal-placental unit is required for its production (Lasley et al., 1990; Stabenfeldt et al., 1991). Despite this the method is of no use for early diagnosis.

14.2.2.4. Early pregnancy factor

An early pregnancy factor (EPF) has been identified, using a rosette inhibition test in several animals as early as 6 h post coitum (Shaw and Morton, 1980). An equine EPF has been identified from 2 days post ovulation. After embryo transfer or embryonic death, EPF declines to non-pregnancy levels within 2 days. As such, this test has the potential to provide a useful tool not only for detecting pregnancy, but also for monitoring in vivo viability of equine embryos and for detecting embryonic death (Gidley-Baird and O'Neill, 1982; Takagi et al., 1998). Early detection of pregnancy failure would allow the mare to be prepared for covering on her next natural oestrus or at an artificially accelerated return to oestrus. As yet, the test is not commercially available. If diagnosis of fertilization ever becomes routinely possible as early as 6 h post coitum, this will allow the mare to be re-covered at that same oestrus giving her a second chance to conceive.

14.2.3. Urine tests

Urine tests, though not very popular, do have their uses in non-lactating mares where rectal palpation or blood sampling proves difficult (Cox, 1971; Evans et al., 1984).

14.2.3.1. Oestrogens

Oestrogens in the mare, being of a relatively small molecular weight, are the only reproductive hormones capable of passing unaltered through the kidney's filtration system and can, therefore, be detected in the urine. Other hormones, such as eCG, are too large and, therefore, cannot be detected in urine. As with plasma concentrations, pregnancy detection is possible from about day 90, although accuracy of diagnosis at this stage is low. By day 150, accuracy improves significantly (greater than 95%; Boyd, 1979).

14.2.4. Milk tests

Plasma hormone concentrations are most commonly used, but there may be occasions in lactating mares when obtaining a milk sample is easier and less stressful.

14.2.4.1. Progesterone

Progesterone is the major hormone which can be isolated in milk for use as a pregnancy test. Milk progesterone concentrations increase in parallel with plasma concentrations. Elevated levels are observed after oestrus; a subsequent decline after 10–15 days is indicative of no pregnancy, but continued elevated levels indicate pregnancy.

14.2.4.2. Oestrogens

Conjugated oestrogens can also be identified in milk, the concentration pattern again correlating closely to that observed in plasma and urine samples (Sist, 1987; Raeside *et al.*, 1991).

14.2.5. Faeces test

Unconjugated oestrogens have been isolated in the faeces of pregnant mares from day 120 and so may be used as a late pregnancy test. Such a test is particularly useful for feral mares and zoo equids (Bamberg *et al.*, 1984; Sist, 1987; Celebi and Demirel, 2003).

14.2.6. Ultrasonic pregnancy detection

In recent years ultrasonic techniques have revolutionized the detection of pregnancy in many animals. Ultrasonic detectors are based on the principle that ultrasonic sound waves are absorbed or reflected by the objects they hit. The relative amount of absorption or reflection depends on the density and movement of the object and can be transduced into a visual image. This method has the advantage of giving an immediate result and of usually being done on the stud, as the equipment is fully portable. More detailed accounts on the general reproductive use of ultrasound are given in Sections 11.2.6 and 13.4 and in reviews by other authors (Chevalier and Palmer, 1982; Allen and Goddard, 1984; Ginther, 1984; McKinnon, 1993; Sertich, 1998).

14.2.6.1. Doppler system

Fraser *et al.* (1973) initially developed the use of the Doppler ultrasound. This machine enables movement, and hence fetal heartbeat, to be detected along with uterine arterial blood flow. The ultrasonic signal is emitted from, and received by, a transducer placed on the mare's abdomen or within her rectum via a rectal probe, and is transduced into an audible sound. The fetal heart beat can be heard from day 42 of gestation onwards as a distinct beat, accelerated in comparison to the mare,

though it is not consistently heard until day 120. The enhanced blood flow through the pregnant mare's uterine artery can also be heard at the same time. It is present as a distinct whooshing noise at a slower beat than the fetal heart. This characteristic blood flow through the pregnant uterus is diagnostic in itself.

This method is very accurate at day 120, but accuracy in early pregnancy is not guaranteed and thus it cannot be successfully used within the timescale required for the mare to be returned to the stallion at her next oestrus. The timespan of use is similar to that of the test for eCG, but it has the obvious advantage of detecting a viable foal (Fraser *et al.*, 1973; Mitchell, 1973). More recently, colour Doppler ultrasonography has been used to monitor ovarian and uterine blood flow with the possibility of using it as a way to predict pregnancies at risk of abortion (Bollwein *et al.*, 2004).

14.2.6.2. Visual echography (ultrasonic pregnancy detection)

Hackeloer (1977) developed the use of visual ultrasonic echography. The reflected ultrasonic waves in this system are transduced into a visual picture on a visual display unit (VDU). Examination is per rectum by means of a probe carrying the ultrasonic emitter and transducer (Figs 11.11 and 14.2). The uterus is examined and within it the embryonic vesicle (Roberts, 1986a; McKinnon and Carnevale, 1993; McKinnon *et al.*, 1993). The embryonic vesicle can be detected from days 11–12 onwards as a discrete spherical sack (Ginther, 1984). This

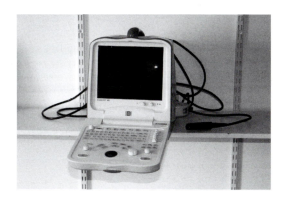

Fig. 14.2. The transducer probe (on the right) is placed in the rectum of the mare and angled down towards the uterus. The image produced is displayed on a VDU.

typical spherical nature of the equine trophoblast and its characteristic position, after day 17, at the junction of the uterine horn and body is fortunate and, as in the case of the human, allows detection at an early stage with an accuracy in excess of 98% (Ginther, 1989b; Hallowell, 1989). This method can also be used in the accurate detection of twins in early pregnancy, in addition to uterine cysts, fluid accumulation, sacculations, etc. or at a later date fetal viability.

The image produced by ultrasonic echography illustrates hard structures, such as bone, which reflect sound waves, as white, and fluid, which absorbs sound waves, as black, with variations in between.

The size of the embryonic vesicle (conceptus) at various stages of early pregnancy is given in Table 14.1. Pregnancy can be first detected at days 11–12, at which stage only the embryonic vesicle can be identified. At this stage, the conceptus is mobile and migrates within the uterus, making detection more difficult. In addition, there is a higher natural risk of embryo mortality in day-11 embryos than to older ones. By days 17–18 the conceptus has become 'fixed', normally at the junction of the uterine body and uterine horn, and so identification is easier (Ginther, 1983a,b; J. Newcombe, Wales, 2007, personal communication). By day 20 the embryo itself can be identified within the embryonic vesicle (Fig. 14.3). Day 24 heralds the first detection of the fetal heart beat (Allen and Goddard, 1984; Ginther,

1984) and roughly between days 55 and 90 fetal sex may be first determined (Curran and Ginther, 1989; Mari et al., 2002).

In practice most studs scan for pregnancy at day 18, which is the most accurate time and early enough to allow arrangements to be made for re-covering. An additional scan at day 40, after the period of highest risk, may be considered. In the Thoroughbred industry where there is a high incidence of twins, initial scanning is normally carried out at days 11–12, in order to identify twins and manage the mare accordingly (Section 14.3). The mare is then re-scanned again at days 18–20 and again at day 40.

14.2.7. Fetal electrocardiography

A fetal heart electrocardiogram may be obtained from electrodes strategically placed on the mare's body, which pick up the fetal electrical heart impulses. The read-out given can be used to detect the presence of a fetal heartbeat and also any abnormalities (Colles et al., 1978; Rossdale and Ricketts, 1980; Roberts, 1986a). Its popularity as a tool for diagnosing pregnancy is limited, due to the complication of setting it up, and it can only be used in late pregnancy. However, it does have its uses in detecting and monitoring fetal stress, or cardiac abnormalities, especially near to parturition or during a difficult delivery. It can also confirm fetal viability and the presence of twins (Parkes and Colles, 1977; Buss et al., 1980).

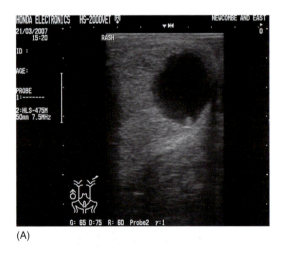

(A)

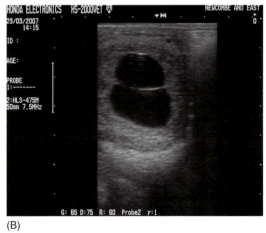

(B)

Fig. 14.3. A typical scanning photograph showing (A) a day 20 conceptus, the embryo is seen at the bottom of the black spherical conceptus; (B) unilateral (in the same horn) twins at day 16, the two black spherical concepti can be seen to be rather squashed together. (Photograph courtesy of John Newcombe.)

14.3. Management of Twin Pregnancies

The conception of twins is a significant and increasing problem in pregnant mare management. As discussed previously (see Section 5.4.3), the mare is monocotous, the uterus being unable to adequately support two pregnancies. Twin pregnancies rarely survive to term, most commonly resulting in abortion in mid- to late pregnancy (9–10 months; Rossdale, 1987; Ball, 1993b). They are the most common cause of non-infectious abortion (Roberts and Myhre, 1983; Macpherson and Reimer, 2000), accounting for 20–30% of all occurrences (Torbeck, 1986). Only 9% of twin pregnancies survive to term; 64.5% of cases result in two dead foals, 21% in one live foal and 14.5% result in two live foals (Card, 2000). Rates of twinning differ significantly between breeds but, in the Thoroughbred twins have been reported to be as high as 16.2% (Newcombe, 1995; Davies Morel et al., 2005). As such, it is advantageous to identify and manage twin pregnancies early on, especially if re-covering is planned. The advent of ultrasonic scanning has significantly helped the early identification of twins.

There are four main management practices used to reduce the incidence of twins: monitor ovulation; wait and see; manually reduce; or treat with prostaglandin F2α (PGF2α; Card, 2000; Macpherson and Reimer, 2000). Historically, the incidence of twinning was reduced by monitoring ovarian activity using rectal palpation, and withholding covering from mares with more than one large follicle. The mare would then be covered on the next natural or the next artificially advanced oestrus. This successfully reduced twinning rates within a population, but with it conception rates declined and the time interval between parturition and successful covering increased (Miller and Woods, 1988; Pugh and Schumacher, 1990). In order for these drawbacks to be addressed, identification and treatment of actual twin pregnancies, rather than potential twin pregnancies, is required. Naturally, in excess of 83% of twin pregnancies are spontaneously reduced to singles around the time of embryonic fixation (day 18; Ginther, 1987; Ginther and Bergfelt, 1988). So, one option is to monitor the pregnancy and observe if natural reduction occurs. If not induced abortion at a later stage may be advocated. The advent of scanning now allows such monitoring to occur easily. An alternative to natural reduction is to manually reduce. Manual reduction of twins to a single has been reported to be up to 96% successful between days 13 and 16 (Pascoe et al., 1987a,b; Ginther, 1989a,b). Manual reduction involves the manual squeezing of the smallest embryo, identified by ultrasound, either between the thumb and forefinger or by using the scanner probe to push the conceptus against the uterine wall and pelvis until the vesicle ruptures (Ginther, 1987; Pascoe et al., 1987a; Rossdale, 1987). This is best done prior to fixation (day 18) so is normally carried out at initial scanning days 11–12 (Ginther, 1989b). After fixation, the manual reduction of bilateral twins (one in each horn) can still be very successful, but reduction of unilateral twins (both in the same horn) runs a higher risk of loosing the whole pregnancy (Ginther, 1989a,b). Other methods of manual reduction can be used including transvaginal ultrasound-guided aspiration, which is reported to have a 70% success rate in eliminating just one embryo between days 16 and 25. Use after day 40 significantly increases the chance of both embryos dying (Macpherson and Reimer, 2000; Mari et al., 2004). At a later stage of pregnancy (after day 40) ultrasound-guided allantocentesis and transabdominal fetal cardiac puncture have been used but have not proved as successful as early manual reduction (Rantanen and Kinkaid, 1989; Card, 2000). An alternative to manual reduction of one twin is to artificially induce abortion of the whole pregnancy and re-cover the mare at the next advanced oestrus. Abortion and subsequent return to oestrus and ovulation can be induced using a single injection of PGF$_{2\alpha}$, multiple injections may be required later on in pregnancy. Abortion can be induced prior to the next expected oestrus, i.e. before day 21 of pregnancy. This is often done at the time of first scanning, which may be as early as day 11, in which case the mare's return to pregnancy post abortion may be delayed by only 10 days or so. However, with the success of ultrasonic-guided manual reduction of twins, abortion of the whole pregnancy at this stage is rarely an option chosen. Alternatively, the pregnancy may be allowed to progress longer in the hope that natural reduction may occur before PGF$_{2\alpha}$ is required. However, if a rapid return to oestrus is desired then PGF$_{2\alpha}$ must be administered prior to the development of the endometrial cups at day 40.

14.4. Embryonic Loss

Mares suffer from relatively high rates of early embryonic death (EED), particularly in the first 40 days of pregnancy. The reasons for, and causes of, embryonic loss and the subsequent failure to produce a foal are discussed in detail in Chapter 19 (this volume). In an attempt to reduce embryonic loss, some studs routinely place their mares on daily progesterone supplementation therapy in the form of injections or oral progesterone treatment for the first 100 days of pregnancy. From day 100 onwards, placental progesterone alone should be adequate. The use of such routine progesterone supplementation is questioned by some, as there is little evidence that it prevents embryonic mortality in the majority of pregnancies and it prevents the expression of the warning signs of a return to oestrus should embryo mortality occur (Darenius *et al.*, 1987; Holtan, 1993).

14.5. Pregnant Mare Management

The general management of the pregnant mare is important in order to ensure that the fetus is given every advantage and that the mare's future reproductive capacity is maintained. A fit condition throughout pregnancy is advantageous allowing a mare to foal with ease and be in optimum physical condition to embark on another pregnancy as soon as possible post partum. An understanding of the developmental changes of both the mare and the fetus during gestation is essential in order to gear her management towards optimum reproductive performance. These developmental changes are detailed in Chapter 5 (this volume).

There is little specific information on the environmental factors that affect fetal well-being. However, good management of the pregnant mare is clearly one, and can be reviewed under six main headings: exercise; nutrition; parasite control; vaccination; teeth and feet care.

14.5.1. Exercise

Exercise and nutrition are closely related and together are the major determining factors of body condition. As with most things, it is the extremes that can prove harmful, and so in the case of exercise a happy medium is to be aimed for (Rossdale and Ricketts, 1980). The absolute level of exercise depends on the individual mare and her history. A moderate exercise regime may be provided in the form of self-exercise by regular turnout in an open field or gentle ridden exercise in early pregnancy (Fig. 14.4). Mares that are used to being ridden can be ridden with increasing gentleness up until 6 months of pregnancy. In the last trimester of pregnancy, forced exercise should cease and ideally be replaced by group pasturing. Mares turned out in groups tend to exercise more effectively than those turned out alone (Kiley-Worthington and Wood-Gush, 1987).

Fig. 14.4. Exercise is essential in order to maintain, a mare's fitness – body condition score of 3 – and to prevent circulatory problems in the later stages of pregnancy. (Photograph courtesy of Tom James, Penpontbren Stud.)

Gentle exercise promotes and enhances the circulatory system. As discussed in Section 5.4.4 the fetus depends entirely upon its dam for its nutrient intake and waste output. This transport system is provided by the blood circulatory system, in particular the utero-ovarian artery and vein. Blood circulation through this system is enhanced by exercise, and hence oxygen and nutrient supply to the fetus. Exercise reduces water retention, or oedema, often associated with mares that are kept standing inside for prolonged periods of time, especially in late pregnancy. Exercise helps to maintain body condition and reduce obesity; hence the chance of complications at parturition is reduced and the maintenance of muscle tone will enhance delivery (Huff *et al.*, 1985).

Exercise must be consistent and not exhaustive, as this has been associated with high abortion rates. Excessive exercise may also cause stress, as may sudden movements, travelling, sale rings and even low-flying aircraft. Stress is known to increase abortion rates especially in early pregnancy – around day 40, as well as during the last 6–8 weeks of pregnancy.

14.5.2. Nutrition

The nutritional requirements of a pregnant mare vary according to whether she has a foal at foot and is lactating, or whether she was previously barren and so has only to satisfy the requirements for her own maintenance plus fetal growth (Hintz, 1993a). The length of lactation in the mare depends today on our interference, but normally runs for 6 months. It is, therefore, during this time that the two groups of mares with or without foals at foot vary in their requirements. In an ideal system, and providing there were enough mares to justify it, these two classes of mares should be kept and fed separately. During pregnancy, a mare's body condition and weight should be carefully monitored to ensure that neither excess weight is gained, nor does she have to mobilize her own body reserves to supplement inadequate nutrition. Her feeding management should vary accordingly. The use of a weigh pad (Fig. 14.5) is ideal to monitor weight on a regular basis, though a weigh band can still give an accurate indication of weight change. Assessment of body condition by eye and feel can also be undertaken (Fig. 11.1), a condition score of 3 should be maintained throughout pregnancy.

The following discussion concentrates on the nutritional management of the non-lactating mare,

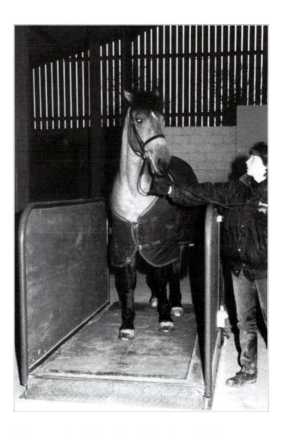

Fig. 14.5. A weigh pad is the ideal method of monitoring a mare's weight, especially when used in conjunction with assessment of body condition by eye and feel.

as the management of the lactating mare is addressed more fully in Chapter 16 (this volume). The general principles of nutrition apply to feeding the mare as to any formulation of a horse ration. The important nutrients are protein, energy, vitamins and minerals, and they should all be balanced according to need. During pregnancy, requirements do not significantly increase until the last 3 months when they should be adequate to allow a 14.5% increase in mare body weight. During early pregnancy, the mare's weight gain should be minimal, though, as she progresses into the last trimester (last 100 days), she should be expected to gain in the order of 0.25 kg day^{-1} (Frape, 1998; Card, 2000). This level of weight gain ensures that it is due to an increase in fetal weight and not to excessive deposition of internal body fat. Over-fat condition in mares during late pregnancy causes excessive pressure on the internal organs at parturition, as well as limiting uterine size and so reducing fetal birth weight and possibly post-natal viability. Limited fat deposition in late pregnancy is, however, desirable to act as a temporary store for emergency mobilization during early lactation. Failure to gain the required weight may mean that the mare's limited fat reserves of pregnancy are already being mobilized, reducing the energy available to the fetus and hence its growth *in utero*. However, it has been reported that mares restricted to up to 55% of National Research Council requirements (Tables 14.3 and 14.4) give birth to normal birth weight foals. This demonstrates that healthy mares, though they may deplete their own body reserves, possess the ability to compensate for low nutritional intake without significant detriment to the foal (Frape, 1998).

As in the case of other animals, feed must be of good quality with adequate roughage to aid digestion and prevent the development of vices through boredom (Winskill *et al.*, 1996). Any good stud will, as a matter of routine, have all batches of hay, haylage, etc. analysed to ascertain the dry matter (DM) content, protein, energy, vitamins and mineral concentrations. This allows accurate balancing of feeds available and identifies the appropriate supplements needed. The quality of feed is also important; conditions such as fungal contamination can cause abortion (Pugh and Schumacher, 1990). General poor nutrition has been associated with prolonged gestation, developmental abnormalities and decreased birth weights. These problems are exacerbated if low nutrition levels are evident in late pregnancy (Comline *et al.*, 1975; Silver *et al.*, 1979). Excess body weight has been associated with uterine inertia and dystocia (Varner, 1983) though this is disputed by others (Henneke *et al.*, 1984).

14.5.2.1. Protein

For a horse weighing 500 kg or more, levels of crude protein (CP) in the order of 630 g CP day^{-1} are required in early pregnancy (Frape, 1998; National Research Council, 2007). However, as pregnancy progresses into the last 90 days protein intake needs to increase in parallel with requirements. Levels of protein in the order of 759 g CP day^{-1} are required in the last 3 months of pregnancy (Tables 14.3 and 14.4). This level may be achieved by feeding good fresh grasses; dried grass

or hay tends to lose protein in the drying process. Legume hay – for example lucerne – tends to have a higher protein content; even after drying, and so may be adequate for the late pregnant mare.

Protein can be supplemented by the addition of animal products or plant products. Appropriate animal by-products include fishmeal, bone meal, etc. Such products tend to be expensive and have now been banned in some countries, in the light of the bovine spongiform encephalitis (BSE) scare in cattle. The other alternatives are plant products, such as soybean meal, linseed meal, etc. These tend to be cheaper and more popular.

The total protein content of a diet is not the only important factor in satisfying the protein requirements; protein quality is also important. The component parts of all proteins are amino acids, some of which are essential and others of which can be manufactured by the body from other amino acids. The latter are termed non-essential amino acids. Certain protein-rich supplements may be lacking in specific essential amino acids, so that even though the total protein content is high, its use to the body is limited. Barley, oats and linseed are all high in total protein content but are lacking in one or more essential amino acid. Soybean meal, on the other hand, contains all the essential amino acids required for the development of the fetus and is, therefore, a very useful protein supplement for use in late pregnant mares.

14.5.2.2. Energy

Energy requirements increase significantly in late pregnancy but are also important during early pregnancy, their deficiency being implicated as a cause of embryo mortality (Potter *et al.*, 1987). Again, the energy demand of mares in good condition in early pregnancy may be met by good-quality forage. For a 500 kg horse, levels of 16.7 Mcal day^{-1} are required (National Research Council, 2007). However, if the mare is a poor doer, or being ridden, her energy intake may need to be supplemented. Her energy intake also needs to increase in the last trimester of pregnancy. This may be achieved in theory by an increase in hay intake. However, in late pregnancy the increase in uterine size begins to limit the capacity of the digestive tract. Energy levels of 21.4 Mcal day^{-1} are required during this period. Good-quality feeds, low in bulk but high in nutrient value, are, therefore, advised. In late pregnancy, the mare's roughage intake should, therefore, be reduced and partly replaced by increasing levels of energy-rich concentrates. Care should be taken to ensure that the required protein intake is still maintained (Tables 14.2–14.4; Pugh and Schumacher, 1990; Frape, 1998; National Research Council, 2007).

14.5.2.3. Vitamins and minerals

Vitamins and minerals are classified as micronutrients. The specific effects of deficiencies in many micronutrients on the pregnant mare, and indeed on general equine welfare, are as yet unknown.

However, the importance of calcium (Ca), phosphorous (P) and vitamin A is appreciated. The Ca/P ratio is of special importance and the involvement of both minerals in bone growth is well documented (Frape, 1998). In the pregnant mare, these micronutrients are important not only to the mare herself, but also to the fetus. Ca and P are normally stored within the bones, much of which are a temporary store and can be mobilized to satisfy demands elsewhere. If the pregnant mare's dietary intake of Ca or P is inadequate, especially during late pregnancy, the mare will mobilize her own stores from within her bones to satisfy the fetal demand. If the dietary

Table 14.2. The average composition of feeds commonly used in horse diets. (From National Research Council, 2007.)

Feed	DM (%)	DE (Mcal/kg)	CP (%)	Lysine (%)	Fat (%)	Fibre (%)	Ca (%)	P (%)	Mg (%)	K (%)	Vitamin A (IU kg⁻¹)
Concentrates											
Barley (rolled)	91.0	3.67	12.4	0.45	2.2	20.8	0.06	0.39	0.14	0.56	817
Sugarbeet (molassed pulp)	88.0	2.84	10.0	0.42	1.1	44.4	0.89	0.09	0.23	1.11	88
Canola meal (mechanical extract)	90.3	2.94	37.8	2.12	5.4	29.8	0.75	1.10	0.53	1.41	–
Maize grain (cracked dry)	88.1	3.88	9.4	0.27	4.2	9.5	0.04	0.3	0.12	0.42	–
Linseed (meal, solvent)	90.3	2.85	32.6	1.20	1.7	36.1	0.40	0.83	0.55	1.22	–
Oats (grain, rolled)	90.0	3.27	13.2	0.55	5.1	30.0	0.11	0.40	0.16	0.52	44
Sorghum (grain, dry rolled)	88.6	3.75	11.6	0.28	3.1	10.9	0.07	0.35	0.17	0.47	468
Soybean (meal, solvent 48% CP)	89.5	3.73	53.8	3.38	1.1	9.8	0.35	0.70	0.29	2.41	–
Wheat bran	89.1	3.22	17.3	0.70	4.3	42.5	0.13	1.18	0.53	1.32	1,048
Wheat grain (rolled)	89.4	3.83	14.2	0.40	2.3	13.4	0.05	0.43	0.15	0.50	–
Forages											
Lucerne	90.3	2.43	19.2	0.83	2.5	41.6	1.47	0.28	0.29	2.37	41,900
Grass (pasture)	20.1	2.39	26.5	0.92	2.7	45.8	0.56	0.44	0.20	3.36	–
Hay (grass, mature)	84.4	2.04	10.8	0.38	2.0	69.1	0.47	0.26	0.18	1.97	–
Hay (legume, mature)	83.8	2.21	17.8	0.89	1.6	50.9	1.22	0.28	0.27	2.38	–
Silage (grass, mature)	38.7	1.98	12.7	0.43	3.0	66.6	0.56	0.31	0.20	2.42	–
Silage (legume, mature)	42.6	2.19	20.3	0.87	2.1	50.0	1.30	0.33	0.26	2.87	–

DM, dry matter; DE, digestible energy; Ca, calcium; P, phosphorous; Mg, magnesium; K, potassium.

Table 14.3. The daily nutrient requirements of pregnant mares of varying weights. (From National Research Council, 2007.)

Stage of pregnancy	Non-pregnant weight (kg)	Daily gain (kg)	Actual weight (kg)	DE (Mcal)	CP (g)	Lysine (g)	Ca (g)	P (g)	Mg (g)	K (g)	Vitamin A (mg)
<5 months	200	–	200	6.7	252	10.8	8.0	5.6	3.0	10.0	8.0
	500	–	500	16.7	630	27.1	20.0	14.0	7.5	25.0	20.0
	900	–	900	30.0	1134	48.8	36.0	25.2	13.5	45.0	36.0
8 months	200	0.13	209	7.4	304	13.1	11.2	8.0	3.0	10.0	8.0
	500	0.32	523	18.5	759	32.7	28.0	20.0	7.6	25.0	20.0
	900	0.57	942	33.3	1367	58.8	50.4	36.0	13.7	45.0	36.0
11 months	200	0.26	226	8.6	357	15.4	14.4	10.5	3.1	10.3	8.0
	500	0.65	566	21.4	893	38.4	36.0	26.3	7.7	25.9	20.0
	900	1.17	1019	38.5	1607	69.1	64.8	47.3	13.8	46.5	36.0

DE, digestible energy; Ca, calcium; P, phosphorous; Mg, magnesium; K, potassium.

Table 14.4. The expected feed consumption of pregnant mares (percentage of body weight). (From National Research Council, 1989.)

Pregnant mares	Forage	Concentrate	Total
Maintenance	1.5–2.0	0–0.5	1.5–2.0
Mares in late gestation	1.0–1.5	0.5–1.0	1.5–2.0

deficiency is great, then her bones will suffer, become brittle and possibly unable to take the strain of the increased weight in late pregnancy or the stresses of parturition. The foals of such mares can also suffer from deformities in tissue and bone growth and general ill thrift at birth.

For a 500 kg mare, the levels of Ca and P required in the first 5 months of pregnancy are in the order of 20 and 14 g day^{-1}, respectively. During the last 3 months the demand increases from 28 g day^{-1} and 20 g day^{-1} in month 8 to 36 g day^{-1} and 26.3 g day^{-1}, respectively at term (Table 14.3). In the tenth month 25.3 mg Ca kg^{-1} of the mare's body weight are deposited in the fetus daily (Frape, 1998; National Research Council, 2007).

Not only are the absolute levels of these two minerals important but also their relative amounts. Excess P interferes with Ca absorption and leads to an effect similar to Ca deficiency. A ratio of Ca/P of between 1:1 and 6:1 is considered acceptable. Legume hay is a good source of Ca and P and supplementation of these minerals should not be required for mares fed legume hay *ad libitum* (Hintz, 1993a; National Research Council, 2007). It is important to know that straights, in the form of grains, tend to be relatively high in P, hence feeding these grains to late pregnant mares, when concentrate intake is increased, may require Ca to be supplemented. Ca may be supplemented in the form of ground limestone flour or milk pellets, which are more readily available in today's market. The high concentration of P in bran largely precludes its use in late pregnancy, except perhaps in small quantities as a laxative.

Other minerals should also be supplemented, though their exact function is unclear: for example, salt, potassium, magnesium, zinc, cobalt, iodine, manganese, copper, etc. (Ott and Asquith, 1986; Ott, 2001). These may be easily supplemented by one of the commercially available vitamin and mineral blocks providing the mares with free access.

Vitamin A, as mentioned, is also important, especially as it is an essential component of epithelium.

As such, it is important in reproductive function, cell regeneration and development. Adequate vitamin A levels are best ensured by feeding fresh green forage (Tables 14.2 and 14.3); mares with no access to fresh pasture must, therefore, be supplemented (Frape, 1998). Excess supplementation as well as deficiency of minerals and vitamin A has been implicated in abnormalities associated with bone growth, such as developmental orthopaedic disease, angular limb deformities and epiphysitis; so feeding the correct levels is important (Beard and Knight, 1992).

14.5.2.4. Water

As with all equine rationing, water is an essential, but often forgotten component. The late pregnant mare, kept at an ambient temperature of 20°C, requires large amounts of water, up to 50 l day^{-1} (6–7 l per 100 kg body weight day^{-1}), depending on the DM content of her ration (Huff *et al.*, 1985). This water must be clean, fresh and available at all times to allow consumption in small but frequent amounts.

14.5.3. Parasite control

Parasite control is of significant importance. A high parasite count is often the reason why some mares appear as poor doers, and parasites cause a large quantity of the food fed to be wasted. If a mare's internal worm burden is excessive, the damage caused may become permanent, condemning that mare to being a bad doer for the rest of her life or may even be a cause of death. Strongyles and ascarids are the main parasites associated with disease in mares and foals (Shideler, 1993d). Pinworms, tapeworms and bots may be present but rarely result in life-threatening conditions (Card, 2000). As with all equids, worming should take place regularly during spring and summer. The ideal interval between worming is often quoted as 12 weeks, however, if good rotational, mixed grazing management and worm count monitoring is employed this interval can be much longer. Less frequent worming is required in winter once the frosts appear.

The development of resistance to wormers (anthelmintic treatments) can be reduced by rotating the wormer types used, i.e. those based on the thiabendazole group, followed by those based on the pyrantel embonate group. Care must be taken in worming pregnant mares, as not all wormers are suitable. Products based upon benzimidazoles, fenbendazole, pyrantel pamoate and ivermectin are considered safe (Varner, 1983; Card, 2000); others may not be suitable for very early pregnant mares, within the first 12 weeks as their use has been associated with a risk of abortion. At the other end of pregnancy, organophosphate wormers used to control bot flies are not recommended, as they may disrupt and trigger smooth muscle contraction, and may induce late abortions due to uterine contractions. In general, it is recommended that no wormers be used in the last month of pregnancy, due to the risk of inducing premature delivery (Varner, 1983). However, some advocate worming immediately prior to parturition to ensure that the foal is not exposed to a high parasite burden at birth (Shideler, 1993d; Card, 2000). The importance of clean grazing must not be overlooked and a combination of worming treatment, worm count monitoring, clean grazing and manure removal gives the best results (Herd, 1987a,b).

Delousing powder may be administered to pregnant mares, but its use should be avoided in late pregnancy. Ideally, pregnant mares should not be allowed to get into such a condition that such treatment is required.

14.5.4. Vaccinations

The vaccination programme required by a mare depends upon the endemic diseases prevalent and hence the country in which she lives. In the UK, vaccination against tetanus and influenza is automatic and is normally administered 4 weeks prior to parturition to allow the mare's titre of antibodies to be raised adequately to ensure transfer to the colostrum and hence to the foal immediately post partum (Varner, 1983; Pugh and Schumacher, 1990; Robinson et al., 1993).

Equine rhinopneumonitis (equine herpesvirus type 1 (EHV 1)) is a problem in the USA and is becoming increasingly so in the UK. This infection causes abortion, normally in the last trimester, in up to 70% of infected mares. If an outbreak is suspected, or routine protection is required, vaccination can be administered in months 5, 7 and 9 of pregnancy (see Section 19.2.2.5.7; Witherspoon, 1984; Mumford et al., 1996).

Equine viral arteritis (EVA) also causes abortion and a modified live vaccine is now available. EVA has been a problem in parts of Europe and USA, for a while and unfortunately is increasingly common in UK. Vaccination is available and may be given to pregnant mares but not in the last 2 months of pregnancy (Timoney and McCollum, 1997). Testing for, and confirmation of, disease-free status is often required by studs before a mare will be accepted for covering in accordance with Horse Race Betting Levy Board Codes of Practice (see Sections 12.5.2 and 19.2.2.5.3).

In other parts of the world routine vaccination for eastern and western equine encephalomyelitis, equine pneumonitis (EHV4), rabies and Potomac horse fever may be considered. With the exception of rabies vaccination, all can be administered in the last 4 weeks of pregnancy to confer protection on the foal via colostrum. Vaccination for strangles, botulism, anthrax, salmonella typhimurium and leptospirosis is possible but only advised in areas of particular risk (Wilson, 1987; Sprouse et al., 1989; Card, 2000).

14.5.5. Teeth care

The teeth of the pregnant mare should not be neglected. Regular teeth rasping ensures that all the plates are level and can efficiently grind food during mastication, enhancing digestion and maximizing the nutrient value of the food. This is of utmost importance when her system is under stress – for example, during late pregnancy.

14.5.6. Feet care

The majority of brood mares are unshod; even so regular trimming should occur. Poor feet cause pain, and this pain can be exacerbated with the increased weight burden of late pregnancy. Such mares will be reluctant to exercise themselves, resulting in problems as previously discussed. Some mares are turned to stud with musculo-skeletal problems and, as such, may need orthopaedic shoeing, especially in late pregnancy. Even so shod mares must have shoes removed prior to parturition to prevent accidental damage to the foal.

14.6. Conclusion

Accurate pregnancy detection is a major key to early pregnant mare management. Once pregnancy has been confirmed mare management can be geared towards ensuring that pregnancy is stress-free, optimizing the chances of a healthy foal and a mare in the appropriate condition for lactation and subsequent re-covering.

15 Management of the Mare at Parturition

15.1. Introduction

Gestation in the mare lasts on average 330–336 days, though considerable variation is evident (Bos and Van der May, 1980; Rossdale and Ricketts, 1980; Badi *et al.*, 1981; Davies Morel *et al.*, 2002). The physical process and endocrine control of parturition have already been detailed in Chapters 7 and 8 (this volume). This chapter will consider solely the management of the mare at parturition, including its artificial induction and dystocia.

15.2. Prepartum Management

Approximately 6 weeks before the mare's estimated date of delivery, she should be introduced to the foaling unit or the yard at which she is to foal. This begins the gradual familiarization of the mare to the surroundings in which she will foal and be kept immediately post partum, so reducing the stress of any sudden changes. Such familiarization allows her to become accustomed to particular management practices, especially if she is to foal away from home. Changes in feed, exercise, housing and routine can be introduced in plenty of time to allow a regular management system to be developed prior to foaling.

If the mare is to foal away from home, a period of 6 weeks is also required to allow her immune system to raise the necessary antibodies against any challenges present in her new foaling environment. This will not only provide protection for the mare herself, but will also allow the antibodies to pass into the colostrum and so provide the foal with immediate protection at birth. Bearing this in mind, it is advised that all mares should be vaccinated with either her annual boosters or new vaccination programme during this 6-week period. The vaccinations used depend upon the country of residence, prevalent diseases and types and ages of mares as discussed in Section 14.5.4 (Golnik, 1992; Card, 2000).

Exercise is very important. Regular free exercise in a paddock or field will be adequate for most mares and will help maintain her fitness for foaling and reduce the chances of oedema (fluid retention) in the legs. A slightly laxative diet may be advised, as many mares suffer from constipation in late pregnancy, especially if exercise is limited. To this end, components such as bran or fresh carrots, etc. may be added in small quantities, but care must be taken not to upset the overall nutritional balance of the diet, especially the calcium:phosphorous (Ca:P) ratio, which is very important at this stage. Lastly, but by no means least, clean and fresh water must be available at all times. Many mares need large quantities of water (500 kg mare will need 27–35 l day^{-1}) in late pregnancy (National Research Council, 2007).

Throughout the preparation of the mare for foaling, she should be observed for the characteristic signs of imminent parturition: increase in mammary-gland size; the secretion of milk; and general relaxation of the abdominal, pelvic and perineal area (see Section 7.2).

By this time, except in the case of an emergency, it will have been decided whether the mare is to give birth naturally or if it is to be induced. If birth is to occur naturally then, as soon as any signs of imminent parturition are noticed, the mare must be put into her foaling box, if she is to foal inside, or into a small quiet paddock if she is to foal out. She should then be monitored closely.

If she is to foal inside, the box provided must be at least 5 × 5 m, with good ventilation, but draught-free. Traditionally, the floor covering would have been a deep bed of straw, which provides a soft, warm, dust-free surface on to which the foal can be born (Fig. 15.1). More recently, other alternatives, such as rubber matting, are increasingly popular, though expensive, they provide a clean, insulated and dust-free floor that can be easily washed and disinfected.

The foaling box should be free of any protrusions, which may cause damage to the mare or foal. Ideally, it should have rounded corners to reduce

Fig. 15.1. The traditional floor covering for the foaling box is straw that provides a good, deep, soft, bed for the foal to be born on to.

the risk of the mare getting cast. Hay nets should not be used, as the foal can get itself caught up in anything left dangling. Hay should be fed off the ground. The use of high hay racks avoids the wastage of feeding off the floor but does run the risk of the mare and foal getting seeds in their eyes and ears. In an ideal purpose-built unit each box should have two doors, one to the outside for horse access and another facing into a central sitting area for human access and viewing (Fig. 15.2). Closed-circuit television is also a good method of viewing mares with minimal disturbance. The provision of

Fig. 15.2. An ideal arrangement of the foaling unit allows the foaling boxes to be viewed from a central sitting area. (From The National Stud, Newmarket, UK.)

radiant-heat lamps in each box is an advantage for weak foals.

Once the mare has settled into her foaling quarters, it is a case of careful watching and patient waiting. Careful observation can minimize the time from the first signs of trouble to action and can, therefore, be crucial in saving lives.

15.3. Induction of Parturition

The mare shows a much wider variation in gestation length than other farm animals in which induction of parturition is practised. As discussed in Section 8.1, the natural length of gestation is affected by several factors. When considering artificially inducing parturition, it is essential that the date is as close as possible to the estimated delivery date. Premature induction will result in foals with all the classical symptoms of prematurity, including breathing difficulties, being late to stand and a delay in the normal post-partum adaptation mechanisms. Such foals may, if they survive, suffer long-term ill effects and there will be a dramatic increase in labour and veterinary expense. Induction of parturition in horses should, therefore, be carried out with great care, and its value, as a routine technique is questionable.

Initially it was thought that induction would be useful as an aid to management, to ensure that all facilities and staff are available and ready when required, particularly if limited experienced labour is available and only a few mares involved. However, high foal mortality rates mean that induction is now largely limited to emergencies, such as prolonged gestation, preparturient colic, pelvic injuries, ventral rupture, previous premature placental separation, pending rupture of the prepubic tendon or painful skeletal or arthritic conditions which can be unbearable in late pregnancy. It may also be considered if the mare has a history of difficult foalings, premature placental separation, uterine inertia, inability to strain effectively, also if she has pelvic abnormalities or injures; in all these cases induction of parturition allows the organization of expert help to be on hand at parturition to assist both mare and foal. Finally, it may be used as a research tool.

15.3.1. Fetal maturity

As indicated, the timing of induction in relation to the expected natural delivery date and fetal maturity is crucial to maximize chances of fetal/foal survival. In horses, parturition is related to fetal development (see Section 8.2; Ousey et al., 2004), especially maturity of the adrenal cortex and its ability to secrete corticosteroids. In the 5 days preceding natural parturition, fetal cortisol levels are seen to increase significantly and are required for the final maturation of major organ systems (Rossdale et al., 1997). In order for the fetus to survive parturition, it must have been exposed to these elevated cortisol levels; hence induction of parturition prior to this runs increasingly higher risks to the foal (Chavatte et al., 1997b). As gestation lengths are so variable in mares, it is not possible to use these as more than a rough guide to determine the expected parturition date; hence fetal maturity needs to be determined by an alternative means. The normal signs of parturition (see Section 7.2) can again give an indication but are not accurate enough to use to time artificial induction (Le Blanc, 1997). However, changes in secretions of the mammary gland are a good indicator. Ca concentrations ≥ 4.0 mg ml^{-1} are reported to indicate a mature fetus, concentrations ≤ 1.2 mg ml^{-1} indicate an immature fetus (Leadon et al., 1984; Ousey et al., 1984; Ley et al., 1993). An inverse relationship between sodium (Na) and potassium (K) is also evident. As parturition approaches Na concentration declines and K concentration increases. When Na < K, then the induction of parturition is reported to be successful. Based upon this, Ousey et al. (1984) devised a scoring system to determine when induction of parturition is safe (Table 15.1).

Haematological assessment of the fetus has also been suggested as a means of determining the safety of induction (Jeffcote et al., 1982; Rossdale et al., 1984) as well as rectal palpation, vaginal examination, amniocentesis, fetal electrocardiography, fetal eye diameter and length and aortic diameter (determined by ultrasonography), but when compared to

Table 15.1. Scoring system used to determine when the induction of parturition is safe. A total score of ≥ 35 suggests that safe induction of parturition is possible. (From Ousey et al., 1984.)

Colostrum concentrations			
Calcium (mg dl^{-1})	Sodium (MEQ l^{-1})	Potassium (MEQ l^{-1})	Points for each electrolyte
≥ 40	≤ 30	≥ 35	15
≥ 28	≤ 50	≥ 30	10
≥ 20	≤ 80	≥ 20	5

Ousey's scoring system they all have limitations (Reef *et al.*, 1995; Le Blanc, 1997; Turner *et al.*, 2006).

15.3.2. Methods of induction

Once fetal maturity has been determined, there are several methods that may be employed to induce parturition (Hillman and Lesser, 1980); these will be discussed in turn.

15.3.2.1. Corticosteroids

In ruminants, the most successful means of inducing parturition is the use of corticosteroids (Adams and Wagner, 1970). Foaling can similarly be induced in large mares using 100 mg dexamethasone, a corticosteroid administered daily for 4 days from day 321 of gestation. Parturition is reported to occur within 1 week and to result in live and healthy foals (Alm *et al.*, 1974, 1975). However, other workers, using pony mares, report less success with slow, protracted, difficult labours, stillbirths and placental retention (Rossdale and Jeffcote, 1975; First and Alm, 1977; Van Niekerk and Morgenthal, 1982).

15.3.2.2. Progesterone

Progesterone administration over a period of 4 days in late pregnancy, and its subsequent withdrawal, will induce parturition over approximately 1 week (Alm *et al.*, 1975). Such use of progesterone does not result in parturition in ruminants.

Both these methods are interesting as they are an indication of the possible differences in endocrine control of parturition when compared to ruminants (see Section 8.2). However, they have the disadvantage of being relatively inaccurate in the timing of the reaction to their administration, which can be over a period of 1 week, and the number of stillbirths. They are, therefore, not used commercially. A more accurate and more immediate induction agent is required.

15.3.2.3. Prostaglandins

The use of prostaglandins gives a more immediate result than progesterone and corticosteroids. Prostaglandin analogues are normally used; for example, 250–1000 µg fluprostenol, administered intramuscularly, is sufficient to cause parturition within 2 h in a mare at full term. The mare will show the initial signs of parturition (stage 1) within

30 min. The foal is usually born within 2 h and the afterbirth appears 2 h later. Complications, such as insufficient cervical dilation and hence rupture and decrease in foal viability, have been reported (Rossdale *et al.*, 1979a; Ousey *et al.*, 1984; Ley, 1989). In the natural course of events, the rise in prostaglandin coincides with the delivery of the foal. Fluprostenol presumably imitates this. Its use too early, therefore, has little effect, as elevated levels at this time are out of synchrony with other endocrine changes (Alm *et al.*, 1975; Cooper, 1979; Van Niekerk and Morgenthal, 1976).

15.3.2.4. Oxytocin

A further agent – and the one most commonly used today to induce parturition – is oxytocin. In the natural course of events oxytocin plays a central role in parturition and concentrations rise markedly especially during the second stage of labour, causing the rapid uterine myometrial contractions associated with birth. Initial research work used high levels of oxytocin (60–120 IU) in a single injection, but these were associated with cervical rupture and reduced foal viability, a likely consequence of the sudden onset of myometrial contractions, bypassing the natural more gradual build-up of events. More recently, low levels of oxytocin, <20 IU, have been used, administered intravenously over time (usually 3 h) either via multiple injections or infusion over time in physiologic saline and this results in parturition beginning within 20 min (Pashan and Allen, 1979b; Pashan, 1982). The exact amount of oxytocin required depends on the endocrine balance within the mare at administration (Pashan, 1980). It is known that, in many mammals, there is a synergistic relationship between oxytocin and prostaglandins and that such an association does occur in mares. Oxytocin administration near term, therefore, results in an immediate release of prostaglandins, equivalent to the natural release. Oestrogen can be used in combination with oxytocin as part of an induction regime and is reported to aid cervical dilation and hence ease the process of parturition (Hillman and Ganjam, 1979).

It cannot be emphasized enough that if induction of parturition is being considered, accurate records of the mare's date of service and expected delivery date are very important, along with close observation for signs of parturition and a means of determining fetal maturity. Inappropriate use of induction agents has disastrous consequences. Most of the danger is to the fetus rather that the dam, though the risk to both increases with the asynchrony between induction and the natural course of events. Induction of parturition in mares remains a risky business and must be used bearing this in mind.

15.4. Management of the Mare at Parturition

Whether the mare delivers naturally or is to be artificially induced, her management should be very similar. The main difference is that the time of delivery with induction will be known and, therefore, preparation can be better timed and organized. A mare may foal in a specifically built foaling unit or outside. Whichever system is chosen, and there are advantages in both, then the principles of the stages of labour and their management will be the same. Foaling mares outside is increasingly popular, as the risk of disease is lower and the system is much nearer the natural situation. Foaling outside is normally restricted to pony or cob-type hardy mares or multiparous mares foaling later on in the season. Early foalers, maiden or difficult mares and those of great value are normally foaled inside, allowing closer observation.

15.4.1. Foaling kit

In readiness for foaling, the following equipment should be organized:

- Veterinary telephone number – ideally they should also be warned beforehand;
- Mare halter and lead rope;
- Towels;
- Bucket;
- Soap or antibacterial wash;
- Cotton wool;
- Access to warm water;
- Obstetric lubricant;
- Obstetric ropes;
- Sharp knife or scalpel;
- Radiant-heat lamp;
- Antiseptic spray/navel dressing – e.g. 0.5% chlorohexidine;
- Feeding bottles for milk/colostrum;
- Gastrointestinal feeding tube;
- Access to colostrum, e.g. frozen colostrum.

15.4.2. First stage of labour

First-stage labour requires no special measures except close observation for the start of the second stage and

for the identification of any potential problems. Once first-stage labour has been diagnosed, the mare's tail should be bandaged up and the perineal area thoroughly washed. Some mares may loose milk prior to, or during, the first stage of delivery (Fig. 7.2). This milk is valuable colostrum, of which there is a finite amount. If a mare shows considerable milk loss prior to parturition, she should be milked lightly and the colostrum collected into a clean sterile container. As soon as the foal is born this can be bottle-fed to the foal, or tubed, if necessary, to ensure that valuable antibodies are received.

At this stage an episiotomy (cutting of the vulva and perineal area) should be performed, if required. Unfortunately, today many mares routinely undergo Caslick's operations (see Section 1.3.1; Fig. 1.9), due to poor perineal conformation. Such mares, along with those that are naturally small, will need an episiotomy to allow passage of the foal (Fig. 15.3).

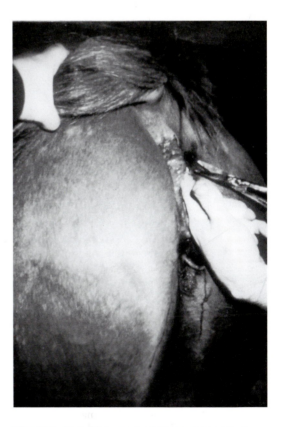

Fig. 15.3. An episiotomy should be performed by the beginning of stage 2 labour. This is routinely required in the case of mares having undergone a Caslick's operation.

During the first stages of labour the mare will seem restless and may repeatedly get up and lie down. She may well loose her appetite, sweat profusely and appear uneasy, glancing at her flanks and grimacing; she may also dig up her bedding.

This continual moving around is thought to help to position the foal within the birth canal. The mare may well show signs of discomfort followed by quiet. Thoroughbreds are notorious for this type of behaviour (Jeffcote and Rossdale, 1979). Whatever, her discomfort will increase with the frequency of contractions, culminating in the breaking of the waters (release of allantoic fluid) at the cervical star. Excessively prolonged first-stage labour may be a sign of problems, especially if the mare seems to be very distressed. It is very difficult to state how long first-stage labour should last and at what stage you should call for assistance, as some mares will naturally show several false starts in the days preceding birth. However, as a general rule, the assistance of a veterinary surgeon should be sort if the mare seems to be in prolonged discomfort, showing considerable agitation and profuse sweating, and before she is in any danger of becoming exhausted.

15.4.3. Second stage of labour

The management of the second stage of labour is more important and is marked by the breakage of the chorio-allantois at the cervical star and the resultant release of allantoic fluid. In 90% of cases the mare will now take up a recumbent position, the most efficient for straining (Fig. 15.4).

At this stage the amniotic sack should be evident as a white membrane bulging through the mare's vulva (Fig. 7.7) and a brief internal examination may be made to ensure that the foal is presented correctly (Fig. 7.8).

If the foal's forelegs, one leg slightly in advance of the other, and muzzle can be felt within the vagina, the mare should be left alone to deliver naturally (Figs 7.9 and 7.13). If there are problems, assistance should now be called. Care should be taken not to rupture the amnion during this process. Ideally it should be left to break naturally. When it does the colour of the amniotic fluid should be noted for evidence of meconium staining – dark brown/green coloration – which is indicative of fetal stress. If this is the case, then delivery of the foal should be speeded up by traction as soon as the foal's head appears (Figs 7.11 and 15.5).

If all is well, all attendants should now leave the box and allow the mare to foal unaided but observed from a discrete distance. Most mares lie down during second-stage labour, as this is the most efficient position for voluntary straining (see Section 7.3.2). Plenty of room is required to allow the mare to stretch out fully during straining.

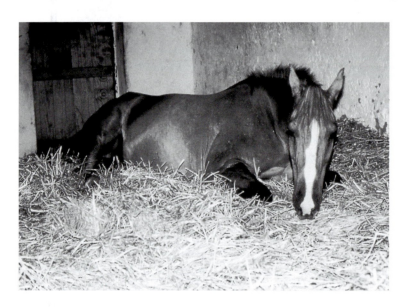

Fig. 15.4. During second-stage labour the mare invariably takes up a recumbent position, the most effective for straining. (Photograph courtesy of Steve Rufus.)

Fig. 15.5. Gentle traction may be used to aid in the final stages of delivery, especially if the mare is showing signs of exhaustion. (Photograph courtesy of Steve Rufus.)

The vast majority of foalings, up to 90%, require no outside interference from man (Vandeplassche, 1993). Occasionally, however, things do go wrong and it is as well to be prepared for, and have an understanding of, such eventualities. In such cases, prompt action can often save the life of both foal and mare. Foaling abnormalities are considered in detail at the end of this chapter.

Second-stage labour should last on average 15 min (range 5–30 min). Mares foaling for the first time tend to have a longer second-stage labour and so do mares that have had a hard first-stage labour.

15.4.4. Immediately post delivery

After delivery the foal should be left with its hind legs still within its mother and the umbilical cord intact (Fig. 7.12), the umbilical cord must be allowed to break naturally to minimize blood loss (Fig. 15.6; Rossdale, 1967). The foal lying with its legs within the vulva of the dam appears to have a tranquillizing effect on the mare. As a result, most mares are reluctant to get up immediately, though they may turn to lick the foal. A mare may remain recumbent for 30 min post partum. This should be encouraged, as it allows initial recovery of the tract and reduces the inspiration of air and, therefore, bacteria through the still relaxed vulva (Fig. 15.7). Such contamination of the vagina increases the chance of post-partum endometritis or acute metritis, and delays uterine involution and any return to

oestrus. Temporary Michel clips may be used after stage 3 to hold the dorsal (upper) vulval lips together and reduce air inspiration. These clips can easily be removed at the normal day 2–3 post-partum veterinary examination. They can then be replaced with a Caslick operation if required. This

Fig. 15.6. Very soon after parturition the healthy foal should be alert and sternally recumbent. The umbilical cord should be left to break naturally. This normally occurs as the foal makes its first attempts to rise to its feet. (Photograph courtesy of Steve Rufus.)

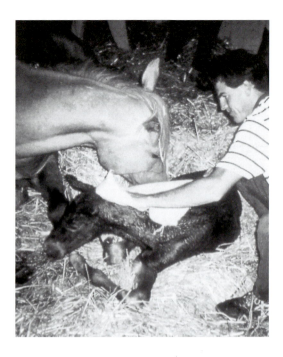

Fig. 15.7. The foal may be dried off immediately after delivery. At the same time the heart rate of the foal can be checked by placing a hand on the side of the foal's thorax. The mare should be encouraged to remain recumbent for a while after birth in order to reduce the inspiration of air and, therefore, bacteria in through the relaxed vulva. (Photograph courtesy of Steve Rufus.)

period, immediately post partum, is very important and marks the beginning of maternal foal bonding and recognition. Minimal interference is required in order to maximize the chances of a good mare–foal bond developing. Interference, especially if the mare is stressed or a maiden, may cause her to get up and paw the ground, appear disorientated and confused at some danger to the foal.

If the foal has shown signs of distress and is limp or weak at delivery, the amnion should be broken immediately and the foal's head lifted to aid breathing (Fig. 15.8). Occasionally, the umbilical cord does not break after birth, despite drying up and constriction at the foal's abdomen. In such cases, it may be broken by a sharp pull while placing the other hand on the foal's abdomen.

Immediately post delivery the foal can be dried off (Fig. 15.7) and the severed umbilical cord must be dressed with an antiseptic agent such as 0.5% chlorohexidine to prevent infection. Iodine was traditionally used as a navel dressing. This is effective in preventing infection, but may cause sloughing of skin cells causing sores and the possible reopening of the infection site. The milder, but just as effective, chlorohexidine is now advised. At this stage the foal's heart rate may also be checked by placing a hand on the thorax (Fig. 15.7). It may also be weighed; the birth weight of most normal foals is 10% of their expected mature weight. Details of the parameters required to be met by the

Fig. 15.8. The amnion may be removed from the foal's head and the nasal passages cleared to aid initial breathing. (Photograph courtesy of Steve Rufus.)

foal at set stages in the first few hours of life are given in Chapter 16 (this volume).

15.4.5. Third stage

During third-stage labour, the mare will appear restless again, similar to that during the first stage. The average duration of this third stage is 60 min but there is wide variation. Occasionally, the placenta may be expelled immediately after, or with, the foal and still attached to the umbilical cord. At the other extreme, it may take several hours. If the placenta is not expelled before the mare stands, it may be tied up to prevent the mare standing on it and ripping it out prematurely. The extra weight provided by tying up also encourages its expulsion (Figs 15.9 and 15.10). Third-stage labour in excess of 10 h is indicative of problems and will be discussed in the following section.

As soon as this third stage has been completed and the placenta expelled, it should be removed from the box and examined for completeness (Fig. 7.14). An effective method of detecting holes and, therefore, any missing fragments, is to tie off both uterine horn ends of the placenta and fill the

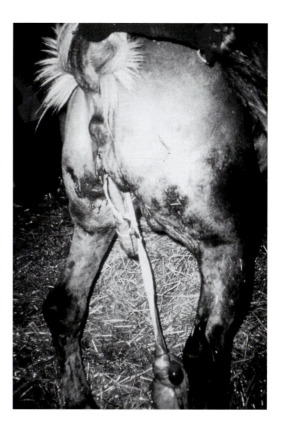

Fig. 15.10. Uterine contractions of the third stage of labour plus vasoconstriction of the placental blood vessels should result in expulsion of the placenta. Accidental tearing of the membranes can lead to partial placental retention, which must be treated immediately. (Photograph courtesy of Steve Rufus.)

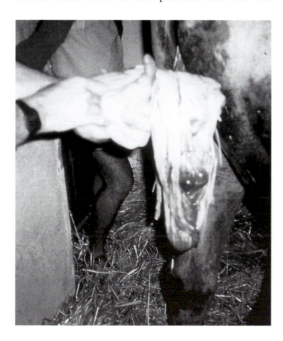

Fig. 15.9. The allantochorion (placenta) can be tied up to prevent the mare from standing on it and ripping it out prematurely or partial placental retention. (Photograph courtesy of Steve Rufus.)

placenta with water through the cervical star. Leakage of water indicates a break, which should be examined to ensure that no membranes are missing. If this is suspected, then a veterinary surgeon should be called to ensure their complete removal. Placental retention, of even just a fragment, can lead to acute metritis, septicaemia and eventual death if not treated as a matter of urgency (see Section 19.2.2.5.3.4). The temptation to pull the placenta to try and release it must be resisted, as this will invariably lead to rupturing of the placental membranes and the danger of fragments being retained. Only a very small fraction is required to set up a septicaemic reaction.

During this period, the bond between mare and foal starts to develop and can be irretrievably damaged by man's interference, however well

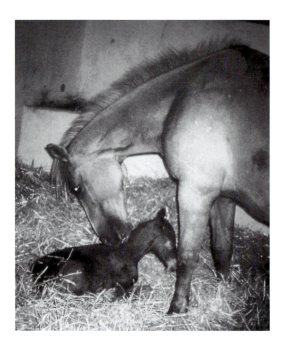

Fig. 15.11. During the third stage of labour the mare and foal should be left undisturbed to start the process of bonding. There is no need to drag the foal around to the head of the mare as she will invariably reach around herself and move towards the foal without assistance. (Photograph courtesy of Steve Rufus.)

intentioned (Fig. 15.11). Damage to newborn foals by their mothers is very rare, and if it does occur, it is usually a result of her being disturbed or flustered by unwanted human interference. Occasionally, mares will nibble or gently bite their foals to encourage them to move. Within reason this may be allowed, but, if the dam is too aggressive a muzzle may be used. Very occasionally, usually in maiden mares, real aggression towards the foal may be evident involving kicking and vicious attack. Such mares may be tranquillized for a short period of time. Tranquillizers may also be used on mares that will not stand to allow the foal to suckle. After a while, most mares will get used to the foal and interference will not be required. Use of the twitch should be avoided for the first 24 h post partum as this is thought to increase the risk of internal haemorrhage.

Foals that appear to be weak and unable to get adequate colostrum from the mare may be bottle-fed or tubed. This ensures that they receive sufficient antibodies as soon as possible.

Enemas used to be routinely administered to foals, using either medical paraffin or warm soapy water. This treatment is no longer deemed necessary unless the foal shows signs of meconium retention after 24 h. Foals should be given the opportunity to suckle and pass meconium naturally before the decision is made to use an enema. The enema technique causes the foal unnecessary stress and rough handling at a very early age when delicate adaptation to the extra-uterine environment is still occurring.

The mare should be given a feed about an hour after the birth. A light, easily digested and slightly laxative feed is best, plus fresh hay and water. Water should be given only under supervision initially, unless automatic water feeders are installed, to minimize the risk of the foal drowning.

15.5. Foaling Abnormalities

Foaling abnormalities or dystocia are relatively rare, especially in multiparous mares (mares having foaled before) with less than 4% experiencing problems (Vandeplassche, 1987). Dystocia can be classified as a condition or complication, associated with parturition that prevents a natural birth. It can be divided into fetal dystocia, resulting from fetal complications, or maternal dystocia, resulting from maternal problems. Dystocia often causes a delay in parturition, which may result in a reduction in the oxygen intake by the foal, due to partial placental breakdown or constriction of the umbilical blood supply. This may result in the birth of a weak foal requiring careful intensive post-natal care and/or more permanent problems, including brain damage and death in the more extreme cases. Delay in parturition due to dystocia may result in other complications further hampering progress. The uterus may close around the fetus as it dries out and the lubricating effect of the allantoic fluid diminishes. As a result, manipulation of the foal becomes increasingly difficult. Further drying out of the tract due to delay hinders passage through the birth canal and thus makes any successful manipulation even harder.

If dystocia does occur, there are three main actions that can be taken: manipulation and traction; Caesarean; and fetotomy. Initially the least drastic option of manipulation, or mutation, of the foal (manual correction of the fetal position) followed by traction, or pulling, of the foal should be considered. However, more drastic action may be required, such as a Caesarean (Vandeplassche *et al.*, 1979a) especially in cases where extensive manipulation would be needed. A Caesarean runs higher risk to both mare and foal and must be conducted under some form of anaesthetic. Mortality rates for foals born by Caesarean are approximately 10%. The mortality rates for mares delivering by Caesarean are lower than this, especially if she is operated upon before labour has progressed too far (Vandeplassche *et al.*, 1979a). A Caesarean involves either a ventral midline incision with the mare in a dorsal recumbent position (laid on her back; Juzawiak *et al.*, 1990), or flank incision if she is standing. Apart from the mortality risk, mares that have undergone Caesareans have the added risk of developing adhesions and so lowered fertility in the future (Stashak and Vandeplassche, 1993). Adhesions form between internal structures that have become damaged, often by exposure to air. If the adhered structures include parts of the reproductive tract, there is a risk of interference with their function. Infection is also another potential problem, though with the advances in modern medicine this is nowhere near the risk that it used to be (Vandeplassche *et al.*, 1979a). A final alternative, in response to severe cases of dystocia, is fetotomy (Vandeplassche, 1987). This is performed on foals that have died *in utero* and involves the dividing up of the foal's body within the mare. A Caesarean is a possibility in such cases but involves a higher mortality risk to the mare. Therefore, if the foal is already known to be dead, fetotomy is often the preferred option.

15.5.1. Fetal dystocia

Fetal dystocia is caused almost exclusively by malpresentations *in utero*, making normal unaided birth impossible. Dystocia may be caused by the foal being in one of the many positions, and the ease of correction and likely outcome depends on its position and the skill of the manipulator (Vandeplassche, 1993; Blanchard, 1995). Some of the more common positions and their treatment are given below.

15.5.1.1. Forward positions

One or more forelegs may be flexed or folded under the foal whose head is in the normal position within the birth canal. As discussed previously (see Section 7.3.2), the widest cross section presented as the foal passes through the pelvic cavity is the thorax area; therefore, in this position, it includes the flexed forelegs, making passage impossible (Fig. 15.12).

The head of the foal may be flexed back. In this position, the nose of the foal will not be felt in the birth canal. This again presents a much wider maximum cross section than the mare can deliver naturally (Fig. 15.13).

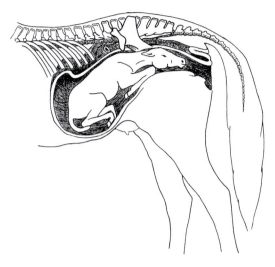

Fig. 15.12. Carpel flexion – flexion of one (unilateral) or both (bilateral) forelegs at delivery significantly increases the cross section across the foal's thorax and, therefore, makes the passage of the foal very difficult.

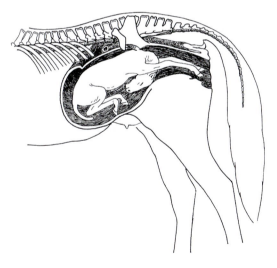

Fig. 15.13. If only the forefeet are presented in the birth canal the head may be flexed back, again presenting such a wide cross section of foal that it is very difficult for the mare to foal naturally.

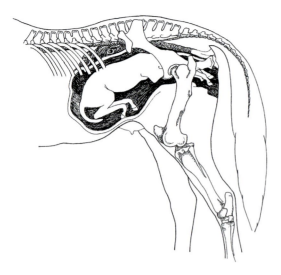

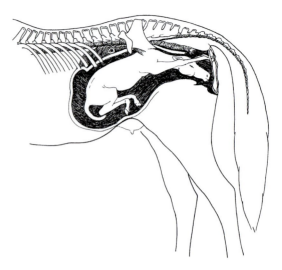

Fig. 15.14. A significant misalignment of the hoof of one leg and the fetlock of the other may be due to one elbow being flexed and becoming lodged at the pelvic brim.

Fig. 15.15. The forelegs may be positioned over the foal's head. Not only does this increase the cross-sectional diameter of the foal but also creates the risk of a rectal vaginal fissure.

The foal's legs may be more misaligned than the normal hoof-to-fetlock alignment. One leg may become caught on the pelvic brim at the point of the elbow, preventing the foal's delivery (Fig. 15.14).

The foal's forelegs may lie over its head and be lodged behind its ears. This presents a cross section too large for natural passage and runs the risk of rectal vaginal fissure if the hoof penetrates the roof of the vagina and into the rectum (Fig. 15.15).

All four of these positions can, in theory, be corrected reasonably easily by pushing the foal back into the uterus and manipulating it so it is presented in a normal position, followed by traction to aid the mare in its delivery. In practice, the strength of the uterine contractions can make this quite difficult. It must also be remembered that traction has to be applied in a curved manner to ease the foal along the birth canal as dictated by the pelvic anatomy.

Other more complicated positions are seen, which require veterinary assistance, and in the worst scenario, a Caesarean may be required, especially if straining by the mare has caused damage to the cervix, vagina or uterus. The foal may have both its head and neck turned back presenting just two forelegs (a more extreme version of that shown in Fig. 15.12), or the presentation of

the head and forelegs is correct but the hind legs are also being presented at the same time (Fig. 15.16), putting four legs in the birth canal. Four legs are also presented first in the crosswise position (Fig. 15.17).

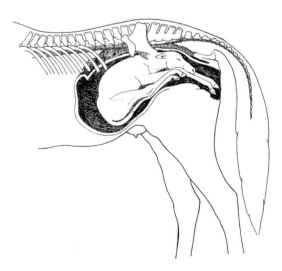

Fig. 15.16. If both forefeet and hind feet are felt within the birth canal, veterinary assistance should be called immediately, as this position can prove very difficult to correct, especially if the mare has progressed far into labour.

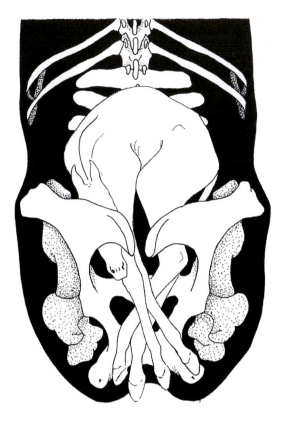

Fig. 15.17. A crosswise presentation also presents four feet first. This can, in theory, be corrected by pushing the hind legs back into the uterus.

Finally, the foal may be presented in a ventral position, in which case the arch of the foal's back bones does not allow expulsion in the required curved manner. This position must be corrected by rotation of the foal before delivery is possible (Fig. 15.18).

15.5.1.2. Backward presentation

Backward positions may be evident with the back legs presented first, a position that need not be manipulated *in utero* but must be carefully watched during labour. In such cases, there is a danger that the umbilical cord may become trapped between the abdomen of the foal and the pelvic brim, starving the foal of oxygen. There is also the danger that the foal may drown by inspiring allantoic fluid *in utero*. Foals in this position must, therefore, be pulled quickly and the amniotic sack removed from the muzzle immediately to allow extra-uterine breathing as soon as possible.

A more complicated backward position is indicated when no back legs are presented and only the tail can be felt. This position is known as a breech and veterinary assistance will be required to deliver such a foal (Fig. 15.19).

Other positions, variations of the above, may be found; those discussed are those most commonly observed.

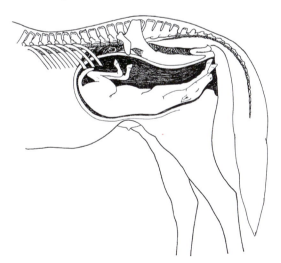

Fig. 15.18. Ventral position of the foal. The foal must be rotated into a normal position before delivery is possible.

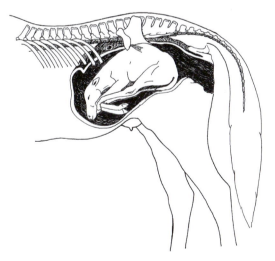

Fig. 15.19. If no legs can be felt within the birth canal and only the tail and rump is presented, this is a true breech position and will require veterinary assistance.

15.5.2. Maternal dystocia

Maternal dystocia is the failure of natural delivery as a result of a maternal complication. These complications may be found in isolation or in combination (Blanchard, 1995). Some of the more common ones will be discussed.

Placenta previa is evident as the protrusion of the intact red allantochorionic membrane through the mare's vulva, due to its failure to rupture at the cervical star. This may be due to extra thick placental membranes, in which case manual breakage of the protruding allantochorion will allow parturition to progress normally. However, failure of the allantochorion to rupture at the cervical star is more likely to be due to rupture elsewhere, indicating premature separation of the placenta from the maternal epithelium a much more serious condition. As this separation is one of the triggers for the foal's first breath there is a danger of suffocation and so the allantochorion must be cut immediately and the mare helped to foal. The condition may also be indicative of placentitis (placental infection), which, if long-standing, is likely to have led to fetal brain damage or stress due to excessive nutrition in late pregnancy (Vandeplassche, 1980, 1993; Asbury and Le Blanc, 1993).

Small pelvic openings may also cause problems. Restriction in this area may be the result of an accident or fracture or caused by malnutrition during the mare's early developmental stages of life. In such cases a Caesarean is often the only course of action that can be taken. It is questionable whether such mares should be used for breeding.

Uterine problems may also cause dystocia. Uterine inertia is a more common condition, and may often accompany fetal dystocia. In such cases, the uterine myometrium becomes exhausted due to excessive straining, especially during the second stage of labour. The condition can be accentuated by age or previous uterine infections and may result in uterine rupture or placental retention. Uterine inertia may not occur to the same extent across all the myometrium. In such cases, tight rings of muscle contraction may occur. As a result, there is a danger that the foal may be crushed or strangled. In most cases of uterine inertia, traction to aid the mare, along with an oxytocin injection to encourage myometrial contraction, is all that is required. Occasionally a Caesarean may be advised (Vandeplassche, 1980, 1993; Lopate et al., 2003). Rupture of the uterine wall occasionally may be seen and is most common due to: mutation or traction; excessive straining especially during second-stage labour; or significant intrauterine movement (Wheat and Meager, 1972; Patel and Lofstedt, 1986; Honnas et al., 1988). Many of these uterine problems are exacerbated in multiparous older mares.

15.5.3. Abnormal conditions

Abnormal conditions associated with parturition do not normally directly prevent normal delivery but may indirectly cause concern (Lofstedt, 1993; Blanchard, 1995). These conditions include problems that become evident prior to labour, such as ruptured prepubic tendon, uterine prolapse and uterine torsion. The prepubic tendon is the tendon sheet attached to the abdominal muscles and so supports the entire abdominal contents including the uterus. Its rupture results in loss of support for the abdominal organs and death is the normal outcome (Fig. 15.20; Jackson, 1982).

Uterine prolapse results in the inversion of the vagina, and often part of the uterus through the vulva, due to incompetent uterine support, possibly associated with old age. In such cases, the uterus can be replaced manually after the administration of an anaesthetic or relaxant. This reduces the mare's straining and allows the uterus to be replaced and the vulva to be sutured to prevent a reoccurrence (Brewer and Kliest, 1963; Vandeplassche, 1975; Breen and Bowman, 1994).

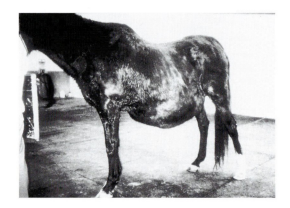

Fig. 15.20. Prepubic tendon rupture may occur in late pregnancy, resulting in the loss of all abdominal support. The prognosis is not good, death normally results. (Photograph courtesy of Julie Baumber.)

One of the more rare conditions is uterine torsion, where the uterus has become twisted, often as the result of a fall or weakened broad ligaments (Pascoe *et al.*, 1981). The twist, which is often towards the cervix end of the uterus, prevents any delivery through the birth canal. The prognosis for such cases is normally 50% chance of survival but some success has been reported by enlisting the help of gravity and rolling the mare over rapidly, or by using manual surgical correction via a flank incision or via the rectum (Bowen *et al.*,1976; Vasey, 1993; Lopate *et al.*, 2003). Often a Caesarean is required (Bowen *et al.*, 1976; Vandeplassche, 1980; Pascoe *et al.*, 1981).

Hydrops amnion and hydrops allantois are occasionally observed problems in the mare near parturition. Both cases involve excessive fluid accumulation in either the amnion or allantois and may occur separately or together. The reason for the condition is unclear though it normally develops from 7 months of pregnancy onwards; treatment is via abortion or induction of parturition if the mare is approaching parturition. Prognosis for the mare is good but foals are commonly significantly compromised and/or abnormal (Vandeplassche *et al.*, 1976).

Some conditions do occur during parturition itself – for example, intestinal rupture which is associated with the feeding of large, infrequent meals during late pregnancy (Littlejohn and Ritchie, 1975; Fisher and Phillips, 1986; Lopate *et al.*, 2003), and maybe the result of a weakness arising from previous damage.

Finally, some conditions may not become evident until after parturition but are a result of the forces of delivery. For example, haemorrhage, both internal, often the cause of sudden death post partum, and the less drastic external, may be a result of labour but not detected until after delivery (Rooney, 1964; Pascoe, 1979b; Lopate *et al.*, 2003). Rectal vaginal fistulas or perineal lacerations may be evident, resulting from the foal's hoof puncturing the roof of the vagina and passing through the floor of the rectum, opening up a cloaca as the foal is delivered (Fig. 15.21). Short fissures may be sutured, but longer ones cause significant problems, largely due to inaccessibility (Lopate *et al.*, 2003).

Delayed involution may be apparent after parturition. Under normal conditions, within 2 h the uterus should have shrunk to one-half of its fully expanded state and continued low-grade myometrial contraction will have expelled most of the fluid and

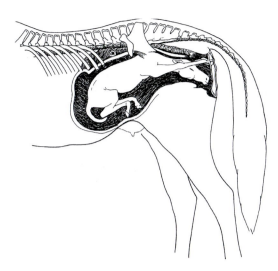

Fig. 15.21. During foaling the foal's feet may pass through the vagina wall and up into the rectum, forming a rectal vaginal fissure.

bacteria remaining after parturition. By day 7 post partum, it should be only two to three times the size evident in a barren mare, returning fully to its normal size by day 30 post partum. Recovery of the uterine endometrium takes approximately 14 days (see Section 16.5.2). Evidence of a dilated cervix or enlarged uterus at day 30 indicates problems, possibly associated with retention of placental membranes or dystocia. Uterine involution can be encouraged in such mares by daily infusion of oxytocin and antibiotics, plus gentle exercise (Blanchard and Varner, 1993a). Exogenous oxytocin may be used not only to provide the extra impetus for placental expulsion but also the following uterine involution (Threlfall, 1993; Haffner *et al.*, 1998).

Finally, hypocalcaemia, though not common in mares, if evident can be quite successfully treated by administration of calcium borogluconate post partum. The incidence of hypocalceamia is higher in mares suffering from dystocia and after Caesarean section.

15.5.3.1. *Retained placenta*

Retention of the placental membranes may cause serious complications and is one of the more common complications post partum (Vandeplassche *et al.*, 1971; Held, 1987). Normally, the placenta reduces in size gradually, as its blood flow reduces;

it, therefore, shrinks and separates away from the endometrium and is delivered by the final contractions of the uterus in third-stage labour. Retention of the placental membranes will result in infection that, if left untreated, will prove fatal. Retention of the afterbirth for more than 10 h indicates problems. Retention is normally in the previously non-gravid uterine horn, especially after dystocia or Caesarean section. It may be due to a hormonal imbalance resulting in inadequate oxytocin and, therefore, reduced muscle activity or due to selenium deficiency or Ca:P imbalance (Lopate *et al.*, 2003). Treatment can be in one of the several ways, including manual removal and/or oxytocin treatment. Manual removal must be carried out with great care and, if attempted, it must be certain that the entire placenta is removed. After initial attempts, an antibiotic pessary can be inserted and the rest of the membranes removed a few days later. It is imperative that the entire placenta is removed as soon as possible. If the whole of the placenta has been retained, then iodine solution can be pumped in through the cervical star to fill up the allantoic sack. It may take 9–11 l for an effect to be obtained. After about 5 min the mare will be seen to strain against the filled placenta and so aid expulsion. This method can be used in conjunction with oxytocin treatment, though the pressure on the uterus itself induces an elevation in endogenous oxytocin (Threlfall, 1993). Other workers have had some success by injecting collagenase into the placenta via the umbilical cord in an attempt to speed up separation from the uterine endometrium (Haffner *et al.*, 1998).

15.6. Conclusion

Management of the foaling mare is of utmost importance in ensuring the birth of a healthy foal. Inappropriate management, or the failure to call in professional help when required, can have disastrous consequences. Induction may be used as an aid to the management of the foaling mare, but must be used with great caution, as inappropriate timing can have fatal consequences.

16 Management of the Lactating Mare and Young Foal

16.1. Introduction

Correct management of the lactating mare and young foal is crucial for long-term foal survival and the rapid return of the mare to oestrus with successful re-covering. As far as the foal is concerned, the adaptation from intra- to extra-uterine environment is the most crucial stage. The subject of neonatal complications and disease is vast and beyond the scope of this book. This chapter will, therefore, concentrate on the normal foal in order that abnormal foals can be identified. More specific detailed texts should be consulted on the problems that may be encountered in the neonate (Koterba, 1990; Madigan, 1990; Adams, 1993a,b; McClure, 1993; Reef, 1993; Roberts, 1993; Seltzer *et al.*, 1993; Traub-Dargatz, 1993a,b; Vaala, 1993; Welsch, 1993; Knottenbelt *et al.*, 2004).

16.2. Foal Adaptive Period

Immediately post partum the foal has to undergo substantial anatomical, functional and biochemical adaptive changes in order to survive in the extra-uterine environment. In the normal foal, adaptive changes can be identified until puberty and even up until the achievement of mature size. However, in considering the true adaptive period for survival outside the uterus, the first 4 days of life are the most crucial. It is within this period of time that the majority of adaptive problems can be identified and, hopefully, rectified. If the foal satisfies all the normal criteria at this age, it has a very good chance of survival.

In a normal birth, the foal is born on its side, lying with its hocks still within its mother and the umbilical cord intact (Fig. 15.8). The newborn foal may be assessed within 3 min of birth, on its appearance, pulse (rate), grimace (response to stimuli), activity (muscle tone) and respiratory rate and scored using the (APGAR) scoring system as normal, moderately depressed or markedly depressed

(Table 16.1; Madigan, 1990; Le Blanc, 1997). The long-term prognosis for the foal is dependent upon this classification.

16.2.1. Anatomical adaptation

Anatomical adaptations can be considered as the milestones that the foal should achieve within set periods of time in order to survive. These parameters can quite easily be observed by the foaling attendant and used to indicate foal viability and identify problems.

Passage through the birth canal compresses the foal's thorax and helps to expel excess fluid from airways allowing the normal foal to breath within 30 s of final delivery. It may take a few sharp intakes of breath as its muzzle first reaches the air during passage through the birth canal, but a rhythm is normally established within 1 min of final delivery. If breathing does not occur within 3 min, there are serious consequences for the foal (Rose, 1988). This initial breathing rhythm is normally a steady 60–70 breaths min^{-1} with a tidal volume (volume of air inspired per breath) of 520 ml, resulting in a minute volume (volume of air inspired min^{-1}) of 35 l (Koterba, 1990). This initiation of breathing results in a significant increase in blood oxygen levels. During this first minute of life, the foal's heart rate should be in the order of 40–80 beats min^{-1}. This can be measured by placing a hand on the left side of the chest near the heart. The foal's body temperature should be 37.5–38.5°C (Koterba, 1990; Traub-Dargatz, 1993a; Vaala, 1993). The foal's mucous membranes should appear pink and moist and have a refill time of 1–2 s within the first 2 h. Within a few hours the respiratory rate declines to 35 breaths min^{-1} with a tidal volume of 550 ml and a minute volume of 20 l and may stay at this relatively high rate compared to adults (8–12 min^{-1}) for a few days (Rossdale and Ricketts, 1980). The heart rate rapidly increases to 120–140 beats min^{-1}, when the foal moves in its attempts to stand, plateauing

Table 16.1. The APGAR scoring system to aid in the classification of newborn foals, score 7–8 normal, score 4–6 moderate depression and score 0–3 markedly depressed. (From Madigan, 1990.)

Parameter	Score		
	0	1	2
Appearance(A)	Recumband and lifeless	Some attempts to move	Significant attempts to sit up (sternal recumbancy)
Pulse rate (P)	Absent	$<60\,min^{-1}$	$60\,min^{-1}$
Respiration (R)	Absent	Slow, irregular	$60\,min^{-1}$ regular
Activity (Muscle tone) (A)	Limp	Some flexion of extremities	Sternal position
Grimace (responce to nasal stimuli) (G)	No response to stimulation	Grimace, slight rejection on stimulation	Cough or sneeze on stimulation

at 80–100 within the first week (Lombard *et al.*, 1984; Lombard, 1990).

The significant increase in blood oxygen concentrations as a result of breathing activates the first reflexes and muscle movements. Within 5 min of birth, the normal foal should be in a sternal recumbency position (Fig. 15.6). It will respond to pain and begin to show evidence of the reflexes associated with rising to its feet, in the form of raising its head, extending its forelimbs, blinking and possibly a whinny (Madigan, 1990; Traub-Dargatz, 1993a). It will also demonstrate the suckling reflex if offered a finger or bottle.

The next major event is the breaking of the umbilical cord, on average 5–9 min after birth. It is considered important that the cord is left to break naturally, as premature breakage is reported to result in the loss of up to 1.5 l of blood by preventing drainage from the placenta (Rossdale and Mahafrey, 1958), though work done by Doarn *et al.* (1985, 1987) failed to support such a blood loss. Breakage naturally occurs at a constriction about 3 cm from the abdomen of the foal. The umbilical artery and vein have thinner walls in this area and so collapse and constrict naturally at this point, as the pressure of blood circulating within these vessels declines, allowing a clean break, with minimal blood loss and reducing the chance of infection. The mare may be recumbent post partum for up to 40 min and therefore, the cord is usually broken by the movements of the foal in its first attempts to rise to its feet, the increased tension resulting in the cord breaking at the constriction, the weakest point. Once the cord has severed, the navel must be dressed. Traditionally, iodine-based preparations have been used. More recent work suggests that treating with iodine causes a sloughing of skin cells and a risk of reopening the navel to infection (O'Grady, 1995). It is suggested that 0.5% chlorohexidine is a better alternative.

Around the time of umbilical cord breakage, the foal makes concerted efforts to raise itself to its feet (Fig. 16.1). The exertion of standing causes the heart rate to increase to up to 150 beats min^{-1}. Heart rate continues to fluctuate with activity around a normal resting heart rate of 80–100 beats min^{-1} evident in the first few days. When attempting to stand, the series of movements is the same as in the mature horse; stretching forward of the head and neck, extending the forelegs; flexing the hind legs and raising the front end first off the ground, followed by the hind quarters. Many initial, unsuccessful attempts are made, which form part of the process of developing reflexes and muscle coordination and control. At this stage, the foal is at risk of damage from projecting objects such as buckets, hay racks, automatic water feeders, etc. Successful standing normally occurs in ponies within 35 min post partum but may take up to half an hour longer in Thoroughbred foals (Fig. 16.2). Failure to stand within 2 h is indicative of a problem and veterinary assistance should be sought (Rossdale, 1967; Jeffcote, 1972; Madigan, 1990). This remarkable ability to stand and walk so quickly is the result of the evolutionary development evident in plain dwelling animals, such as the horse, to enable them to flee from potential predators as soon as possible after birth.

Though the suckling reflex is developed within 5 min post partum, successful suckling can obviously not occur until after standing and locating the udder. The actual reflexes involved in suckling are elicited by contact with soft warm surfaces; hence, foals are seen to suckle and nuzzle their dam's flanks while searching for the udder. As soon as the foal stands, it demonstrates directional movement towards the udder, located by the dark and warmth (Fig. 16.2). This process of locating the mare's udder can be easily disrupted by human

Fig. 16.1. Soon after the end of second-stage labour the foal makes concerted efforts to rise to its feet, at this time the umbilical cord breaks. (Photograph courtesy of Stephen Rufus.)

Fig. 16.2. Successful standing in the pony foal takes on average 35 min compared to up to 1 h in the Thoroughbred. The foal then searches for the udder; the suckling reflex is already present and is elicited by contact with soft warm dark areas. (Photograph courtesy of Stephen Rufus.)

interference. It is very tempting to try and help a foal and frustrating to watch it suckling at the hock or chest and seemingly unable to locate the teat. However, it is much better to resist the temptation to interfere. The mare will normally assist the foal by gently nudging it and moving her hind leg away from her body to allow the foal easier access. Occasionally, primiparous (maiden) mares may need to be held to allow the foal to reach the udder, appearing ticklish and initially objecting to the foal's attentions. However, she will soon settle down and should be left alone. In ponies, successful suckling is achieved normally within 65 min; Thoroughbreds are again slower, taking 30 min or so longer (Fig. 16.3; Rossdale, 1967). At suckling, a real affinity develops between mother and foal, which develops into a very strong bond. Human interference may well disrupt this bonding process (Chavatte, 1991). The foal, throughout the first few days of life, will suckle at 15–30 min intervals; later on the interval period lengthens. If during the first few days of life a foal is not seen to suckle for 3 h or so, problems should be suspected.

During the first 12 h the foal should be seen to pass meconium, its first bowel movement. It may well be passed earlier than this and is sometimes seen within a few minutes of the first feed. Meconium consists of bowel glandular secretions collected during the foal's inter-uterine life, along with digested amniotic fluid and cell debris, which are passed through the foal's digestive tract *in utero*. Meconium is stored in the colon, caecum and rectum ready for expulsion after birth. Premature expulsion may occur under stressful conditions during, or immediately prior to, delivery. Meconium staining of the amniotic fluid or the perineum area of the foal is, therefore, indicative of fetal or foal stress. Meconium should be brown to greenish brown in colour and is usually all expelled within the first 2 days. Meconium is followed by the characteristically coloured yellow milk dung, which indicates correct gut function (Rossdale, 1967). The routine use of enemas is advocated, by some, 12–18 h post partum (Madigan, 1990). However, their repeated use can irritate the mucosal lining of the gut. Routine enemas are becoming less popular as the adverse effect of such stresses on the newborn foal is increasingly understood. Colt foals should urinate for the first time within 5–6 h, whereas filly foals urinate later, on average at 10–11 h (Jeffcote, 1972; Roberts, 1975). Regular urination of small volumes of near colourless urine should be observed in the normal foal (Rossdale, 1967; Traub-Dargatz, 1993a).

16.2.2. Functional adaptation

This section will consider how the functions of the pulmonary, cardiovascular, temperature control, immune and renal systems adapt to accommodate the change from the intra- to extra-uterine environment. The change to their functions is reflected in the observed anatomical changes previously discussed.

16.2.2.1. Pulmonary ventilation

Successful gaseous transfer within the lungs across the air–blood interface depends upon their functional and structural maturity. One of the major events is the laying down of surfactant. Surfactant, a complex lipoprotein, provides the alveoli with a surface film, so reducing the surface tension and increasing the efficiency of gaseous exchange and reducing lung collapse (Kullander *et al.*, 1975). Surfactant maturation occurs in the last third of pregnancy, but particularly around day 300 onwards and may not be complete until after delivery (Pattle *et al.*, 1975). Maximum respiratory efficiency is not possible until superfactant development has been completed, and this creates a problem with premature foals.

The foal takes its first breath while *in utero* as practice for post-partum functioning of the muscles involved in respiration. Fluid collected within the lungs during pregnancy is expelled by compression of the thorax during delivery and by evaporation and spluttering during early breaths. The first proper extra-uterine breaths and the subsequent breathing rhythm are stimulated by low oxygen (anoxia) and

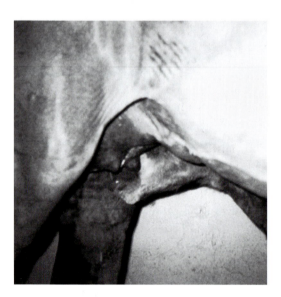

Fig. 16.3. In ponies suckling normally occurs within 65 min, whereas in Thoroughbreds it normally takes 1.5 h.

high carbon dioxide plasma concentrations. Cold shock from the atmosphere and tactile stimulus, such as rubbing and mare's licking, all encourage initial breathing. The first breath should occur within 30 s of the foal's hips appearing through the birth canal (Dawes et al., 1972; Gillespie, 1975; Stewart et al., 1984; Ousey et al., 1991). After birth, as a breathing rhythm is established, the alveoli continue to expand and develop, induced by lung expansion and stretching of the bronchi (Vaala, 1993). At birth, the efficiency of gaseous exchange is low – hence the rapid breathing rate. As the bronchi are stretched and alveolar development continues, so increasing in surface area/air ratio, more efficient gaseous transfer is achieved and the breathing rate declines (Gillespie, 1975). Additionally, calcification of the initially compliant (soft) chest wall makes breathing again more effective with time (Koterba and Kosch, 1987).

16.2.2.2. Cardiac and circulatory systems

In utero the placenta acts as the 'lungs' in being the major site of oxygen and carbon dioxide exchange, as well as nutrient uptake. In order that the placenta is supplied, blood must pass from the pulmonary artery via the ductus arteriosus to the aorta, so bypassing the pulmonary system (the lungs) and passing directly to the placenta (Fig. 16.4). Only a small supply of blood to the pulmonary system is required, just enough for pulmonary growth and development. Additionally, the foramen ovule allows blood to pass from the right atrium directly to the left atrium and hence immediately around the body via the aorta, rather than to the right ventricle. Blood enters the placenta via the two umbilical arteries and leaves via the single umbilical vein, to pass to the liver and then, back to the right side of the heart. The bypassing of the lungs is aided by the relatively high pulmonary vascular resistance compared to the systemic resistance. Both ventricles work in parallel, with the right ventricle dominating in size and output (MacDonald et al., 1988; Vaala, 1993).

Immediately post partum, the circulatory system of the foal must change dramatically to redirect blood through the pulmonary system to the lungs, and away from the umbilical system to the placenta (Fig. 16.5). The trigger for this change is unclear, but a decrease in pulmonary resistance plays a role. As the foal takes its first few breaths the collapsed lungs inflate, stretching the alveoli and rapidly reducing pulmonary resistance, resulting in increasing blood perfusion of the lungs (Kullander et al., 1975). As pulmonary resistance declines, blood is drawn up directly through the pulmonary artery to the lungs and not across the ductus arteriosus to the aorta. As more blood is drawn away from the right-hand side of the heart and more blood enters the left side of the heart from the pulmonary vein, blood pressure in the left-hand side of the heart becomes greater than that in the right, as a result of this differentiation in blood pressure the foramen ovule closes. Some blood leaving the left-hand side of the heart to the aorta may still continue to pass through the ductus arteriosus to the pulmonary system to recirculate through the lungs and vice versa. This continues until the closure of the ductus arteriosus at 2–4 days of age (Rossdale, 1967; Button, 1987; Lombard, 1990). The trigger for closure of the ductus arteriosus is unclear but is thought to be associated with increasing plasma oxygen concentrations and decreasing tissue concentrations of prostaglandins (Vaala, 1993). Until its complete closure, it may reopen in response to stress or hypoxemia (Cottrill et al., 1987). Delayed closure is often associated with Caesarean section births or induced parturition as final preparation of the ductus arteriosus has not been allowed to occur (Machida et al., 1998). In the newborn foal blood should now be pumped from the right side of the heart via the pulmonary artery to the lungs for oxygenation and back via the pulmonary vein to the heart for circulation around the body (Figs 16.4 and 16.5).

Many newborn foals initially suffer from arrhythmia, irregular heartbeat, but this soon settles down naturally (Yamamoto et al., 1992). Many foals show signs of asphyxia during the second stage of labour evident as a blue tongue and mucous membranes of the eyes caused by a reduction in blood flow and, therefore, oxygen to the head. Such constriction of the head, neck and chest during passage through the pelvis is of no long-term significance providing the foal continues to be delivered normally and parturition is not delayed. After birth, the mucous membranes may remain blue/grey in colour for a short while, but should be the normal pink colour within 2 h.

At birth, the red blood cell count is elevated compared ($9–13 \times 10^{12}$ l^{-1}) to that later on in life (7.5–10.5×10^{12} l^{-1}), though haemoglobin levels are similar to that seen in adults. This is unusual as in most mammals haemoglobin levels are elevated in the newborn (Knottenbelt et al., 2004). Elevated red

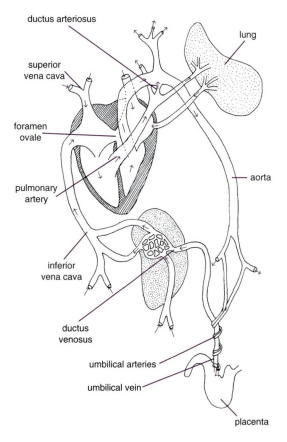

Fig. 16.4. The fetal circulatory system *in utero*. The blood passes through two openings, the foramen ovule between the left- and right-hand side of the heart and the ductus arteriosus, hence by passing the lungs.

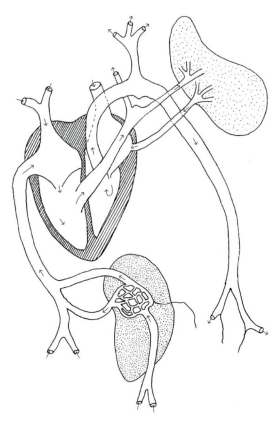

Fig. 16.5. The circulatory system of a normal foal post partum. The foramen ovule and the ductus arteriosus close within a few days of birth.

blood cell counts are thought to be due to fetal stress during birth, as levels are further elevated in foals born with difficulty. Within 2 h of birth, red blood cell counts decline and white blood cell counts rise to normal levels (Table 16.2; Chavatte *et al.*, 1991).

16.2.2.3. Temperature

The foal is born with a well-developed temperature-control mechanism, unlike many other mammals, especially primates which cannot effectively control their body temperature for several weeks after birth. At birth, the foal can maintain a steady body temperature of 37–37.5°C (100°F; Rossdale, 1968), which increases to 38–38.5°C within 1 h despite a cold environment. This is due to the high metabolic

rate of newborn foals (200 W m⁻²), which is three times that of a 2-day-old foal (Ousey, 1997). A body temperature below 37°C or above 40°C is a cause for concern. The exact mechanism by which it maintains this steady body temperature is unclear but is likely to involve the ability to shiver, which is first evident in newborn foals during the first 3 h post partum. In addition, heat produced by muscular activity and the strain of the foal's first movements is likely to contribute to maintaining body temperature, along with the foal's insulating layers of fat and its hair coat. Unlike the human baby, the foal does not have brown adipose tissue. Its ability to shiver earlier in life negates the requirement for brown fat, heat-producing tissue. The presence of brown fat is associated with neonates unable to shiver and those that have less fine control over their body temperature (Ousey *et al.*, 1991). Hypothermia can occur in new-

Table 16.2. The major haematological and biochemical parameters for foals from parturition to 7 days of age. (From Irvine, 1984b; Stewart *et al.*, 1984; Vivette *et al.*, 1990; Knottenbelt *et al.*, 2004.)

	Abbreviation	Units	Birth	24 h	7 days+
Haematology					
Haemoglobin	Hb	g l^{-1}	120–180	130–155	115–175
Haematocrit	PCV	l l^{-1}	0.40–0.52	0.34–0.46	31–40
Erythrocytes	RBC	×10^{12} l^{-1}	9.0–13.0	8.0–11.0	7.5–10.5
Leukocytes	WBC	×10^{9} l^{-1}	5.5–11.5	–	7.0–12.0
Lymphocytes	L	×10^{9} l^{-1}	–	1.8–3.0	2.0–4.0
Metabolites					
Glucose	Gluc	mg 100 ml^{-1}	50–70	100–110	100–110
Lactate	Lact	mg l^{-1}	–	7–14	–
Fibrinogen	Fibrin	g l^{-1}	<2	2.0–3.0	1.6–2.8
Total protein	TP	g l^{-1}	45–47	–	60–65
Electrolytes					
Calcium	Ca	mmol l^{-1}	–	2.5–4.0	–
Phosphates	P	mmol l^{-1}	–	2.2–5.2	–
Iron	Fe	mmol l^{-1}	72–88	18–63	18–54
Blood gases					
Bicarbonate		mmol l^{-1}	23	27	24
Carbon dioxide	pCO$_2$	mmol l^{-1}	–	37–50	–
Oxygen	pO$_2$	mmol l^{-1}	77	75–98	75–106
pH			7.36	7.39	7.43
Minerals					
Copper	Cu	μmol l^{-1}	–	9–12	–
Selenium	Se	μmol l^{-1}	–	1.2–1.6	–
Hormones					
Cortisol		ng ml^{-1}	120–140	60	30
Thyroxine	T$_4$	nmol l^{-1}	6–10	–	8–20

born foals, its onset may be rapid and can result from infection, as well as dystocia and a cold environment. Hypothermia may cause hypoxemia and acidosis, causing an attempted reversion to fetal circulating patterns and altered gastrointestinal function.

16.2.2.4. Immune status

As discussed in Section 5.4.1.4, the equine placenta is epitheliochorial and, as such, presents a considerable barrier to the passage of blood components from mother to fetus *in utero*, especially those of large molecular size, such as immunoglobulins (antibodies). In the foal, therefore, the attainment of immunoglobulins *in utero* is limited and so colostrum is vitally important for achieving adequate immunity for survival in the extra-uterine environment. At birth, the foal is plunged from

sterile conditions into an environment of varying immunological challenge. The foal's system is perfectly capable of meeting this challenge by producing its own antibodies over time, but is born immunologically naive (without antibodies) apart from a small concentration of immunoglobulin M (IgM) and so has no 'safety net' to protect it until it has produced enough antibodies to protect itself. This 'safety net' is provided by the immunoglobulins in colostrum. Equine immunoglobulins can be subdivided into IgG, IgM and IgA. The most predominant in colostrum is IgG and it is these that are most evident in the circulation of the young foal (McGuire and Crawford, 1973). Adult levels of immunoglobulins are not immediately evident in the foal; these are reached only after the foal starts to actively produce its own immunoglobulins. For the first 24 h post partum, enterocyte cells lining the foal's small intestine are able to absorb by pino-cytosis large protein molecules such as immu-noglobulins (Jeffcote, 1987). The ability to absorb whole proteins is seemingly enhanced by other components of colostrum and controlled in part by cortisol, though the exact mechanisms are unclear. Over time, the enterocytes are replaced by cells incapable of absorbing proteins. It is, therefore, essential that newborn foals receive colostrum, 500 ml at least, within the first 24 h of life, and preferably within the first 12 h, when absorption is most efficient (Pearson et al., 1984; Jeffcote, 1987). This ensures that the foal obtains maximum pro-tection from infection via maternal antibodies. Several tests to ascertain the immunological (IgG) status of the foal are available (Bertone and Jones, 1988; Le Blanc, 1990; Madigan, 1990). The amount of colostrum depends on the size of foal and concentrate of immunoglobulins in the colos-trum. It is generally agreed that foal IgG serum concentrations of 4–8 g l^{-1} are appropriate. Colo-strum may also be tested and it has been suggested that a specific gravity of >1060 is indicative of an IgG concentration of >30 g l^{-1} which, in a 50 kg foal with approximately 5 l blood volume, will, if the foal ingests 800 ml, result in a foal serum concentration of >5 g l^{-1}. This is the minimum requirement and so ideally the foal should ingest more to bring its IgG levels up to 10 g l^{-1} by 24 h (Le Blanc et al., 1986; Stoneham, 1991). If the colostrum is of poor quality then the foal will need to ingest more, and vice versa if the colostrum has a higher IgG concentration.

The foal's own immune system does start to function to a very limited extent during pregnancy; hence, a small concentration of IgM is evident at birth, but it does not reach maximum capacity until 3–4 months of age (Cullinane et al., 2001). Colostrum, therefore, provides protection in the interim until the foal's own immune system is fully functioning (Rouse and Ingram, 1970). If the mare is immunized during late pregnancy, then the anti-bodies raised pass to her colostrum and are avail-able to the foal, providing it with essential temporary protection. For this reason immuniza-tion and introduction to the foaling environment is advised 4–6 weeks prior to expected delivery.

16.2.2.5. Renal function

Newborn foals will initially urinate frequently (around 150 ml^{-1} kg^{-1} day^{-1}), producing urine that is more dilute than adult horses (Brewer et al., 1991). The dilute nature of the foal's urine is indicative of the fact that the fetal kidney is not mature at birth. This reduced ability to concentrate urine can be of concern in dehydrated foals and has implications on the use of renally excreted drugs, including antibiotics (Holdstock et al., 1998). This is another reason to be cautious about the prophy-lactic use of antibiotics in newborn foals.

16.2.3. Biochemical adaptation

The foal's metabolic system at birth undergoes dra-matic alterations from a dependent to an independ-ent status (Ousey et al., 1991). While in utero, it is dependent entirely upon the maternal system via the placenta; post partum this dependency is removed and replaced by reliance upon the pulmo-nary and gastrointestinal systems.

At birth, the foal goes through a transitional period after the severing of the maternal connection and before suckling. This period of time is one of consider-able stress and exertion, for which energy is required, provided by liver glycogen stores laid down during the later stages of gestation; the equine fetus only stores limited glycogen within the brain. Mobilization of glycogen, i.e. its conversion to glucose, is via the process of glucogenesis, one of the major enzymes in this pathway is glucose-6-phosphate which the liver only produces after birth. Hence, glycogen reserves can only be mobilized post partum. Full glycogenic ability is not reached until 1 month post partum (Ousey et al., 1991). Glucose levels can be measured in the plasma of newborn foals and used to indicate the availability of these glycogen stores. However, these stores are finite and are quickly depleted in cases of stress/hypoxemia, etc. Immediately post partum, glucose concentrations should be in the order of 50–70 mg 100 ml^{-1} blood. Levels lower than 50 mg 100 ml^{-1} indicate hypoglycaemia. Once the foal has suckled, glucose levels increase and in a normal foal that is 36 h old they reach values of 100–110 mg 100 ml^{-1} blood.

Bicarbonate levels rise steadily over the first 36 h of life from 23 mmol l^{-1} evident at birth to 27 mmol l^{-1} (Jones and Rolph, 1985; Fowden et al., 1991). Lactate concentrations also raise immediately post partum and decline to normal adult levels within 36 h (7–14 mg l^{-1}). The initial increase in lactate coin-cides with a fall in venous pH (7.4 post partum to 7.35 at 30 min) and may be as a result of the energy demands during the transition period. This fall in pH rectifies itself within 12 h when pH increases to 7.39 (Stewart *et al.*, 1984). Table 16.2 illustrates the major haematological and biochemical parameters for foals from parturition to 7 days of age.

Hepatic function is thought to be good at birth but does not fully function until 4–6 weeks of age, unlike pancreatic function which appears to be fully functional at birth (Knottenbelt *et al.*, 2004).

Endocrine function can also be indicated via blood sampling. Two hormones of particular interest are cortisol and thyroxine. In the normal fetus adrenal cortex activity significantly increases in the last 4 days of gestation; cortisol concentrations rising to 70–80 ng ml^{-1}. In the first few hours of life this significantly increases again to 120–140 ng ml^{-1} before declining to 60 ng ml^{-1} within 6 h of birth, finally declining to normal basal levels, 30 ng ml^{-1}, within 3 days (Silver and Fowden, 1994). Such a pattern of cortisol release is not evident in premature foals, in which cortisol concentrations may not reach above 30 ng ml^{-1} and reaction to adrenocorticotropic hormone (ACTH) is very poor (Silver *et al.*, 1984).

The newborn foal has higher circulating concentrations of thyroid hormones, T_3 and T_4, than most other domestic animals. At birth, these may be up to 10–20 times of that seen in adults (T_3 3.36 ± 0.65, T_4 8.05 ± 2.09 nmol l^{-1}). Concentration then drops in the first few days but rises again (T_3 0.86 ± 0.4, T_4 14.34 ± 6.7 nmol^{-1}) 4–6 days post partum, levels then drop again over the next 3 months (Irvine, 1984b; Vivette *et al.*, 1990; Knottenbelt *et al.*, 2004). Thyroid hormones are known to be involved in many physiological functions and also integrally linked to metabolic rate, this is noteworthy, as discussed earlier (see Section 16.2.2.3) the metabolic rate of the newborn foal is particularly high.

The significant changes in both cortisol and thyroxine in the later stages of gestation and during early life make them very likely candidates as the main drivers of neonatal adaptation.

The biochemical changes apparent in the newborn and very young foal can only be assessed via blood sampling and vary considerably. They must, therefore, be viewed with a certain amount of caution when used as a diagnostic aid (Jones and Rolph, 1985; Fowden *et al.*, 1991).

16.2.4. Post partum foal examination

Within 1 h of birth, it should be evident whether or not the foal is adapting to extra-uterine environment

appropriately though there is variation between foals as to when various milestones are met. If problems are suspected a veterinary surgeon should be called. Assessment of the following parameters, some of which have already been detailed, will give a very good indication of foal well-being.

Foal:

- Heart rate;
- Respiration;
- Ability to stand;
- Vigour;
- Ability to suckle;
- Straight legs;
- Body weight;
- General demeanour.

Placenta:

- Weight;
- Integrity;
- Abnormalities.

Mammary-gland function:

- Colostrum quantity and quality.

16.3. Early Foal Management: The First 6 Weeks

Management of the lactating mare and foal depends to a certain extent on whether she is to be returned to the stallion or not. If the mare is to foal at stud, her early management and that of the foal will largely be determined by the general stud management and practice. As such, management has a significant effect on the foal's long-term prospects; this is an area that should be discussed during initial stud selection.

Immediately after foaling and foal examination the following management procedures may be carried out:

Navel dressing
Administration of broad spectrum antibiotics (penicillin/streptomycin) – a precautionary method taken by some
Administration of vitamin supplements
Blood testing
   Haematology and blood biochemistry (see Table 16.2)
   Isoerythrolysis test – mare–foal compatibility test
   Immune status test/immunoglobulin uptake
Enema – still practised routinely by some, but not advised unless meconium retention is suspected
Vaccination – particularly for tetanus, is practised by some but not advised if the mare is adequately protected

Apart from this, the mare and foal should be left in peace, with regular, unobtrusive observation. A radiant-heat lamp may be used to provide warmth for the foal. As indicated, antibiotics may be given as a precautionary measure; however, indiscriminate use is not encouraged, due to the potential for bacterial resistance, disruption in the colonization of the small intestine and caecum, diarrhoea and selection for salmonella antibiotic resistant organisms (Madigan, 1990). Enemas are routinely administered by some. If used, they are best administered during the first passage of meconium or at 12 h. Administration too early may have no effect if no meconium has yet passed beyond the pelvic inlet (Madigan, 1990). No controlled trials appear to have been carried out on the use of enemas and many now confine their use to foals with meconium retention after 12 h.

16.3.1. Exercise

For the first 3 days of life the eyesight of the foal is not good enough to safely allow it out of the stable or small foaling paddock. The foal is born with poor eye reflexes and low corneal response which gradually improves over the first few days of life (Knottenbelt *et al.*, 2004). After 3 days it should have developed adequate appreciation of distance and depth to be turned out with its dam for an hour or two during the day, providing the weather is good (Fig. 16.6).

It is best to avoid turning out young foals if it is wet or very cold and windy. Such weather will easily soak the foal through and, as both mare and foal will be reluctant to move around in such weather, it defeats one of the main objectives of turning them out, that of exercise (Back *et al.*, 1999). The paddock provided for foals should be small; half an acre is ideal. It should have strong, well-constructed fences, ideally post and three rails with no wire. There should be no protruding objects, old machinery, wire, holes, low branches, etc. as these can prove death traps to young foals still not that sure on their feet. An alternative system used in Europe, especially France, is electric fence paddocks. Providing mares are accustomed to electric fences, large fields can be subdivided up into small paddocks with electric tape, allowing association with other mares and foals but security in the first week (Fig. 16.7).

Water should be provided in a bucket. Streams or large water troughs can also prove lethal for a very young foal. The paddock should have plenty of good grass, as this will encourage the mare to eat, which is especially important in mares whose appetite has declined after parturition.

Persuading the foal to leave its stable for the first time can be a challenge, but should be made as free

Fig. 16.6. After 3 days a foal's eyesight should have developed enough to allow it to be turned out with its mother.

Chapter 16

Fig. 16.7. Large fields can be divided up into small single mare and foal paddocks using electric fence tape, providing contact with other mares and foals along with security from them during early life.

from trauma as possible; otherwise nervousness will be perpetuated. During the first 3 days of life the foal should have been handled gently, stroked all over and got used to having arms put around it. When the big day comes, therefore, it should be used to human contact. The best day to turn a foal out for the first time is a nice sunny day, but not too hot or flies will be a problem, and bright sunlight may discourage the foal from leaving the much darker stable environment. There should be at least two handlers. The mare should be led ahead slowly by one handler and another should cradle the foal in his/her arms, one arm behind its hindquarters and the other around its chest, and encourage it to follow its dam (Fig. 16.8).

Some foals will follow easily; others prove more difficult. A foal should never be pulled from the head by means of a halter, as this may seriously damage its neck and head. A soft twisted cloth, bandage or thick rope can be put around its neck initially which can later be replaced with a soft leather or webbing halter. Leather is preferred, as it will stretch and eventually break under strain. Some people like to leave head collars on foals while they are out; this can be very convenient for catching them and gives them time to get used to them. However, the collar must be very well fitting

to ensure that it will not get caught on anything or allow the foal to catch its feet in it (Fig. 16.9).

16.3.2. Handling

Initial handling in the first few days before turn-out should consist of gentle stroking over the whole body and general familiarization to humans. Once a foal can be led, it must start to learn how to be tied. This is best done by using a round pole with no projections, so that the foal cannot get itself twisted up or caught on fences. A rope can then be attached to its head collar and on to the pole. As mentioned earlier, there is a risk of damage if the foal is pulled by a rope attached to its head. Hence, an alternative is to loop the rope around the foal's girth and up through its head collar and on to the pole. This method of restraint means that all the pull is taken on the girth and not the foal's head. However, foals will soon learn that they cannot escape and that it is easier to stand still.

Once the foal has learnt to accept tying up, the general stroking and handling can progress on to grooming and attention to feet and eventually travelling. These are particularly important if you intend to show the foal. Grooming can develop

Fig. 16.8. The foal may be encouraged to lead for the first few times by cradling the foal in your arms, one arm behind its hindquarters and the other around its chest. (Photograph courtesy of Angela Stanfield.)

Fig. 16.9. A leather halter is preferred as a first time halter as it will stretch and eventually break under strain.

slowly and the foal will soon come to enjoy it, providing all progression is done slowly and patiently (Fig. 16.10).

Providing the weather is good, a foal can be bathed – again of great use if it is to be shown. The weather must be warm and it should not be bathed very early or late in the day, to avoid it catching a chill.

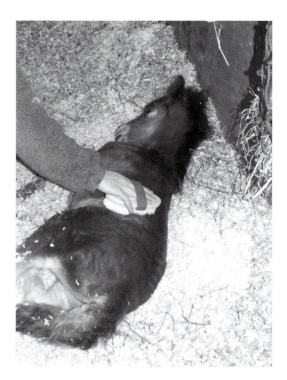

Fig. 16.10. A foal should learn to be groomed as one of its first lessons.

Fig. 16.11. Foals are very inquisitive and so care must be taken is what is left in their stables or field.

Introduction to a trailer or lorry can also be done in the first 6 weeks of life. The mare can be used to encourage the foal and many take to it easily, providing the mother is a good loader. If she is not, there is danger of the foal picking up her bad habits or her fear. In such cases, leaving the trailer in the foal's paddock with the door open and feed in the top end can encourage it to investigate and get used to going in and out at will. If this is done, it must be watched at all times to ensure that the foal does not hurt itself. Once the mare and foal have been successfully loaded and unloaded a couple of times they can be taken for a short ride. A foal will sometimes travel better if there is no central partition dividing the trailer. If there is a top door, it should be closed or a bar or cover used to prevent the foal from trying to escape over the tailboard if it panics.

16.3.3. Feet care

The foal's feet should need little attention in early life unless they have a significant deformity. Nevertheless, picking up the feet, picking out the hooves and grooming the legs should be done regularly, and these, along with ensuring a general acquaintance with the blacksmith when the mare's feet are attended

to, will ease work on foal's feet later on. Regular inspection of the feet will allow examination for injury and damage, and light trimming every 6–8 weeks from 3 months onwards is advised.

16.3.4. Behaviour

During the first 6 weeks of life, the foal shows quite significant development in behaviour and social interaction. Initially the foal's whole world and social experience revolves around just its mother. This includes play (Fig. 16.12), which may consist of rubbing her mane and tail and kicking. Through this, it begins to learn how far it can push it before being reprimanded and so what is acceptable and what is not. Once the foal has developed more steadiness on its feet, normally after about a week, it will start to explore further away from its mother, but never straying far. Over the next few weeks the circle gets bigger and it spends more time away from its mother, investigating and playing alone (Figs. 16.12 and 16.13).

If at this stage the foal has access to other foals, it will begin to interact with them and play will gradually include them rather than its mother. By 8 weeks it spends up to 50% of its time playing with other foals and only 10% playing around its

Fig. 16.12. Play is a very important aspect of a foal. Early behaviour and through play, it will explore its environment and capabilities. (From Penpontbren Welsh Cob Stud.)

Fig. 16.13. The foal's circle of play becomes increasingly larger over the first few weeks of life, as it gains independence. (From Penpontbren Welsh Cob Stud.)

mother. If, however, the foal has no contact with others, it will play with its mother much longer and may try to play with other older horses present or even dogs or other animals regularly in its company. If its mother is particularly possessive, or shy,

these characteristics can be passed on to the foal and it will not integrate as well with other foals (Carson and Wood-Gush, 1983b).

Apart from play, the foal spends a significant amount of time lying down resting (Fig. 16.14).

Fig. 16.14. Apart from play and suckling, the foal spends a large proportion of his time lying down resting.

These are normally short periods of rest, particularly in warm sunlight, between periods of play.

The remainder of its time is spent suckling. These periods of suckling in the first week are short and may occur as often as every 15 min. With time, the intervals between sucklings become longer – 35 times day^{-1} by week 10 – but the periods of time spent suckling and hence the intake increases (Fig. 16.15; Carson and Wood-Gush, 1983a).

16.3.5. Nutritional requirements and the introduction to solids

The newborn foal requires 120–150 kcal day^{-1} energy and 5.5–6.0 g kg^{-1} day^{-1} protein (Knottenbelt

Fig. 16.15. As the foal grows up it spends less time suckling its mother, the intervals between suckling getting larger.

et al., 2004). In early life milk will satisfy these requirements; however, the foal will soon need to supplement this and so may be first seen investigating concentrate feeds, and even ingesting some, as early as 3 days of age. Investigation of the mare's feed at an early age is to be encouraged, as the mare's milk is naturally short of iron and copper (Cu), which invariably causes anaemia in very young foals. Cu and iron are vitally associated with red blood cell function and haemoglobin levels. Adequate levels of Cu and iron can be achieved by the foal picking at the mare's feed (Fig. 16.16). Anaemia may persist in foals too weak to nibble hay or concentrates until they are treated or are able to eat. Foals may also be observed nibbling the mare's dung in the first 5 weeks or so. The reason for this is unclear but is maybe a means of addressing mineral deficiency or introducing bacteria and protozoa into the gut in order to set up the appropriate microflora for digestion. This coprophagic behaviour does, however, run the risk of parasite ingestion by the foal and so worming of mare's with young foals at foot is particularly important.

In theory, the mare's milk up until peak lactation, on average week 8, provides the vast majority of the nutrients required by the developing foal. During this time, however, gradual investigation and ingestion of its dam's feed provides an increasing amount of nutrients, but this is not significant until after weeks 6–8. The amount of extra creep feed that a foal will require in the first few weeks depends largely on the mare's milk yield. In addi-

Fig. 16.16. A foal will soon be seen investigating its mother's feed. This is to be encouraged, as it provides the foal with essential copper and iron.

tion to concentrates, the foal must be introduced to roughage in the form of grass or hay, as a diet of concentrates and milk alone can cause diarrhoea.

Creep feed can be introduced as an optional extra as early as week 1, but the foal should never be forced to eat it. The progression from milk to solid food must be gradual and can start slowly at an early age. Many propriety creep feeds are available on the market, specially formulated to ensure the foal gets the adequate nutrients for a healthy start in life. If you mix your own, there are a few considerations to bear in mind. Protein should be relatively high in foal diets (up to $6\,g^{-1}\,kg^{-1}\,day^{-1}$) when compared to adult diets. In particular, these proteins should be digestible and contain the ten essential amino acids for horses: lysine, methionine, leucine, isoleucine, histidine, arginine, tryptophan, valine, phenylalanine and threonine. Many legumes, grains and pulses tend to lack lysine, and soybean meal and linseed meal can be added, within reason, as good suppliers of lysine. Other dietary components of special importance in growing animals are calcium (Ca) and phosphorous (P). A ratio of these two minerals of 2:1 should be aimed for. These minerals are essential for healthy growth of bones, cartilage, tendons and joints. Excess Ca or P can cause problems. Excess P causes Ca to be mobilized from the foal's bones in order to maintain the ideal 2:1 ratio, causing bone weaknesses and epiphysitis, so delaying growth. Many people supplement Ca in the form of limestone flour or equivalent as many legumes, grains and pulses have a relatively high concentration of P. Low Cu concentrations are reported to be associated with angular limb deformities and high zinc concentrations with developmental orthopaedic disease (DOD).

In addition to concentrates, foals should have access to fresh green forage. Lucerne (alfalfa) is good, as it is relatively high in digestible protein and Ca. Hay may be fed, but it must be of a good quality, with no evidence of dust, mould, dampness, etc. Best of all is free access to fresh grass, which provides an *ad libitum* supply of continually fresh material.

Once the foal starts to pick at its mother's food, care must be taken that, when its intake becomes significant, she receives enough to meet her own requirements. Foal and mare should now be fed individually. By 3–4 months the foal should be eating $1\,kg\,day^{-1}$ about 0.3–0.5% of body weight Careful monitoring of the foal's feed is required to prevent obesity and resultant conditions such as developmental orthopaedic disease (DOD; Ralston, 1997; Coleman *et al.*, 1999).

Water intake should not be ignored during this period of early life. In the first 3 weeks the foal's fluid intake will be satisfied by milk. However, increasingly as it takes in more solid food it will require access to clean water. At 4–6 weeks the foal will require approximately 4 l water day^{-1} (approximately 2 l per 100 kg body weight day^{-1}; Martin *et al.*, 1992).

16.3.6. Dentition

Providing the teeth of the foal erupt as expected and at the correct angle, there is no need to do anything with the teeth in the first 6 weeks. Most foals are either born with the central incisors or they erupt within 8–9 days, the middle incisors should then erupt at 4–6 weeks (Table 16.3).

16.3.7. Immunization and parasite control

Foals are often routinely immunized against tetanus in the first few days of life. However, colostrum from a suitably immunized mare is a much more effective method of providing protection against tetanus and numerous other diseases. Work suggests that immunization of very young foals, born to mares that are adequately protected by vaccination, may even have a detrimental effect on their long-term protective response to subsequent immunization (Van Maanen *et al.*, 1992). In some countries, it may be worth considering immunization against strangles or rabies depending on the prevalence of the disease. Foals are also at greater risk of botulism and rotovirus than adults, vaccination of foals is not recommended but the best protected is conferred by vaccination of the mare in the eighth, ninth and tenth month of pregnancy. Worming can be first done at 7 days of age, though ideally it would delayed until 6–10 weeks of age, good managerial control of parasites should be practised in order to minimize the need for anthelthintics. The foal should then follow a regular worming regime, worming against *Strongyloides westeri* in both the mare and young foal is particularly important (Clayton, 1978; Craig *et al.*, 1993; Shideler, 1993d). By 6 weeks of age ascarids also become a potential problem (30–60% of foals being infested) and so appropriate wormers should be selected (Knottenbelt and Pascoe, 2003).

16.4. The Orphan Foal and Those Requiring Additional Support

The inability of a dam to bring up her foal due to her death, illness or injury can be a serious setback for the foal. The newborn foal depends on its mother for a variety of things including colostrum, nourishment via milk, maternal affection and psychological development. Any replacement dam has to satisfy all these needs. Survival rates in orphan foals were at one stage very low. However, research and development have now allowed the needs of the foal when changing from newborn to weanling to be better understood and the management of the orphan foal geared to suit these needs.

The treatment required by the foal depends upon when the problem occurred, under what conditions and the reason for additional support. For example, foals that have received no colostrum or a very limited amount need to be identified and fed colostrum immediately. If the mare died during parturition colostrum may be milked from her and fed to the foal. Foals that have been orphaned after 48 h or so may have been able to suckle and may have received enough colostrum for adequate protection. Colostrum may be collected and frozen for up to 5 years, though it is suggested that some degeneration starts to occur after 12 months. Some studs routinely collect and freeze extra colostrum from mares that produce excess or from those that lose their foals. Colostrum substitutes are available; colostrum from other animals was traditionally used but is now frowned upon, due to digestive

Table 16.3. The ages of eruption of equine teeth. (From Tremaine, H., Wales, 2002, personal communication.)

Tooth	Deciduous	Permanent
1st incisor	<1 week	2.5 years
2nd incisor	4–6 weeks	3.5 years
3rd incisor	6–9 months	4.5 years
Canine	–	4–5 years
Wolf tooth	5–6 months	
1st cheek tooth	Birth–2 weeks	2.5 years
2nd cheek tooth	Birth–2 weeks	3 years
3rd cheek tooth	Birth–2 weeks	4 years
4th cheek tooth	–	9–12 years
5th cheek tooth	–	2 years
6th cheek tooth	–	3.5–4 years

upsets arising from differences in composition. Alternatively, serum transfusions may be given to provide the foal with immediate protection.

In addition to classifying a foal according to its colostrum intake, its state of health and strength should also be assessed. Thus, any assessment should compare the parameters of the orphan foal against those expected of a normal foal. A difficult birth due to dystocia, in addition to the death of the dam, will result in an orphaned foal requiring immediate intensive care, such as respiratory support; stomach tubing with colostrum; antibiotic treatment, precautionary against infection; multivitamin injection; oxygen replenishment or even blood transfusion. On the other hand, a strong orphan may readily suck from a bottle and not require additional treatment.

Apart from death, a dam may for some reason be unable to raise her foal herself. The reasons for such failure are important when deciding upon the appropriate management. A mare may be unable to feed her foal through inadequate milk production, due to mastitis or a physical abnormality. In the case of mastitis, the inability to produce milk may be permanent or temporary. If temporary, the foal may only need supplementary feed to keep it going until the infection has cleared. If a mare seems physically unable to produce enough milk, her foal may be supplemented to a certain extent but left hungry enough to encourage it to suckle his mother. This suckling may then stimulate her mammary glands to increase their milk production. Occasionally, foals may need supplementing just for the first few days of life, especially if they are premature, so giving the mammary gland time to catch up and secrete adequate milk. In such cases, the foal need not be removed from its mother, allowing the mother–foal bonding to continue: psychological damage to the foal is, thereby, limited.

Occasionally, if the birth has been particularly stressful or in the case of some maiden mares, the foal may be rejected by its mother. This rejection may vary from disinterest to physical attack. Such mares may be sedated temporarily or the procedures used for fostering foals, discussed later, can be applied. Rejection is often only temporary, the mare subsequently accepting the foal. During such incidences of weak mare–foal bonding, human interference must be minimized, as any external stimulus will only serve to worsen the situation. Intervention must be used only if the foal's wellbeing is at risk.

The vast majority of orphan, or bottle-fed, foals are at a disadvantage as far as health status is concerned. Health care for such foals is, therefore, very important. All the routine parameters for the newborn foal should be checked at birth. It is also advisable to check temperature and heart rate twice daily for the first few weeks to allow rapid identification of problems, so that immediate treatment or further investigation can be commenced. The foal may be helped by multivitamin and antibiotic injections.

Orphan foals and those requiring additional support are more susceptible to the ordinary infections that normal foals take in their stride. They should be watched carefully and kept in a scrupulously clean environment. Diarrhoea is a common complaint of bottle-fed foals, especially if fed artificial diets. The diarrhoea may be a direct result of the diet but may also be a sign of infection. Persistence after 24 h should be considered as serious, as dehydration can bring a foal down very quickly. Respiratory diseases can also be a problem in orphan foals, so their housing should be draught- and damp-free while providing good ventilation. Meconium retention may also be a problem, especially in foals that have not received adequate colostrum, as colostrum acts as a laxative. An enema may be required and the foal watched for subsequent constipation.

As discussed, the loss of a mare deprives the foal of nutritional and psychological security. A foal can survive without the psychological stimulus of a mother, though this may affect his long-term behaviour. It is not, however, able to survive without nutrition. There are several ways of providing that nutrition.

16.4.1. Fostering

The fostering of a foal on to a mare that has lost her own foal is the best solution. This provides the foal with a source of nutrients and psychological security, and provides the mare with a substitute for the lost foal. The mare used should be as close in her stage of lactation as possible to the foal's natural mother. This is ideal as the nutritional components of the mare's milk vary with time of lactation and are coordinated with the foal's developing requirements (Chapter 9, this volume).

If no such foster-mother is available, but there is a mare that has lost her foal in the last few weeks, then she can be used as a companion to the foal and will be an invaluable support to its psychological development. In such a case, the foal may be supplied with artificial foal milk, designed for the very young foal, in addition to being allowed to suckle its foster-mother. If the foal shows signs of poor health, it can be prevented from suckling its foster-mother and fed specially formulated foal diet for its age, but it will still have her psychological support.

Within the UK and in other countries, 'foaling banks' exist which arrange the teaming up of orphan foals and mares. Multiple suckling – the introduction of an orphan foal to a mare that has plenty of milk but with her own foal still at foot – is not very successful in horses. This technique is successfully practised with cattle, where milk yields are artificially high. In horses multiple suckling tends to lead to two poor foals and may lead to resentment between them. In North America nurse mares are available as foster-mothers. These mares are exceptionally good milkers and are kept primarily for leasing out as foster-mothers, after their own foals have been weaned early.

Once a foster-mother has been found, it can be quite tricky persuading her to accept the orphan foal. She is much more likely to accept a foreigner if it smells of her or her own foal. Tricks such as rubbing the placenta of the foster-mother over the foal or skinning the dead foal and placing the skin over the orphan work quite well, but depend upon you, the mare and foal being all together at the right time. Later on, the orphan foal can be rubbed with the mare's urine, faeces or milk, especially in the head, neck, back and navel region, again to mask its own foreign smell. Other tricks are used, such as rubbing strong smelling substances on the mare's nose and over the foal in an attempt to disguise its smell.

Once prepared the foal should be introduced to the mare with great care. One person must hold the mare and at least one hold the foal. They should be introduced to each other at the mare's head end and her reaction noted very carefully. If all goes well and she shows no objection, the foal can be allowed to slowly explore around the mare, making sure that plenty of assistance is available if the mare should object. The foal can then be introduced to the mare's udder and allowed to suckle, providing again the mare shows no signs of objection. Rarely will things go according to plan and the mare will often initially show signs of annoyance or objection to the orphan, in which case it should be removed and reintroduced to her head end again. Eventually, most mares will let the foal

suckle and once this has occurred it can be left in a stable with her unrestrained for a short period of time. The area used should be relatively confined so that they remain in close contact with each other and the foal does not become isolated. Throughout this period, observation at all time is very important and immediate action taken to remove the foal if the mare starts to object. Slow, patient progress will pay off and, once the foal has suckled several times, the mare will rarely object to it. After a couple of days, they can be turned out into a small paddock alone to help develop the bond between them over a distance before introducing them to the melee of other mares and foals.

Problems with fostering do occur. Some mares, regardless of all persuasion, will not accept a foal and it should be given up as a bad job before humans and horses lose patience and the foal experiences yet another rejection. There are, however, various techniques to help persuade reluctant foster-mothers, including the use of a crate in which the mare is held such that she is unable to turn around. The sides of the crate should be solid with a hole at the end nearest the udder through which the foal, which is outside the crate, can suckle. This can work quite well and after the foal has suckled several times the mare can be removed from the crate and she will often accept the foal. Other tricks are used to elicit the mare's protective response: the introduction of a dog or another mare within sight can induce a protective response in the mare towards the foal, which often leads to acceptance, but the applicability of such means depends on the individuals concerned.

If all else fails, a nanny goat has been advocated as a foster-mother. The goat can be placed on an elevated platform for suckling, and its continual presence provides company for the foal. However, care must be taken, as goat's milk is not of the same composition as horse's milk. It has two-thirds the sugar content and three times the fat content of mare's milk and like cow's milk tends to cause gastrointestinal upsets especially associated with gas retention (see Chapter 9, this volume).

16.4.2. Artificial diets

If it is not possible to find a foster-mother or a foal requires supplementary feeding, there are specifically formulated equine milk substitutes available on the market (Table 16.4). These are formulated to mimic the mare's natural milk components

Table 16.4. An example composition of a foal milk replacer for use with orphan foals or those that require additional nutritional support. (From Frape, 1998.)

Component	Percentage
Glucose	20.0
Fat-filled powder	5.0
(20% fat)	
Spray-dried skimmed	
Milk powder	40.0
Spray-dried whey powder	32.7
High-grade fat[a]	1.0
Dicalcium phosphate	1.0
Sodium chloride	0.2
Vitamins/trace elements[b]	0.1
Disperse in clean water at a rate of 175g l⁻¹	
May also be pelleted and mixed with a stud mix as a weaning feed for orphan foals	

[a]High-quality tallow and lard, including dispersing agent. Stabilized vegetable oil could be alternatively added at the time of mixing.
[b]To provide vitamins A, D_3, E, K_3, riboflavin, thiamine, nicotinic acid, pantothenic acid, folic acid, cyanocobalamin, iron, copper, manganese, zinc, iodine and selenium.

(Frape, 1998). All these dry powders must be mixed under sterile conditions and fed at 37.5°C. Other formulas are used, based on cow's milk or dried cow's milk, with added components to make up the shortfalls in cow's milk compared to mare's milk. These are not very successful, though still popular, and often result in gastrointestinal upsets, causing diarrhoea, dehydration and if not rectified, a rapid decline in foal development and growth. If diarrhoea does occur, then milk substitutes should be replaced by a 50% glucose-electrolyte solution, made up in sterile water, for 1–2 days, with the slow reintroduction of the milk substitute. Ensuring regular small feeds are fed rather than fewer large ones can reduce the incidence of diarrhoea. A young foal will naturally suckle up to five times h⁻¹ (Chavatte, 1991). It is, therefore, necessary that frequent feeding is gradually reduced from 2 hourly feeds during the first few days to every 2h after 2 weeks of age (Carson and Wood-Gush, 1983b; Naylor and Bell, 1987). Specially designed mare's milk substitutes are widely available nowadays and there is no excuse for feeding other formulas except in real emergencies (Green, 1993; Frape, 1998).

During the first few days of life, the orphan foal should also be given a broad spectrum antibiotic plus vitamin injection, to give it an extra boost over what is to the foal, a very stressful time. The foal should be watched carefully and help called at any signs of trouble. Orphan foals tend to go down hill more rapidly than those with their mothers. For the first few weeks the orphan foal will need to receive its milk via a bottle, and a large plastic squash bottle with a lamb teat works quite well. It is very difficult to prescribe how much and how often it should be fed, as there is no hard and fast rule. However, as a rough guide 110–220g of milk should be fed every hour in the first week to ensure a daily intake of 120–150kcal digestible energy (DE). Over the first few weeks daily intake should increase to 10–15l, and with time the frequency of feeding can decrease (week 2 every 2h, week 3 every 3h, week 4 every 4h). After 4 weeks it may be fed just four times day⁻¹. The exact timing and amount depends on the introduction and acceptance of milk in a bucket and/or concentrate feed. After about days 3–5, made-up milk powder can be placed in a bucket and hung at a level giving easy access for the foal. This will allow it to wean itself slowly off the bottle and on to the bucket. Some foals feed from the bucket very quickly; others need more encouragement. It must be ensured that the milk is always clean and fresh; otherwise the foal will be discouraged from taking it. Warm milk and encouragement by licking off human fingers can help a reluctant foal to accept bucket milk (Green, 1993; Frape, 1998).

Some foals, instead of moving over to bucket feeding, are kept on bottle feeds by use of an automatic feeder. The more sophisticated automatic feeders mix and regulate the amount and temperature of the milk delivered, and the foal sucks through a conveniently placed teat in its pen wall. Less sophisticated machines require manual refilling with premixed milk. Either, if accepted, reduces the labour, especially at night if there are several orphan foals to be fed. It is very important that such feeders are regularly stripped down and sterilized, as bacteria build up rapidly in such an ideal environment and can cause the foal immense problems.

16.4.3. Introduction of solids

Many breeders recommend the introduction of solids as early as 1 week of age, though the foal's intake at this time will be very limited. Introduction at an early age gives the foal time to become accustomed to solid feed and gradually wean itself off milk. This progression from milk to solids can be helped by introducing foal milk pellets as a half-way house. As soon as the foal is eating a regular amount of concentrates, then the milk pellets fed can be reduced. *Ad libitum* access to fresh green grass, or lucerne as a substitute, will also encourage a foal to eat. By 2 weeks of age, the foal should be turned out, at least for part of the day, with a companion, weather permitting, into a small safe paddock. This will introduce it to fresh grass and allow it room to stretch its legs, and experiment with movement and investigate its new environment.

By 2 months of age, most orphan foals are consuming enough concentrates to allow them to be weaned off milk. The exact time will depend on the foal's well-being, and final weaning should not be attempted unless the foal is fit and healthy, as it will always result in a slight setback. Orphan foals can be fed normal foal creep feeds as discussed earlier (Section 16.3.5).

By 3 months, the foal should be completely weaned off milk and fed on a ration of concentrate to supplement fresh grass and forage. With orphan foals in particular, a close watch should be placed on the foal's body weight and its diet should be changed to accommodate any significant rise or fall in condition. As with normal foals under- or over-condition can be detrimental to growth and development (Ralston, 1997; Coleman *et al.*, 1999).

16.4.4. Developmental orthopaedic disease

DOD is the generalized term given to disturbances in skeletal growth and development of foals and young stock, such as angular limb deformities, contracted tendons, incorrect calcification of bone, bone and joint inflammation (osteochondritis dissecans (OCD)), epiphysitis, vertebral abnormalities (wobblers syndrome) and general abnormalities in bone and joint structure and development (Coleman *et al.*, 1999). There are several reasons for DOD including inherited conditions, limb trauma from excessive work on growing limbs, endocrine dysfunction, toxicity, fast growth rates and malnutrition (Thompson, 1995). The last two are due to incorrect feeding which can have several effects. First, general overfeeding of both energy and protein results in overweight, this leads to excessive strain on young, still growing limbs and joints, causing deformities. Second, incorrect feeding, both too much and too little, of specific nutrients can have specific effects. For example, excessive

energy appears to have a direct effect on the hormone regulation of bone growth and development (Thompson *et al.*, 1988). This effect is made worse if animals are fed large carbohydrate (glucose) meals infrequently, i.e. feeding once a day is more likely to cause DOD than three times per day (Raub *et al.*, 1989). Minerals such as Ca and P are also important, deficiencies of either will cause DOD; additionally, the ratio of Ca to P needs to be correct. If P levels are relatively high this imbalance will tie up Ca reducing the amount available for bone growth and, therefore, cause DOD, even if Ca levels in the diet appear correct. Finally, Cu is important for bone growth and development, inadequate levels again leading to DOD. It is therefore very important that youngsters are fed a correctly rationed quality diet in several meals per day and that exercise is provided (ideally turnout) but not excessively so in order to prevent limb trauma.

16.4.5. Discipline and the orphan foal

One of the problems encountered more often with orphan foals than those brought up by their dams is bad behaviour. The mare's presence acts to discipline the foal, teaching it respect for others, both horses and humans, and passing on, it is hoped, good behavioural characteristics. Hand-reared foals can become over friendly with, and lack respect for, human beings, treating them much as they would other foals and horses with nipping, kicking, chasing, etc. If this is allowed to get out of hand, such animals can be extremely hard to handle in later life.

To avoid this, orphan foals should be allowed sight of other horses as soon as possible. Direct contact may be dangerous at such a young age but sight and smell will help. By the time they are 1–2 months old they should have an established gentle companion – one that is placid and does not bully the foal; Shetland ponies and/or donkeys are often used. By 4–5 months old, they should be able to cope with other weanlings and can be run out together, allowing play and social interaction with their fellows. Discipline while in the company of humans should continue to be strict and consistent.

16.5. Mare Management in Early Lactation: the First 6 Weeks

Immediately post partum, the mare should be left in peace to bond with her foal. She should have a good supply of hay or fresh grass plus clean water available to her. After an hour or so, she may be given a small nutritious feed. She should then be just discretely observed from a distance (Asbury, 1993). Apart from being involved in the management practices applied to the foal, it is important that, during the first few weeks, she is watched carefully to ensure that she has not suffered any detrimental long-term effects of the birth. In general, however, a lactating mare should be managed in much the same way as other stock with common sense and an eye for potential problems.

16.5.1. Milk production

At about 72 h after delivery, the mammary gland produces predominantly milk rather than colostrum. Details of milk composition and quality plus further details of the physiology of lactation are given in Chapters 9 and 10 (this volume). During the start of lactation, it is important that the mare looks healthy and fit in herself, as any infections or disease will easily pass on to the foal. The general well-being of the foal should also be used as an indication of milk production. If the mare is not producing enough milk, the foal will appear tucked up, the mare's teats may be sore from continual unsuccessful suckling and she may begin to object to the foal suckling, risking the development of a perpetual situation making the condition worse. Low milk yields may be due to a physical inability, low nutritional intake, poor body condition or mastitis (McCue, 1993).

Occasionally, immediately post partum, the mare fails to produce any milk at all. This is due usually to either a failure in the milk ejection reflex or in the production of milk. Failure of milk ejection reflex is most common, and is thought to be due to high circulating adrenalin levels as a result of stress and anxiety, causing vasoconstriction and hence inhibits the action of oxytocin on the udder. Injection of oxytocin and/or warm compresses applied to the udder, along with a quiet calm atmosphere, will often rectify the situation. Failure of the udder to produce any milk, i.e. galactopoiesis or lactogenesis failure, is a serious condition, about which little can be done. Many foals born to such mares have to be classified as orphan foals and brought up accordingly. Such mares should not be bred again, as the condition invariably repeats itself (McCue, 1993).

Teat abnormalities can also cause problems. Inverted nipples or conical nipples make it very hard for the foal to suckle, though the condition may not affect milk production itself. Supernumerary teats may also be present, but are rare, and do not seem to affect milk ejection from the normal teats. Fluid collection occasionally occurs within the udder of high-yielding mares. Udder oedema, as it is termed, results in fluid accumulation around the udder and along the abdomen. As a result the udder becomes too painful to allow the foal to suckle and the condition predisposes the mammary gland to infections.

There is a relatively rare condition, called neonatal erythrolysis, in which the mare becomes isoimmunized against the foal; i.e. she has raised antibodies against the foal. This results in a fatal reaction within the foal if it ingests the mare's colostrum. Alternative colostrum must be fed for at least 36 h post partum until the gastrointestinal system loses the ability to absorb whole protein molecules. From this time onwards, the foal can suckle normally.

Mastitis, an infection of the mammary gland, is relatively rare in the mare compared to the cow. It is characterized by a hot, swollen and painful udder, with oedematous swellings developing along the abdomen and up between the hind legs. Milk secreted tends to be thick and clotted and should not be fed to foals. If mastitis does occur, it is most often evident post weaning, especially if the mare is still producing large amounts of milk. Mares that have had foals prematurely weaned are, therefore, at particular risk. Treatment is similar to that in cows, by the administration of antibiotics directly into the mammary gland via an intramammary tube inserted into the streak canal and repeated for several days. Alternatively, intramuscular injection of systemic antibiotics can be used, as many mares violently object to having a mastitic mammary gland touched.

16.5.2. Uterine involution and breeding on the first oestrus post partum (foal heat)

Many studs perform a routine internal examination within 3 days of parturition to identify problems and check that uterine involution is progressing appropriately. Any problems identified at this stage can be treated in time for either covering at the foal heat (first oestrus post partum), or, if not, by her next return to service.

The rate of uterine contractility and involution to its pre-pregnancy state has a significant bearing on the conception rates at covering on the foal heat. Within 7 days the uterus should have returned to two to three times its size in the barren mare and, by days 30–32 both the uterine body and horns should be back to their normal pre-gravid size (Blanchard and Varner, 1993a). In addition, recovery and return to pre-gravid state of the endometrium is required; this is normally completed by day 14 (Gygax et al., 1979; Blanchard and Varner, 1993a). Clearance of all uterine fluid discharge should also have occurred by day 15 (McKinnon et al., 1988a). In order for conception rates to be maximized, full uterine involution must have occurred. However, the mare invariably returns to oestrus and ovulation between 4 and 10 days post partum, and, as such, she may be covered before full uterine involution has occurred, as a result conception rates are often poor (Belling, 1984). A correlation between uterine involution and conception rates is clearly documented, and mares that return to oestrus and ovulation 10 days post partum, or later, have significantly better conception rates to that oestrus, compared to those returning nearer to day 5 post partum (Loy, 1980). The time interval of 10 days prior to ovulation, plus 5 days for the embryo to reach the uterus, allows 15 days for the uterus to recover prior to receiving another embryo. As has been indicated, by day 15 uterine fluid clearance and endometrium recovery should be complete.

Ideally, mares returning to oestrus prior to day 10 post partum should not be covered but left until their second oestrus post partum. However, the industry demands foals are born as early in the year as possible and that a mare should produce one foal per year. The mare's 11-month gestation makes this difficult to achieve. In practice, there are several management techniques that can be used to help achieve these aims. The main aim is to manipulate the timing of the first or second ovulation and oestrus, in order to allow a 15-day recovery period prior to the arrival of the embryo in the uterus. The mare's foal heat can be delayed using progesterone supplementation for 10 days post partum followed by withdrawal. This will induce a mare to ovulate within 3–4 days. Thus, the minimum 15-day recovery period is achieved. Alternatively, the mare may be allowed to show her normal post-partum oestrus and ovulation. If this occurs prior to day 10 she is not covered but her next ovulation is advanced by the use of prostaglandin F2α 6–10 days after the

foal heat ovulation, again ensuring at least a 15-day recovery period. Both these methods allow longer time for uterine involution and hence result in better conception rates (Blanchard and Varner, 1993a). In feral populations, conception and foaling rates to foal heat are in excess of 90%; this is thought to be due to exercise and so turning out mares as soon as possible post partum is advised in order to encourage uterine involution, especially uterine fluid clearance (Lowis and Hyland, 1991; McDonnell, 2000a; Blanchard et al., 2004). More recent work by Ishii et al. (2001) indicates close proximity to a stallion as well as warmer environmental temperature is associated with a reduced risk of post-coital endometritis and better conception rates to foal heat covering.

16.5.3. Nutrition

One of the specific areas to note in the management of the lactating mare is her nutrition. A lactating mare has higher nutritional requirements than any other equid, even one in heavy work (Doreau et al., 1988). As parturition approaches, her nutritional demand increases to meet the demands of the growing fetus. After parturition, the mare continues to provide all the nutrients for that foal, but now the foal is extra-uterine and is larger. The supply of nutrients via milk is less efficient than nutrient transfer via the placenta. The efficiency of energy transfer in milk is only 60%, i.e. 40% of the energy intake does not appear as energy in the milk (Table 14.2, 16.6). After parturition the mare's nutrient requirements increase by 70–75% during early lactation and by 50% in later lactation to make up in part for this decrease in efficiency (McCue, 1993). At peak lactation, a mare may produce up to 3% of her body weight as milk.

16.5.3.1. Protein and energy

Protein and energy are important components of a lactating mare's ration. If her diet is nutritionally low in protein, she will not have enough to satisfy the demands of her milk production. Milk production will then decline (Martin et al., 1991). Low dietary energy will result in mobilization of the mare's own body reserves in order to try and maintain production. If low dietary energy persists, milk yield will decline and the mare will become emaciated (Pagan and Hintz, 1986). Some weight loss, especially in mares that milk well, is to be expected, but should be minimized by appropriate feeding. Weight loss in mares that are to be returned to the stallion can affect conception rates (Sutton *et al.*, 1977; Gill *et al.*, 1983). Hence, the protein and energy content of a lactating mare's ration are very important to her reproductive efficiency. For a 500 kg mare, DE intakes of 31.7 Mcal day^{-1} are required in the first month. Similarly crude protein intakes of 1535 g day^{-1} are required in the first month in order to ensure adequate nutrition for maximum milk production (Tables 14.2, 16.5 and 16.6; Frape, 1998; National Research Council, 2007).

16.5.3.2. Calcium and phosphorus

In addition to protein and energy, Ca and P intakes are also very important. For a 500 kg mare average Ca intakes of 59.1 g day^{-1} in the first month of lactation are required to ensure that the foal obtains adequate Ca for bone and tendon growth and development. As previously discussed, the ratio of Ca to P is important; excess P causes a drain of Ca from the mare's bones in an attempt to redress the balance. P intakes of 38.3 g day^{-1} in the first month are required in early lactating mares of 500 kg.

In order to satisfy the demands for milk production the mare will require concentrate feed in addition to good-quality forage. P and protein tend to be deficient with regard to lactating mares, even in very good-quality forage diets. The concentrate ration should address this but consist of no more than 50% of the total diet. Not only should protein quantity be high to account for the inefficiency in conversion of nutrient protein to milk protein, but protein quality is also important (Frape, 1998). Lysine is often a limiting essential amino acid in conventional grain and grass forage diets. This can be addressed by the inclusion of lucerne, or soybean meal (Frape, 1998).

As stressed in earlier discussions, it is imperative that all home-grown or bought-in straights and forages are analysed. An appropriate ration for a lactating mare can be ensured using this information and analysis of commercial concentrates (Tables 14.2, 16.5 and 16.6).

16.5.3.3. Vitamins

Extra vitamins and minerals are also required during lactation. Deficiency will cause lacklustre and ill thrift in both the mare and foal and inefficient use of other nutrients. Vitamins A and D are of

Table 16.5. Daily nutrient requirements of lactating mares of varying weights. (From National Research Council, 2007.)

Lactating mares	Weight (kg)	Milk production (kg day^{-1})	DE (Mcal)	CP (g)	Lysine (g)	Ca (g)	P (g)	Mg (g)	K (g)	Vitamin A (10^3 IU)
1 month	200	6.52	12.7	614	33.9	23.6	15.3	4.5	19.1	10.0
	500	16.3	31.7	1535	84.8	59.1	38.3	11.2	47.8	25.0
	900	29.34	54.4	2763	152.6	106.4	68.9	20.1	86.1	45.0
2 months	200	6.48	12.7	612	33.8	23.6	15.2	4.5	19.1	10.0
	500	16.2	31.7	1530	84.4	58.9	38.1	11.1	47.7	25.0
	900	29.16	54.3	2754	152.0	106.0	68.6	20.1	85.8	45.0
3 months	200	5.98	12.2	587	32.1	22.4	14.4	4.3	18.4	10.0
	500	14.95	30.6	1468	80.3	55.9	36.0	10.9	45.9	25.0
	900	26.91	52.4	2642	144.5	100.6	64.9	19.6	82.7	45.0
4 months	200	5.42	11.8	559	30.3	16.7	10.5	4.2	14.3	10.0
	500	13.55	29.4	1398	75.7	41.7	26.2	10.5	35.8	25.0
	900	24.39	50.3	2516	136.2	75.0	47.1	19.0	64.5	45.0
5 months	200	4.88	11.3	532	28.5	15.8	9.9	4.1	13.9	10.0
	500	12.2	28.3	1330	71.2	39.5	24.7	10.2	34.8	25.0
	900	21.96	48.3	2394	128.2	71.1	44.4	18.4	62.6	45.0
6 months	200	4.36	10.0	506	26.8	15.0	9.3	3.5	13.5	10.0
	500	10.9	27.2	1265	66.9	37.4	23.2	8.7	33.7	25.0
	900	19.62	46.3	2277	120.5	97.4	41.8	15.7	60.7	45.0

DE = digestible energy; CP = crude protein; Ca = calcium; P = phosphorous; Mg = magnesium; K = potassium.

Table 16.6. The expected feed consumption by lactating mares (percentage of body weight). (From National Research Council, 1989.)

Mares	Forage	Concentrates	Total
Early lactation	1.0–2.0	1.0–2.0	2.0–3.0
Late lactation	1.0–2.0	0.5–1.5	2.0–2.5

particular importance. Vitamin A is available in fresh green forage and vitamin D from exposure to sunlight (Frape, 1998; Duren and Crandell, 2001).

16.5.3.4. Water

Access to a clean, reliable water source is particularly important for the lactating mare, whose water requirement is as high as those of any other equine. A 500 kg mare in mid-lactation in an ambient temperature of 20°C may require as much as 78 l day^{-1} (11–14 l 100 kg^{-1} body weight day^{-1}; National Research Council, 2007).

It is not appropriate to be too prescriptive on exactly what a lactating mare requires, as requirements will vary with individuals. Mares that produce more milk will probably require more feed and will probably require higher levels of concentrate to meet their demand. However, some mares seem to be better doers and more efficient as milk producers. In such mares, obesity may prove to be a problem and, therefore, limiting their concentrate intake should help. The yardstick to go by is the body condition of the mare; she should be fit not fat, with a body condition score of 3. From 6 weeks of lactation onwards, the foal, as far as nutrition is concerned, becomes increasingly independent. The mare's milk yield then decreases; her nutrition management must, therefore, reflect this change in demand.

16.5.4. Exercise

The importance of exercise in uterine recovery is not to be underestimated. Increasing exercise in the first few days post partum is linked to accelerated uterine involution and hence better conception rates at the foal heat. It is interesting to note that conception rates to the foal heat in feral ponies are normal and in excess of 90% (Camillo *et al.*, 1997; McDonnell, 2000a). Confinement of domesticated mares may, therefore, be a contributory cause of decreasing foal heat conception rates (Lowis and Hyland, 1991; Blanchard *et al.*, 2004).

16.5.5. Immunization, parasite control, dentition and feet care

No specific immunization of mares that were appropriately vaccinated before parturition, apart from annual boosters, should be required at this stage. Worming should be carried out in the first week, especially against strongyles and ascarids, which are particularly significant in mares and young foals but also strongylosis, a problem in mares but not in foals. From then on, a normal regular worming regime should be adhered to. Teeth and feet care should not be neglected and their care continued as normal.

16.6. Foal Management: 6 Weeks of Age Onwards

From 6 weeks onwards, the foal becomes increasingly independent of its mother.

16.6.1. Exercise

Continued turnout is essential to help muscle coordination and development, fitness, gastrointestinal and cardiovascular system function and independence. Ideally, mares and foals should be turned out together in mixed gender groups to help the foal's development of social awareness and appreciation of hierarchy. Group turnout is an ideal starting point for gradual weaning systems.

16.6.2. Handling

The foal's handling at this stage should develop on from that started in the first 6 weeks of life, remembering patience and reward. Halter breaking should have been started by now (Fig. 16.17). Late halter breaking can lead to confrontation. Leading lessons should develop and the foal should learn to lead without resistance. The process can be aided by a rope around the foal's hindquarters that can be pulled to encourage it to walk forward. Well before weaning, the foal should be happy to be led without resistance or fuss, both behind its mother and away from her. This can only be achieved by continual and patient training, using short and frequent lessons (Fig. 16.18). The foal should also become further accustomed to travelling, leading on to travelling alone.

Fig. 16.17. By 6 weeks of age, halter breaking should have been started. (From Penpontbren Welsh Cob Stud.)

Fig. 16.18. The foal should learn to lead without resistance from an early age.

16.6.3. Feet care

Foot problems can be identified and possible correction considered within the foal's first year of life. In addition to regular trimming and handling, this can ensure that minor faults and problems can be identified before training begins. Over zealous attack of a foal's feet, in attempts to correct leg problems, should be avoided as it exacerbates existing problems. Indeed, if left alone many such deformities often prove to be self-correcting.

Corrective trimming should be done only by a trained and experienced farrier or veterinary surgeon. Deformities, such as an incorrect hoof/pastern angle, can be corrected within reason by specific trimming to change the length of the horse's heel. Toes, out or in, result in uneven wear of the hoof wall and corrective and compensatory trimming can alleviate the problem. Excessively long toes or the opposite – club foot – can be corrected by ensuring that any trimming done is adequate, not overzealous.

16.6.4. Behaviour

From 6 weeks of age the foal continues to develop its independent traits, spending more and more time away from its mother, playing and interacting with other foals. This is invaluable in developing social awareness and learning about hierarchies within a group, in preparation for survival alone with its peers and without its mother for protection.

16.6.5. Nutrition

From about 8 weeks of age, a creep feed becomes increasingly important to the foal as a source of nutrients. From this stage onwards milk quality declines as an encouragement to the foal to seek nutrients elsewhere. The quality of creep feed fed must be carefully assessed to ensure that it provides all the nutrients required for optimum growth and development. However, it should be borne in mind that optimum growth is required not maximum growth. As a rough guide, a foal destined to make 150–160 cm may gain up to 2 kg day^{-1} but by 1 year of age it should not weigh in excess of 80% of its expected mature weight. Excess weight gain causes strain on muscles, tendons, joints, the circulatory system, etc. It is especially important when these structures are still developing that undue stress does not cause permanent deformity.

The protein requirement of the foal for growth is high and it becomes the first limiting factor as far as nutrient supplied by milk is concerned. Creep feeds have been discussed previously (see Section 16.3.5), but for the older foal the feed should contain 20% protein, at 4 months, a foal destined to make a mature weight of 500 kg requires 669 g CP day^{-1} in a highly digestible form, in addition they require 39.1 g day^{-1} Ca and 21.7 g day^{-1} of P, though a Ca:P ratio within the range of 1:1 and 3:1 is acceptable. The importance of zinc and copper is increasingly evident.

As far as quantity of feed is concerned 0.5 kg day^{-1} at 2 months is adequate, gradually increasing as weaning approaches. As a guide foals of 3 months of age should be consuming 0.3 kg 100 kg^{-1} body weight day^{-1} which is normally about 1 kg day^{-1} or 0.25–1% of body weight (Frape, 1998).

It is important that the mare does not have access to the foal's creep feed, as this may discourage the foal from feeding. Specially designed creep feeders permit the foal to feed alone without danger that the mare can have access to the feed. The amount of feed per foal should also be controlled and, if several are run together, it is best if they are fed individually. This may prove difficult but will ensure that each foal is fed according to need and monitored against weight gain. Free access allows greedy foals to gorge themselves at the expense of smaller, less-dominant individuals.

If foals are fed outside in a creep feeder, then food should be checked regularly to avoid mould developing. Free access to fresh grass or lucerne is essential and access to a mineral supplement is good practice.

Again water must not be forgotten and foals of 8–10 weeks in age require at least 5 l water day^{-1}. This will increase in warm environments and with exercise and the dry matter content of feed (Martin et al., 1992).

16.6.6. Dentition

By 6 weeks the foal's first and second incisors should have erupted. These are then followed by the wolf-teeth at 5–6 months and then the third deciduous incisors at 6–9 months (Table 16.3). Attention to teeth beyond familiarization with opening the mouth to allow the teeth to be viewed should not be required at this stage.

16.6.7. Immunization and parasite control

Immunization against tetanus and influenza can be instigated at 5–6 months of age, similarly vaccination against strangles, rabies, equine viral encephalitis and African horse sickness may be considered at this time depending on the prevalence of the diseases, though the side effects of immunization

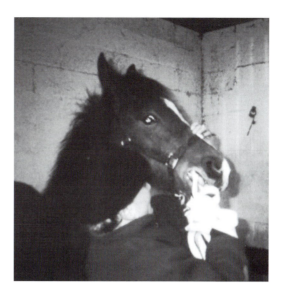

Fig. 16.19. Regular worming should be established by 2 months of age and continue as a routine for the rest of the foal's life.

such as abscessed vaccination site with strangles inoculation make some unpopular as part of a general routine.

A regular worming regime should be established from 2 months of age. Wormers against strongyles and ascarids are particularly important, as these specifically affect young horses (Fig. 16.19; Rossdale and Ricketts, 1980).

16.7. Management of the Lactating Mare: 6 Weeks Post Partum Onwards

From 6 weeks post partum, lactation yield decreases, along with the quality of milk produced. The mare's udder shrinks as milk demand reduces.

16.7.1. Nutrition

As the mare's milk yield decreases, so do her nutrient requirements. Concentrates should then be slowly reduced to ensure that she does not become obese. For a 500 kg mare protein intakes can be gradually reduced to 1468 g day^{-1} and DE to 30.6 Mcal day^{-1} by 3 months post partum. Ca and P are still important, intakes of 55.9 and 36.0 g day^{-1}, respectively, being required (Tables 16.5 and 16.6). Particular attention should be paid to the mare's body condition, as mares often put on extra weight in the summer months as the demand for milk declines and before the demands of the fetus *in utero* become significant. It is not, however, appropriate to put a mare in foal on a strict reducing diet later on in the autumn in an attempt to lose excess weight gained during the summer months.

16.7.2. Exercise

Exercise at this stage is essential to build up the mare's fitness again. Indeed exercise, along with a decrease in nutrient intake, helps to dry up milk production. The mare and the foal should be turned out for as long as possible, ideally day and night, providing the weather is good. If the weather is too hot it may be appropriate to bring them in during the day to avoid the flies and be turned out at night.

16.7.3. Immunization, parasite control, teeth and feet care

Worming regime, immunization programmes, attention to teeth and feet should be maintained as normal and not neglected, to ensure that the mare remains in optimum condition, especially if she is in foal again.

16.8. Conclusion

In summary, the management of the lactating mare and foal is particularly critical during the first few hours post partum. Once this initial critical adaptation of the foal to the extra-uterine environment has successfully been achieved, the likelihood of the foal surviving is high. From this stage onwards, concentration should be given to the management of the mare so that she can be successfully covered as soon as possible, and to the management of the foal so that it grows into a healthy, well-disciplined individual.

17 Weaning and Management of Young Stock

17.1. Introduction

Correct weaning and young stock management is critical in ensuring the foal's long-term good health, physical growth and development, psychological development, social interaction with fellow equids and humans and long-term productivity. The following account will consider weaning management and the management of young stock separately.

17.2. Weaning

Weaning is essential to allow the mare's mammary gland to recover in order to ensure an adequate milk supply for the forthcoming, new foal. Naturally, the foal would be weaned at 9–10 months of age, giving the mare at least a month to recover before the birth of the new foal. By 9 months of age the foal will normally be consuming a large quantity of solid food, with minimal reliance on milk. The resultant dry period, after weaning and before the new lactation, allows the mare's system to recover, concentrate on supporting the foal *in utero* and replenish body reserves.

17.2.1. The timing of weaning

Naturally, during the last 3 months of lactation, the foal, now over 6 months of age, derives most of his nutrients from grass and herbage. It is, therefore, quite possible, with careful management and the provision of a concentrate supplement feed (creep feed) to wean foals at 6 months, providing they are consuming adequate amounts of feed (Coleman *et al.*, 1999). Weaning at 6 months is, therefore, the practice in most studs. However, it should not be considered unless it is certain that the foal is in good physical condition as well as taking in adequate concentrates. The time of weaning will also depend upon the mare's behaviour, month of parturition and the dependence of the foal upon the mare (Apter and Householder, 1996). Early

weaning, as soon as 4 months of age, may be practised if the mare is suffering. This requires planning and good management; but, providing the foal is well prepared, introduced to solids in good time and is in good physical condition, it should not suffer significantly as a result. Early weaning can cause complications, as such, foals cannot be run out with other foals, which will either not yet be weaned or will dominate the foal and not allow it adequate access to concentrate feeds. An alternative companion needs to be provided to give psychological security and development; a small, quiet donkey or pony can be ideal.

17.2.2. Weaning stress

Weaning is a very stressful process, both physically and psychologically (Apter and Householder, 1996). The foal will be separated from its dam, milk will be eliminated from its diet, it will be introduced to strange horses and there will be more handling and contact with humans. Careful management, however, can ease these stresses and thus reduce the stress of weaning.

The physical stress of weaning can be reduced by ensuring that the solid food intake of the foal prior to weaning is adequate to minimize any setback due to the sudden removal of milk from the diet. Sudden changes in diet at all ages can cause digestive upsets and in foals they can additionally lead to growth and developmental retardation (Fig. 17.1; Warren *et al.*, 1998a,b). To ensure a gradual change, some studs advocate milking the mare for a few days after weaning and feeding the milk to the foal along with its concentrate diet. This is also reported to reduce the risk of mastitis but is time-consuming and some mares will object violently to being milked by hand. Alternatively, foals can be fed milk pellets for a while after weaning. Such artificial inclusion of milk into a weanling's diet, however, defeats one of the main objects of weaning – that of removing milk from the diet. In addition,

Fig. 17.1. Intake of solid food must be adequate prior to weaning so as to minimize the upset to the digestive system as a result of the change from a liquid, milk-based to solid, concentrate- and forage-based diet.

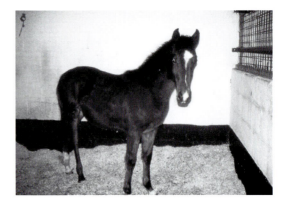

Fig. 17.2. Traditionally foals are weaned by sudden removal of the mare, leaving the foal in a secure stable.

if the foal has naturally significantly reduced its intake of milk prior to weaning, the addition of milk post weaning will be a retrograde step and may adversely affect newly established gut microflora and so cause digestive upsets.

A foal being considered for weaning must be in good health. Any animals showing signs of illness, such as runny nose, coughing, listlessness, starry coat, diarrhoea, etc. must not be weaned until their condition improves. Young animals can suffer quite dramatically from seemingly small problems, resulting in considerable setbacks to their development, with possible permanent damage. If in doubt, it is advisable to call a veterinary surgeon. In exceptional cases, and only under veterinary supervision, foals suffering from ill health may be weaned, as some medicines are easier to administer and are more effective in a foal not on a milk diet. Psychological stress can be reduced by introducing the foal to his post-weaning companions and regular handling prior to removal from his mother (see Chapter 16, this volume).

17.2.3. Methods of weaning

Plans for weaning foals should be considered well in advance of the actual event. There are four main types of weaning: sudden or abrupt; gradual; interval or paddock; and weaning in pairs. The method employed is often dictated by the facilities available, the numbers of foals and young stock and also personal preference. Traditionally, foals were weaned suddenly and individually by removing the mare abruptly and leaving the foal in a stable or loose box out of earshot of its mother (Fig. 17.2). More recently, other methods have been advocated, which are based on a more gradual removal of the mare or the introduction of substitute companions.

17.2.3.1. Sudden or abrupt weaning

Sudden or abrupt weaning involves the abrupt separation of the mare and foal. If this system of weaning is to be employed, then a safe and secure stable is required for the foal. It must be free from any projections likely to cause damage; water buckets should either be fixed to the wall or not left unattended. Hay should be fed in a hay rack off the ground – a hay net is not advised as the foal may strangle itself; hay on the floor is better but can be wasted. The bed should be deep, ideally made of straw, providing good protection as the foal launches around the box. The stable door should be secure, with an upper grill or metal-mesh door as well as a solid upper door (Fig. 17.3).

At weaning, the mare is abruptly removed from the loose box, leaving the foal behind. The foal should by now be accustomed to handling and can be held in the stable while the mare is removed. She must be kept moving even though she is likely to be very reluctant; the quicker she is removed and with

Fig. 17.3. The stable used to leave the foal in, after removal of the mare, should be very secure and have an upper grill or mesh as well as a solid upper door.

the least fuss, the better. As soon as she is out of the stable, both solid doors top and bottom should be shut and a light left on in the stable. The foal should be relatively safe under these conditions for a short while until the mare has been attended to (Heleski *et al.*, 2002).

The mare should be taken to a field out of ear-shot of the foal, with limited grass cover. The field should be secure, with safe boundaries. Some mares are very disturbed for the first few hours and can easily damage themselves by careering around; others appear to consider weaning a relief. The mare should be watched until she has settled down and started to graze. The foal should then be checked. It should be given water, as it will have invariably worked itself into a good lather. Hay should also be made available *ad libitum,* along with a small feed fed as soon as it has calmed down. The foal should remain in the box for the first few days to allow it to get used to life alone. A large stable is, therefore, advantageous. For these first few days the upper mesh door should be closed at all times to prevent the foal attempting to jump out and yet still provide ventilation. Foals are notoriously unaware of danger and will launch themselves at obstacles that an adult horse would not dream of attempting. They are, therefore, very prone to damage and extra care should be taken to

avoid potential hazards: prevention is infinitely better than cure. The foal should be handled and mucked out regularly.

These first few days are very stressful and the foal is susceptible to physical damage and disease, and this is one of the main disadvantages of this system. After a few days, providing the foal is calm, it may be turned out for short periods of the day with a companion in a small secure paddock. The length of turnout can be increased gradually to all day and night if appropriate. In many systems, foals are still brought in at night right through until the following spring, as weaning does not occur until late summer or autumn.

During this time, the mare should not be neglected. Her udder may start to show signs of tenderness and discomfort due to the increase in milk pressure. The milking out of a small amount of milk daily for the first few days is advocated by some in order to reduce the pressure and hence the chance of masti-tis. The amount of milk removed daily should only be small and should gradually be reduced over 5 days. The removal of too much milk will only serve to prolong the problem. If milk build-up within the udder leads to excessive pressure, infection and mastitis can result. Post-weaning mares must, there-fore, be watched for such problems, especially in the first few days. Mastitis in mares is relatively rare but, if it occurs, can prove fatal if not appropriately treated. Antibiotic treatment is usually successful, though there is always the danger of the infected half of the udder being lost.

17.2.3.2. Gradual weaning

Gradual weaning is a newer and increasingly popular method, as it attempts to reduce the stresses of sudden weaning. It can be practised in yards with single mares or groups. As with abrupt weaning, if two stables are to be used for the mare and foal, they must be safe and secure and ideally have an interconnecting barred window. More commonly, adjacent paddocks are used and, providing these are well fenced and secure, there should be no problems. It is advised that the fencing should be post and rail, rather than wire, to reduce the chance of injury. If two paddocks are to be used, it is normal to select the more lush pasture for the foal, as its requirements will be greatest and eating will provide a distraction. Initially, the mare and foal are turned into the separate paddocks or stables for a short period of time, half an hour or so. Over the next couple of weeks this time of separation increases, until they are turned out separately all the time. The close proximity of the foal to the mare allows physical contact and interaction but does not allow suckling. Independence and a reduction in the reliance on milk are developed over a period of time so the stress of abrupt complete

separation is much reduced. An additional advantage is that, as the time of separation is gradually increased post-weaning mastitis is not a problem (Fig. 17.4).

17.2.3.3. Paddock or interval weaning

Paddock weaning requires careful planning and is not possible with single foals or with foals of vastly differing ages. Ideally, the foals should be born in batches within 2 weeks of one another and brought up together in the same paddock after the first couple of weeks (Fig. 17.5). This allows them to become accustomed to each other and a hierarchy is developed while their mothers are still around to dilute any aggression. In such systems some mares may be seen allowing a foal other than their own to suckle, providing her own foal is not also demanding milk. Near the projected time of weaning, all foals should be checked for physical condition and adequate solid food intake. In an ideal system on large studs, there will also be other batches of younger foals born later and following behind, so any foals not ready for weaning in one batch can be transferred to the next one, giving them 2 more weeks or so to become prepared.

Once all the foals are ready for weaning, the most dominant mare or the dam with the most independent foal can be removed on one day, followed by the next dominant the next day, etc. until all have been removed. Occasionally, a gentle dry mare may be

Fig. 17.4. Gradual separation of mare and foal over time can help minimize the stress of weaning.

Fig. 17.5. In gradual weaning system, groups of mares and foals of similar ages are run together in preparation for the gradual removal of the mares at weaning.

introduced as a companion. The mares must be taken well away from the foals and out of earshot. In this system, the foal that has lost his mother soon forgets, due to the solace and security of its fellow foals and the other mares. Mares should again be turned out on to poorer grazing and be watched for mastitis.

17.2.3.4. Weaning in pairs

A final alternative, not used widely in practice, but an area of some research, is paired weaning. Based upon the sudden weaning system, foals are weaned abruptly in pairs rather than singles, two foals being left in a stable together (Hoffman *et al.*, 1995). However, this system does require the two foals to be 'weaned' from each other at a later stage, which may be as stressful as initial sudden weaning alone would have been.

17.2.4. Variable stresses of weaning systems

As indicated, gradual and paddock weaning are considered the least stressful, as the foals are solaced by familiar surroundings and companions and any change is gradual (De Ribeaux, 1994). This is supported in work by McCall *et al.* (1985, 1987) who, using vocalization as an indicator of stress, suggested the following ranking of weaning methods, based upon the decreasing stress involved:

sudden weaning with no creep feed; sudden weaning with creep feed; gradual weaning with no creep feed; and gradual weaning with creep feed. Paired and paddock weaning are advocated by some as less stressful; however, they require the foal to be 'weaned' from its other foal companions at a later stage, prolonging the stress. This is of particular concern in paired weaning systems. Indeed, work by Hoffman *et al.* (1995) and Malinowski *et al.* (1990) indicated that paired sudden weaning was in fact more stressful than single sudden weaning, as measured by plasma cortisol levels and foal behaviour. Malinowski *et al.* (1990) also indicated that lymphocyte proliferation to a challenge of concanavalin A was suppressed in foals weaned in pairs compared to singles. This would indicate a lower disease resistance. This suppression, however, could be improved by human contact. Paddock-weaning systems appear less stressful but are not possible with a single, or a few foals, as age-group batches are required. In addition to the immediate physical and psychological stress, recent work by Nicol *et al.* (2002), among others, has suggested that weaning stress has an additional long-term effect, predisposing horses to gastric ulcers and as a result depress growth rates. Though this is not supported by all workers (Rogers *et al.*, 2004), who report no difference in daily weight gain between foals abruptly or gradually weaned. A link

Fig. 17.6. After weaning, mares can be turned out into a paddock with limited grass cover to help dry up their milk production.

between stable weaning and confinement post weaning and the development of stereotypic behaviour has also been demonstrated (Heleski *et al.*, 2002; Waters *et al.*, 2002). Diet at weaning may also be linked to stress levels; creep feed and fibre diets being less stressful than high sugar and starch diets (Nicol *et al.*, 2005). It is increasingly evident that careful management of weaning is required in order to minimize any effects on growth and development, especially that of bone potentially predisposing to injury (Coleman *et al.*, 1999; Fletcher *et al.*, 2000). However, weaning need not always lead to problems, and with good management, maintenance of familiar surroundings and routines, gradual preparation in diet and handling for the event, stress can be minimized as can any setback in growth, even in foals weaned as early as 4 months (Warren *et al.*, 1998). Physical damage to the foal can also be minimized by ensuring that it is in safe and secure surroundings and health promoted by exercise and access to sunlight and fresh air (Holland *et al.*, 1997).

17.2.5. Post-weaning care of the mare

As discussed, the mare's udder should be watched for evidence of tenderness due to milk accumulation and possible mastitis (Perkins and Threlfall, 1993). In order to minimize this, post weaning she should be turned out into a paddock with limited grass cover and no supplementary feed (Fig. 17.6). This limits her nutritional intake and forces her to exercise in order to obtain what grass she eats. Exercise helps to relieve pressure within the udder and utilizes nutrients that would otherwise be directed towards milk production. Her paddock should be secure and free of hazards, as some mares may be initially quite agitated due to the absence of the foal. The mare usually recovers more quickly from the separation than the foal and, in some cases, may even seem relieved to be free of the extra burden (Holland *et al.*, 1997).

17.3. Management of Young Stock

The management of young stock in itself could be the title of a book; hence, this section will only attempt to give a summary of the principles and main points to consider. The more specialized texts available should be consulted for in-depth details of the systems that can be used (Jones, 1978; Equine Research, 1982; Coldrey and Coldrey, 1990; Knowles, 1993; Britton, 1998; Lorch, 1998).

17.3.1. Nutrition

Significant foal development occurs in its first year of life, though it continues until maturity at 4–5 years of age (Table 17.1). Nutrition during this period is, therefore, of the utmost impor-

Table 17.1. Percentage of mature body weight and of wither height attained at different ages in various breeds. (From Hintz, 1980 cited in Frape, 2004.)

Age/Breed	6 months		12 months		18 months	
	Weight	Height	Weight	Height	Weight	Height
Shetland pony	52	86	73	94	83	97
Quarter Horse	44	84	66	91	80	95
Anglo-Arab	45	83	67	92	81	95
Arabian	46	84	66	91	80	95
Thoroughbred	46	84	66	90	80	95
Percheron	40	79	59	89	74	92

tance. Nutrition must satisfy the body's requirements for growth and development (Table 17.2), but not to excess, as this may cause obesity, which in turn exerts extra strain on young limbs, tendons and the circulatory system, leading to conditions such as developmental orthopaedic disease (DOD), osteochondritis dissecans (OCDs) and epiphysitis (Ralston, 1997; Frape, 2004). The direct links between nutrition, growth and development, and subsequent performance (athletic and conformational) are increasingly evident (Ott and Asquith, 1986).

Table 17.2. Daily nutrient requirements of young stock of varying weights. (From National Research Council, 2007.)

Animal	Weight (kg)	Mature weight (kg)	Daily gain (kg)	DE (Mcal)	Crude protein (g)	Lysine (g)	Ca (g)	P (g)	Mg (g)	K (g)	Vitamin A (10 IU)
Maintenance	200	200		6.7	252	10.8	8	5.6	3.0	10.0	8
	500	500		16.7	630	27.1	20	14	7.5	25.0	20
	900	900		30.0	1134	48.8	36	25.2	13.5	45.0	36
Growing horses											
Weanling 4 months	67	200	0.34	5.3	268	11.5	15.6	8.7	1.4	4.4	3.4
	168	500	0.84	13.3	669	28.8	39.1	21.7	3.6	10.9	8.4
	303	900	1.52	23.9	1204	51.8	70.3	39.1	6.4	19.7	15.2
Weanling 6 months											
	86	200	0.29	6.2	270	11.6	15.5	8.6	1.7	5.2	4.3
	216	500	0.72	15.5	676	29.1	38.6	21.5	4.1	13.0	10.8
	389	900	1.3	28.0	1217	52.3	69.5	38.7	7.5	23.3	19.4
Yearling 12 months											
	128	200	0.18	7.5	338	14.5	15.1	8.4	2.2	7.0	6.4
	321	500	0.45	18.8	846	36.4	37.7	20.9	5.4	17.4	16.1
	578	900	0.82	33.8	1522	65.5	67.8	37.7	9.7	31.4	28.9
Yearling 18 months (not in training)											
	155	200	0.11	7.7	320	13.7	14.8	8.2	2.5	8.1	7.7
	387	500	0.29	19.2	799	34.4	37	20.6	6.2	20.2	19.4
	697	900	0.51	34.6	1438	61.8	66.7	37.1	11.1	36.4	34.9
Two-year old 24 months (not in training)											
	172	200	0.07	7.5	308	13.2	14.7	8.1	2.7	8.8	8.6
	429	500	0.18	18.7	770	33.1	36.7	20.4	6.7	22.0	21.5
	773	900	0.32	33.7	1386	59.6	66.0	36.7	12.0	39.6	38.6

DE = digestible energy; CP = crude protein; Ca = calcium; P = phosphorous; Mg = magnesium; K = potassium.

Prior to weaning, the foal's digestive system progressively adapts to the digestion of solids and roughage. At the time of weaning, the transition from fluids to solids is complete. In general, nutrient deficiencies in young stock are more critical than in older animals, and so rations for weanlings should be designed with particular care (Table 17.2; Traub-Dargatz, 1993a; Ralston, 1997). As a general guide, a weanling should be fed 3–4 kg day^{-1} split into 3–4 feeds (Ralston, 1997).

17.3.1.1. Energy

The precise energy, protein, vitamin and mineral requirements vary significantly with rate of growth and expected mature weight. The specific requirements for growth are being superimposed upon those for maintenance (Tables 17.2, 17.3; Ott and Asquith, 1986). Energy is the most important component of a youngster's diet. Low energy levels depress growth and, at the extreme, can cause stunting even if the other components of the diet are appropriate. For a 4-month-old youngster expected to reach 500 kg mature weight energy intakes of 13.3 Mcal day^{-1} are required. This increases at 12 months of age to 18.8 Mcal day^{-1} (Tables 17.2, 17.3; National Research Council, 2007). These relatively high energy levels can be obtained by feeding high concentrate diets; however, an adequate intake of roughage must also be maintained. A ratio of 4:6 roughage: concentrates is ideal if the roughage has a digestible energy (DE) content of 2.0 Mcal kg^{-1} dry matter (DM). This can be achieved by feeding legumes such as lucerne (alfalfa). Most other roughages have energy concentrations

Table 17.3. Expected feed consumption by young stock (percentage of body weight). (From National Research Council, 1989.)

Animal	Forage	Concentrate	Total
Maintenance Young horses	1.5–2.0	0–0.5	1.5–2.0
Nursing foals 3 months	0	1.0–2.0	1.0–2.0
Weanling foal 6 months	0.5–1.0	1.5–3.0	2.0–3.5
Yearling foal 12 months	1.0–1.5	1.0–2.0	2.0–3.0
Yearling 18 months	1.0–1.5	1.0–1.5	2.0–2.5
Two-year old 24 months	1.0–1.5	1.0–1.5	1.75–2.5

less than 1.0 Mcal kg^{-1} and so either the amount of concentrate fed or the energy concentration of the concentrate feed should be increased to compensate. If the amount of concentrate is increased it should not represent more than 70% of a 12-month-old youngster's diet (Tables 14.1, 17.2, 17.3; Ott, 2001; National Research Council, 2007).

Within these guidelines the body condition of the youngster should also be monitored and the level of feed should be altered to account for any overweight or underweight. Inappropriate energy levels are a prime cause of obesity in young stock and are also thought to be contributory to conditions such as epiphysitis, structural deficiencies, contracted tendons, etc. in the growing horse (Ralston, 1997; Warren et al., 1998a,b; Coleman et al., 1999).

17.3.1.2. Protein

Protein is a necessity for bone growth and development, in addition to its involvement in energy utilization. For 4-month-old youngster with an expected mature weight of 500 kg, protein intakes of 669 g day^{-1} increasing to 846 g day^{-1} at 12 months of age are recommended (Tables 17.2, 17.3; National Research Council, 2007). The quality of the protein provided, in the form of amino acid composition, is again just as important as the quantity. The adequate supply of essential amino acids is extremely important to young stock. Lysine, usually the most limiting amino acid in most diets, is especially important in youngsters (Ott, 2001; Stainer et al., 2001). Hence, even if the total protein content of the diet is appropriate, the horse may still suffer from protein deficiency due to the lack of an essential amino acid. Feeding quality protein straights can alleviate this. Soybean meal is a good source of protein, being highly palatable and high in lysine, and it is a popular choice as a component of a youngster's diet.

As mentioned, protein is involved in the utilization of energy within the diet. Inadequate protein concentrations are associated with signs of energy deficiency, even if there are adequate DE levels within the diet fed. It is, therefore, usually recommended that protein levels slightly above requirements are fed in order to ensure that energy utilization is optimized (Tables 14.1, 17.2, 17.3).

17.3.1.3. Vitamins and minerals

As might be expected, vitamin and mineral intakes in growing youngsters are of importance. Vitamin and mineral deficiencies within the component feeds of a diet and deficiencies in feedstuffs are due to the environmental conditions under which they are grown, particularly soil deficiencies. Ideally, all rations fed to youngsters should be analysed for the major minerals, at least for calcium (Ca), phosphorus (P), sodium (Na) and chlorine (Cl) and, if necessary, an appropriate supplementary feed or a general vitamin and mineral supplement fed as routine to ensure no deficiencies arise. However, general supplements should be used with care as they may complicate imbalances if adequate levels of specific minerals were originally present within the ration (Frape, 1998; Duren and Crandell, 2001).

17.3.1.4. Calcium and phosphorus

Ca and P are essential for bone, tendon and cartilage growth and are, therefore, extremely important in a youngster's diet. For a 4-month-old youngster with an expected 500 kg mature weight respective daily intakes for Ca and P should be 39.1 g and 21.7 g. In practice, levels slightly above this are often fed to allow for a margin for error. Requirements decrease to 37.7 g and 20.9 g, respectively, for youngsters of 12 months of age (Table 17.2, 17.3; National Research Council, 2007). However, as discussed previously, the ratio of Ca:P is as important as their specific concentrations; excessive Ca reduces P availability and so causes signs of P deficiency, excessive P causes mobilization of Ca from bone. A ratio around of 1.5:1 is ideal for 6-month-old youngsters, though some variation either side of this can be tolerated. The older animal is able to tolerate more variation, but even so the Ca:P ratio should not exceed 3:1.

17.3.1.5. Sodium chloride

Youngster's rations should contain 0.9% salt, though the actual demand and intake of salt will be affected by work and environmental conditions. A salt lick or general vitamin and mineral supplement containing salt can be made available to youngsters on an *ad libitum* basis to allow them to supplement their salt intake as and when required.

17.3.1.6. Trace minerals

Other minerals required by the growing horse include selenium, iodine, iron, copper, cobalt, manganese, potassium, magnesium, molybdenum, sulfur and fluorine. The requirements are, however, low – hence the term, trace minerals. Most feedstuffs will provide adequate levels of these, though, certain feedstuffs may be deficient in some trace minerals due to deficiencies within the soil in which they were grown, for example, selenium is deficient in parts of mid-Wales in the UK, in areas around the Great Lakes in the USA, and also in parts of New Zealand. If horses are fed exclusively on feeds grown within these areas, they will show signs of deficiencies. Most commercial diets are a combination of feedstuffs from numerous locations and, as such, normally contain adequate trace elements. Problems may be encountered in horses fed purely on home-grown and mixed rations in deficient areas. Ideally feeds should be analysed and appropriate supplements used.

17.3.1.7. Water

Water must not be forgotten as an essential component of any diet, a continual supply of clean fresh water is, therefore, essential. The estimated daily water intake for a yearling, weighing about 300 kg in an ambient temperature of 20°C, is reported to be 17–21 l day^{-1} (6–7 l per 100 kg body weight day^{-1}; National Research Council, 2007).

17.3.2. Housing

The nutritional requirements of any youngster are partly determined by its environment and housing. The higher a horse's maintenance requirement, the higher his overall nutritional demand. If maintenance requirements are increased by adverse environmental conditions, i.e. wind, rain, cold, etc. then the overall ration must be increased to compensate. If this is not done, then the growth and development of the horse will suffer. In order to minimize maintenance requirements and feed costs, many youngsters are housed for part, if not all, of the day, especially in adverse weather conditions. This also helps prevent damage and injury to the horse. However, such management is not advantageous to the psychological development of the youngster. Isolation can lead to behavioural problems and such horses are also prone to obesity. An alternative, that goes some way to solving the problem of isolation is to keep youngsters together in groups in a barn (Fig. 17.7). Youngsters are successfully kept in this way providing time is allowed for them to adjust to each other and to develop a

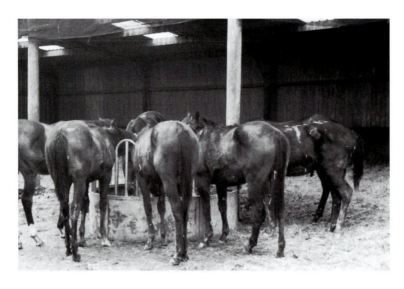

Fig. 17.7. As a halfway house between stabling youngsters and turnout all day is barning youngsters.

hierarchy, preferably in a large open field, before being confined within the barn. This system does run a higher risk of injury to weanlings from close contact with fellows, and animals kept in this way must be watched closely for signs of bullying, and appropriate action taken. Ideally, weather permitting, weanlings should be turned out all the time with supplemented feed. The benefits of social interaction, sunlight and *ad libitum* exercise should not be underestimated.

17.3.3. Exercise

As previously mentioned, exercise and social interaction with other horses are essential for a youngster's physical and psychological development (Back *et al.*, 1999). Ideally, youngsters should be reared at pasture and this is becoming increasingly popular. Such animals can exercise at will, reducing the strain on growing bones, allowing muscles and tendons to develop and grow in strength and size. Social interaction with other horses allows a respect for fellows and reduces boredom and the associated bad habits of confinement (Fig. 17.8). Environmental conditions are often the reason for restricted turnout. Conditions such as heavy rain, snowfall and driving wind or very hot weather may justify youngster's being brought in. The only major disadvantage of pasture-kept youngsters is the difficulty in making sure that all have an adequate intake of concentrates. Most studs feed a

standard ration to all pasture-kept youngsters all together and any individuals that appear to be in under- or over-condition are fed separately.

17.3.4. Handling

Before weaning all foals should be halter broken and be used to being handled (Fig. 17.9). It is much easier to teach foals basic manners at this age than face a stroppy and potentially dangerous yearling.

A well-handled foal that is used to grooming, feet trimming, clipping, boxing, etc. will be much easier to train in later life. Such an animal will have developed a confidence and respect that can be built upon in later training schedules. It is also much more likely to concentrate on the lesson in hand and, therefore, be easier to train, as it has confidence in its relationship with his handler.

When handling any youngsters, it must be remembered that they can be very unpredictable. By nature, they are often flighty and there is no guarantee that they will always react to things in the same way. As they develop, their interaction with and, therefore, their reaction to, their environment changes. Every precaution should be taken to ensure that, if the unexpected occurs, no one is hurt and minimum panic ensues. A youngster is very impressionable and panic or nervousness within the human handling can easily be sensed by, and transferred to, the youngster. This can become

Fig. 17.8. Ideally youngsters should be reared at pasture, allowing them to develop social interaction and respect for others, as well as providing exercise, to the advantage of bone, muscle and tendon development. (From Derwen International Welsh Cob Stud.)

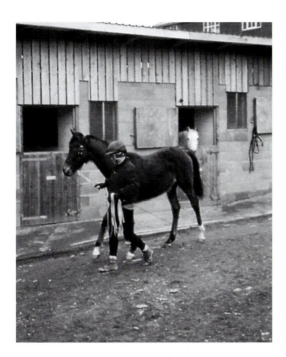

Fig. 17.9. Before foals are weaned, they should be halter broken and used to being handled, which considerably eases their management post weaning.

self-perpetuating and result in an owner too frightened to handle the youngster and a horse that is a bag of nerves and increasingly unpredictable in its behaviour. A complete outsider is often required to break such a downward spiral.

The youngster should always be treated with constant discipline. It gains security from predictable, consistent handling, even though it will try to push the limits of behaviour. Bad habits and behaviour should be disciplined immediately with the voice. Physical punishment is not normally necessary, except in extremes, and if not carried out with care, can degenerate into a confrontational situation.

Youngsters should be taught not to bite, nip, push, barge, etc. A foal should not be allowed to play with handlers. It is great to teach a foal to 'shake hands', but, when it starts to kick out in front as a yearling or older horse, this can be very dangerous. In such a scenario, the horse cannot really be blamed, as it will find it very confusing to now be told off for doing something it was once praised for. Charging and pushing handlers, especially when leaving a stable or going through a gate, are potentially very dangerous. The foal should be taught to let the handler go first and, if it barges, it should be halted with a short sharp tug on the lead rope. If that does not work, then a series of harder tugs should be given. One continual tug is not advised, as this will encourage the horse to fix his neck and head and use his strength against his handler in the future.

Praise is as important as discipline and, whenever a youngster does well or as he is told, it should be praised by a pat and also by voice. Throughout its handling, it should get used to the human voice and certain clear commands. Initially 'Whoa' is very useful to get the horse to stand still and can be used in later lunging lessons. This is one of the most important discipline lessons to learn. As long as a horse will stand still when told, you will always have control over it. There are other commands for other actions, e.g. come to call, walk, trot on, etc. as appropriate. A horse will get used to any commands, providing they are clearly and consistently used.

Initially, a youngster will be very exuberant and full of energy and at this stage, it may be appropriate to overlook minor bad behaviour until the basics have been achieved. It can be very demoralizing for a youngster – and handler – if it is always being disciplined and can seemingly do nothing right. The less dangerous undesirable behaviour can be disciplined at a later stage providing it does not get out of hand. The handler must use his common sense and assess the situation.

Beyond the basics, further handling and training will depend on the youngster's destination in life and can be geared towards a particular aim, for example, racing, breeding, showing, hacking, etc. It is beyond the scope of this book to discuss the specific training of youngsters for specific disciplines.

Whatever the ultimate destiny for the youngster, exercise is of particular importance. At about a year of age, forced exercise can be introduced in the form of leading out or lunging and provides the grounding for further training, especially in riding horses. Regular exercise also provides conditioning and fittening of horses, ensuring that when training begins in earnest an initial period of fittening is not required. Training of unfit horses runs the risk of strains and sprains of unconditioned muscles and tendons, leading to a waste of time in resting injuries. A youngster that does not have enough free exercise can be very difficult to train, as he will spend much of the time careering around before work can commence in earnest. Such animals are also very difficult to lunge or put in a horse walker and run a high risk of injury. A combination of free and forced exercise allows agility and physical ability to develop, along with mental stimulation. Development of both mental and physical faculties is essential as a highly sophisticated state is required in many equine disciplines.

17.4. Conclusion

Considerable thought should be invested into the method of weaning and the handling of young stock. Management at this young age can have long-term repercussions on the foal's physical and psychological development and so affect its ability to fulfil its potential in later life.

18 Stallion Management

18.1. Introduction

The management of the stallion should not be neglected in the enthusiasm to obtain optimum mare and foal management. Management of the stallion has already been considered in some depth in Chapters 11, 12 and 13 (this volume). However, it is also worth considering the introduction of the stallion to his work as a breeding animal and the general training and management principles that should be borne in mind when keeping stallions.

18.2. Early General Training

Early training, well in advance of the stallion's first introduction to a mare, is very important to reduce the chances of injury to both horses and handlers. It will also reduce the risk of him developing potentially dangerous bad habits.

It is very important that a stallion is taught discipline and respect from a young age. Once a good grounding has been established, this can be built upon. It is near impossible to start disciplining a 3- to 4-year-old stallion without the considerable risk of injury, and it will inevitably lead to conflict and not respect. In training a stallion the handler's attitude and competence are of extreme importance, as rough handling and incorrect and inconsistent training can cause many problems, from poor breeding behaviour and performance to dangerous vices (McDonnell, 2000b).

Once basic discipline has been achieved, including the acceptance of the handler, bridle, leading, obeying voice commands to halt, walk on, back up, along with boxing, shoeing, veterinary inspection and general handling, he may be trained further in a specific area or discipline or maintained at this level for breeding. Further training for riding or driving is advantageous, as it provides the stallion with another constructive outlet for his energies other than covering. This advanced level of discipline normally leads to greater respect and subsequently an easier stallion to handle.

18.3. Restraint

There are several means of restraint that can be used to control a stallion. The method used depends on the stallion's age and temperament, the facilities available and the handler's personal preference. The effect of the handler on the behaviour of the stallion cannot be overemphasized. A nervous and insecure handler will transfer these feelings to the stallion which, in picking them up, is more likely to act uncharacteristically and unexpectedly. It is especially important that the handler dealing with young stallions is calm and confident and has had plenty of experience. The overuse of restraint or punishment to compensate for nervousness is a trap that can, all too easily, be fallen into. If a stallion needs to be reprimanded, it should be immediate, quick and effective. Continuous ineffective, half-hearted attempts, often due to a lack of experience or confidence on behalf of the handler, lead to resentment of the handler by the stallion.

Stallions are by nature proud and courageous, attributes much to be admired. They must be treated with respect in order to maintain these attributes and channel them into a safe expression and not into conflict (Pickett and Voss, 1975b).

Ideally the stallion should be restrained for day-to-day management by a good strong leather or webbing head collar or halter, and lead rope. This must be checked regularly as, due to his strength, a stallion may break away easily from inadequate restraint and has the potential to wreak havoc in a yard. He should have been taught acceptance of the halter and leading at a young age so that this is not a problem. However, it is inevitable that at an older age he will become more boisterous and, if he has been inappropriately trained in early life, he may need a more substantial means of restraint. This may consist of a halter plus chain, so arranged over the nose that a pull on the chain applies pressure to the nose and provides extra restraint on the stallion. A more severe method along these lines is to pass the chain through the stallion's mouth or under his chin. These should be done with care, as they are potentially very dangerous to the stallion's mouth and nose if extreme pressure is applied.

Many stallions are restrained, especially for covering, by means of a snaffle or stallion bit and, again, a chain may be attached in one of several positions. It may be passed through the left-hand ring and attached to the right-hand ring (Fig. 18.1); attached to a separate chain that is passed through both rings; or, more severely, passed through the left-hand ring, then through the right-hand ring and passed back and attached to the left-hand ring. This arrangement of the stallion chain passed under the chin is popular but is thought by some to encourage rearing, and so an alternative is to pass the stallion chain over the nose, looped through the noseband. This is reported to discourage rearing by encouraging the head to come down when pressure is applied (Fig. 18.2). The more severe forms of restraint should only be used as a last resort.

The effective use of the handler's voice also should not be underestimated. A clear, confident voice command is just as effective as a physical reprimand to a well-trained stallion. A halter for everyday use and a snaffle bridle for covering, along with the effective use of the voice, are all a well-trained stallion should require (Figs 18.3 and 18.4). Whatever restraint is chosen ideally a stallion should have different tack for varying occasions, so that he knows what is expected of him by the restraint used.

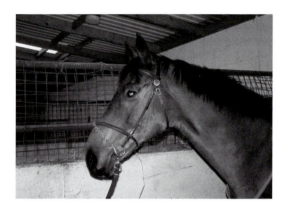

Fig. 18.2. This alternative form of stallion restraint with the chain passing over the nose is thought to discourage rearing, as pressure applied encourages the stallion's head to come down.

Fig. 18.3. All a well-behaved stallion should require for restraint on an everyday basis is a halter. (From Paith Welsh Cob Stud.)

18.4. Introducing the Stallion to Covering

A young stallion should not be expected to cover mares until he is at least 3-years old. Work by Johnson *et al.* (1991) indicates that a 3-year old stallion is capable of covering mares successfully in his first season, but only a few. A 4-year-old is capable of covering a full book of mares (50 season[-1]) but his fertility rates and libido cannot be

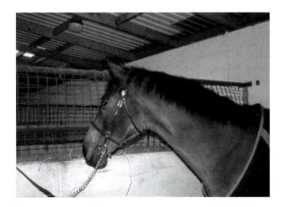

Fig. 18.1. This form of stallion restraint, with the chain passed under the chin, is commonly used but thought by some to encourage rearing.

Fig. 18.4. A snaffle bridle, along with effective use of the voice, is all a well-mannered stallion should require when covering a mare.

expected to be consistent. By 5 years of age, he should have reached his full reproductive potential, which, for most stallions, is 50–100 mares season^{-1} and up to three mares day^{-1} with rest periods. The workload that he is capable of at 5 years of age is likely to be that which can be expected of him at least until his twenties, baring unforeseen circumstances. It is advised, therefore, that during his first season a stallion should be limited to 15 mares or so spread out over the season and he should not be expected to cover more than one per day. The owner of a young stallion should also be prepared to cancel further nominations in the first season if the stallion is showing signs of losing interest and lacks libido or gets injured. All mares for young stallions should be individually picked, as only mature mares of quiet disposition well in oestrus should be chosen. Such mares are often offered on a 'no foal, no fee' basis, as even if a semen evaluation has been conducted, the stallion has as yet no proven fertility record.

A stallion's first covering should be with an experienced handler who knows the stallion well. Even if the stallion is eventually to be used in pasture breeding, or other non-in-hand breeding situations, it is advisable that his first cover is in hand, or in more controlled conditions, to ensure that evasive action can be taken in the event of emergencies. This first cover is extremely important and its success can seriously affect a stallion's long-term ability and behaviour. It is essential that the mare to be covered is experienced, quiet and in full oestrus. Maiden mares are not advisable as they themselves may be unpredictable and it is enough of a job watching an inexperienced and, therefore, unpredictable stallion, without having the added complication of an unpredictable mare. Ideally, the mare should be slightly smaller than the stallion, making it easier for him to mount (Pickett, 1993d).

The stallion and mare should be prepared for covering as detailed in Section 13.5, though the omission of washing of the genitals is practised by some for the first few covers (Samper, 2000). The stallion should be familiar with the covering area, having been introduced to it beforehand, and extra care should be taken to ensure that the floor is non-slip and that there are no protrusions that may injure him or cause him to fall.

The handler should be experienced in the normal sequence of events when mating mature stallions, as ultimately the novice will need to behave in a similar fashion. An inexperienced stallion cannot be expected to conform immediately. At the first covering, it is best to more or less allow him his head. As discussed in Section 13.6, excessive interference by humans discourages the stallion, and at this stage excessive guidance or discipline should be avoided and the stallion allowed to gain confidence in his ability, before he is taught manners as well. However, potentially dangerous habits, such as kicking or biting, should be corrected immediately, as, if allowed to persist, they could render the stallion unusable.

At his first few coverings, the stallion may need prolonged teasing and may mount the mare several times before ejaculation is achieved. He should not be hurried or forced in any way, as this will only serve to upset him, put him off his stride and result in long-term problems. If he seems unable to ejaculate properly, he should be taken away and returned to his box and either tried again with the same mare later on in the day or, better still, with another mare. There is no reason why such hiccups in the first few covers should have any effect on his long-term performance. This first season is all about building up the stallion's confidence in his ability and gradually instilling manners for the sake of safety. It is a gentle balancing act between the two aims, and the rate of progress very much depends on the individual stallion. You must always be prepared to suspend all attempts at covering if he has a bad experience and start again at the beginning to restore his confidence. A bad experience may mean he develops an aversion to a particular type of mare, i.e. colour, size, age, etc. or even a permanent reduction in his libido.

The aim of the stallion's first season is to ensure that he associates his new job with pleasure, in a calm and secure atmosphere, so that he will be able to deal with the occasional not so cooperative mare in later life. It is essential that everything is carried out calmly and that any incident is dealt confidently. Panic and insecurity in the handler will affect the stallion's attitude and performance. Confidence in his handler and surroundings can only serve to enhance his own self-confidence and, therefore, his ability.

18.5. General Stallion Management

The general management of the stallion is extremely important to ensure that he is fit, able and willing to do his job. It also helps to prevent the acquisition of bad habits and increases safety. The main areas of stallion management that need to be considered are housing, exercise, nutrition, feet care, dental care, vaccination and worming programmes.

18.5.1. Housing

Naturally, a stallion would roam wide areas of land, migrating over new pasture with his mares. He could, therefore, exercise himself at will and was always provided with fresh clean grazing. Domestication has largely put paid to this, except in some pasture breeding systems. A stallion's management needs to compensate him for this loss and so optimize his welfare and performance.

Ideally, a stallion should be turned out in a large paddock, but this is not always possible. Climate and limited grazing in many areas precludes the use of all year turnout.

Most stallions are confined to a stable for some period of time at least. This stable must be large, at least 5 × 5 m for a 150 cm horse, and be light and airy. It should have a good strong secure door, with top and bottom sections, plus a top grid that can be shut to provide extra safety but still allow ventilation. Details, such as the stallion's pedigree, can be displayed, adding interest, especially on yards where visitors are catered for (Fig. 18.5). The stable should have a tie ring at the back, to which the stallion should be tied (racked up) as part of his general daily routine, normally when his stable is mucked out and water and hay replenished. It is a good idea to rack stallions up routinely as it makes them easier to handle if visiting mares are around in the yard, and allows them to be easily caught if required. A consistent routine enhances their discipline.

A stallion's stable, as with all horses, should be kept clean and free of flies. Regular cleaning of water troughs and feed mangers is essential.

In order to reduce boredom and the development of vices, such as crib-biting, weaving, stable-walking, etc. the stable door should overlook a busy part of the yard but not too close to any mares. Chains, plastic bottles and swedes hung from the ceiling, or a football or even a cat have been successfully used to provide the stallion with entertainment and, therefore, reduce boredom. The box should have easy access to a paddock, which is normally exclusively for his use. Two acres per stallion is ideal and allows plenty of room for exercise. The cost of fencing can be minimized, as just this paddock need have strong high stallion-proof fencing. For larger stallions, fencing should be post and rail, at least 2 m high. An electric fence, run along the top or projecting into the field about 15 cm

Fig. 18.5. A stallion box at the National Stud, Newmarket, showing the stallion's pedigree displayed on the inside of the upper doors.

from the top of the fence, may be added to provide extra security (Fig. 18.6).

The paddock has to be cleared of manure at regular intervals to ensure it remains clean and reduces the parasite burden. It may also be provided with a field shelter, to give the stallion protection in inclement weather.

Research indicates that the housing of stallions has a direct effect upon reproductive performance. Housing stallions in close proximity to other stallions

Fig. 18.6. Two-metre-high post and rail fencing is ideal for a stallion's paddock. An electric fence running along the top or inside the fencing provides extra security. (From The National Stud, Newmarket, UK.)

may mimic the natural bachelor herd scenario (McDonnell and Murray, 1995). Stallions in such bachelor groups have lower testosterone levels and, therefore, a lower libido than harem stallions. Presumably, it is the social interaction with mares in the absence of other stallions that results in elevated testosterone concentrations. Pasture-bred stallions are reported to have higher libido and exhibit higher fertility rates than stabled, in-hand-bred stallions (McDonnell and Murray, 1995). In general it appears advantageous for stallions to be turned out under natural daylight as much as possible, with plenty of exercise and, within the constraints of safety, any social interactions should be with mares rather than other stallions, at least during the breeding season (McDonnell, 2000b).

18.5.2. Exercise

Exercise, as for all horses, is essential for the physical and psychological well-being of the stallion. It helps to reduce boredom and maintain basic fitness and muscle tone. Fitness is especially important, as stallions undergo short, sharp periods of extreme exercise when covering. Exercise improves the cardiovascular system and reduces the chances of conditions such as azoturia (tying up) and improves general well-being (Dinger and Noiles, 1986a). Exercise also helps aid digestion and promotes a healthy appetite. It can either be free or forced, i.e. turned out or ridden/lunged. Many stallions are unbroken, in which case the only option is free exercise. As such, they should be turned out for as long as possible each day (Fig. 18.7).

Free exercise is fine for those willing to exercise themselves. However, some refuse to move around the paddock, or at the other extreme charge around like mad and pace the fence, spending no time grazing and so lose condition. The exercise of these stallions has to be controlled by means of forced exercise, which also improves discipline, especially in nervous and highly strung animals. Riding and lunging are popular forms of forced exercise. Other forms include swimming (Fig. 18.8), which is particularly beneficial to the cardiovascular system and for lame horses. Treadmills or horse walkers provide an effective means of forced exercise but should be restricted to stallions accustomed to them (Fig. 18.9).

Some stallions, especially native types, can be turned out with mares and foals. This system has the added advantage that mares returning to oestrus after unsuccessful covering can be detected and re-covered by the stallion turned out with them (Fig. 18.10). This system should be confined to use with well-behaved, older stallions, which are well into their working season.

Fig. 18.7. Turnout into a field provides ideal exercise for a stallion, weather permitting.

Fig. 18.8. Swimming provides good exercise, especially for the cardiovascular system and for stallions with lameness problems.

Fig. 18.9. Treadmills or horse walkers are a good means of forced exercise but should be restricted to stallions that are used to them.

Exercise must be closely monitored along with nutrition to ensure that the stallion remains in body condition score 3. He must not be over exerted or he will not have enough energy for the real job in hand.

18.5.3. Nutrition

A properly balanced diet is essential for a stallion's well-being. Each stallion should be fed individually according to his size, condition, workload,

Fig. 18.10. Stallions of a quiet disposition can be turned out with mares, especially towards the end of the season. This relieves boredom in the stallion and allows mares returning to service to be recovered. (From Derwen International Stud.)

temperament, etc. He should be in a condition score of 3 and his feed should be carefully monitored throughout the year in order to maintain this. One of the problems encountered with stallions is obesity. Good nutrition and exercise management can prevent this. During the breeding season, the workload, in nutritional terms, of a stallion with a full book of mares, is as great as that of a performance horse. As a general rule, a stallion should have a daily intake of 2–3% of body weight; at least 50% of this should be of good-quality roughage (Hintz, 1993b). Young growing stallions should be

fed a slightly higher proportion of concentrates, i.e. a ratio of 60:40 of concentrates/roughage.

18.5.3.1. Protein and energy

For a 500 kg mature working stallion, a protein daily intake of 789 g is recommended with higher levels for young stallions (Hintz, 1983, 1993b; Hurtgen, 2000; National Research Council, 2007). Daily energy levels of 21.8 Mcal, similar to those for horses in heavy work, are recommended (Tables 14.2, 18.1, 18.2).

Table 18.1. Daily nutrient requirements of stallions of varying weights. (From National Research Council, 2007.)

Animal	Weight (kg)	DE (Mcal)	Crude protein (g)	Lysine (g)	Ca (g)	P (g)	Mg (g)	K (g)	Vitamin A (10³ IU)
Stallion (working)									
	200	8.7	316	13.6	12	7.2	3.8	11.4	8
	500	21.8	789	33.9	30	18	9.5	28.5	20
	900	39.2	1421	61.1	54	32.4	17.1	51.3	36
Stallion (non-breeding)									
	200	7.3	288	12.4	8	5.6	3.0	10	8
	500	18.2	720	31	20	14	7.5	25	20
	900	32.7	1296	55.7	36	25.2	13.5	45	36

DE = digestible energy; Ca = calcium; P = phosphorous; Mg = magnesium; K = potassium.

Chapter 18

Table 18.2. Expected feed consumption by stallions (percentage of body weight). (From National Research Council,1989.)

Animal	Forage	Concentrate	Total
Mature stallion	1.0–2.5	1.0–1.5	2.0–3.0
Young stallion	0.75–2.25	1.25–1.75	2.0–3.0

18.5.3.2. Vitamins and minerals

Many breeders feed a vitamin and mineral supplement on a free access basis to stallions, regardless of feed analysis. This is not always necessary, but can be used as a precaution. The only vitamin that is likely to be short in a well-balanced diet is vitamin A. However, the inclusion of roughage in the diet in the form of leafy green forages, which are high in vitamin A, helps to address this potential shortfall (Hintz, 1993b, Hurtgen, 2000). There is no research which indicates that any single nutrient can improve sperm quality or quantity (Steiner, 2000).

Inappropriate nutrition is one of the major causes of low libido and poor reproductive performance. Correct monitoring of a stallion's condition and adjustment of nutrition and exercise, accordingly, cannot be overemphasized. However, sudden changes to feeding immediately prior to the breeding season can have as detrimental an effect on performance as over- or under-nutrition per se (Hintz, 1993b).

18.5.3.3. Water

As with all horses access to clean reliable water source is essential. A working stallion, of 500 kg in weight in an ambient temperature of 20°C, may well require in excess of 50 l day^{-1} (10 l 100 kg^{-1} body weight day^{-1}) compared to a similar stallion, not in work, i.e. maintenance requirement only, which would require only 25 l day^{-1} (5 l 100 kg^{-1} body weight day^{-1}; Frape, 2004; National Research Council, 2007).

18.5.4. Feet care

The feet of a stallion should never be neglected. Lameness can severely reduce and restrict the stallion's ability to cover. This is especially evident in hindlimb lameness and is often first evident as uncharacteristically low libido. A stallion should ideally not be shod during the breeding season, as shoes can inflict more damage than unshod feet.

Regular, 6–8 week, trimming should be carried out to ensure that the feet remain clean, and uncracked. Any problems should be dealt with immediately to avoid long-term complications. Regular turnout is conducive to good-quality hooves, again reinforcing the ideal of keeping stallions in paddocks for at least part of the day (Hurtgen, 2000).

18.5.5. Dental care

Uneven, rough teeth, along with lesions or abscesses, can reduce a stallion's appetite due to pain. They can also cause food to pass into the stomach without full mastication, reducing the efficiency of digestion. If a stallion starts to lose condition for no obvious reason, one of the first things to check is his teeth and mouth. In any case, a stallion's mouth should be checked annually to see whether rasping is required.

18.5.6. Vaccination

Vaccination is an essential part of a preventative medicine routine that should be developed and implemented regularly for a stallion throughout his life. The vaccinations required depend upon the country in which the stallion resides and on the prevalence of various infections. In the UK, stallions should have up-to-date influenza and tetanus inoculations. Vaccination against equine herpes virus 1 (EHV1) is becoming more popular in UK, and is also available now for equine viral arteritis (EVA; Timoney and McCollum, 1997). In other parts of the world, vaccination for rabies, botulism, eastern and western encephalomyelitis, African horse sickness and Potomac horse fever may be considered; vaccination against strangles is possible but not popular due to side effects (see Section 19.2.2.5.7). In general, any vaccination is advised to be administered at least 60 days before breeding starts to ensure that any resultant fever does not affect breeding performance (Hurtgen, 2000; Steiner, 2000).

18.5.7. Swabbing

Swabbing of the stallion's genitalia to test for pathogenic bacteria, as detailed in Section 12.5.2, is increasingly practised. An annual series of swabs is a compulsory requirement for Thoroughbreds and is increasingly demanded by other breed societies. Swabs are normally taken from the prepuce, urethral fossa and sheath at the beginning of each season and if a problem is suspected.

18.5.8. Parasite control

Worming is another essential part of a preventative-medicine routine. As with all wormers in all horses, the key to success is regular use in combination with good grazing management and rotation of the product to ensure that resistance does not develop. A high worm count, like any parasitic infection, causes listlessness and, hence, low libido and reduced reproductive performance. Worming and/or faecal egg count monitoring should be carried out regularly, especially throughout the spring, summer and autumn. Some wormers are themselves reported to cause listlessness and reduced libido but only for a few days after administration. Bearing this in mind some studs try to organize their worming regime so that stallions are not wormed at the height of their covering season.

18.6. Stallion Vices

Largely due to the manner of current management practices, many stallions are in danger of developing bad habits, or vices, due to boredom. Prevention is infinitely better than cure and it is, therefore, essential that the stallion's management is geared appropriately (McBride and Hemmings, 2005). A regular routine of work, exercise, feeding, etc. and a stable in an area of the yard where activity can be observed goes a long way to achieving this. A frustrated and bored stallion releases his energies and tensions in the only way possible to him, by developing vices. These vices can be harmful to the stallion himself, dangerous to the handler and may also affect his reproductive performance. If prevention has failed and vices have developed there are certain practices that can help control them and/or their effects.

18.6.1. Crib-biting and wind-sucking

Crib-biting and wind-sucking are primarily caused by boredom and are related, often developing the one from the other. During crib-biting the horse bites part of the stable structure or other convenient object (Fig. 18.11). It is thought to develop from the horse's natural urge to eat or graze regularly. The condition is exacerbated when feed is delivered in small concentrate meals with limited roughage. Increasing roughage and allowing *ad libitum* availability will certainly reduce the chance of this vice developing and will help alleviate the condition in sufferers. The habit can be discouraged by removing all objects that

Fig. 18.11. Telltale signs of crib-biting in a horse; protruding surfaces, commonly stable doors, showing signs of teeth marks.

can be grasped by the teeth, or by painting structures with Cribox or an equivalent foul-tasting substance as a deterrent. Wind-sucking can develop on from crib-biting and in this more serious condition, the horse, while grasping the projecting structure, arches his neck and gulps in air. If the habit is allowed to continue unchecked, it can lead to colic and reduced appetite, as well as excessive wear and tear on the upper incisor teeth. A muzzle can be used to prevent both vices. A cribbing strap, placed around the horse's throat, preventing the stallion from tensing the neck muscles used in wind-sucking, may also help (Fig. 18.12).

Fig. 18.12. A cribbing strap may be used to prevent the stallion from tensing the neck muscles required for wind-sucking.

A more extreme method of curing wind-sucking is severing of the neck muscles attaching the hyoid bone to the base of the tongue. Alternatively, the nerves serving these muscles can be severed. Such a procedure, in the majority of cases, will effect a cure, and in the remainder, considerable improvement is obtained, but these are drastic solutions to a problem that is largely avoidable with appropriate management (McGreevy *et al.*, 1995).

18.6.2. Weaving

Weaving involves the lateral swaying of the horse's head and neck rhythmically from side to side, often over the stable door. This can cause damage to the forelegs as the horse's weight is repeatedly shifted from side to side. The condition is thought to develop from the horse's natural urge to move continuously, associated with grazing over large tracts of lend. A chronic weaver may weave himself to the point of exhaustion. The condition can be alleviated by anti-weave bars over the lower stable door (Fig. 18.13).

Unfortunately, chronic weavers will continue to weave within their boxes. Furthermore, stable-walking may develop, in which the stallion continually paces around his box, seemingly chasing his tail. There is little that can be done to cure this behaviour except to turn the horse out into a paddock to relieve the boredom, but such horses often then fence walk and usually revert to stable-walking as soon as they are stabled again (Mills and Nankervis, 1999).

Fig. 18.13. During the times that the top door of the stable is open, anti-weave bars may be used to discourage the stallion from weaving over the bottom door.

18.6.3. Self-mutilation

This vice, unlike the others, is not normally an expression of boredom, but nevertheless can be extremely distressing to the stallion and his owner. The stallion bites his own legs, shoulders and chest, causing himself some considerable damage. It is normally particularly evident after mating. If evident only then, thorough washing of the stallion after dismount reduces expression of the vice, which in this case is thought to be due to the smell of the mare. If, however, the stallion is a habitual self-mutilator, there is very little that can be done to cure him, though the use of a muzzle or cradle can prevent him inflicting damage.

18.6.4. Masturbation

Masturbation by a stallion is considered by some to be a further vice, thought to originate from boredom, especially if sexually frustrated. However, it is evident that feral and wild ponies also demonstrate such behaviour. It has been suggested, therefore, that it is a natural behaviour, rather than a problem, and any problem lies with human perception and potential embarrassment. The old-fashioned use of penile rings, etc. is now frowned upon and unnecessary. Masturbation is expressed by the stallion rubbing the extended penis along the under side of his abdomen. In extreme cases masturbation may result in ejaculation and concern over the loss of valuable sperm. Providing the stallion's workload is not too high, such behaviour should not affect his fertility rates, indeed increasing his workload may go some way to reducing it (Pickett, 1993d; McDonnell, 2000b).

18.6.5. Aggressive behaviour

Some stallions develop extremely aggressive behaviour and become a danger to both handlers and mares (McDonnell, 2000b). Occasionally, this behaviour is associated with certain conditions or restraint, or is directed towards certain people and can be averted by avoiding such situations. However, more often than not, it is expressed generally and is due to mismanagement during formative years. If the behaviour is beyond control, the stallion can be gelded. In 95% of cases, gelding significantly reduces aggressive tendencies. If the stallion must remain entire, then certain measures can, and should, be used to protect handlers and mares.

He should be muzzled, to prevent savaging mares during covering, and controlled by a pole attached to his bit, giving his handler more control and the only option with some stallions may be artificial insemination. In deciding whether to continue to use an aggressive stallion, it must be certain that his behaviour is management-induced and not inherited, as it is very important that such behaviour is not perpetuated in subsequent generations.

Rearing and striking out with the front feet constitute a relatively common vice, though potentially very dangerous to handlers. This should be corrected, especially in young stallions, where the vice can be cured. To avoid being kicked, the handler should always stand to one side of the stallion, and never in front. A long lead rein should be used so that contact can still be maintained from a distance if the stallion rears. As soon as the stallion starts to rear, his lead rein should be jerked sharply, along with a verbal reprimand. Backing the stallion at the first signs of rearing can also help to avert the situation. The use of a chain under the chin is one of the more popular forms of stallion restraint, but has been reported to be associated with a higher incidence of rearing and hence may be best avoided.

Finally, biting is another relatively common vice in stallions. This should be corrected at a young age by a short, sharp jerk on the lead rein, or sharp tap on the muzzle and verbal reprimand as punishment. If allowed to continue, a stallion can become almost impossible to handle. Some stallions only bite in certain situations, such as when they are eating, after mating, when they are being groomed, or when fed by hand. Such situations should, therefore, be reduced to a minimum, and any handlers warned of the problem.

18.7. Conclusion

Stallion management from a very early age has important implications for reproductive ability and behaviour. Many problems encountered in stallions, which either do not perform to their full potential or exhibit antisocial behaviour, stem from mismanagement at an early age. One of the major problems encountered with stallions is boredom, due to confinement and isolation. This can directly affect libido, performance and other behavioural characteristics. Unfortunately, this often becomes a self-perpetuating downward spiral, which is very difficult to break. Many of the problems encountered in the management of stallions can be averted by consistent discipline and alleviating boredom by providing turnout, social interaction and activity.

19 Infertility

19.1. Introduction

Infertility is a vast subject and this chapter can only be an introduction providing a basis from which further information can be sought. Infertility may have its root cause in either the stallion or the mare, and each of these aspects will be discussed in turn.

Expected fertility rates vary enormously and are affected, among other things, by management. As an average within the UK, 40–80% of mares covered deliver a live foal (Osborne, 1975; Sullivan *et al.*, 1975; Baker *et al.*, 1993). The definition of fertility is however variable and may be expressed in many ways: percentage of mares pregnant per mating, percentage of mares pregnant per oestrous period, percentage of mares pregnant at the end of the season (often from coverings on numerous oestrous periods), percentage of mares producing a live foal, etc. It is, therefore, very difficult to make comparisons between reported fertility rates. Regardless of the precise definition of fertility, the ability of a mare to produce a live offspring the following year is the main consideration, failure to do so can be due to extrinsic or intrinsic factors, and this is the case with both the mare and the stallion. Extrinsic is the term given to external factors affecting reproductive performance, whereas intrinsic is the term given to internal, usually physiological, factors. When considering the mare and the stallion, these two areas will be dealt with in turn.

19.2. Mare Infertility

On average 40–80% of services result in a live foal. This figure, however, varies significantly with the population of mares (Osborne, 1975; Sullivan *et al.*, 1975; Mahon and Cunningham, 1982; Baker *et al.*, 1993). Human interference in the reproduction of the horse has also had a significant effect on reproductive performance. Live foal rates for wild horses and ponies are in the region of 95% compared with 60% for hand-bred mares (Bristol,

1987). Embryo mortality or early embryonic death (EED) is held responsible for a significant amount of the apparent infertility. Rates of 5–14% have been reported for in-hand mating, compared to 1–2.5% for free mating, though higher rates of EED (up to 50% for normal mares) have been reported by others (Baker *et al.*, 1993). As most EED occurs prior to day 40 (Ginther, 1985) most measurements of infertility will also include this.

Reproductive performance has been largely ignored in the improvement of the equine. This is in contrast with other farm livestock, where reproductive efficiency is of utmost importance. It is evident that barrenness in a mare at the end of the season is not necessarily due to pathological infertility but is also a consequence of other environmental and managerial influences and the natural tendency in many mares to take a season off breeding. Some mares are consistently barren in early life but successfully breed later on. Failure to produce an offspring in a particular year is, therefore, a complicated problem and is not to be confused with infertility and EED. Before the reasons for a mare failing to produce a foal are discussed in detail, the following glossary should be noted to prevent confusion in terms.

Fertile	– able to produce a live foal
Infertility	– a temporary inability to reproduce
Barrenness	– lack of a pregnancy at the end of the season, but perfectly capable of producing a foal, as demonstrated in previous years
Subfertility	– inability to reproduce at full potential may be temporary or permanent
Sterility	– a permanent inability to reproduce
Embryo mortality or death (EED)	– embryo loss prior to day 40, often occurring between scanning at days 15 and 40
Abortion	– fetal death after day 40
Stillborn	– fetal death after day 300
Fertilization rate	– number of ova fertilized per ovulation (85–90%; Pycock, 2000)

| Pregnancy rate | – number of mares pregnant on a specified day, expressed per oestrous cycle (45–60%) or per breeding season (75–90%; Pycock, 2000) |
| Live foal rate | – number of mares foaling per number of mares bred over the season (50–85%; Pycock, 2000). Arguably the best indicator of reproductive performance |

In the following account, the term reproductive performance will be used, as it encompasses all the above definitions. In many texts, the strict definitions of infertility, EED and abortion are not adhered to and make it very hard to distinguish precisely what is responsible for reported changes in reproductive performance. In the mare, reproductive performance does not just depend upon successful gamete production, as is the case with the stallion, but also upon an appropriate environment for fertilization, the free-living embryo, implantation, placentation and subsequent parturition. As such, the following sections will include consideration of fertilization failure, EED and abortion as appropriate. The failure of a mare to produce a foal at the end of a season may have numerous causes and can be divided into extrinsic and intrinsic factors.

19.2.1. Extrinsic factors affecting reproductive performance in the mare

Extrinsic factors affecting reproductive performance in the mare may be considered to include: lack of use; subfertile or infertile stallion; poor stallion management; poor mare management; and the artificially imposed breeding season.

19.2.1.1. Lack of use

A mare may not be covered in a particular season, due to design or unavoidable circumstances. If she foals late in one season she may not be re-covered, in order to allow her to return to foaling earlier in the year. Disease or infection may preclude a mare from use in a particular year, either because she herself is not in a fit condition to successfully carry a foal or there is a danger of systemic or venereal disease transfer. In the case of a performance horse she may not be bred in a particular year due to work commitments, though advances in embryo transfer now potentially allow such mares to foal via use of a recipient mare (Chapter 21, this volume).

19.2.1.2. Subfertile/infertile stallion

Half the responsibility for the success or failure of a covering lies with both the stallion and the mare. If a mare is covered by an infertile or subfertile stallion her chances of producing a foal are significantly reduced through no fault of her own. The causes of infertility in the stallion will be discussed in Section 19.3. Mares can only be expected to perform to their full reproductive potential if they are covered by a stallion whose semen meets minimum requirements. A stallion must also be physically capable of covering a mare effectively; a good semen evaluation in the absence of the ability or willingness to cover is of no use in the natural service of a mare. To a certain extent, this problem may be overcome by the use of artificial insemination (AI), but in such cases, it must be certain that the stallion's lack of libido or ability is not due to a potentially heritable fault.

19.2.1.3. Poor stallion management

Stallion management has been considered in detail in Chapters 12, 13, 18 (this volume) and will also be covered later in this chapter. It is evident that, if any aspect of a stallion's management, especially his covering management, is not correct, then there is the potential for his fertility rates to be affected. Management in the earlier formative years also has a significant effect on a stallion's libido and hence reproductive performance.

The imposition of an artificial breeding season causes problems with reproductive performance. Even with the use of artificial lights, a stallion's performance during the months of December to February is lower than during his true breeding season. So fertility rates for mares covered within this period of time cannot be expected to meet normal expectations.

Behavioural abnormalities, often associated with inappropriate management, such as failure to obtain or maintain an erection, incomplete intromission and ejaculation failure, may cause apparently low fertilization rates as can the incorrect detection of successful ejaculation (Kenney, 1975a; Pickett and Voss, 1975b).

19.2.1.4. Poor mare management

Mare management has been discussed in detail in Chapters 12, 13 and 14 (this volume). Any deficiencies or inadequacies in brood-mare management can lead to poor reproductive performance. Of specific significance is covering management, especially oestrus and ovulation detection. Better, more experienced management and the use of veterinary diagnosis and hormonal manipulation of the cycle are seen to significantly improve reproductive performance. Stress associated with handling and transportation is suggested to be associated with EED (Osborne, 1975), possibly via changes in plasma cortisol (Bacus *et al.*, 1990) and progesterone concentrations (Van Niekerk and Morgenthal, 1982). Finally, nutritional stress is also reported to affect reproductive success, again mainly via EED and abortion (Henneke *et al.*, 1984; Potter *et al.*, 1987; Ball, 1993a).

19.2.1.5. Imposed breeding season

As detailed previously and in Section 12.6.1, there is considerable pressure for foals to be born as soon as possible after 1 January. The methods by which the mare's breeding season may be manipulated in order to achieve this are also discussed in Section 12.6.1. Regardless of the treatment used, pregnancy rates outside of the natural breeding season are never as high as normal expectations. As such, the continued imposition of an arbitrary breeding season places unfair constraints upon a mare's potential reproductive performance.

19.2.2. Intrinsic factors affecting reproductive performance in the mare

Intrinsic factors affecting reproductive performance in the mare may include: age, chromosomal, hormonal, pituitary, ovarian, Fallopian tube, uterine, cervical, vaginal and vulval abnormalities and infections. All these will be discussed in turn; many, however, are closely interrelated.

19.2.2.1. Age

Age is reported to have the most significant bearing on reproductive performance (McDowell *et al.*, 1988; Barbacini *et al.*, 1999). In general, fertility decreases with age and EED increases (Ball *et al.*, 1989; Ball, 1993a; Carnevale *et al.*, 1993). Baker

et al. (1993) indicated that young mares have 90–95% fertilization rates with EED up to 50%. However, the fertilization rates of older mares decline to 85–90% and EED may reach 100%. It has been suggested that this decrease in fertility may in part be due to a increase in the transit time (greater than 4 h) for sperm to reach the oviduct (Scott *et al.*, 2000), making the timing of covering in older mares more crucial. This would also in part explain the observation that significantly more cleaved ova can be collected at day 15 in 2–10-year old compared to 20-year-old mares. Additionally, older mares yield more embryos with morphological abnormalities (Carnevale and Ginther, 1992; Carnevale *et al.*, 1993) and demonstrate more multiple ovulations and so an increased likelihood of multiple pregnancies (Davies Morel and O'Sullivan, 2001), thus further adding to the high apparent infertility rates in older mares. Evidence would also suggest that the occurrence of anovulatory oestrus is greater in mares over 20 years (Vanderwall *et al.*, 1993) and that placental development, particularly blood supply to the microcotyledons is adversely affected by old age (Bracher *et al.*, 1996; Merkt *et al.*, 2000; Allen, 2001). Despite reducing reproductive performance, providing a mare is in good physical condition, she may breed successfully well into her 20s. However, welfare considerations may preclude breeding a mare of advanced age. In such cases, embryo transfer may prove a viable alternative to putting an older mare through the stresses of carrying a pregnancy to term (Chapter 21, this volume; Carnevale and Ginther, 1992).

19.2.2.2. Chromosomal abnormalities

The normal chromosomal complement for the equine is 64 (32 pairs), the female complement being denoted as 64XX. Various variations on the normal complement include 63XO (a female with a single X chromosome), termed Turners syndrome, one of the more common chromosomal abnormalities. Such individuals are characterized by small rudimentary ovaries, flaccid poorly developed uterus, no ovarian activity and, therefore, permanent anoestrus. Such mares tend also to be short in stature (Chandley *et al.*, 1975; Hughes *et al.*, 1975). Mosaic chromosomal configuration may occur as 64XX complement in some cells, and 63XO in others; such individuals demonstrate erratic oestrous cycles with no ovulation (Chandley *et al.*, 1975). Very rarely, chromosomal complements of 64XX/65XXY are found; such horses are termed intersex (Dunn *et al.*, 1981; Halnan, 1985; Kubien and Tischner, 2002). Positive diagnosis of chromosomal abnormalities is only possible by karyotyping (genetic mapping) via cytogenic analysis of blood samples, though they may be indirectly indicated physiologically (Ricketts, 1975b, 1978; Halnan, 1985; Bowling *et al.*, 1987; Bowling and Hughes, 1993). Newcombe (2000) suggests that chromosomal abnormalities incompatible with life are a significant cause of EED in the mare.

It is interesting to note that inbreeding is often practised within the equine industry in an attempt to fix traits within a population. Such levels of inbreeding would cause significant fertility and congenital problems to other mammals; however, this does not seem to be such a big problem in horses (Mahon and Cunningham, 1982).

19.2.2.3. *Hormonal abnormalities*

As discussed in Chapter 3 (this volume), the control of the mare's reproduction is a finely balanced cascade and interrelationship of hormones involving the hypothalamic–pituitary–ovarian axis. Abnormalities/inefficiencies in any of these centres can cause imbalance throughout the whole axis. The majority of hormonal deficiencies are associated with pituitary abnormalities – for example, Cushing's syndrome (Pycock, 2000). Complete failure, or neoplasia, of the pituitary is relatively rare in the horse, but temporary malfunction may occur, especially in association with the transitional period at the beginning or end of the breeding season. Mares during this transitional period tend to suffer from prolonged or persistent oestrus (nymphomania), prolonged dioestrus (persistent corpus luteum (CL)), silent ovulations (failure to exhibit oestrus despite ovulation), split oestrus (oestrus over a period of up to 3 weeks with possibly a quiescent period in the middle), anovulatory follicles (follicles that luteinize but fail to ovulate), etc. Similar problems may be seen in post-partum mares due to lactational anoestrus. In such cases, it is evident that a period of time is required to allow the mare's system to re-establish regular 21-day cycles.

Diagnosis of hormonal abnormality is initially via a mare's behaviour and the seeming inability to detect oestrus or conversely, apparent continual oestrus. Diagnosis of the cause is helped via scanning and rectal palpation, by which ovarian activity can be monitored. The incidence of hormonal deficiencies or abnormalities is particularly evident today, as continued attempts are made to breed

mares earlier in the season. As discussed in Section 12.6.1, the use of exogenous hormonal treatments and/or light treatment does successfully advance ovulation and oestrus within the year, but does not eliminate the transition period, which may still be associated with problems.

Pituitary or hypothalamic tumours are rare in mares; they are associated with muscle wasting, hypoglycaemia, docility, alopecia, blindness and uncoordinated movement in addition to prolonged anoestrus.

Hormonal deficiencies during pregnancy may also result in reproductive failure. In particular, progesterone insufficiency may result in EED or abortion, dependent upon when it occurs (Ginther, 1985; Morgenthal and Van Niekerk, 1991). Exogenous progesterone supplementation has proved successful in problem mares, and it is routinely used in some stud practices, as a safe guard, though such use is not normally justified with normal mares (Ganjam *et al.*, 1975; Pycock, 2000).

19.2.2.4. *Physical abnormalities*

19.2.2.4.1. OVARIAN ABNORMALITIES Occasionally ovaries may be absent due to surgical intervention, or chromosomal abnormality. Inactive ovaries and ovulation failure are often observed in mares and are exacerbated by the imposition of an arbitrary breeding season (Kenney *et al.*, 1979).

19.2.2.4.1.1. **Follicular atresia** Follicular atresia is responsible for some incidences of ovulation failure. In such cases, a group of follicles will develop normally, to about 3 cm in diameter, but there is a failure in the emergence of a dominant follicle, which would be expected to develop further. Conditions such as ovarian hypoplasia, granulosa cell tumours, ovarian cysts, uterine infections and malnutrition have all been implicated in follicular atresia (Bosu, 1982; Pugh, 1985; Bosu and Smith, 1993). The best cure appears to be time, especially in mares encountering problems during the transitional stage of the breeding season. Often succeeding cycles will not demonstrate the condition.

19.2.2.4.1.2. **Corporal lutea persistence and failure** CL persistence and conversely failure are also causes of reproductive failure in the mare, manifesting themselves as long or short oestrous cycles, respectively. Failure of the CL is less evident in the mare than persistence. However, CL failure is implicated in experiments using progesterone supplementation to prevent abortion (Ganjam *et al.*, 1975; Allen, 1993). The effect of progesterone insufficiency has also been considered in Section 19.2.2.3. The presence of a persistent CL is more common in mares and is an important cause of anoestrus. The normal lifespan of a CL is 14 days, after which, in the absence of a pregnancy, the luteolytic hormone prostaglandin F2α (PGF2α), secreted by the uterine endometrium, takes effect. A persistent CL is presumably, therefore, a result of failure in the release of PGF2α, or the ability of the CL to react appropriately. The presence of such conditions in a mare is implicated by the lack of oestrous behaviour, and is confirmed by scanning or rectal palpation. The failure of the luteolytic message may be linked to uterine infection, rendering the uterine endometrium unable to produce PGF2α. EED and associated pseudopregnancy are other potential causes of a persistent CL. Treatment with exogenous PGF2α is normally successful (Kenney *et al.*, 1975b; Pycock, 2000).

19.2.2.4.1.3. Anovulatory follicles
Anovulatory follicles (haemorrhagic, persistent or luteinized unruptured follicles) can be a cause of anoestrus. They occur most commonly in the transition period into and out of the breeding season. Anovulatory follicles are characterized as large follicles, which

fail to rupture and ovulate. Instead, over time they fill with blood and persist as haematomas, possibly over a number of cycles. Their presence is further complicated by their ability to secrete progesterone – hence their alternative name of luteinized unruptured follicles and their similarity to functional CL. Differentiation between the two can only be made at scanning by variations in their echogenic characteristics (Ginther and Pierson, 1989; Pycock, 2000).

19.2.2.4.1.4. Ova fossa cysts
Ova fossa cysts, sometimes termed cystadenomas, are reported, especially in older mares. They appear to be associated with the epithelium of the fimbrae and may cause blockage of the ovulation fossa and, therefore, a disruption of ova release. They are often evident as a bundle of cysts, similar to a bunch of grapes, near the ovulation fossa. In the extreme, they may also interfere with the blood supply to the rectum (Prickett, 1966; Stabenfeldt, 1979).

19.2.2.4.1.5. Granulosa (theca) cell tumours
Granulosa theca cell tumours are the most common tumour within the equine ovary and an important cause of anoestrus (Sundberg *et al.*, 1977; Rambegs *et al.*, 2003). They normally affect mares between the ages of 5 and 7, and are usually associated with a single ovary. The ovaries are either polycystic or large solid structures and may weigh up to 8 kg (Fig. 19.1; Norris *et al.*, 1968; Stabenfeldt *et al.*, 1979).

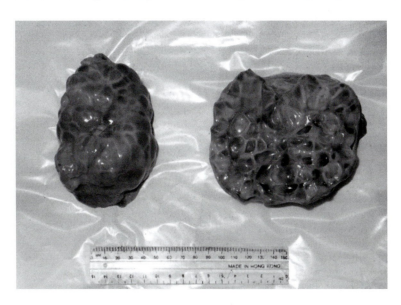

Fig. 19.1. Granulosa theca cell tumour in the mare evident in the polycystic form.

The symptoms demonstrated by such mares depend on the hormones secreted by the tumours. Oestrogen, the most common, may result in nymphomaniac behaviour (prolonged oestrus) in the absence of ovulation. Testosterone-producing cysts result in stallion-like behaviour and muscular development. Elevated inhibin concentrations are thought to cause the characteristically small contralateral ovary due to negative feedback effects (Stabenfeldt *et al.*, 1979; Piquette *et al.*, 1990). Removal of the affected ovary often allows the resumption of normal reproductive activity by the remaining ovary (Meager, 1978).

19.2.2.4.1.6. **Ovarian teratomas** Ovarian teratomas, arising from germ cells and containing hair, teeth, bone, cysts, etc., have been reported but very rarely occur (Rossdale and Ricketts, 1980; Hughes, 1993) and, are non-secretary and benign. They are unilateral in occurrence, allowing the other ovary to function normally; pregnancy rates may or may not be affected significantly. Parovarian or paroophoron cysts are also reported, but rarely cause problems.

19.2.2.4.1.7. **Hypoplasia** Ovarian hypoplasia (underdevelopment) of the ovary is a further cause of anoestrus. It is characterized by small, immature ovaries with no ovarian activity or increase in ovarian size within the breeding season. Hypoplasia is usually bilateral and is often associated with chromosomal or hormonal abnormalities as previously discussed.

19.2.2.4.1.8. **Cystic ovaries** The term cystic ovary implies the presence of fluid-filled structures within the ovarian stroma, which are hormonally active. Such structures are reported in cattle but are not reported not to occur in mares (Pycock, 2000).

19.2.2.4.1.9. **Other abnormalities** Dysgerminomas (malignant tumours of cells resembling primordial cells), abscesses and haematomas (overfilling of the follicular cavity with blood post ovulation) are also reported to occur but rarely (Meuten and Rendano, 1978; Bosu *et al.*, 1982; Neely, 1983; Bosu and Smith, 1993).

19.2.2.4.1.10. **Multiple ovulation** Multiple ovulations, and the resulting multiple pregnancies, have been considered previously in Sections 3.2.5 and 14.3. However, they warrant mentioning here as a major cause of EED and abortion, hence

reproductive failure. The mare is monocotous, and, as such, she is unable to satisfactorily support more than one fetus *in utero*. Fetal restriction results, causing either spontaneous abortion of one or both fetuses or the mummification of the smaller twin (McDowell *et al.*, 1988). Twins are, therefore, not desirable and management in general is geared towards avoiding them (see Section 14.3). Twinning is an inherited trait and can be avoided or an awareness of its possibility gained by studying a mare's breeding history and any previous incidents of twinning (Jeffcote and Whitwell, 1973; Ginther, 1982; Miller and Woods, 1988).

19.2.2.4.2. FALLOPIAN TUBE ABNORMALITIES The extent of follicular tube abnormality is disputed. Arthur (1958), using slaughterhouse material, demonstrated that the incidence of such abnormalities was very low. However, Vandeplassche and Henry (1977) demonstrated a 40% incidence, normally involving adhesions of the infundibulum to other parts of the reproductive tract. Collagenous masses that may occlude the lumen of the follicular tubule have been more recently documented (Lui *et al.*, 1990). Rarely, an ovarian cyst may be seen to block the entry to the Fallopian tube at the infundibulum. Tumours of the Fallopian tube are extremely rare (Allen, 1979).

19.2.2.4.3. UTERINE ABNORMALITIES Uterine abnormalities due to congenital defects or infections are relatively well understood in the mare, compared to abnormalities of the remainder of the tract. The development of techniques such as ultrasonic scanning, endoscopy and uterine biopsy has significantly advanced our understanding of uterine physiology and pathology (Kenney, 1975b; Brook, 1993).

19.2.2.4.3.1. **Endometriosis, or chronic non-infective degenerative endometritis** Endometriosis is caused by degeneration, rather than by infection of the endometrium, and may be classified as infiltrative or degenerative. Infiltrative endometriosis may be a result of changes within the uterus due to a busy breeding career, and is associated with a natural increase in leucocyte response to the normal bacterial challenge post coitum (Van Furth *et al.*, 1972; Ricketts, 1978). Degenerative endometriosis is a degeneration of the endometrial glands rendering the uterus incapable of supporting a pregnancy. It is associated with EED and is often the result of repeated gestations, especially in mares with a history of uterine infections. Degeneration of the endometrial glands results in a failure to return to normal post partum, leaving lymph-filled lesions (Walter *et al.*, 2001). Recent work suggests that endometriosis does not affect the ability of the uterus to react to a bacterial challenge (phagocytic activity of polymorphonuclear neutrophilic granulocytes, PMN; Zerbe *et al.*, 2004). Treatment may be attempted by the stimulation of growth of new healthy endometrium using mechanical or chemical curettage (see Section 19.2.2.4.3.6.1). However, such treatment is not very successful and runs the risk of further uterine damage. The prognosis for such mares is poor (Witherspoon, 1972; Bergman and Kenney, 1975; Kenney and Ganjam, 1975; Asbury and Lyle, 1993).

19.2.2.4.3.2. Hyperplasia Uterine hyperplasia (overdevelopment) is characterized by an overdevelopment of the uterus for the reproductive stage of the mare, or a failure to recover from a previous event such as pregnancy. Again, the endometrial glands are most significantly affected. As indicated, the condition is often a result of delayed involution post partum, the uterine endometrium failing to return to normal within the time expected (see Section 16.5.2; Ricketts, 1975a; Kenney, 1978; McKinnon, 1987c; McKinnon *et al.*, 1988a). Hyperplasia may also be a result of EED or abortion or as a consequence of hormonal imbalance, often resulting from hormone-secreting tumours. Hyperplasia is normally a temporary condition, which can be reversed by reproductive rest or hormonal treatment (Bosu *et al.*, 1982; Van Camp, 1993).

19.2.2.4.3.3. Hypoplasia Uterine hypoplasia (underdevelopment) is characterized by an inability to develop adequately in order to maintain a pregnancy. The endometrial glands are most significantly affected, tending to be very small and so incapable of adaptation to support a pregnancy. As a result, even if fertilization does occur, the EED rate is high. Covering mares too close to puberty is associated with high rates of EED due to hypoplasia, simply because the uterine development to date is inadequate. The actual age at which the uterus is fully mature depends very largely on the individual mare, 18 months to 4 years is considered acceptable. Hypoplasia at 4-years old or over is indicative of a problem, which is likely to be permanent, and may be associated with chromosomal or hormonal abnormalities (Ricketts, 1975a,b).

19.2.2.4.3.4. Uterine atrophy Mares with uterine atrophy or senility are normally characterized as repeat breeders with high rates of EED (Greenhof and Kenney, 1975; Kenney, 1978; Van Camp, 1993). Uterine atrophy is caused by a decrease in the number of endometrial glands, due to atrophy or an inability to regenerate themselves. It is often associated with chromosomal intersex conditions, ovarian incompetence or progressive wear and tear in multiparous mares. It is also reported to have a greater occurrence late in the breeding season, presumably due to a decline in oestrus and ovarian activity. Generally, such late-season atrophy is of little concern, but evidence of it occurring early in the season may be indicative of a permanent problem and effect on reproductive performance. This condition is normally irreversible.

19.2.2.4.3.5. Uterine fibrosis Uterine fibrosis (periglandular fibrosis) is a degenerative uterine change, most commonly found in old multiparous mares; it is characterized by fibrotic changes around the endometrial glands forming glandular nests. As a result, the secretions of the endometrial glands decrease, the glands dilate, increasing the incidence of uterine cysts and resulting in increasing EED due to a disruption of embryo mobility, or abortion in late pregnancy due to restricted placental size (Van Camp, 1993).

19.2.2.4.3.6. Uterine luminal cysts Uterine luminal or endometrial cysts are the most common form of uterine lesion (Eilts *et al.*, 1995; Stanton *et al.*, 2004). They are generally thin-walled, greater than 3 cm in diameter, filled with lymph and may occur singularly or in multiples (Fig. 19.2).

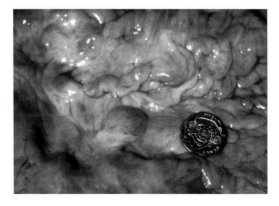

Fig. 19.2. A single luminal cyst within the uterus. They may also be present in multiples. Luminal cysts do not normally cause excessive problems unless evident in large numbers where they may interfere with embryo mobility and subsequent implantation.

They are particularly evident in mares, of 10-years old or over. Their effect on reproductive performance is disputed. If they are present in any number, they are likely to interfere with embryonic mobility increasing EED (McDowell *et al.*, 1988) and also reduce the uterine surface area available for placental attachment, increasing abortion rates (Curnow, 1991). Treatment may be attempted by puncturing the cysts via curettage, endoscopic manipulation or thermocautery, though they may subsequently recur (Mather *et al.*, 1979; Neely, 1983; Wilson, 1985; McDowell *et al.*, 1988; Pycock, 2000).

19.2.2.4.3.6.1. Uterine curettage Uterine curettage was traditionally used as a treatment for a whole range of conditions that resulted in damage to the uterine endometrium. It works on the principle of mechanical or chemical irritation of the endometrium, the rationale being that irritation stimulates and initiates a cleansing and regeneration process within the uterine endometrium and a mobilization of neutrophils to the affected site. Mechanical curettage involves physically scraping the entire surface of the endometrium, using a cutting edge mounted upon a long shaft that is passed through the cervix into the uterus. Chemical curettage involves the infusion of a chemical irritant, e.g. povidone-iodine or kerosene, which is reported to have a similar effect (Bracher, 1992). Curettage was once very popular but has largely been discredited as ineffective, with the potential to cause excessive scar tissue, haemorrhage and uterine adhesions.

19.2.2.4.3.7. Ventral uterine dilation Ventral uterine dilation or sacculation is caused by uterine myometrial atrophy, normally in the base of one uterine horn, forming an out folding or sacculation, which often collects fluid. This is again more common in older multiparous mares, due to a weakening of the myometrium. It often occurs at the implantation site, and may be caused by a gradual weakening of the wall in an area of repeated excessive stretching. Treatment is, unfortunately, relatively unsuccessful, but some beneficial results have been reported using oxytocin, or oxytocin in combination with warm saline lavage (Kenney and Ganjam, 1975; Neely, 1983). The fluid accumulation is normally of greatest concern making such mares susceptible to chronic endometritis and pyrometra (Kenney and Ganjam, 1975; Brinsko *et al.*, 1990).

19.2.2.4.3.8. Uterine adhesions Uterine adhesions are present as single or multiple bands or sheets of tissue within or across the lumen of the uterus and are the result of uterine trauma from dystocia, intrauterine infusion, severe endometritis or after treatment with caustic solutions. Their effect upon fertility depends upon their extent, but they may disrupt embryo mobility, restrict placental attachment, or even cause post-partum problems, such as placental and fluid retention, leading to endometritis. As such, they may be associated with EED or abortion. Attempts may be made to remove or break adhesions manually via an endoscope and biopsy forceps or via electrocautery (McKinnon, 1987a,b; Van Camp, 1993).

19.2.2.4.3.9. Uterine neoplasia Neoplasia or tumours within the uterus are very rare, leiomyoma are the most common and can be evident as single or multiple nodules and can cause persistent haemorrhage (Quinn and Woodford, 2005). Treatment may be attempted by surgery or endoscopy with some success (Bostock and Owen, 1975; Kenney, 1978).

19.2.2.4.3.10. Lymphatic lacunae Poor uterine lymphatic drainage may result in oedema within the uterine wall causing continuation of the normal ultrasonic 'chart wheel' appearance of the uterus beyond ovulation. Normally such oedema disappears soon after ovulation via absorption into the lymphatic system. Lymph is pumped along the lymph vessels by mild rhythmic contractions of the uterine myometrium but sometimes, especially in older multiparous mares, uterine myometrial contractility is impaired and so uterine oedema is not dissipated and leads to a doughy thick-walled uterus, which is unable to sustain a pregnancy; pregnancy normally fails at 12–20 weeks (Le Blanc *et al.*, 1994, 2004).

19.2.2.4.3.11. Foreign bodies Very occasionally foreign bodies, such as fetal bone, tips of uterine swabs, frozen semen straws, etc., are found within the uterus resulting in chronic endometritis. Their removal followed by treatment and recovery time normally restores reproductive performance (Ginther and Pierson, 1984c; Pycock, 2000).

19.2.2.4.4. CERVICAL ABNORMALITIES Cervical abnormalities normally arise from damage at parturition. Lacerations or injuries to the cervix often do not heal properly severely affecting the dynamic properties of the cervix which allow it to vary from a tight seal to wide dilation at parturition; often causing adhesions, which may block the entrance to the uterus through the cervix, or cause cervical incompetence. This will inhibit sperm deposition, allow infection to enter the uterus and prevent natural drainage of the uterus (Sertich, 1993). As discussed in some detail in Sections 1.3 and 1.5, the cervix naturally forms the final seal protecting the upper reproductive tract from infection and so such mares are predisposed to uterine infection, fertilization failure and EED. Minor adhesions may be treated by physically cutting or electrocauterizing the scar tissue and inserting a plastic tube to prevent reoccurrence, and lacerations can be surgically corrected (Brown et al., 1984; Aanes, 1993). Excess adhesions and interruption of the normal cervical relaxation at oestrus can prevent the natural drainage of uterine secretion at oestrus, increasing the chance of endometritis. The prognosis in most cases, however, is not good. Neoplasms of the cervix are very rare (Sertich, 1993). Inherited cervical incompetence though has been reported in pony mares (Lieux, 1972; Brown, 1984).

19.2.2.4.5. VAGINAL ABNORMALITIES Vaginal abnormalities have several causes. Among these is damage at parturition, often a result of fetal malpresentations. Superficial damage will correct and heal naturally, though there is the risk of adhesions. Severe adhesions may cause the mare pain at subsequent coverings. In the extreme, rectal vaginal fissures may be opened up by the foal's foot passing through the top of the vagina and into the rectum during parturition. The prognosis in such cases depends on the length of the opening formed, but can be very poor when substantial rectal vaginal fissures occur (Section 15.5.3; Spensley and Markel, 1993).

Two of the most common vaginal abnormalities are associated with poor perineal conformation: pneumovagina and urovagina. Pneumovagina (inspiration of air and bacteria into the vagina) predisposes the mare to endometritis and so is a common cause of infertility, especially in Thoroughbred mares, and is due to the incompetence of the vestibular and vulval seals and associated poor perineal conformation (see Section 1.3). This can be alleviated quite successfully by a Caslick's vulvoplasty operation. Full details of vulval incompetence, its consequence and treatment have already been given in Section 1.3.

Urovagina (urine pooling within the vagina) results from weakness within the vaginal walls and a collection of urine within the resultant sacculation often around the fornix and so urine can easily pass through the cervix into the uterus. As such, it is associated with infection (endometritis, cervicitis and vaginitis) and hence adversely affects reproductive performance. The condition is normally observed in older multiparous mares, with pendulous reproductive tracts, due to continual stretching and weakening with successive pregnancies. Occasionally, it may also be seen as a temporary phenomenon at foal heat, but in most circumstances it will have rectified itself by the second heat post partum. If it is evident, it is essential that the mare is not covered, as there is an increased chance of post-coital endometritis. Treatment using oxytocin has proved reasonably successful (Monin, 1972; Brown et al., 1978).

19.2.2.4.5.1. Persistent hymen Occasionally, a persistent hymen may be evident. The hymen divides the anterior and posterior vagina and occasionally does not break in early life before first service and so may be evident in maiden mares as a white/blue membrane possibly pushing through the vulva. A persistent hymen will impede natural drainage and, as such, when fillies reach puberty the secretions of the reproductive tract, associated with oestrous cycles, build up behind the hymen causing it to bulge through the vulva. Simply manually breaking the hymen will allow the fluids to drain and the mare's subsequent fertility should be unaffected. If the hymen is not broken prior to the first service it may tear causing the development of scar tissue.

19.2.2.4.6. VULVAL ABNORMALITIES Vulval abnormalities most commonly involve inappropriate perineal conformation, as discussed in detail in Section 1.3. The resulting vulval-seal incompetence increases the chance of infection entering the reproductive tract. Lacerations of the vulva occurring at parturition or due to accidental injury can again compromise vulval-seal competence, due to incorrect healing and adhesion formation. Failure to cut a Caslick's vulvoplasty prior to covering or parturition will also cause tearing and will predispose to adhesion formation, fibrosis and inappropriate vulval healing (Aanes, 1969, 1973, 1993).

Haemorrhage of the vulval lips may be evident, caused by the bursting of varicose veins. This has minimal direct effect on reproductive ability but may cause discomfort at breeding. Neoplasms of the vulva are reported, most commonly melanomas, originating in the pigment-producing cells of the skin, especially prevalent in grey mares. These tumours can spread from the perineal area around the anus and eventually throughout the whole of the body. Squamous-cell carcinoma, normally associated with the penis, may also be seen on the vulval lips. Finally, enlarged clitorises, sometimes in the form of a vestigial penis, may be observed and are associated with chromosomal abnormalities and such animals are sterile.

19.2.2.5. Infectious infertility

19.2.2.5.1. OVARIAN INFECTIONS As far as infection or disease is concerned, the ovary is essentially unaffected and the vast majority of ovarian abnormalities, and hence, ovarian infertility, are not the result of pathogenic agents.

19.2.2.5.2. FALLOPIAN TUBE INFECTIONS (SALPINGITIS) Salpingitis, inflammation of the Fallopian tubes or saplings, is rarely seen; however, it may occur as a consequence of endometritis. Complete blockage of the Fallopian tubes is rare, but inflammation can interrupt the process of fertilization, the passage of ova towards the utero-tubular junction and sperm movement towards the ampulla. Infertility or subfertility may result. Occasionally, infection may cause inflammation of the valve at the utero-tubular junction, affecting the passage of sperm and/or fertilized ova.

19.2.2.5.3. UTERINE INFECTIONS One of the major causes of infertility in the mare is endometritis (Bennett, 1987). Endometritis, inflammation of the uterine endometrium, is often caused by infection by venereal and opportunistic bacteria but may also be due to non-infectious degenerative endometritis and/or persistent post-coital endometritis. The main consequence of endometritis is a uterine environment hostile to embryo survival and implantation, resulting in EED and abortion. Endometritis is evident in four forms: acute endometritis; chronic endometritis; acute metritis; and pyrometra. These will be discussed in turn later in the chapter.

There are several factors that predispose the mare's tract to infections, including immunological, physiological or endocrinal deficiencies, which may be inherited, leading to a predisposition to endometritis.

Unfortunately, the mare's reproductive tract is not well designed for the easy removal of infective organisms or the resulting exudate. Infections are difficult, therefore, for the mare's system to eliminate naturally and can easily develop into chronic infections. Chronic infections that are not identified can cause serious problems to the mare if not treated in good time. Temporary infertility is nearly always evident with endometritis, and, if the infection/damage is great, permanent reduction in reproductive performance will result. Bacterial infection is nearly always introduced at covering or by inadequate hygiene precautions during internal examination or immediately post partum. It is, therefore, most important that during covering and manipulation or examination of the mare's tract, strict hygiene precautions are adhered to (see Section 13.5.1).

One of the major problems with uterine infections is that they may remain undetected for prolonged periods of time, thus not only reducing the mare's reproductive performance, but also risking transfer to the stallion and hence to other mares. Regular swabbing is not only compulsory in many studs, but is good practice in order to ensure that all chronic endometritis infections and latent asymptomatic infections are identified and treated immediately.

Endometritis is often characterized by excess mucus, which may be seen exuding from the vulva, high leucocyte counts and increased uterine blood flow. Oedema (fluid accumulation) can be identified at scanning and the uterus at rectal palpation can be felt as large, flaccid and doughy. The mare may also show shortened oestrous cycles, due to the irritation of the uterine wall resulting in premature CL regression.

19.2.2.5.3.1. Potential endometritis-causing bacteria As indicated, endometritis is primarily caused by bacterial infection. There are six major bacteria causal to endometritis, with up to 15 different bacteria identified in some cases (Pycock, 2000). The six major bacteria will be considered and can be classified as opportunistic or venereal.

19.2.2.5.3.1.1. Opportunistic bacteria Opportunistic bacteria are those which are common within the environment. They often have no effect, but rapidly invade a micro-environment once the opportunity arises. In the mare this often occurs after a disruption to the natural microfloral balance due to antibiotic treatment, stress, excessive use of soaps or antiseptics, etc. Disruption to the natural balance leaves a space into which opportunistic bacteria invade and then populate. Opportunistic bacteria can be introduced quite easily but especially at covering, internal examination, AI, foaling, etc. (Allen and Pycock, 1989; Samper and Tibary, 2006). As such, they are potential causers of acute endometritis, especially in compromised or susceptible mares (Le Blanc *et al.*, 1991). There are three main opportunistic bacteria of concern.

Streptococcus zooepidemicus is implicated in 75% of acute endometritis cases, particularly during the initial stages. They are spherical bacteria, found normally in chain formation, often in the intestine and mucous membranes. *Streptococcus* is classified into two subgroups: alpha and beta. *S. zooepidemicus* is a beta *streptococcus* and, as such, causes the destruction of red blood cells and has a major role in initiating infection of the mare's cervix and uterus (Hughes and Loy, 1969). It may also promote the proliferation of other bacteria within the tract (Le Blanc *et al.*, 1991; Asbury and Lyle, 1993).

Hemolytic Escherichia coli is a rod-shaped aerobic bacterium that is found either alone or in short chains. It is the second most common cause of uterine infection. It is naturally found in the intestine and is associated in particular with faecal contamination. It can cause not only acute endometritis but also severe systemic infection, which can prove fatal (Allen and Pycock, 1989; Asbury and Lyle, 1993).

Staphylococcus aureus is a less-common cause of endometritis. It is a spherical or oval bacterium, normally evident in clusters and found associated with skin and mucous membranes. Under suitable conditions, such as the disruption of the natural microflora, ill health or stress, the bacteria will invade the reproductive tract of the mare (Allen and Pycock, 1989; Asbury and Lyle, 1993).

19.2.2.5.3.1.2. Venereal disease bacteria Venereal disease bacteria are those that are transferred solely via the venereal route, i.e. they are present within the semen and the reproductive tract of the mare and stallion and are capable of producing endometritis in both the normal and susceptible mare. They may also be present in apparently asymptomatic animals – in particular the stallion which rarely shows symptoms. There are three main venereal disease bacteria of concern.

Taylorella equigenitalis is an extremely contagious bacterium and is the causal agent of contagious equine metritis (CEM; Tainturier, 1981). It was first isolated in Newmarket, UK, by Crowhurst (1977), where it spread rapidly and widely, due to the reluctance of infected-mare owners not to present their mares for service. The bacterium is initially rod-shaped and becomes spherical with age. The stallion is seemingly not affected by the bacterium, but is the prime means by which it is spread from mare to mare. In the mare, the typical symptoms of acute endometritis are seen, characterized by uterine, cervical and vaginal inflammation along with copious grey discharge within 2–5 days of infection; she may then appear to recover but remains a carrier. In rarer instances, the mare may not show any clinical symptoms but still be a carrier capable of infecting a stallion. At the other extreme, the infection may develop on to give chronic endometritis.

Klebsiella pneumoniae is an encapsulated rod-shaped bacterium, associated with acute and chronic endometritis. The ones of particular concern are capsular types 1, 2 and 5 (Pycock, 2000). The bacteria are endemic and widespread, but diagnosis is reasonably accurate via cervical-uterine swabbing. Unfortunately, the bacteria are relatively insensitive to antibiotics and antiseptic washing agents (Crouch *et al.*, 1972).

Pseudomonas aeroginosa, a slender rod bacterium with rounded ends and flagella, is found widely within the environment. However, some strains of *P. aeroginosa* are causal to endometritis and may be isolated in stallion's semen or in swabs taken from the urethral fossa, but clinical symptoms are rarely evident. In the mare, *P. aeroginosa* causes a greenish-blue or yellowish-green exudate, which appears to be more prevalent in older mares. It is relatively resistant to antibiotics and antiseptics, so early diagnosis and cessation of natural cover is the best course of action (Hughes *et al.*, 1966; Hughes and Loy, 1975).

19.2.2.5.3.1.3. Diagnosis Due to the highly contagious nature of venereal disease endometritis, diagnosis and prevention are very important. Diagnosis of acute endometritis may be obvious due to exudates, or via identification of inflammation via scanning, rectal palpation or endoscopy. Once inflammation has been diagnosed, the causal agent, i.e. infective or not, needs to be identified.

Bacterial infections may be identified by swabbing of the reproductive tract. It is normal and recommended practice (Horse Race Betting Levy Board, 2008) that swabs are taken from the uterus, cervix, clitoris and urethra opening. Uterine swabbing should be carried out, using a guarded swab to prevent contamination en route and through an open cervix during oestrus (Fig. 19.3; Greenhof and Kenney, 1975). The other swabs may be taken throughout the mare's oestrous cycle. The resultant swabs are plated out and incubated under varying conditions, anaerobic, aerobic and microphillic, etc., to aid bacterial identification (Ricketts *et al.*, 1993). Fungal infections may also be identified in a similar manner. The use of swabbing is a widespread and often compulsory practice. Some breed societies have successfully used it to eradicate specific causes of infection in many areas worldwide. In particular, within the UK, the Horse Race Betting Levy Board publishes Annual Codes of Practice, which are used worldwide and have resulted in a near eradication of *T. equigenitalis* (CEM) from the UK as well as significantly reducing the incidence of venereal diseases caused by *K. pneumoniae* and *P. aeroginosa*, it also produces guidelines on equine herpes virus (EHV) and equine viral arteritis (EVA; Horse Race Betting Levy Board, 2008). These Codes of Practice are reviewed annually and detail the number and type of swabs that need to be taken for different classes of mare. CEM is now a notifiable disease in UK and codes of practice have been laid down for exportation and importation of stock and, in cases of suspected CEM abortion (Platt *et al.*, 1978; Crowhurst *et al.*, 1979; Rossdale *et al.*, 1979b).

Uterine aspirations and washings may also be collected, especially if purulent material and fluid are present. Culturing of the washings allows bacteria to be identified (Freeman and Johnston, 1987).

19.2.2.5.3.2. Acute endometritis
Acute endometritis is invariably a result of either significant bacterial challenge, by venereal or opportunistic bacteria or a persistent acute reaction to covering. If infective, acute endometritis develops rapidly, giving immediate symptoms of exudate or pus and irregular oestrous cycles. Internally, it causes deep haemorrhage and degeneration of luminal epithelial cells and, in severe cases, degeneration of the deeper stroma cells, leading to areas of missing endometrium. This may lead to hypertrophy and abscessed uterine glands (Rooney, 1970).

19.2.2.5.3.2.1. Acute infective endometritis Acute infective endometritis is a major cause of infertility in the mare, providing a hostile environment for both sperm and embryo survival. Bacteria are introduced into the system at covering, both natural or AI, or at veterinary inspection. It is now accepted that some degree of acute endometritis is evident after all coverings regardless of the extent of

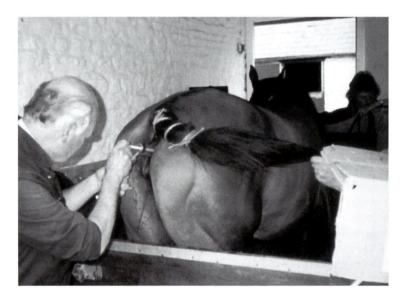

Fig. 19.3. Many studs require all mares to be swabbed prior to service. The swabs can be taken from clitoral sinuses and fossa along with the urethral opening, endometrium and cervix.

bacterial invasion (see Section 19.2.2.5.3.2.2); however, introduction of bacteria causes a significant uncharacteristic inflammatory reaction.

Treatment for general acute endometritis begins with identification of the infective agent and targeted use of local antibiotics, systemic antibiotics or uterine lavage (Threlfall, 1979). Local antibiotics are applied via infusion, using an indwelling catheter passed through the cervix and placed into the uterus. The end of the catheter is looped into two ram's-horn shapes, which help keep the catheter in place, and allows repeated infusions without the need to change and reintroduce the catheter (Figs 19.4 and 19.5). This reduces the risk of introducing more opportunistic bacteria via the technique itself into what is already a compromised system.

The catheter is most easily introduced through a relaxed cervix at oestrus. However, this is not always possible, as endometritis often leads to anoestrus and, hence, a tight cervix. Treatment with PGF2α is, therefore, sometimes required to bring the mare back into oestrus sooner, so reducing the delay until treatment can commence. PGF2α may be administered as a series of injections to provide a series of short cycles and, therefore, accelerate a series of treatments. Infusion times depend upon the severity of the infection and may vary from daily infusions for 3–5 days to 15 days. Such antibiotic treatment must be used with care, as some antibiotics may cause necrosis or erosion of the endometrium. Bacterial resistance is increasingly a problem, so identification of the causal bacterium and use of a specific target antibiotic are very important. Excessive antibiotic use may allow fungal infections to develop which will themselves require treatment (Asbury, 1987; Asbury and Lyle, 1993). Systemic antibiotics have been used, but evidence for their success is inconclusive. They have been advocated for use in conjunction with local antibiotics (Brown et al., 1984).

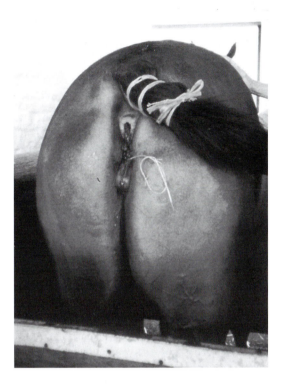

Fig. 19.5. An infusion catheter *in situ* allowing repeated treatment of mares with antibiotics in cases of endometritis. (Photograph courtesy of Elm Stud, Vikki Kingston.)

Uterine lavage, using 1–2 l of saline, is increasingly popular. Lavage has been demonstrated not only to remove debris and exudate but also to encourage neutrophil release to the infection site. The washings may also be used to identify causal agents. The extent and regularity of lavage again depends upon the severity of the condition (Asbury, 1990). Uterine infusion or lavage with chemical

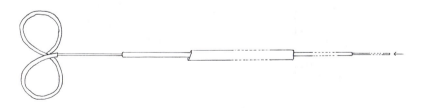

Fig. 19.4. An infusion catheter for the treatment of mares with antibiotic solution in cases of endometritis.

irritants or disinfectants, as a form of chemical curettage (see Section 19.2.2.4.3.6.1), have been advocated. However, results are variable and such treatments should be used with great care. Povidone-iodine has been used with some success (Asbury and Lyle, 1993) as has plasma infusion (Asbury, 1984; Pascoe, 1995). In addition to treating the infection any physical abnormalities that may be predisposing the mare to infection should be corrected, i.e. via Caslick's vulvoplasty, a Pouret operation, removal of adhesions, etc. (Caslick, 1937; Pouret, 1982).

In the treatment of endometritis, topical treatment of vaginal or clitoral infections should be considered. This involves cleansing the whole area with a non-antiseptic soap for *K. pneumoniae* and *P. aeroginosa*, or chlorohexidine for *T. equigenitalis*, followed by topical application of antibiotic creams. Unfortunately, *K. pneumoniae* and *P. aeroginosa* are particularly difficult to eliminate, in which case clitorectomy may be considered (Pycock, 2000). Clitorectomy, removal of the clitoris, is more widely practised in the USA than in UK and Europe.

19.2.2.5.3.2.2. Post-coital acute endometritis

Post-coital acute endometritis is the specific term given to acute uterine inflammation resulting from covering. As indicated previously, this is evident, to some extent, in all mares but is especially evident in susceptible mares. The persistence of post-coital endometritis is encouraged by several factors, including general stress, a decline in the mare's general well-being and cervical, vaginal or vulval abnormalities. However, even in the absence of these predisposing factors, acute endometritis may still persist, and it is evident that some mares are inherently more susceptible than others. Initially, it was thought that the acute response in some mares was due to an immunological incompetence, rendering the mare unable to react to, and so eliminate, bacteria introduced at covering (Hughes and Loy, 1969). More recent research would indicate that there is no significant difference in immunoglobulin release, or the functional ability of neutrophils in normal and susceptible mares post coitum (Allen and Pycock, 1989), and that an inflammatory response is evident even with the insemination of semen without bacterial contamination (Kotilainen et al., 1994). It is now thought that acute post-coital endometritis is due to a reduction in the physical ability of susceptible mares to drain fluid from the tract. This is likely to involve defective myometrial contractility,

but the causes of such a defect are as yet unclear (Troedsson, 1999; Knutti et al., 2000; Watson, 2000; Rigby et al., 2001; Campbell and England, 2006). Whatever the cause the inflammation within the uterus provides a non-ideal environment reducing sperm motility (Alghamdi et al., 2001) and providing an inappropriate environment for the survival of the conceptus.

Post-coital prophylactic measures are often employed in such mares to reduce the incidence of inflammation by assisting uterine exudate clearance. Those sperm required for fertilization reach the Fallopian tube within 2–4 h of ejaculation and the fertilized ovum does not arrive in the uterus until day 5 (Bader, 1982). In theory, therefore, uterine treatment is safe within these time limits. In practice, however, due to the rapid rise in progesterone post ovulation and along with it a natural decline in uterine myometrial contractility (Stecco et al., 2003), it is best not to attempt treatment after 48 h post ovulation. Treatment can be via uterine lavage using a saline plus antibiotic solution, which successfully removes uterine fluid and debris, enhancing neutrophil function and antibiotic efficiency. Lavage also stimulates uterine contractility and encourages the release of fresh neutrophils through irritation of the endometrium (Pycock, 2000; Knutti et al., 2000; Card, 2005). As such, it is often the treatment of choice for such mares. Oxytocin may also be used, both alone and in combination with lavage, again to encourage myometrial activity, and hence fluid clearance (Allen, 1991; Le Blanc, 1994; Pycock and Newcombe, 1996; Campbell and England, 2002; Vanderwall and Woods, 2003). PGF2α, by virtue of its similar action on uterine myometrial contractility, has also been used successfully (Combs et al., 1996). These systems may be supplemented by AI with semen extended with antibiotic extenders. The addition of antibiotics significantly reduces any bacterial challenge and the use of AI reduces the total number of sperm introduced into the uterus and further reduces the inflammatory response (Davies Morel, 1999; Nikolakopoulos and Watson, 2000; Sinnemaa et al., 2003). For the same reason such mares should ideally only be covered once (Kenney and Ganjam, 1975; Kenney et al., 1975a). It is also increasingly evident that teasing plays a role in encouraging uterine myometrial contractility. Work by Madill et al. (2000) indicates that uterine contractility is greatest in mares that are teased and increasingly less in mares that are covered by AI, than mares with sight of a stallion and finally mares with just sound of a stallion call.

19.2.2.5.3.3. Chronic endometritis

Chronic endometritis may be more accurately divided into chronic infective endometritis and chronic non-infective degenerative endometritis. It has recently been suggested that chronic non-infective degenerative endometritis should now be termed endometriosis.

Chronic infective endometritis can arise from an untreated or inappropriately treated, acute uterine infection, or due to a mare's inability to satisfactorily combat the initial infection. As with acute endometritis this may be due to fungal and yeast infection as well as bacterial. The condition is more often found in older multiparous mares, especially with poor perineal conformation where the breakdown in uterine defence mechanisms and possibly poor uterine myometrial contractility results in an inability to respond to introduced infection and may also have allowed normal genital bacterial flora to contaminate the uterus. Such infection is often long term, but not so evident as a dramatic inflammatory response. It can be extremely damaging to the endometrium, causing degeneration and necrosis resulting in permanent infertility. Treatment, though not that successful, is as indicated for acute endometritis but with particular use of infusion and lavage. Large-volume infusion with a broad-spectrum antibiotic is advised, as often wide ranges of bacteria are present (Pycock, 2000). Similarly, antimycotic agents are infused for fungal infections. Lavage using isotonic saline, followed by antibiotic and/or plasma infusion, is reported to be successful (Asbury and Lyle, 1993). At breeding, such mares should be treated in a similar manner to those susceptible to acute post-coital endometritis (Asbury and Lyle, 1993).

Chronic non-infective degenerative endometritis, more recently named endometriosis, has been previously discussed in Section 19.2.2.4.3.1.

19.2.2.5.3.4. Acute metritis

Acute metritis is potentially the most serious uterine infection. It is associated with a massive contamination of the whole uterus as a result of trauma, often associated with parturition involving retained placental or fetal tissue, or bacterial infection introduced via air inspired post partum or via hands used to aid parturition. Occasionally, it may be evident post coitum. Decomposition of retained tissue encourages rapid bacterial growth along with toxin production. The inflammation of the entire uterus then favours the passage of toxins into the main circulation, resulting in toxaemia and, potentially, death.

Prevention is infinitely better than cure; and absolute hygiene at parturition, plus the complete expulsion of all placental and fetal tissue post partum is essential. Treatment must be immediate and normally involves large-volume lavage and possibly oxytocin to encourage uterine contraction and thus the flushing out of the uterine contents. Lavage should then continue until recovered fluids are relatively clear. Recovery is not possible until the source of the toxaemia is removed (Blanchard and Varner, 1993a; Threlfall, 1993). The prognosis is often poor and, even if the toxaemia is successfully resolved, long-term lameness from laminitis may result (Eustace, 1992; Pycock, 2000).

19.2.2.5.3.5. Pyrometra

Pyrometra is characterized by fluid accumulation in a large, pendulous uterus. In time the uterine walls may become leathery, tough and fibrous, due to continual infection. Such mares may appear healthy in themselves, but often do not show oestrous cycles due to the inability of the uterus to produce PGF2α. Pyrometra may be associated with a blockage of the uterus, fibrosis, adhesions, etc. resulting in a build-up of exudate within the uterus, with no normal drainage. It is often due to infection, but not necessarily so. Treatment normally involves drainage, followed by antibiotic infusion or lavage, but the prognosis for a breeding career is often poor. If infection is not evident and breeding is not required, such mares may not require treatment if they show no signs of discomfort. However, the presence of infection poses problems and, if left untreated, infective pyrometra may develop into septicaemia (Ricketts, 1978; Hughes *et al.*, 1979).

19.2.2.5.4. CERVICAL INFECTIONS

Cervicitis, inflammation of the cervix, is usually associated with, and often precedes, endometritis. Such infection causes inflammation and possible pus accumulation (Sertich, 1993).

19.2.2.5.5. VAGINAL INFECTIONS

Vaginal infections are often a prelude to endometritis, especially in mares suffering from poor perineal conformation. Alternatively, it may be caused by chemical irritation of the vagina, such as antimicrobial agents used in examination. These can also result in vaginal necrosis, which may also result from damage in cases of dystocia (Le Blanc *et al.*, 2004). Systemic and topical antibiotic treatment is often successful and the prognosis, providing the infection is not long term or developed into necrosis, is good.

19.2.2.5.6. VULVAL INFECTIONS Equine coital exanthema, genital horse pox, evident as vesiculation and ulceration of the vulval lips or penis, is caused by equine herpes virus 3 (EHV3; see Section 19.2.2.5.7). Infection causes pain at covering and can, therefore, affect reproductive performance. It is sexually transmitted and symptomless carriers are reported. There is no direct affect on reproductive performance, but in order to prevent transfer natural covering must cease. Treatment with antibacterial creams or powders prevents secondary infections and helps the natural healing process (Pascoe *et al.*, 1968, 1969; Gibbs *et al.*, 1972).

19.2.2.5.7. VIRAL INFECTIONS The incidence of viral abortion is 1–5%, mainly occurring in late pregnancy. There are two main viruses that have a major effect on reproductive performance in the mare: equine arteritis virus (EAV), the causal agent for EVA and EHV3, the causal agent for coital exanthema.

EAV is of reproductive significance as it causes abortion in mares (Timoney and McCollum, 1987; Castillo-Olivares *et al.*, 2003). The virus can be spread via the venereal route (natural covering and AI) but also via the respiratory route and from the placenta of aborting mares and the urine of infected animals (Acland, 1993). EVA is evident worldwide with the exception of Japan and Iceland and until relatively recently the UK (Wood *et al.*, 1995; Samper and Tibary, 2006). Many countries have strict regulations to limit its importation and spread. Stallions are the major route of infection as they can become asymptomatic carriers; the asymptomatic carrier state does not exist in mares and geldings which only shed the virus during the initial infective phase. Mares and geldings eliminate the virus within 60 days but remain seropositive due to the previous infection; similarly, all stallions become seropositive but 30–60% do not eliminate the virus and become persistently infected, the virus lodging in the accessory glands and then being shed in semen (Glaser *et al.*, 1997). Seropositive stallions, which it is suggested can number up to 80% in some countries, can, therefore, be classified as shedders and non-shedders and it is the seropositive shedder stallions that are a risk to mares. The

carrier state and the shedding of the virus through semen are testosterone-dependent, therefore, gelding of a stallion removes the risk, similarly it has been reported that shedding stallions treated with gonadotrophin-releasing hormone (GnRH) antagonist stopped shedding the virus, though a return to the shedding status was resumed after the end of treatment (Fortier *et al.*, 2002). Shedding stallions can be classified as short-term shedders (only excreting the virus in the initial infective period, as seen in mares and geldings), long-term shedders (excreting the virus for 3–9 months) or chronic persistent shedders (which will permanently excrete the virus). It is these later stallions that are the biggest risk, 85–100% of seronegative mares mated by a seropositive shedder stallion will become infected (Samper and Tibary, 2006), whether mating be via natural service or AI. Infected mares may not show clinical signs but shed the virus for the first 60 days or so via nasopharyngeal secretions, urine and the infected placenta if abortion occurs. Infection does not affect fertility per se but can cause abortion, usually in months 3–10 of pregnancy, due to severe oedema and necrosis of the endometrium (Clayton, 1986). Mares and geldings normally recover spontaneously and so treatment, apart from supportive care, is not required and is largely unsuccessful. There is no treatment for carrier stallions. Prevention and management of control measures is, therefore, very important. The status of all mares and stallions should be ascertained by blood sampling, strict hygiene precautions should be practiced and ideally shedding stallions should not be used for covering, though it may be acceptable to use them to mate seropositive mares. Mares and stallions can be vaccinated with a modified live vaccine at least 3 weeks prior to mating (Parlevliet and Samper, 2000), which will give protection for up to 2 years. It is important, however, and required by some breed societies that all animals are blood-tested prior to vaccination to certify that their subsequent seropositive status is due to the vaccination and not infection (Timoney and McCollum, 1987, 1988, 1997; Timoney *et al.*, 1988). In the UK, EVA is a notifiable disease under certain circumstances and, as such, is now included in the Horse Race Betting and Levy Board Codes of Practice.

There are four main strains of EHV, EHV1, 2, 3 and 4. EHV3 is the causal agent for equine coital exanthema, which is primarily transmitted venerally, but may also be transmitted by infected equipment, i.e. AI, gynecological examination, etc. (Rathor, 1989). It causes blister-like lesions 5–7 days post infection, both on the perineal area of the mare and the penis and prepuce of the stallion. These resolve within 3–4 weeks leaving scars. Though the direct effect on fertility is minimal, covering during the active phase may cause discomfort and bleeding and is not advised in order to prevent transmission (Samper and Tibary, 2006). EHV1 and 4 are of more limited concern with regard to reproductive performance but they can cause abortion. In particular, EHV1 is the causal agent for rhinopneumonitis abortion. Virus transfer is via the respiratory route, from birth fluids, soiled bedding, placental tissue, etc. It may also be found in semen (Acland, 1993; Davies Morel, 1999). The virus causes placental separation resulting in fetal suffocation and abortion with 96% occurring in the last 4 months of pregnancy. It can have a devastating effect causing abortion storms in mares plus neonatal losses. As with EVA, EHV1 is now included in the Horse Race Betting and Levy Board Codes of Practice and is as yet of minor concern in UK.

There are other viruses whose major effects are not on fertility but may be a minor cause of reduced reproductive function, for example, West Nile virus a mosquito-borne virus that primarily causes encephalitis (inflammation of the brain) and/or meningitis (inflammation of the lining of the brain and spinal cord) but indirectly affects reproductive ability (Bunning et al., 2002; Long et al., 2002).

19.2.2.5.8. PROTOZOA INFECTION Dourine, caused by *Trypanosoma equiperidum*, a sexually transmitted protozoa, is now eradicated from the UK, most of Europe and North America but is still prevalent in many temperate countries including Africa, South and Central America, the Middle East and Asia. It causes intermittent fever, depression and progressive loss of body condition along with vaginal and vulval infection and inflammation along with discharge. Infected horses also develop characteristic subcutaneous lesions, areas of thickened skin. If left untreated it will develop systematically to form raised rings within the mare's coat along with depigmentation of the genitals plus fever and death in 50–75% of cases (Brun et al., 1988; Brown, 1999).

Piroplasmosis, caused by the haemoparasite *Babesia equi* or *Babesia caballi*, may also be a potential risk to mares and stallions. Present world-wide, except for UK, Ireland, Japan, the USA, Australia and Canada, it is most often transmitted by ticks as a blood-borne protozoa; however there is a chance of transfer of infection to mares if semen of an infected stallion becomes contaminated with blood (Samper and Tibary, 2006).

19.2.2.5.9. FUNGAL AND YEAST INFECTION Mycotic or fungal infections can potentially be transferred venerally and can potentially cause endometritis (Dascanio et al., 2000). The most commonly isolated fungi include *Chlamydia* spp. and *Microplasma* spp. *Micoplasma* spp., in particular, have been associated with endometritis, abortion and balanitis in the stallion (Bermudez et al., 1987). *Chlamydia* spp. have been reported to cause salpingitis and reduced fertility and abortion (Herfen et al., 1999). Yeasts may also cause problems to both mares and stallions and be transferred venereally; these include *Candida* spp. and *Aspergillus* spp. Though present occasionally in semen the greatest risk is transfer at AI, if strict hygiene procedures are not adhered to. All these infective agents, as with bacterial infections, disrupt the ability of the uterus to support a developing embryo and, if present later in pregnancy, can cause abortion via placentitis and occasionally fetal infection. The majority of fungal abortions occur around 10 months of pregnancy (Platt, 1975; Acland, 1993). Treatment is via infusion of antimycotic agents such as povidone-iodine, nystatin or lufenuron but the success rate is low (Hess et al., 2002). Acidic agents, such as vinegar and acetic acid, have been used with some success (Pycock, 2000). If the mycotic growth cannot be arrested the prognosis is hopeless.

19.2.2.6. Fetal congenital deformities

Many fetal developmental deformities have been reported (Leipold and Dennis, 1993). Many of these are not compatible with fetal life, and so cause abortion. Chromosomal defects also occur, leading to EED rather than abortion (Blue, 1981; Ricketts et al., 2003).

19.3. Stallion Infertility

On average, a stallion might be expected to cover one to two mares per day during the breeding season, with a rest day every 7–10 days. This gives reasonable results, with fertilization rates of about

60% (Pickett and Voss, 1975b). These figures are only an average and are affected by many things as discussed for the mare (see Section 19.2) including the type and condition of mares presented to a stallion and also by the characteristics of the individual stallion. Some stallions are capable of much heavier workloads and some struggle with less (Pickett and Shiner, 1994). It is one of the major responsibilities of the stallion manager to be aware of the limitations of their individual stallion and to work within these constraints (Kenney, 1990).

Research into the causes of infertility in the stallion is as yet limited, due to previous concentration on the mare. However, in any breeding programme, 50% of the outcome is determined by the stallion, and as such, deserves fair discussion. The difficulty in obtaining standard figures for fertility and the reluctance of the majority of stallion owners to select for reproductive performance and to assess for breeding soundness have led to relatively low average fertility rates, which have shown little, if any, improvement in recent years. Before discussing the subject further the following glossary should be noted to prevent confusion in terms.

Sterility   – permanent inability to reproduce
Infertility   – temporary inability to reproduce
Subfertility   – inability, either temporary or permanent, to reproduce at full potential
Impotency   – temporary or permanent inability to ejaculate semen; sperm capable of fertilizing an ovum may, however, be produced

19.3.1. Extrinsic factors affecting reproductive efficiency in the stallion

Extrinsic factors affecting the reproductive efficiency of a stallion include: lack of use, the presentation of subfertile or infertile mares, poor mare management, poor stallion management and the imposition of an artificial breeding season. Each of these factors will be discussed in turn in the context of reproductive efficiency/infertility. It will be noted that many aspects have already been discussed in previous chapters, especially those concerning, management and so such details will not be repeated here.

19.3.1.1. Lack of use

Reproductive efficiency in any animal reflects its use. A stallion may not be used in a particular year

by design, due to financial or management considerations. Disease may also preclude a stallion from use for part or all of a season, due to the risk of direct disease transfer to mares in the case of venereal or contagious diseases, or may limit his ability to perform, in the case of non-contagious disease. Alternatively, the stallion may have suffered from disease or infection during the previous year and is not to be used in the following season in order to allow full recovery or because the long-term effects of disease on his reproductive performance deem it inappropriate to use him until he is fully recovered. Diseases of the stallion's reproductive tract will be discussed later under intrinsic factors.

Finally, semen evaluation is part of good practice and should be carried out regularly at the beginning of each season. Poor semen quality may lead to the stallion being taken out of use until the cause has been isolated and the problem solved.

19.3.1.2. Subfertile or infertile mare

Both mare and stallion are equally responsible for fertilization. A stallion is, only as good as the mare he is to cover, and vice versa. It is essential that any mare presented to a stallion is capable of reproducing and does not suffer from any of the factors affecting reproduction that have been discussed previously in the sections on mare infertility. If the mare herself is subfertile or infertile, lack of success cannot be blamed on the stallion.

19.3.1.3. Poor mare management

Mare management is discussed in detail in Chapters 12–16 (this volume). Inappropriate management will adversely affect the mare's ability to conceive and, therefore, the apparent fertility rates of the stallion that covered her. The most important management area as far as stallion reproductive efficiency is concerned is during the time of covering. Service at an inappropriate time due to a failure to detect oestrus and ovulation accurately will obviously be reflected in poor fertility rates. Failure to detect oestrus is usually due to either prolonged dioestrus preventing oestrus from being displayed or infrequent or inaccurate teasing, along with lack of records and mare observation. In such cases, veterinary examination by means of rectal palpation and/or scanning have been shown to significantly improve fertility rates by allowing more accurate detection of ovulation.

19.3.1.4. Poor stallion management

Stallion management is discussed in detail in Chapters 12, 13 and 18 (this volume). All aspects of a stallion's management will affect his ability to cover mares successfully. Stallion management, as far as it directly affects reproductive efficiency, can be subdivided into the following.

19.3.1.4.1. EXCESS WORKLOAD As discussed previously (see Section 18.5), the amount of work or number of mares a stallion may be expected to successfully cover during a season is highly variable (Pickett *et al.*, 1975c; Pickett and Shiner, 1994). It is one of the responsibilities of the stallion manager to know the capabilities of his/her stallions. The workload of a stallion depends upon the ability of his testis to produce sperm and epididymal sperm storage reserves. Among other things, this is a function of testis size, which can be assessed by calipers or ultrasonically (Fig. 19.6; Love *et al.*, 1991).

Stallions with large testis have larger daily sperm outputs and can cope with a heavier workload than stallions with smaller testis. Testis size is also a function of age, which has an important bearing on fertility, especially at the extremes of youth and old age (Johnson and Neaves, 1981; Pickett *et al.*, 1989; Pickett and Shiner, 1994). It is a good idea, especially with new stallions, to carry out a full

semen analysis to give a guide to daily sperm production (see Section 20.4).

Sperm concentrations are usually in the range of $100–300 \times 10^6$ sperm ml^{-1} and for successful fertilization $300–500 \times 10^6$ sperm are required (Pickett and Voss, 1972). On average, 50–60% of sperm produced can be classified as normal progressively motile sperm capable of fertilizing an ova (Kenney, 1975a; Pickett and Voss, 1975a). The average daily sperm production for a stallion is $0.6–6 \times 10^9$ depending upon season, environment, age, etc. (Pickett and Voss, 1972; Lopate *et al.*, 2003). From these figures, it is apparent that, in theory, the average number of successful services a stallion could be expected to perform per day is one to three. There are, however, other considerations to take into account when looking at workloads.

It is interesting to note that total sperm production per week is the same, regardless of whether a stallion is used daily or on alternate days. However, use on a daily basis results in a lower concentration of sperm ml^{-1} (Pickett *et al.*, 1975a). This may be of no consequence in stallions with high daily sperm-production rates, as concentrations will still be acceptable, but daily use of stallions with lower daily sperm-production figures may have a detrimental effect on fertility rates.

Excessive workloads may also result in a lack of libido. As a result, the stallion will be slow to breed

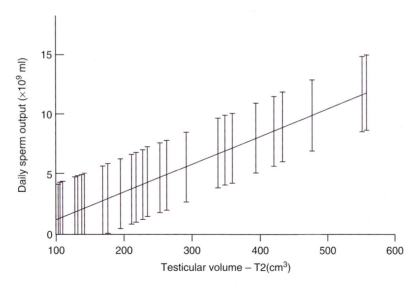

Fig. 19.6. The relationship between testicular volume, and daily sperm output. (From Love *et al.*, 1991.)

or may even fail to ejaculate. In such cases, it is best to take the stallion out of work for a short period of time, and reintroduce him a week or so later. If libido is still low, it may well be indicative of further problems.

In an ideal world, a stallion would be used just once a day and given regular periods of rest of 1–2 days every 10 days or so, this is particularly important in young stallions. However, there are many pressures, not least financial, that entice stallion managers to increase workloads. Due to individual variation, some stallions can cope with more than one mare per day or periods of excessive use in a busy season, provided they are given adequate periods of rest.

19.3.1.4.2. TRAINING MANAGEMENT As discussed in Section 18.4, early training is of utmost importance in the long-term ability of a stallion to perform to his full potential. A stallion brought up in a relaxed, unstressed environment with consistent and fair discipline and respect is much more likely to perform to his full potential in later life.

One of the major problems encountered as a carry-over effect from early life, which is often unappreciated, is stallion isolation. Many managers isolate stallions to ensure the safety of personnel, other stock on the yard and the stallions themselves. This is a self-perpetuating problem, and, as such, treatment often results in boredom, vices and excessive excitability and unpredictability, which in turn results in further isolation in the interests of safety. A happy medium between safety and stallion participation in the general yard activities has to be achieved.

19.3.1.4.3. BREEDING DISCOMFORT Full physical examination of the stallion is essential before purchase to ensure that no abnormalities are present. Details of this are given in Chapter 11 (this volume). It is also advisable that stallions undergo regular examinations at the beginning of each season to ensure that no problems have arisen since the last season that may cause pain at breeding. Pain associated with the act of covering can cause a permanent reduction in libido. Poor feet care or conditions such as laminitis cause pain on mounting, especially if the problem is in the hind feet. Muscular or skeletal problems, including arthritis, may also limit the stallion's ability to mount, due to pain. Irritation and soreness of the penis or sheath area may also cause pain at covering, especially if smegma has accumulated or soap

or antiseptic wash has not been rinsed thoroughly. Breeding accidents involving inadequate erection at intromission, kicking by a mare and rough handling will discourage a stallion from future covering, as he will associate covering with pain. Finally, when using an artificial vagina (AV), care should be taken that the internal temperature is not too hot, as this will cause pain and will reduce his future willingness, not only to use an AV, but also in natural service.

19.3.1.4.4. NUTRITION As discussed in Section 18.5.3, appropriate nutrition throughout the year is essential in order to ensure that the stallion is in optimum physical condition for the season. Obese or excessively thin stallions suffer from low libido, and nutrition, along with exercise, is a major determinant of body condition. A body condition score of 3 on a scale of 1–5 is to be aimed for.

As far as specific deficiencies are concerned, only limited research has been carried out. It is known that in general severe nutritional deficiency is associated with a delay in puberty, testicular atrophy and a reduction in sperm production. Deficiencies in energy, and, to a lesser extent, protein, have also been associated with low reproductive efficiency (Jainudeen and Hafez, 1993). Severe deficiencies in vitamins A and E and selenium are specifically associated with a reduction in spermatogenesis in other farm animals and have been suggested to have a similar effect in stallions (Ralston *et al.*, 1986). In addition, low dietary intake of copper, iron and/or cobalt results in a reduction in appetite, with accompanying weight loss and anaemia and, via these, a decline in semen quality (Jainudeen and Hafez, 1993).

Obesity will result in a loss of libido and may also cause a reduction in spermatogenesis. Obesity is associated with excess fat deposition within the scrotum, increasing scrotal insulation and hence causing an increase in testicular temperature, with an associated decline in spermatogenic efficiency.

19.3.1.4.5. CHEMICALS AND DRUGS As discussed in Section 12.5.1, it is essential that prior to use, if a stallion has been on any drug regime, time must be allowed for the drug to be eliminated from his system.

Though illegal in many countries and/or competition rules anabolic steroids are sometimes used in an attempt to improve male characteristics – for example, weight gain, muscle growth and performance in young horses. They have also been used in attempts to improve stallion libido. In humans and other animals, such use of anabolic steroids is known to be associated with infertility and a similar association has been indicated in stallions. Anabolic steroids have been reported to result in a decrease of up to 40% in testicular size and weight (Blanchard *et al.*, 1983; Koskinen *et al.*, 1997). Spermatogenesis is also reduced, with fewer sperm per gram of testicular tissue being produced and lower sperm-motility rates (Squires *et al.*, 1982a; Blanchard *et al.*, 1983).

Anabolic steroids are, therefore, not recommended for stallions in breeding work. These drugs not only have an immediate effect, but may also have a long-term effect, certainly until they are completely eliminated from the stallion's system.

Testosterone therapy has been used to improve libido in stallions and is relatively effective. However, it does have serious potential side effects as far as fertility is concerned. Chapter 4 (this volume) outlines the fine control and delicate hormonal balance controlling male reproductive functions. If one of the components of the system is altered, it affects the delicate balance of the whole system; hence, if a stallion is treated with testosterone, this increases circulating levels, which in turn act as a negative feedback on the hypothalamus and pituitary, reducing gonadotrophin (luteinizing hormone (LH) and follicle-stimulating hormone (FSH)) release and hence stimulation of the testis including sperm production (Squires *et al.*, 1981b; 1997). Testosterone therapy is, therefore, associated with low fertility due to reduced sperm counts and hence is not advised for use in stallions in work, unless under veterinary supervision.

In addition, the use of any other drug or treatment that causes a drop in appetite, diarrhoea or lack of condition is ill advised during the breeding season and should only be used under veterinary supervision. Some wormers have also been reported to be associated with a temporary decline in fertility and many breeders arrange their parasite control regimes to ensure that stallions are not treated during the breeding season.

19.3.1.5. Imposed breeding season

Reproductive activity in the stallion, as in the mare, is naturally limited by a breeding season, though with enough encouragement most will cover mares out of season (Pickett and Shiner, 1994). As discussed in Section 4.1, season affects the number of sperm per ejaculate, total sperm number, number of mounts per successful ejaculation and reaction times. As a result, fertilization potential out of season is significantly reduced (Pickett and Voss, 1972). The natural breeding season, with its optimum fertilization rates and libido, is nature's way of ensuring that foals are born during the spring and early summer to maximize their chances of survival.

Unfortunately, this natural breeding season does not coincide with the arbitrary breeding season man has determined in an attempt to achieve foaling as near as possible to 1 January, the official registered birth date of all foals in several breed societies, the Thoroughbred being the most well known. The arbitrary breeding season in the northern hemisphere starts on 15 February, as opposed to the natural breeding season that starts in April/May. In the southern hemisphere the imposed season starts 15 August, as opposed to the natural season in October/November. Stallions are, therefore, expected to cover mares at a time of the year when their libido and fertilization rates are naturally low and when they are unable to perform to their full potential. The adverse effect of season on reproductive efficiency is increasingly evident in older stallions (Johnson and Thompson, 1983).

Occasional use at either end of the non-breeding season can be quite successful, but a full workload can in no way be expected. Exact performance depends on the individual animal, but improved fertilization rates and libido can be obtained by the use of artificial lighting in the stallion's stable, from November onwards, to give 16 h light and 8 h dark, so mimicking the early onset of spring and advancing the breeding season (Clay and Clay, 1992).

19.3.2. Intrinsic factors affecting reproductive performance in the stallion

Intrinsic factors affecting reproductive performance in the stallion include: age, along with chromosomal, hormonal, physical and semen abnormalities. These will be discussed in turn in the context of reproductive performance.

19.3.2.1. Age

Age is an important aspect in considering the potential fertility of a stallion. Young and old stallions may have problems with taking on a full workload with consistent success.

A young stallion is still learning the job and can easily be adversely affected by his handlers and/or management. He may, therefore, be slow to breed, mounting several times per successful ejaculation or even failing to ejaculate, ejaculating prematurely or exhibiting enlargement of the glans penis before intromission. Careful treatment and handling during this period is essential to ensure that any such behavioural problems are not perpetuated (Naden *et al.*, 1990). As far as physical capabilities are concerned puberty (17–22 months) heralds the beginning of sexual activity (Clay and Clay, 1992). Three-year-old stallions may, therefore, be used for covering and are perfectly capable of fertilizing a mare, but they have a limited sperm-producing capacity. By 4 years of age they are capable of producing adequate numbers of sperm to cover as many mares as an adult stallion, but full consistent fertilizing capacity is not attained until 5 years of age on average (Berndston and Jones, 1989; Johnson *et al.*, 1991).

At the other end of the spectrum, old age may be a problem. An age-related decrease in semen quality after 20 years of age has been reported by some (Johnson and Thompson, 1983; Amann, 1993a,b), but this is not supported by other work (Johnson *et al.*, 1991). Conditions such as epididymal fibrosis which reduce epididymal sperm reserves and hence daily sperm production have been linked to old age. However, a decline in fertility is associated with general age-related problems, such as arthritis, many of which cause pain on mounting, a major cause of low libido and, therefore, low fertilization rates. If such problems are encountered they may be alleviated, to a certain extent, by the use of breeding platforms, or AI as well as allowing him extra time. The effect of age is very variable between different stallions, and older stallions should not automatically be precluded from use, as such, animals have had many years in which to prove their worth as far as their own performance and that of their progeny are concerned. Older stallions often tend to be more gentlemanly to handle, know their job well and good to use on maiden, shy or nervous mares, giving them confidence. When using an older stallion, it is particularly important that his semen should be evaluated regularly and monitored closely to allow a reduction in his workload if a decline in semen quality is detected.

19.3.2.2. Chromosomal abnormalities

Chromosomal abnormalities or genetic inadequacies may be the cause of infertility in stallions that otherwise appear fit. These may be associated with semen abnormalities, or more obvious abnormalities of the genitalia. Intersex conditions such as hermaphrodites (both ovarian and testicular tissues are present internally with an intermediate male/female external genitalia) and pseudohermaphrodites (either ovarian or testicular tissue is present internally with an intermediate male/female external genitalia) do occur (Varner and Schumacher, 1999). The most common is male pseudohermaphroditism, where the animal is genetically female but has some rudimentary male and female external genitalia (Keifer, 1976). In addition, genetic chimeras or mosaics (63XO:64XY), male syndrome (64XX) and Klinefelter's syndrome (65XXX) have been reported in stallions but are relatively rare (Halnan and Watson, 1982; Bowling *et al.*, 1987; Bowling, 1996; Makinen *et al.*, 2000). Other genetic abnormalities are associated with cryptorchidism (rig; Section 19.3.2.4.1) and testicular hypoplasia both directly affecting reproductive efficiency (Varner and Schumacher, 1991).

Some genetic deformities may not directly affect reproduction but may preclude the stallion from use, such as umbilical and inguinal hernias. These may well correct themselves naturally but the trait may well be perpetuated in succeeding generations. Other genetic factors causing abnormalities of the reproductive system will be discussed under the specific areas of the tract detailed below.

19.3.2.3. Hormonal abnormalities

As detailed in Chapter 4 (this volume), the endocrine control of the reproduction is governed by a finely balanced system. Circulating concentrations of testosterone have a direct effect on reproductive performance, on both libido and sperm production; low testosterone levels are often blamed for poor fertility rates (Nett, 1993c). Testosterone, human chorionic gonadotrophin and gonadotrophin-releasing hormone therapy have been used, with mixed success, to address this problem. The lack of success may well be due to the fact that depressed pituitary function is the cause of infertility in only 1% of cases (Boyle *et al.*, 1991; Roger and Hughes, 1991). Abnormal hormone levels may be associated with hypothyroidism, resulting in delayed puberty, smaller testes, decreased spermatozoa production and decreased libido. Feminization of the genitalia may also be observed. It has been postulated that changes in thyroid function may be the cause of stallion summer infertility associated with elevated environmental temperatures (Brachen and Wagner, 1983).

19.3.2.4. Physical abnormalities

Numerous abnormalities of the stallion's genitalia have been reported. As with most anatomical abnormalities, they are caused either by disease or are inherited. It is reported that one in five males has an anatomical abnormality, the significance of which varies from life-threatening to a minor flaw, that may be of little consequence as far as reproductive performance is concerned but may still reduce his market value. Only those most commonly encountered will be considered in the following sections.

19.3.2.4.1. CRYPTORCHIDISM A cryptorchid stallion or a rig is an animal in which either one or both of the testes have failed to descend into the scrotum. The passage of the testes from a position next to the kidneys should occur, as a gradual process, *in utero* or during the first few months of life (Fig. 2.9). A cryptorchid stallion may be further classified as illustrated in Figs 19.7 and 19.8 (Cox, 1993a,b).

The failure of testes to descend may be temporary (most will descend within 3 years of birth) or permanent. Evidence suggests that more than 75% of cases, especially temporary retention, involve the right testis (Bishop *et al.*, 1964, 1966). Circumstantial evidence and work reported by Cox (1993a) and Leipold (1986) would suggest that cryptorchidism has a heritable component and has a higher incidence in ponies (particularly temporary retention) and in Quarter horses and Paint horses (particularly permanent retention). The retention of one or both testes results in a significant decline in testes weight in the retained testis, often accompanied by relative increase in epididymis size, even if it does subsequently descend. The size may be reduced by up to 20-fold in the abdominally retained testis; the reduction in size of the inguinally retained testis is not as great but a difference of up to sevenfold has been reported (Bishop *et al.*, 1964). A unilateral cryptorchid is perfectly capable of successfully covering mares, though his total sperm output per ejaculate will be reduced and he will, therefore, be unable to bear a full workload. In practice, however, it is not advised to breed cryptorchids due to the possible heritability of the condition, indeed several breed societies do not allow such animals to be registered for use as a stallion. It is advised that cryptorchids are castrated; however, removal of the retained testis is not without complication, especially in abdominal cryptorchids. Unfortunately, it

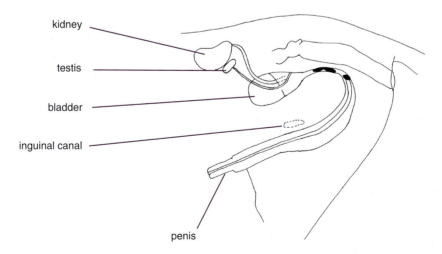

Fig. 19.7. An abdominal cryptorchid stallion is characterized by the testis lying up within the body cavity. In a unilateral abdominal cryptorchid, only one testis has failed to descend; in a bilateral, both remain in the body cavity.

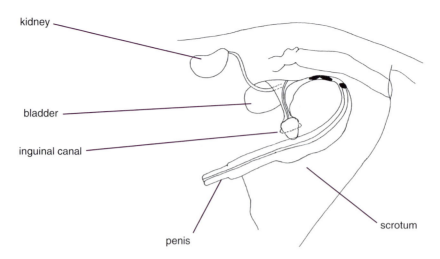

kidney

bladder

inguinal canal

penis

scrotum

Fig. 19.8. An inguinal cryptorchid stallion is characterized by the testis having only partly descended and remaining associated with the inguinal ring. Again, failure of testis descent may be seen in both (bilateral) or only one (unilateral) testis.

is, therefore, not uncommon for such animals to be unilaterally castrated, retaining the non-descended testis, which, though it is not able to produce sperm, will continue to produce testosterone (Cox *et al.*, 1973), and so an animal that outwardly appears to be a gelding will demonstrate stallion-like behaviour. Diagnosis of cryptorchidism is usually via blood test for oestrone sulfate. In animals of 3 years of age or older, oestrone sulfate levels greater than $0.1\,ng\,ml^{-1}$ indicate a retained testis. In younger horses oestrone sulfate concentrations are less accurate, an alternative is to measure the release of testosterone in response to a challenge of 6000 IU human chorionic gonadotrophin (hCG) injection. The presence of a retained testis is indicated by an increase in plasma concentrations of testosterone; in geldings no such reaction is seen (Lopate *et al.*, 2003b).

19.3.2.4.2. HERNIAS Stallion hernias may be classified in a number of ways (Figs 19.9 and 19.10). All have the potential to effect spermatozoa production, due to an elevation in testicular temperature from the close proximity of the herniated part of the gastrointestinal tract (Cox, 1988).

Testicular hernias may be acquired, usually due to accident or strain (Varner and Schumacher, 1991) or are congenital, due to inherited abnormality (Schneider *et al.*, 1982; Cox, 1993a). Testicular hernias may be further classified as inguinal or

scrotal depending on the extent of herniation. Inguinal hernias result from intestinal tissue passing solely through the inguinal ring (Fig. 19.9; Stashak, 1993). Scrotal hernias result from further herniation where the intestine extends into the scrotum (Fig. 19.10; Varner and Schumacher, 1991). The most common form of hernia is the inguinal, especially in young foals, where large inguinal rings are the prime cause (Wright, 1963). Spontaneous recovery normally occurs within 3–6 months and no long-term detrimental effects have been reported (Varner and Schumacher, 1991). Surgical intervention is sometimes required (Van der Veldon, 1988).

Apart from the mortal risk of intestinal strangulation, the biggest problem associated with testicular hernias is the effect on testicular function, due to elevated temperature from the close proximity of the intestine (Varner and Schumacher, 1991; Cox, 1993a,b). As discussed previously, an increase in testicular temperature has a direct effect on function (Section 2.5).

19.3.2.4.3. TESTICULAR HYPOPLASIA OR DEGENERATION Both hypoplasia and degeneration are the terms given to an underdeveloped and, therefore, under-functioning organ. Hypoplasia is generally the term given to a condition present from birth. Hence, in the case of testicular hypoplasia, the testes, for some reason, have never developed

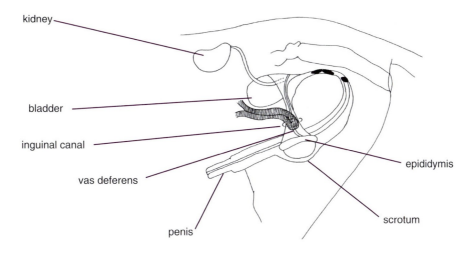

Fig. 19.9. An inguinal hernia in the stallion, in which a loop of the intestine folds through the inguinal ring.

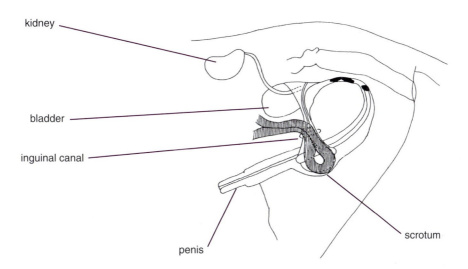

Fig. 19.10. A scrotal hernia in the stallion is a more extreme case of inguinal hernia, in which the loop of intestine has entered the scrotum and there is significant danger of complete ligation of the intestine, necrosis and death.

beyond an immature stage (Roberts, 1986b; Varner and Schumacher, 1991; Varner *et al.*, 1991). Its causes are many, including cryptorchidism and hernias, but also malnutrition, endocrine malfunction, infections, irradiation, toxins, and is often an inherited fault (Roberts, 1986b). The extent of the problem varies considerably from mild, where the testes appear normal, though possibly slightly small, to more severe cases, where the testes are significantly smaller than normal, and if the condition is advanced, the testes may have become hard

due to the overdevelopment of connective tissue (Ladd, 1985). Spermatozoa production depends on the severity of the condition, varying from slight impairment to aspermic (no sperm at all). Any spermatozoa that are ejaculated have a higher incidence of abnormalities. In such cases, the libido of the stallion is often not affected (Varner and Schumacher, 1991; Varner *et al.*, 1991).

Testicular degeneration refers to the condition where testicular development did originally occur to some extent but some subsequent problem has resulted in a degeneration of the tissue. The testes are highly sensitive to extrinsic factors and so testicular degeneration is a major cause of infertility, especially in older stallions. Unlike hypoplasia, degeneration is an acquired condition. Degeneration may be temporary or permanent; it may be unilateral (the cause being localized in origin) or bilateral (a systemic cause; McEntee, 1970). The condition is evident as a shrinking of the testes, often showing small epididymides with a reduced number of spermatozoa within (Watson *et al.*, 1994a). Spermatozoa counts are depressed (Roberts, 1986b), and a decline in spermatozoa output is observed, with an increase in the percentage of morphologically abnormal spermatozoa (Friedman *et al.*, 1991; Blanchard and Varner, 1993b). The causes of testicular degeneration are many and varied; the prime causes being elevated testicular temperature, scrotal/testicular injury especially that associated with haemorrhage, increased scrotal insulation due to scrotal oedema, scrotal dermatitis (McEntee, 1970; Varner and Schumacher, 1991; Blanchard and Varner, 1993b) and cryptorchidism, autoimmune disease (Squires *et al.*, 1982a; Zhang *et al.*, 1990b). More minor causes include toxins, tumours, obstructions of the vas deferens, testicular torsion and old age (Rossdale and Ricketts, 1980; Pickett *et al.*, 1989; Varner and Schumacher, 1991; Varner *et al.*, 1991). Testicular degeneration is, in most cases, reversible, providing that the duration of the problem is limited and the causative problem can be alleviated. However, infective and traumatic degeneration is more likely to be permanent (Burns and Douglas, 1985; Blanchard and Varner, 1993b).

19.3.2.4.4. TESTICULAR TORSION Testicular torsion is the twist or rotation of the testis within the scrotum, the extent of the twist and the resultant effect is variable, though torsion occurs most commonly in younger stallions, those with larger scrotal sacks but small testis. The twist may be complete, i.e. occur through an angle of up to 360°, a condition difficult to detect immediately, as the testes, on cursory examination, would appear to be positioned correctly. More commonly the twist is partial, i.e. through 90–180°, resulting in the epididymis being in a cranial position (towards the stallion's abdomen; Hurtgen, 1987). A minor torsion may be transient and may present just a little pain and a slight decrease in ejaculated spermatozoa concentration. Such torsions may have no long-term effects and may correct themselves (Threlfall *et al.*, 1990).

Major torsion can result in symptoms similar to orchitis (Section 19.3.2.5.1) including acute colic pain, scrotal and testicular swelling and obstruction of the blood supply, which, if present in a chronic case, may lead to degeneration and permanent damage (Kenney, 1975a; Threlfall *et al.*, 1990). There is some dispute as to the effect of the condition on semen quality. It is evident that if degeneration does result then semen quality will suffer.

19.3.2.4.5. TESTICULAR TUMOURS Testicular tumours are rare in horses, though their true exact incidence rate is difficult to ascertain, as the majority of stallions are gelded at a young age. Tumours or neoplasms can largely be divided into two (Bostock and Owen, 1975; Caron *et al.*, 1985): (i) germinal neoplasms, seminomas (most common; Vaillencourt *et al.*, 1979), teratomas (seminiferous tubule in original and so may contain various tissues such as hair, teeth, etc.); and (ii) non-germinal neoplasms, interstitial cell or Sertoli cell tumours (Rahaley *et al.*, 1983; Morse and Whitmore, 1986; Schumacher and Varner, 1993). Such conditions normally occur in older stallions and cryptorchids (Vaillencourt *et al.*, 1979). Often, the condition is not associated with pain or elevated temperatures, but a firm swelling may be felt in the testicular tissue and the testis affected is enlarged. Neoplasms are causes of testicular degeneration and, therefore, associated with depressed spermatozoa counts and high incidence of morphological abnormalities (Hurtgen, 1987).

19.3.2.4.6. VAS DEFERENS AND ACCESSORY GLAND PHYSICAL ABNORMALITIES Physical abnormalities of the vas deferens and the accessory glands are rare but if they occur are invariably associated with a current or previous infection (Section 19.3.2.5) or inherited abnormalities (Varner *et al.*, 1991). Varicocele (enlargement of the spermatic vein) may also be seen, which affects the functioning of the pampiniform plexus and hence affect testicular temperature control. Similarly, verminous granulomas, formed from parasitic migrations, affect testicular blood supply and pampiniform plexus function. Abnormalities associated with infection are manifest as fibrous growths or swellings at inflammation sites, which may cause obstruction and aspermia. Congenital abnormalities may be evident as immature, underdeveloped structures or complete absence. Most abnormalities can be identified by rectal palpation or ultrasonic scanning.

19.3.2.4.7. PENIS AND PREPUCE PHYSICAL ABNOR-MALITIES

Abnormalities of the penis or prepuce are normally associated with trauma or injury. The penis, especially when erect, is very vulnerable to traumatic injury from a kick by an unreceptive mare (Vaughan, 1993). This causes vascular rupture and/or haemorrhage, making the return of the penis to within the prepuce difficult and painful. Haemorrhage of penile blood vessels (penile hematoma) may also occur if a stallion covers a mare with a Caslick's, prior to episiotomy, or if the mare suddenly lunges to one side while being covered. The long-term effects of such trauma will not only depend on the physical recovery of the penis but also on the stallion's psychological recovery. Such trauma can make a stallion, especially a young one, very reluctant to cover a mare again. Damage to the urethra within the penis is evident as blood contamination of semen, termed haemospermia (Schumacher *et al.*, 1995). Congenital conditions such as small or short penis or stricture of the preputial orifice (phimosis) may be seen. Penile paralysis is a further condition often caused by trauma but also by neurological disease and general ill health, as is priapism (persistent erection; Pearson and Weaver, 1978; Lopate *et al.*, 2003).

Tumours of the penis are not often malignant, the most common are squamous cell carcinomas, but lesions may also be due to melanomas, sarcoids and herpes virus (see Section 19.2.2.5.7), which often burst and result in haemorrhage at covering and cause the stallion considerable pain. Blockage of the urethra or vas deferens has been reported, characterized by a normal libido but small-volume aspermatic semen, even though testicular function is normal.

19.3.2.5. Infectious infertility

19.3.2.5.1. TESTICULAR DISEASE AND INFECTION

Infection and/or inflammation of the testes (orchitis) is relatively rare in the stallion and is often associated with testis degeneration (Section 19.3.2.4.3). Orchitis may have a systemic or localized cause. The usual infection entry is via the bloodstream, wounds or ascending infection resulting in elevated testicular temperature and associated decline in spermatogenesis. The magnitude of the decline in fertility rates and the time period reflect the severity of the disease and the duration of the problem. Recovery, when it occurs, will be some-

what delayed, as the prime site of effect, as far as spermatogenesis is concerned, is the germinal cells. The spermatogenic cycle being 56 days, this period of time must be allowed after recovery for semen quality to return to anywhere near normal (Varner *et al.*, 1991).

Systemic disease, causing orchitis, normally results in bilateral inflammation of the testes and epididymis. Bacterial agents causing such orchitis include *Streptococcus equi* (strangles), *S. zooepidemicus*, *K. pneumonia*, *Actinomyce bovis* and *Pseudomonas mallei* (glanders) and possibly *Salmonella abortus equi*. Viral agents may also be a systemic cause of orchitis: these include EVA, equine infectious anaemia, equine influenza and EHV (Rossdale and Ricketts, 1980; Ladd, 1985; Roberts, 1986b; De Vries, 1993; Slusher, 1997). Systemic infections cause chronic, rather than acute, orchitis and have more of a chance of causing low-grade testicular degeneration and, with it, permanently depressed semen quality.

Localized infections may be caused via a wound, often to the scrotum, but may also be caused by ascending infection via the inguinal canal (Varner and Schumacher, 1991). Such infections tend to cause acute orchitis (De Vries, 1993), which may be unilateral or bilateral and present initially as soft, flabby, swollen testes. If the condition persists, chronic orchitis may result. Semen quality will be poor, with a decline in spermatozoa concentrations and an increased incidence of abnormalities (Hurtgen, 1987). The major infective agents associated with localized orchitis are *Staphylococcus* spp., *E. coli*, *S. zooepidemicus* and *S. equi*. In cases of acute orchitis, rises in testicular temperature are also a potential hazard (Blanchard and Varner, 1993b).

Orchitis may also be caused by testicular trauma (one of the commonest problems) or torsion and by parasites. The parasite most often associated is *Strongylus edentatus* larvae (Smith, 1973). These can migrate into the testicular tissue causing orchitis or obstruction of the testicular artery within the pampiniform plexus. This will have an additional detrimental effect upon the efficiency of the countercurrent heat-exchange mechanism (Roberts, 1986b; Varner *et al.*, 1993).

Finally, orchitis may be caused as a result of damage to Sertoli cells and hence the blood testes barrier causing autoimmune orchitis. This will result in an autoimmune response to spermatozoa and testis inflammation (Papa *et al.*, 1990; Zhang *et al.*, 1990b).

19.3.2.5.2. VAS DEFERENS AND ACCESSORY GLAND DISEASE AND INFECTION

Inflammation/infection of the vas deferens and accessory glands is very uncommon but is often accompanied by inflammation of the epididymis and is often the cause of epididymitis. It is noteworthy, as such, infections are often very persistent and so the stallion remains a carrier and of danger to mares he covers. Infection or inflammation of the ampulla gland and seminal vesicles, though rare, is more likely than infection of the prostate and bulbourethral glands (Blanchard *et al.*, 1987) and can be due to ascending/descending infection, blood-borne infection or from surrounding infective tissue. Infective agents include: *Corynebacterium pyrogenes* and *Brucella abortus* as the most common cause of accessory gland infection, but *P. aeroginosa*, *K. pneumoniae*, *Streptococcus* spp. and *Staphylococcus* spp. have also been identified. Infection is often characterized by increased leucocyte concentrations within semen, especially in semen collected after rectal palpation as well as bacterial contamination of semen, both of which affect motility as well as presenting a risk of infection transfer (Diemer *et al.*, 2003). Rectal palpation will also reveal that the seminal vesicles are swollen and painful (Varner *et al.*, 1991; Malmgren, 1992). Treatment is problematic as it is very difficult for systemic antibiotics to reach significant concentrations in the accessory glands in order to have an effect. Flushing and local infusion with an appropriate antibiotic, though not easy (Reinfenrath *et al.*, 1997), gives the best success rates.

19.3.2.5.3. PENIS, PREPUCE AND URETHRAL DISEASE AND INFECTIONS

Infection or inflammation of both the penis (balanitis) and urethra (urethritis) may be due to non-infectious irritation (chemical) or infective agents both often resulting in haemospermia. Contamination of semen with blood not only indicates the risk of possible infection transfer but is also associated with low fertility rates. Infective agents include: parasites – *Habronema larvae* (Habronemiasis or summer sores), Myiasis (fly strike; Hurtgen, 1987; Varner and Schumacher, 1991), *S. edentatus* larvae (Pickett *et al.*, 1981); protozoa – *Trypanosama equiperdum* (dourine; Couto and Hughes, 1993); viruses – equine herpes virus III (coital exanthema or genital horse pox); bacteria – *Streptococcus* spp., *K. pneumoniae*, *P. aeroginosa* and *T. equigenitalis*.

Not only may the penis itself be infected, but it is also the major means by which venereal infection can be passed from the stallion to the mare and vice versa (Parlevliet and Samper, 2000; Samper and Tibary, 2006). The penis has a naturally balanced microflora that causes no problem to him or any mares that he covers. However, if this balance is disturbed, due to systemic infection or disease, general ill health, impaired normal disease resistance or inappropriate use of soaps and detergents for penile washing serious consequences can result (Bowen *et al.*, 1982). Similarly, if he comes into contact with a contaminated mare, his natural microflora balance may be breached and this may allow the invasion of foreign infective agents. The prepuce area protecting the penis then provides an ideal environment in which such organisms can multiply. Contamination of the stallion's penis may result from poor hygiene, especially at covering and veterinary examination. Therefore, regular swabbing of the stallion and all mares to be covered, and the washing of the stallion and mare during the preparation for covering goes a long way to preventing venereal disease transfer.

The stallion is often asymptomatic, failing to demonstrate any clinical signs of infection, but infection may be traced back through symptoms shown by mares he has covered (Samper and Tibary, 2006). Details of the organisms involved and their potential effect on reproductive performance have been given in the previous section dealing with mare infertility (Section 19.2.2.5), as it is in the mare that these infections manifest their symptoms. Isolation and treatment of the stallion is the only course of action when such infections are suspected. Transfer from mare to mare via a stallion in a busy season is very easy and can have disastrous consequences.

Other conditions or infections may be transferred via the stallion at covering as well as cause balanitis; for example, coital exanthema (caused by EHV3), which is often, but not always, characterized by lesions, particularly in the warmer climates of Asia, Africa, South America and South-east Europe (Pascoe and Bagust, 1975; Couto and Hughes, 1993). EVA (Timoney *et al.*, 1988), EHV (EHV4 and possibly EHV1; Pascoe and Bagust, 1975; Jacob *et al.*, 1988), habronemiasis lessons (summer sores; Philpott, 1993) and fungal infections (Zafracas, 1975) are other examples (see also Sections 19.2.2.5.7–19.2.2.5.9).

Complete prevention of venereal disease is difficult, but is aided by adhering to full hygiene precautions prior to covering, though complete disinfection of the stallion's penis is impossible and not advisable. Regular swabbing in accordance with the Horse Race Betting Levy Board guidelines (Horse Race Betting Levy Board, 2008) will greatly increase the chance that any bacteria present are identified so that appropriate treatment can be given. Treatment itself can cause problems, as systemic antibiotics may affect the natural microfloral balance, which will take time to restore. Topical application to such a sensitive area may also cause dryness and cracking, causing pain at covering. Semen, for use with AI, can be treated with an extender containing an appropriate antibiotic. Though not advised, natural service can be risked after thorough washing of the stallion's penis and with uterine lavage of the mare 4–6 h post coitum followed by uterine infusion of an appropriate antibiotic (Samper and Tibary, 2006).

19.3.2.6. Immunological infertility

Semen contains many antigens, including those within seminal plasma and those that are spermatozoa-bound. Under certain conditions an autoimmune response to these antigens may occur, causing the destruction of spermatozoa both within the testis and the female tract (Wright, 1980; Teuscher et al., 1994).

19.3.2.7. Semen abnormalities

Semen abnormalities are discussed in full in Chapter 20 (this volume), along with the evaluation of semen. In summary and as indicated throughout the previous text, most infections and trauma of the male reproductive tract have an adverse effect on sperm production and, hence, fertility. The normal parameters expected of a semen sample are given in Table 20.3.

Infection and/or abnormalities of the reproductive tract may affect any of these parameters but usually cause a reduction in sperm concentrations, inadequate motility and poor longevity. Infection, rather than trauma, is often characterized by high leucocyte counts. If a semen sample does not meet the required parameters, the cause of the problem should be identified before the stallion is used, for his own protection and that of any mare. In addition to poor sperm quality, urospermia (urine contamination) and haemospermia (blood contamination) may be evident (Voss and McKinnon, 1993).

Haemospermia, as mentioned previously, may be caused by a number of infective agents or by physical damage (Schumacher et al., 1995). A red blood cell count >500 ml[-1] or a white blood cell count >1500 ml[-1] will have an affect on fertility. This effect appears to be modulated by the white or red blood cells themselves, rather than via the serum (McKinnon et al., 1988b).

Urospermia has a number of causes including neurological disorders (Leendertse et al., 1990; Mayhew, 1990; Samper, 1995a; Griggers et al., 2001; Lowe, 2001). Stallions suffering from urospermia may appear normal, with no neurological defects and with adequate libido and mating ability. The condition may be continuous, intermittent and unpredictable. Contamination may occur at any time during ejaculation and may be as little as 1 ml or as great as 250 ml (Varner and Schumacher, 1991). Evidence suggests that contamination is not likely to be due to just leakage but to an all-or-nothing effect (Nash et al., 1980). Urine contamination within a semen sample adversely affects spermatozoa motility and their capacity to fertilize an ovum (Hurtgen, 1987; Varner and Schumacher, 1991). Urine contamination may also affect semen pH (Hurtgen, 1987; Samper, 1995a). The severity of the problem is dose-dependent and stallion spermatozoa can tolerate minute amounts of urine without deterioration.

19.4. Conclusion

The causes of infertility, or the failure to produce an offspring, whether on a temporary or permanent basis, are numerous. Some are treatable but, for many, the prognosis for the individual as a breeding animal is poor. It is essential, therefore, that all potential breeding stock are submitted to a thorough examination prior to purchase, in order to ensure that they are capable of fulfilling their reproductive potential.

20 Artificial Insemination

20.1. Introduction

Artificial insemination (AI) was first developed in horses and dogs in the later part of the 19th century, but as the horse was increasingly replaced by the combustion engine (Heape, 1897; Perry, 1968), AI in horses dwindled while its use in other farm livestock, especially cattle, significantly increased. Research specifically into AI in equids is still somewhat limited as historically many breed societies were reluctant to accept for registration progeny conceived in this way. The most noteworthy of these is the Thoroughbred Breeders Association, which will still not accept AI and which has a significant influence on the equine industry. Most of the other breed societies within Britain and worldwide do now accept the progeny of AI, but many set strict regulations, such as a limit on the number of foals that can be registered per stallion per year, the use of semen after the stallion's death, etc. In Europe, Australia, China, South Africa and the USA, equine AI is now widespread. However, it still has some way to go before it reaches the sophistication of cattle AI. More detailed accounts of equine AI may also be found in Brinsko and Varner (1993), Davies Morel (1999), Pickett et al. (2000) and Samper (2000).

20.2. The Uses of AI

There are a variety of reasons why equine AI is practised. Some of these are dependent upon and limited by the regulations set out by the countries and breed societies involved. The reasons for using AI include:

1. Removal of geographical restrictions.
2. Minimization of disease transfer, both venereal and systemic, by the removal of direct contact between the mare and stallion. Semen can still provide a means of transferring venereal disease (VD), e.g. equine viral arteritis (EVA), equine herpes virus (EHV), VD bacteria, etc. but it may also be treated with extenders containing antibiotics to minimize the bacteria content and so reduce the number of potentially pathogenic organisms. Such semen is, therefore, useful for mares that have an increased susceptibility to uterine infection (Clement et al., 1993; Metcalf, 2001).
3. Reduction in injury risk to both handlers and horses by the removal of direct contact between mare and stallion. The risk is further reduced if the stallion can be persuaded to mount a dummy mare.
4. Increasing the number of mares that can be inseminated per ejaculate.
5. Improvement of native stock through semen importation (Sukalic et al., 1982; Ghei et al., 1994).
6. Development of gene banks for future reintroduction of genetic material (Zafracas, 1994).
7. Breeding of difficult mares – those with physical abnormalities, especially caused by accidents, infection, poor perineal conformation, psychological problems, etc. However, care must be taken to ensure that such problems are not heritable.
8. Breeding from difficult stallions – those with physical problems, injury, infection, inadequate semen characteristics, psychological problems, etc. (McDonnell et al., 1991; Love, 1992). As with the mare care must be taken to ensure that such problems are not heritable.
9. Reduction in labour costs (Boyle, 1992).
10. Semen sexing (Johnson et al., 1998; Seidel et al., 1998).

Several concerns over the use of AI have also been expressed, including the reduction of the genetic pool, overemphasis on 'fashionable' strains, the technical skill required and the infection risks if adequate screening is not practised (Davies Morel, 1999).

20.3. Semen Collection

Semen collection can be carried out using one of the several methods. The easiest method is the collection of dismount samples. The drips of semen are collected into a sterile jar from the stallion after withdrawal from the mare. This method is unreliable, the quality of sample is very variable and the majority of the sample is left within the mare. Samples often contain low sperm concentrations and are relatively high in pathogenic organisms. Semen can also be collected from the anterior vagina immediately after mating. However, the semen at collection has already come into contact with the acidic, and hence spermicidal, secretions found within the vagina. The sperm may also become contaminated by pathogenic organisms. Neither of these methods of collection allows the assessment of the total volume of semen produced.

Condoms have been developed for use with horses. They can work very well, but do have a tendency to burst or become dislodged (Perry, 1968; Boyle, 1992).

Finally, the best, and now most commonly used, method of semen collection is the artificial vagina (AV). Usually the stallion is encouraged to ejaculate into the AV by mounting an oestrous jump mare or a dummy. Stallions with musculoskeletal problems can be taught to ejaculate while standing using manual stimulation (McDonnell and Love, 1990; McDonnell and Turner, 1994; Davies Morel, 1999); or alternatively pharmacological ejaculation (chemical stimulation) using xylazine, imipramine or clomipramine plus manual stimulation (McDonnell and Love, 1991; McDonnell and Turner, 1994; Davies Morel, 1999) can be used with some success. The first AV was developed for use with horses in Russia at the beginning of the 20th century. Various models, including the Cambridge, Colorado, Missouri, Nishikawa and Hannover, are now available for use and they are all based on the same principles (Davies Morel, 1999). They provide a warm sterile lumen surrounded by a water jacket, under some pressure, with a collecting vessel at the end, in an attempt to mimic the natural vagina (Figs 20.1 and 20.2).

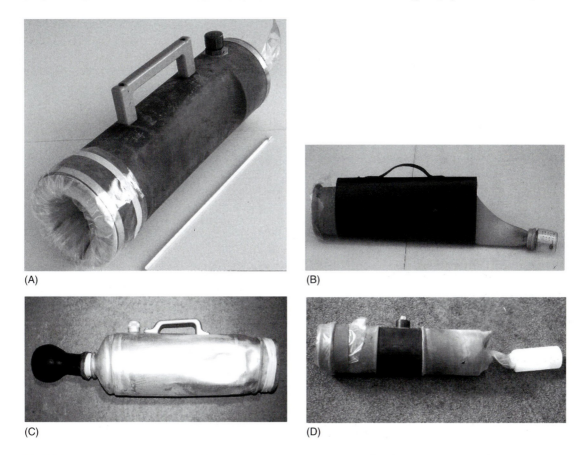

(A)

(B)

(C)

(D)

Fig. 20.1. Various artificial vaginas (AVs) for use in horses: all, except the Nishikawa, have disposable inner liners in place: (A) Cambridge; (B) Missouri; (C) Nishikawa; (D) Hannover.

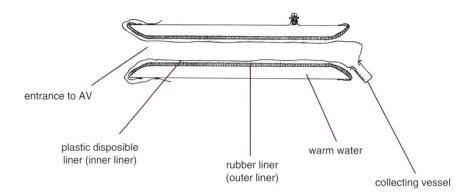

entrance to AV

plastic disposible
liner (inner liner)

rubber liner
(outer liner)

warm water

collecting vessel

Fig. 20.2. A diagrammatic representation of an equine artificial vagina (AV).

Most AVs consist of a solid outer casing with two rubber linings, an outer and an inner. The outer lining and the casing form a jacket, into which warm water and/or air is passed by means of a tap or valve. The temperature inside of the AV should be slightly above body temperature at 44–48°C. The amount of water used must be adequate to ensure that the pressure within the lumen of the AV mimics, as closely as possible, the pressure against the insertion of the penis within the natural vagina. Some models (the Missouri) allow the lumen pressure to be increased by inflating the water jacket; as air is lighter than water, this minimizes the final weight of the AV. The inner lining of the AV is often protected by an additional disposable inner liner, so ensuring sterility. The disposable liner, or in some cases the inner liner itself, is connected to the collecting vessel. Before use, this liner is lubricated with sterile obstetric lubricant to aid the stallion. It is most important that the collecting vessel, as well as the lumen of the AV, is about 44°C during collection, in order to prevent cold shock (see Section 20.6.5.1; Kayser *et al.*, 1992). If the temperature of the AV is too hot, there is a similar detrimental effect on sperm, in addition, there is the risk of discouraging the stallion from using an AV, and also possibly from carrying out natural service. The stallion seems less sensitive to temperature than other farm livestock, but temperatures above 48°C must be avoided (Hillman *et al.*, 1980). Sperm are also susceptible to ultraviolet light and so protection as well as insulation may be provided by enclosing the whole AV and collecting vessel within a protective jacket (Fig. 20.3).

Stallions can be trained relatively easily to use an AV. Initial training uses a jump mare, a mare in

Fig. 20.3. The Cambridge AV, one of the most commonly used, ready for use; to protect the semen from ultraviolet light and to help maintain the required temperature, the end of the AV and collecting vessel can be enclosed within a protective jacket.

oestrus either naturally or induced by treatment with oestradiol, to encourage mating and ejaculation into the AV. The mare is prepared as for normal covering, including swabs in case of accidental covering; she must also be of a quite calm disposition.

Over time most stallions become quite happy to use a dummy (Fig. 20.4), especially if an oestrous mare is in the vicinity. Some stallions are not so keen, usually as a result of low libido, a possible consequence of coming into stallion work late in life after time as a performance horse, or as a result of incorrect AI management in the past. These stallions may never accept a dummy and will always require the extra stimulus of a jump mare.

Fig. 20.4. Many stallions are trained to use a dummy mare. This eliminates the chance of accidental mating of a jump mare. (From Thoroughbred Breeders Equine Fertility Unit, Newmarket, UK.)

A stallion is prepared for semen collection in the same manner as he would for natural covering (see Section 13.5.2).

In readiness for collection all equipment must be at the right temperature for both collection and subsequent handling before the stallion is brought in for collection. Everyone involved must know what is expected of them. Collection of semen always carries a risk, due to the unpredictable nature of stallions, especially when covering. As with in-hand covering, all handlers are advised to wear hard hats, especially the semen collector. Up to three handlers will be needed if a jump mare is to be used; one to hold the stallion, one to collect the semen and one to hold the jump mare. All handlers should stand on the same side of the stallion. This ensures that, in the event of an accident, the stallion can be pulled away by his handler, and minimizes the chance of the semen collector getting kicked. The side used does not affect the sample collected, but once a stallion has got used to semen being collected from one side, it is best to try and stick to that side in future.

The stallion is allowed to mount and the collector diverts the penis towards the AV. The stallion should be allowed to gain intromission and enter the AV of his own free will and not have the AV forced upon him. The AV can be stabilized by being held against the hindquarters of the mare, if present, or the side of the dummy (Fig. 20.5). The occurrence of ejaculation is noted, as in natural covering, by the flagging of the tail or by feeling for the contractions of the urethra and the passage of the semen along the ventral side of the penis.

After collection, the collecting vessel must be carefully removed from the AV and the semen placed in an incubator at 38°C and then evaluated as soon as possible. If it is not possible to carry out semen assessment immediately, it can be extended and stored at 4–5°C for up to 24 h without appreciable reduction in its viability, and a reasonably accurate evaluation can still be obtained (Malmgren *et al.*, 1994; Batellier *et al.*, 2001).

20.4. Semen Evaluation

Prior to evaluation, the gel fraction of semen has to be removed. This gel fraction is the later secretion at ejaculation and has a very low concentration of sperm, which are invariably dead. The gel fraction is, therefore, of no consequence in AI and is removed by either: careful aspiration with a sterile syringe; or filtration through a gauze, or by an in-line filtration system incorporated into the AV; or by careful decanting. Filtration is the most popular method used today and also ensures removal of debris (Davies Morel, 1999).

It is imperative that at all times, including during handling and evaluation, semen is kept warm at 38°C (slightly lower than the temperature actually at collection). Sperm are very susceptible to cold shock (see Section 20.6.5.1) and, if any instruments, slides, microscope stages, etc. are not pre-warmed then results obtained will be misleading. Semen can be evaluated under several categories and not all evaluations involve all the assessments detailed below. It will vary according to what it is hoped to achieve by the assessment and the reasons for it being carried out. It is advisable for all the gross and microscopic assessments (Sections 20.4.1 and 20.4.2) to be carried out before a stallion enters an AI programme for the first time, and it is as well to repeat this at the beginning of every season.

When assessing the reproductive potential of a stallion, it is best to collect more than one sample for assessment. Two possible regimes are recommended: two ejaculates taken 1 h apart followed 3 days later by a single ejaculate for evaluation; or two ejaculates collected 1 h apart, followed by daily collection of samples for evaluation for 6–7 days (Pickett and Voss, 1972; Swierstra *et al.*, 1975). However, these take time and are rarely practical in many systems. In addition, any delays in testing are unpopular, as they delay the start of

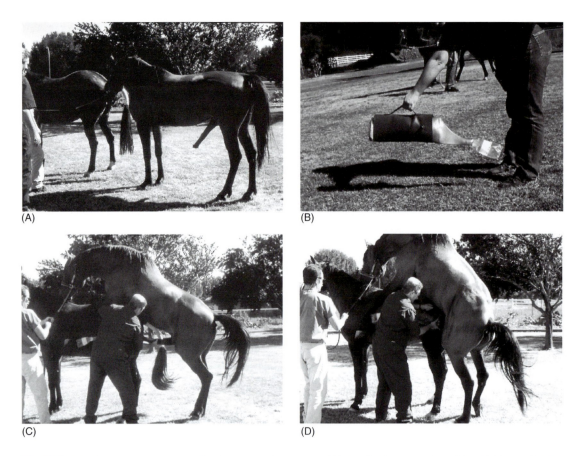

Fig. 20.5. Events involved in semen collection using a jump mare: (A) introduction or teasing of the stallion and jump mare; (B) artificial vagina (AV) ready for use; (C) guiding the stallion's penis into the AV and the correct positioning of handlers on the same side of the mare and stallion; (D) horizontal positioning of the AV to mimic the natural position of the mare's vagina;

the breeding season. Therefore, in practice, most evaluation is carried out on a single ejaculation sample.

In addition to a full evaluation once a year, it is a good practice to perform a basic evaluation on each sample subsequently collected for AI. In this case once the sample has been collected and filtered a small volume (3–5 ml) is removed for evaluation of: appearance, motility, concentration and possibly morphology and the rest prepared for immediate insemination or storage.

20.4.1. Gross evaluation

The sample for gross evaluation is assessed untreated, immediately after collection.

20.4.1.1. *Appearance*

Stallion semen is normally milky white in colour with a thickness equivalent to single cream (Kuklin, 1983; Fayrer-Hosken and Caudle 1989). It should contain no evidence of bloodstaining, urine contamination or clots (Varner and Schumacher, 1991; Samper, 1995a). The normal volume of semen produced by a stallion at each ejaculate varies considerably – 30–250 ml – but on average, most stallions produce 100 ml and, of this, the gel fraction is normally 20–40 ml. However, considerable variation is evident, both between different stallions and as discussed in Chapter 2 (this volume), semen quality and quantity also vary within the breeding season (Pickett *et al.*, 1988; Ricketts, 1993; Davies Morel, 1999).

(E)

(F)

(G)

Fig. 20.5. *Continued* (E) dismount of the stallion and slow removal of the AV, ensuring all semen is collected; (F) after dismount the stallion should be turned away from the mare to reduce the chance of injury; (G) the collected sample. (Photographs courtesy of Julie Baumber and Victor Medina.)

20.4.1.2. pH

Acidity/alkalinity is assessed using a standard pH meter. Acid conditions are known to be spermicidal; elevated pH may also be indicative of extraneous material or infection. A pH of 6.9–7.8 is acceptable, with levels of 7.3–7.7 being best (Pickett and Back, 1973; Fayrer-Hosken and Caudle, 1989; Oba *et al.*, 1993; Pickett, 1993a; Griggers *et al.*, 2001).

20.4.1.3. Motility

A rough estimate of motility can be obtained by placing a drop of raw semen on a warmed microscope slide and viewed under ×10 magnification. It is not possible to track individual sperm but a rough estimate of the percentage of motility can be gained with experience from the characteristics of

the wave-like motion of the sample as the sperm within it move.

20.4.2. Microscopic evaluation

20.4.2.1. Semen extenders for evaluation

Prior to microscopic evaluation the sample is invariably extended, roughly 1:20 semen/extender, primarily to allow individual sperm to be observed, as sperm within raw semen tend to clump together. Extension of the sample also prolongs the life of the sperm providing them with an additional source of energy and substrates for survival while evaluation takes place (Yates and Whitacre, 1988). Addition of the extender must be done immediately post collection, ideally within 2 min, to reduce the chance of obtaining erroneous results from a loss of sperm viability caused by delay. There are

Table 20.1. Examples of extenders used for semen evaluation.

(a) Non-fat dried skimmed milk glucose (NFDSMG) extender, for use in semen evaluation. (From Kenney *et al.*, 1975a.)

Component	Quantity
NFDSM	2.4 g
Glucose	4.0 g
Penicillin (crystalline)	150,000 units
Streptomycin (crystalline)	150,000 µg
Deionized water	made up to 100 ml

(b) Non-fortified skimmed milk extender for use in semen evaluation. (From Varner and Schumacher, 1991.)

Component	Quantity
Non-fortified Skimmed milk	100 ml
Polymixin B sulfate	100,000 units
Heat milk to 92–95°C for 10 min in a double boiler, cool and add the polymixin B sulfate	

Table 20.2. Criteria for grading sperm motility.

Grade	Description
0	Immotile
1	Stationary or weak rotatory movements
2	Backward and forward movement or rotatory movement, but fewer than 50% of cells are progressively motile and there are no waves or currents
3	Progressively, rapid movement of sperm with slow currents, indicating that about 50–75% of the sperm are progressively motile
4	Vigorous, progressive movement with rapid waves, indicating that about 75% of the sperm are progressively motile
5	Very vigorous forward motion with strong, rapid currents, indicating that up to 90% of the sperm are progressively motile

numerous extenders available and used successfully during the evaluation process. In general, these are the same as those used in the preparation of semen for immediate AI and/or chilled semen storage (see Section 20.6.3). The most popular are those based upon either non-fat dried milk solids (NFDSM) or skimmed milk (Table 20.1).

The exact dilution rate required depends upon the concentration of sperm and also the test being performed. Once diluted the sample is viewed under varying conditions under a light microscope with a magnification of ×40 or greater.

20.4.2.2. Motility

Sperm motility is graded on a scale of 0–5 (Table 20.2). Motility must be assessed immediately in order to get an accurate measurement. It may be assessed visually where the motility characteristics of individual sperm can be graded. However, dilution itself may affect motility and so assessment tends to be subjective and dependent on the examiner's previous experience (Jasko *et al.*, 1988).

Due to the subjectivity of microscopic assessment, other methods of assessing motility have been investigated, including the time-lapse darkfield photographic method (Elliott *et al.*, 1973; Van

Huffel *et al.*, 1985), cinematographic techniques (Plewinska-Wierzobska and Bielanski, 1970) and computer analysis (Jasko, 1992; Burns and Reasner, 1995). The correlation between the results obtained by visual and computer analysis is reported to be very good at 0.92 (Bataille *et al.*, 1990; Malmgren, 1997). Computer analyses have now been extended to give a full evaluation, not just of motility (Jasko *et al.*, 1990b). However, the cost of such equipment means that such analyses are largely restricted to research use, though they do provide an automated, rapid, objective result.

Regardless of assessment method used, classification of type of movement, as indicated in Table 20.2 is important. Progressive movement is required; oscillatory movement (movement on the spot) or movement in very tight circles is classified as abnormal. Some evaluators classify sperm motility as a ratio of those showing progressive movement/oscillatory movement. Computer analysis can give detailed information on the velocity and paths of movement of individual sperm.

Semen containing at least 50% progressively motile sperm, which is grade 2–3, can be considered as adequate for AI (Colenbrander *et al.*, 1992; Ricketts, 1993; Davies Morel, 1999). The correlation between motility and fertility is, however, variable and low (0.7; Samper *et al.*, 1991; Jasko *et al.*, 1992b; Heitland *et al.*, 1996). Though when compared with the other parameters evaluated it gives the best correlation, it cannot be relied upon to give an absolute indication.

20.4.2.3. Longevity

Assessment of motility over time may be used as an indication of viability. However, it must be remembered that longevity in a test tube does not necessarily equate to longevity within the mare's uterus (Kenney *et al.*, 1975a). Longevity assessment involves the evaluation of initial motility and then at various intervals after storage at 5°C or 22°C or 37–38°C. For example, if motility is not less than 45% after 3 h or 10% after 8 h at 22°C then the semen may be classified as good enough for AI (Kenney *et al.*, 1975a; Amann, 1988; Amann *et al.*, 1988; Samper, 1995b).

20.4.2.4. Concentration

Sperm concentration is a major determinant of the value of a semen sample and was traditionally assessed using a haemocytometer. A haemocytometer consists of a counting slide with a cover slip under which a set volume of diluted semen can be trapped and viewed. The number of sperm within the sample on the haemocytometer is counted by means of the counting grid. From this figure, and the dilution rate, the concentration of sperm can be calculated. This method gives a very accurate assessment, but is time-consuming and cannot be easily done in the field. Hence the use of manual or automated spectrophotometers has become popular and now form part of complete computerized sperm analysis systems (Davies Morel, 1999). Spectrophotometers can be used anywhere; however, the results obtained can be variable and are only as good as the initial calibration using a haemocytometer. The semen is diluted in the order of 1:30; the diluent used must be optically clear – for example, 10% formalin plus 0.9% saline. The amount of light passing through the sample is then read by the spectrophotometer, from which the concentration of sperm is calculated (Pickett and Back, 1973; Bielanski, 1975; Hurtgen, 1987; Davies Morel, 1999).

The normal range for sperm concentration is $30–600 \times 10^6$ sperm ml^{-1} of undiluted semen (Pickett and Voss, 1972). A total number of $100 \times 10^6–300 \times 10^6$ viable sperm per inseminate is required for acceptable fertilization rates (Kenney *et al.*, 1971). In practice, samples with a concentration of $100–200 \times 10^6$ ml^{-1} are considered acceptable for AI (Pickett *et al.*, 1988; Ricketts, 1993; Davies Morel, 1999).

20.4.2.5. Morphology

Morphological examination of sperm is a common method of trying to assess a stallion's fertility. Individual sperm are examined and the percentage of abnormal sperm in the sample noted. Viewing of individual sperm can be enhanced using a variety of stains, e.g. nigrosin – eosin stain, which stains the heads of dead sperm violet/purple (Dott and Foster, 1972), or by viewing under electron microscopy which allows various sperm components or abnormalities to be specifically identified (Reifenrath, 1994; Samper, 1995a; Pesch *et al.*, 2006). Staining of sperm to assess the integrity of specific areas may also be carried out, e.g. acrosome stains (Oetjen, 1988) immunofluorescence tests (Zhang *et al.*, 1990a) and labelled monoclonal antibodies (Blach *et al.*, 1988). The principles of these methods have been exploited in automated sperm morphometry analysis systems (Gravance *et al.*, 1996). Figure 20.6 illustrates some examples of the abnormalities that may be seen.

Abnormalities can be classified as primary (failure of spermatogenesis, failure of sperm maturation), secondary (sperm damage occurring at ejaculation) and tertiary (inappropriate handling post ejaculation). Failure of spermatogenesis is usually characterized by sperm with two heads, two tails, no mid-piece, no tail, rudimentary tails or excessively coiled tails and indicative of a long-term or even permanent problem. Maturation failure is characterized by the presence of cytoplasmic droplets on the mid-piece of the sperm. As maturation proceeds, these cytoplasmic droplets should progressively move down the tail and disappear. This may well be only a temporary problem, a period of rest rectifying the problem. Damage occurring at ejaculation is normally characterized by tail abnormalities, bends, coils, kinks or swellings, detached heads and tails and protoplasmic droplets. Finally, tertiary or post-ejaculatory damage is manifest as a loss of acrosome, fraying/thickness of the mid-piece and the bursting of sperm heads (Hurtgen, 1987; Varner *et al.*, 1987; Yates and Whitacre, 1988).

The correlation between morphology and fertility is relatively low (0.25–0.5; Van Duijn and Hendrikse, 1968; Hurtgen, 1987). Work has been carried out in an attempt to correlate specific abnormalities to fertility (see Section 20.4.3) but again correlations are poor. Morphology is, therefore, not a very reliable

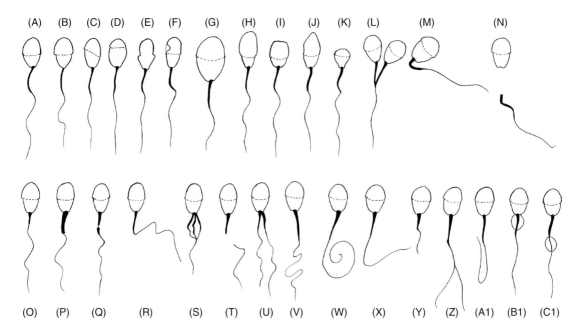

Fig. 20.6. Some examples of the more common abnormalities that may be seen when examining sperm for morphology. Top row from left to right: (A) a normal spermatozoon is followed by the following abnormalities in order. Acrosome defects: (B) swollen, (C) partially lifted, (D) small, (E) lifted, (F) part missing; head defects: (G) big head, (H) elongated, (I) flattened, (J) lanceolated, (K) microhead, (L) double head; neck defects: (M) bent, (N) broken. Second row: tail defects: (O) short, (P) fat, (Q) split/constricted, (R) bent annulus, (S) fibrous, (T) broken, (U) double (V) convoluted, (W) corkscrew, (X) bent, (Y) small, (Z) double, (A1) shoehorn; droplets: (B1) proximal, (C1) distal.

indicator of the fertilizing potential, though, as with motility, in the absence of anything else, it is used.

Semen containing 50% or more morphologically normal sperm is considered appropriate for AI (Fayrer-Hosken and Caudle, 1989; Pickett, 1993a; Davies Morel, 1999).

is acceptable for AI. However, a sample with a lower ratio may be considered to be usable, if it has a high total sperm count, as it can be diluted appropriately and still ensure that the minimum number of live sperm for fertilization are inseminated (Ricketts, 1993; Davies Morel, 1999).

20.4.2.6. Live/Dead ratio

Motility gives an indication of the percentage of live sperm within a sample; however, differential staining of sperm gives a more accurate result. As used to assess morphology staining with nigrosin-eosin stain in equal parts allows the differential staining of dead and live sperm. Dead sperm, as they are permeable to the stain, appear as violet/purple, live sperm remain clear (Dott and Foster, 1972, 1975). Other stains have been used equally successfully, including ethidium bromide and acridine orange fluorescence or H25 fluorescence (Hermenet *et al.*, 1993). The ratio, or percentage, of dead to live sperm is assessed in a number of samples from the collection. A live/dead ratio of 6:4 (60% live sperm)

20.4.2.7. Cytology

Blood cells, leucocytes and erythrocytes can be identified in a semen sample using haematoxylin and eosin or Wrights stain, and viewed under a haemocytometer (Roberts, 1971; Swerczek, 1975). A count of leucocytes greater than $1500\,ml^{-1}$ is indicative of a problem, possibly due to an infection, especially if the pH of the semen is also high. Such a sample would not be appropriate for use in AI (Rossdale and Ricketts, 1980; Ricketts, 1993; Davies Morel, 1999). Erythrocyte concentrations above $500\,ml^{-1}$ are indicative of a problem such as haemorrhage, injury, etc. and would also make a sample inappropriate for use in AI (Rossdale and Ricketts, 1980; Ricketts, 1993; Davies Morel, 1999).

20.4.2.8. Bacteriology

Semen samples will potentially contain bacteria, both pathogenic and non-pathogenic (Madsen and Christensen, 1995). Isolation and identification of pathogenic bacteria, in particular, is required in order to prevent passage of infection at insemination. Bacteria can be identified by direct plating of semen samples on to agar plates or by plating of swabs taken from semen or the genitalia. Incubation under various conditions then allows differentiation of bacteria (Madsen and Christensen, 1995). Long-term or acute infection may also be evident as high leucocyte counts in a semen sample or the evidence of pus. High bacterial counts in general reduce sperm motility (Diemer *et al.*, 2003); this plus the identification of specific bacteria such as *Pseudomonas aerugenosa*, *Taylorella equigenitalis* and *Klebsiella pneumoniae* will preclude a semen sample from use (Fraser, 1986; Scherbarth *et al.*, 1994; Davies Morel, 1999).

A summary of the acceptable semen parameters is given in Table 20.3. All the tests discussed so far are based upon gross or microscopic evaluation of sperm physical characteristics and all have their limitations. The future may, therefore, lie in functional tests (Wilhelm *et al.*, 1996).

20.4.3. Functional tests

Due to the failure to find a physical characteristic of sperm which is consistently and highly correlated with fertility (motility being the best with a correlation of 0.7; Jasko *et al.*, 1992b; Heitland *et al.*, 1996) attention has been drawn to the possibility of

assessing the functional integrity of sperm (Neild *et al.*, 2005b). Many such tests have been investigated in isolation or as part of an accumulative model (Wilhelm *et al.*, 1996). However, many are at the experimental stage and are also costly and labour-intensive and so are not in widespread commercial use at present. They may, however, hold the key to the future of semen evaluation.

20.4.3.1. Biochemical analysis

Biochemical analysis of semen may give indirect information on the quality of the sample. The total number of sperm present within a sample is reflected in the concentration of enzymes and substrates, such as aspartate amino transferase, lactate dehydrogenase (LDH), hyaluronidase, acrosin, adenosine triphosphate (ATP), acetylcarnitine, etc. (Bruns and Casillas, 1990; Castro *et al.*, 1991; Stradaioli *et al.*, 1995). Many of the enzymes, such as LDH, acrosin and hyaluronidase are located in the sperm head, specifically in the acrosome region, or in the mid-piece. As such, they can also be used to indicate sperm viability, as elevated free levels are indicative of sperm damage (Graham *et al.*, 1978), e.g. total acrosin activity is reported to correlate positively with fertilization rates (Reichart *et al.*, 1993; Sharma *et al.*, 1993; Francavilla *et al.*, 1994). Though biochemical analysis can give an indirect indication of sperm numbers and damage, it is as yet very non-specific and it will remain as such until more detailed work has been carried out to ascertain the normal concentration of these enzymes within the sperm and semen.

20.4.3.2. Membrane integrity tests

The integrity of the sperm membrane and internal organelles is a prerequisite for successful fertilization. Several tests have been investigated to evaluate sperm and mitochondria membrane integrity, these include the use of antibodies to various components of the sperm membrane, fluorescent probes or various stains combined with flow cytometry (Magistrini *et al.*, 1997; Silva and Gadella, 2006). Antibodies can be labelled with fluorescent dyes to allow the integrity of the membranes to be assessed visually (Blach *et al.*, 1988; Van Buiten *et al.*, 1989; Amann and Graham, 1993; Casey *et al.*, 1993; Neild *et al.*, 2005a).

Fluorescent probes such as calcein/ethidium homodimer or carboxyfluorescein diacateate/

Table 20.3. The acceptable range for a normal stallion's semen parameters.

Parameter	Acceptable range
Volume of sperm produced	30–250 ml
Sperm concentration	30–600×10^6 ml^{-1}
Morphology	>50% physiologically normal
Live/dead ratio	6.0:4.0
Motility	>50% progressively motile sperm
Longevity at room temperature	45% alive after 3 h
10% alive after 8 h	
pH	6.9–7.8
White blood cells	<1500 ml^{-1}
Red blood cells	<500 ml^{-1}

propidium iodide have been used (Garner *et al.*, 1986; Harrison and Vickers, 1990; Althouse and Hopkins, 1995). Both calcein and carboxyfluorescein diacateate are membrane permeable and so sperm fluoresce green. The counterstains, ethidium homodimer and propidium iodide, are not membrane permeable and can only enter membrane-damaged sperm causing them to fluoresce red. The ratio of green to red fluorescing sperm, therefore, indicates the percentage of sperm with viable membranes. Additionally the acrosome fluoresces more brightly than the rest of the head, allowing its integrity to be specifically assessed. This fluorescent technique is now becoming used in commercial AI laboratories as part of the evaluation procedure particularly for post-thaw samples.

20.4.3.3. Flow cytometry

The use of flow cytometry has been investigated, with both single and double staining to assess the functional capabilities of sperm and mitochondria (Wilhelm *et al.*, 1996; Magistrini *et al.*, 1997; Papaioannou *et al.*, 1997; Love *et al.*, 2003). The flow cytometer then separates out the sperm passed through it on the basis of their fluorescence. So if the stain used specifically attaches to functional mitochondria then the flow cytometer can give a reading for the percentage of sperm with optimum fluorescence, i.e. indicating the percentage of sperm with viable mitochondria and hence functional sperm. The results are encouraging and suggest a significant positive correlation between sperm with optimally functioning mitochondria and viable sperm (Papaioannou *et al.*, 1997; Colenbrander *et al.*, 2003).

20.4.3.4. Filtration assay

Various filters including sephadex beads over glass wool; cotton, with or without sephadex; bovine serum albumin (BSA); cellulose acetate, etc. have quite successfully been used to give an assessment of sperm viability. The passing of semen through such a filter causes the less-viable sperm to be held within the filter, resulting in a filtrate of highly viable sperm. A positive correlation exists, therefore, between the number of sperm in the filtrate and the fertilizing potential of the sample (Strzemienski *et al.*, 1987; Samper and Crabo, 1988; Samper *et al.*, 1988). Such a filter can also be used to remove less-viable sperm and allow the concentration of sperm with high fer-

tilizing potential (Jeyendran *et al.*, 1984; Samper *et al.*, 1988; Casey *et al.*, 1991).

20.4.3.5. Hypo-osmotic stress test

The hypo-osmotic test relies upon the fact that the sperm membrane is semipermeable allowing the selective passage of water through it along an osmotic pressure (OP) gradient. If stallion sperm are placed in a hypo-tonic solution (75–129 mOsm of either NaCl or sodium citrate for 15 min), water passes into the sperm resulting in a ballooning of the sperm head and deformation of the tail (Jeyendran *et al.*, 1984; Neild *et al.*, 1999). As a result the sperm tails show characteristic deformation in the form of bending and coiling (Samper *et al.*, 1991). This effect is only observed in sperm with intact membranes. Significant correlation is reported between the swelling of sperm heads and the characteristic tail coiling with sperm motility and the percentage of sperm successfully penetrating an ovum (Jeyendran *et al.*, 1984; Correa and Zavos, 1994; Kumi-Diaka and Badtram, 1994; Revell and Mrode, 1994; Correa *et al.*, 1997). The test is used in stallions with some success (Zavos and Gregory, 1987; Samper *et al.*, 1991; Lagares *et al.*, 2000) and its use as a potential commercial test is being investigated. Hyper-osmotic tests have also been investigated but with less success (De la Cueva *et al.*, 1997).

20.4.3.6. Cervical mucus penetration test

The ability of sperm to penetrate cervical mucus, the first major biological fluid that the sperm naturally come in contact with, has been used to indicate the viability of sperm from several species. A highly significant, positive correlation between penetration and acrosome integrity and also between penetration and total sperm integrity has been reported (Galli *et al.*, 1991). However, no correlation between penetration and fertility has yet been proved.

20.4.3.7. Oviductal epithelial cell explant test

The effect of sperm on oviductal epithelial cell activity, including protein secretion, is reported to have a high correlation with sperm morphology, and to be a possible prognostic test for *in vitro* fertilization (Thomas *et al.*, 1995; Thomas and Ball, 1996). The number of sperm that bind to explants may also be a possible indicator of sperm viability (Thomas *et al.*, 1995).

20.4.3.8. Zona-free hamster ova penetration assay

The ability of a sperm to penetrate a zona-free hamster oocyte has been used successfully to indicate the viability of human (Yanagimachi *et al.*, 1976; Binor *et al.*, 1980; Overstreet *et al.*, 1980; Hall 1981) and bovine sperm (Amann, 1984). Successful penetration is dependent upon successful completion of capacitation and the acrosome reaction and so gives an assessment of the sperm viability. Alternatively, sperm capacitation and the acrosome reaction can be induced artificially and full penetration and fertilization of the oocyte assessed instead of just attachment. Some success has been achieved with the zona-free hamster oocyte test in stallions (Samper *et al.*, 1989; Zhang *et al.*, 1990a; Padilla *et al.*, 1991), including a correlation with conception rates (Pitra *et al.*, 1985). Incubation of non-capacitated stallion sperm with hamster oocytes results in activation of oocyte chromosomes which has also been suggested as an indicator of sperm viability (Ko and Lee, 1993).

Similarly fertilization of an ovum depends upon an increase in oocyte calcium ion concentrations, which then oscillate at a specific frequency. Recently a sperm-specific oscillation factor phospholipase (PLC zeta) has been isolated in stallion sperm and is reported to correlate well with fertility (Gradil *et al.*, 2006).

20.4.3.9. Heterospermic insemination and competitive fertilization

Pooling of sperm from two males prior to insemination, or incubation with oocytes *in vitro*, allows direct assessment of the relative fertilizing ability of the two populations of sperm. This competitive assessment is termed, heterospermic insemination. In order for the procedure to be successful identification of the source of each sperm is essential. This may be achieved by the use of genetic markers or labelling of sperm (Beatty *et al.*, 1969; Bedford and Overstreet, 1972). Though successful, this technique is of limited practical significance.

20.4.3.10. Hemizona assay

A further development of the zona-free hamster penetration assay and heterospermic insemination is the hemizona assay in which stallion sperm from different stallions are incubated with hemizona from a single oocyte. A significant relationship has been reported between the number of bound sperm for a particular stallion and the probability of pregnancy resulting from insemination with that stallion's semen (Fazeli *et al.*, 1995). Further development of this to assess the zona pellucida binding ability of sperm has also been suggested as a means of assessing sperm viability (Pantke *et al.*, 1992, 1995; Meyers *et al.*, 1995, 1996).

20.4.3.11. DNA analysis

Finally the total amount and structure of sperm DNA may also be assessed to give an indication of viability (Evenson *et al.*, 1995; Kenney *et al.*, 1995; Neild *et al.*, 2005b).

20.5. Sexing Sperm

Many attempts have been made to preselect the sex of offspring both in man and animals. The first sex-selected offspring were rabbits in 1989 (Johnson *et al.*, 1989) and the first reported sex-selected foal was born in 2000 (Buchanan *et al.*, 2000). Preselection of sex in horses is driven by the popularity of different sexes for different disciplines; for example, geldings are most popular as event/leisure horses, colts are most popular for racing, whereas fillies are more popular as polo ponies. Sexing of sperm has been attempted by various methods with differing success in different animals, including the differential staining of the X and Y chromosome (Bhattacharya *et al.*, 1977; Ericsson and Glass, 1982; Windsor *et al.*, 1993), sperm karyotype analysis (Rudak *et al.*, 1978; Amann, 1989) and flow cytometry. Flow cytometry appears to be the method with most potential and has been successfully used with rabbit, bull and stallion sperm (Johnson *et al.*, 1989, 1997, 1998; Cran *et al.*, 1995; Seidel *et al.*, 1998). Recent developments of the flow cytometry method have led to the development of fluorescence-activated cell separation (FACS) which is reported to be very successful and is currently the method of choice for sex selection. FACS works on the basis that X- and Y-bearing chromosomes differ in their DNA content (3.7% difference in horses; Welch and Johnson, 1999). Sperm are diluted and incubated with a fluorescent dye; the amount of dye taken up depends upon the sperm DNA content (i.e. X- or Y-bearing chromosomes).

They are then passed through an argon laser beam which induces fluorescence of the sperm without causing damage. They are then passed through the flow cytometer's high voltage deflection plates which separate them according to their fluorescence (i.e. whether they are X- or Y-bearing sperm; Johnson, 2000; Morris, 2005). The accuracy of sex selection is reported to be 94–96% (Allen, 2005) and good conception rates (up to 60%) have been achieved using fresh or chilled sex-sorted semen (Lindsey *et al.*, 2002, 2005). However, there are two major drawbacks to this technology: first, work with frozen semen provides less encouraging results, with conception rates of around 20% (Lindsey *et al.*, 2002; Lee and Morris, 2005); second, the sorting rate of existing FACS machines. Current FACS machines can sort sperm at 20×10^6 h^{-1} for a limited period (2–3 h). Hence, current technology will only provide doses of 40–60×10^6 which is well below the conventional insemination dose of 300–500×10^6; hence low-dose insemination techniques are required if sex-sorted semen is to be successful (see Section 20.9.2).

20.6. Semen Storage and Use

Once the semen has been evaluated, it can be considered for insemination. Semen can be used in one of five ways: first, used raw and undiluted to immediately inseminate a single or possibly two mares; second, diluted and used immediately for insemination into several mares; third, diluted and kept at room temperature for insemination within 24 h; fourth, diluted and refrigerated for use over the next 48–72 h; or fifth, diluted and frozen for use at a later date. The method used depends on the stud system, the location of the mare(s) to be inseminated and personal preference.

20.6.1. Raw or undiluted fresh semen

Insemination of mares with raw, untreated semen is not usually carried out unless the indication for insemination is due to physical complications on either the mare or stallion side, which precludes natural covering. Insemination with raw semen should be carried out between 30 s and 1 h after collection in order to obtain acceptable results. Insemination with raw semen provides none of the advantages of AI. Even the division of raw semen to allow the insemination of a number of mares with a single ejaculate is less successful than if the division of the sample occurs after the addition of

an extender. The use of an extender is, therefore, recommended (Varner, 1986).

20.6.2. Diluted fresh semen

As with undiluted raw or fresh semen the use of diluted fresh semen defeats many of the objects of AI but does allow several mares to be inseminated per ejaculate from a single stallion with a high workload. Many of the extenders appropriate for use with fresh semen are reviewed below (Section 20.6.3) and also in Aurich (2005).

The semen/extender dilution rate with all insemination is an important consideration. Inadequate dilution will not provide adequate support for sperm. On the other hand, excessive dilution has been associated with depressed fertilization rates (Katila, 1997). A minimum semen/extender ratio of 1:1 has been recommended (Brinsko and Varner, 1992). In general, fresh semen for use is diluted in a ratio of between 1:2 and 1:4 depending upon the concentration of the raw semen sample (Section 20.7).

Work by several authors indicates that conception rates for mares inseminated with extended fresh semen, immediately post extension are similar to that obtained with undiluted raw semen, varying from 60–75% depending upon the extender used (Yi *et al.*, 1983; Jasko *et al.*, 1993b). In general E-Z Mixin appears to be the most successful though many of the other extenders discussed in Section 20.6.3 are successful (Francel *et al.*, 1987; Pickett *et al.*, 1987; Squires *et al.*, 1988).

Due to the limited viability of fresh semen stored at body temperature or non-refrigerated, its use is limited. Stallion semen is, therefore, very often diluted chilled or frozen to allow for at least 2–3 days' storage (see Sections 20.6.5 and 20.6.6).

20.6.3. Extenders for fresh or cooled semen storage and use

The use of extenders allows the number of mares inseminated per ejaculate to be significantly increased. It also prolongs the life of the sperm by providing: additional energy; antioxidants to neutralize toxic by-products of metabolism; and antibiotics to counter infection (Kotilainen *et al.*, 1994; Ball *et al.*, 2001). Extenders are also used to dilute semen prior to evaluation (see Section 20.4.2).

There are a plethora of extenders available; however, most used today are based upon milk products and/or egg yolk, as a source of protein and energy, with the addition of an antibiotic. Several other common components are added in order to maximize sperm survival (Section 20.6.3.5; Douglas-Hamilton, 1984; Province, 1984; Francel *et al.*, 1987; Pickett, 1993b,c; Ball *et al.*, 2001).

In broad terms extenders in use today can, therefore, be divided into milk or milk product-based; cream and gelatin-based; or egg yolk-based.

20.6.3.1. Milk and milk product-based extenders

These extenders include non-fat dried skimmed milk glucose extender (NFDSM-G II) or Kenney, one of the most popular diluents used (Kenney *et al.*, 1975a), along with E-Z Mixin, which is very similar in composition. These extenders are commonly used for extending semen prior to storage, but can also be used for evaluation as they are optically clear, maintain sperm motility and fertility well and are relatively straightforward and cheap to prepare. Although slightly more expensive than the straightforward skimmed milk preparations (Table 20.4c) these extenders are normally the preferred option. The straightforward skimmed milk extender requires the milk to be heated to 92–93°C for 10 min in order to inactivate lactenin, an antistreptococcal agent naturally found in milk which is toxic to equine sperm (Flipse *et al.*, 1954; Householder *et al.*, 1981).

Pregnancy rates with all milk-based extenders are reported to be quite acceptable, 52–62% pregnancy rates with little difference between the different extenders though E-Z Mixin is considered by some to be more successful (Kenney *et al.*, 1975a; Householder

Table 20.4. Milk and milk-based extenders in use today.

(a) E-Z Mixin. (From Province *et al.*, 1984, 1985.)

Component	Quantity
Non-fat dry milk	2.4 g
Glucose monohydrate	4.9 g
Sodium bicarbonate (7.5% solution)	2.0 ml
Polymixin B sulfate (50 mg ml⁻¹)	2.0 ml
Distilled water	92.0 ml
Osmolarity (mOsm kg⁻¹)	375.00 ± 2
pH	6.99 ± 0.02
Mix the liquids first and then add the powders	

(b) Non-fat dried skimmed milk glucose extender II (NFDSM-G II) or Kenney extender. (From Kenney *et al.*, 1975a.)

Component	Quantity
Non-fat dry skimmed milk	2.4 g
Glucose	4.9 g
Gentamicin sulfate (reagent grade)	100.0 mg
8.4% NaHCO$_3$	2.0 ml
Deionized water	92.0 ml
Mix the liquids first before adding the NFDSM to avoid the antibiotic curdling the milk	

(c) Heated skim milk extender. (From Voss and Pickett, 1976.)

Component	Quantity
Skimmed milk	100.0 ml
Heat to 92–93°C for 10 min in a double boiler. Cool to 37°C before use	

(d) Non-fat dried skimmed milk extender glucose I (NFDSM-G I). (From Kenney *et al.*, 1975a.)

Component	Quantity
NFDSM	2.4 g
Glucose	4.0 g
Penicillin (crystalline)	150,000 units
Streptomycin (crystalline)	150,000 µg
Deionized water	made up to 100 ml

(e) Skimmed milk gel extender. (From Voss and Pickett, 1976; Pickett, 1993b; Householder *et al.*, 1981.)

Component	Quantity
Skimmed milk	100.0 ml
Gelatin	1.3 g
Add the gelatin to the skimmed milk and agitate for 1 min	
Heat mixture in boiler for 10 min at 92°C, swirling mixture periodically	

(f) INRA 82. (From Magistrini *et al.*, 1992; Ijaz and Ducharme, 1995.)

Component	Quantity
Glucose	5.0 g
Lactose	300.0 mg
Raffinose	300.0 mg
Trisodium citrate dehydrate	60.0 mg
Potassium citrate	82.0 mg
HEPES	952.0 mg
Penicillin	10.0 IU ml⁻¹
Gentamicin	10.0 µg ml⁻¹
	Continued

Table 20.4. Continued

Water	100.0 ml
UHT skimmed milk	100.0 ml
OP mOsmol kg^{-1}	326.0
pH	7.1

(g) Mare's milk extender. (From Lawson and Davies Morel, 1996.)

Component	Quantity
Mare's milk	100 ml
Osmolarity (mOmol kg^{-1})	303.00
pH	7.07
Heat the mare's milk at 62.8°C for 30 min	

(h) NFS mare's milk. (From Lawson, 1996.)

Component	Quantity
Mare's milk	100.0 ml
Glucose	4.9 g
Streptomycin sulfate (crystalline)	0.1 g
Penicillin (crystalline)	0.1 g
Sterile deionized water	88.0 ml
Osmolarity (mOmol kg^{-1})	307.00
pH	7.14
Heat the mare's milk at 62.8°C for 30 min	

et al., 1981; Province, 1984; Francel *et al.*, 1987). A more recent addition to the range of milk-based extenders is INRA 82 (Table 20.4f), developed and widely used in France (Magistrini *et al.*, 1992), which can be used alone or with added egg yolk (INRA 83-Y; Rota *et al.*, 2004). This extender is reported to maintain sperm motility better than Kenney or E-Z Mixin at 5°C (Ijaz and Ducharme, 1995).

Milk-based extenders are largely based on cow's milk; however, mare's milk has also been successfully used both alone or with glucose (Table 20.4g,h; Kamenev, 1955; Mihailov, 1956; Lawson, 1996; Lawson *et al.*, 1996).

20.6.3.2. Cream-gel-based extenders

The addition of gelatin to an extender is thought to act as a membrane stabilizer so increase pregnancy rates (Table 20.5). The major disadvantage in using cream-gel-based extenders and the reason why, despite their apparent success, they are not widely used as commercial semen extenders, is their difficulty in preparation and the presence of fat glob-

ules which makes them inappropriate for use in microscopic examination.

20.6.3.3. Egg yolk-based extenders

Due to their optical opacity egg yolk-based extenders are used primarily for storage rather than semen evaluation. However, it is evident that egg yolk has some cryopreservation properties particularly above freezing (Bergeron and Manjunath, 2006); it is, therefore, particularly successful as part of an extender for semen destined for chilling or freezing. The specific use of egg yolk as a cryo-protectant will be discussed later under Section 20.6.6. There are many egg yolk-based extenders and many have found popularity for commercial use (Table 20.6; De Vries, 1987; Jasko *et al.*, 1992a; Klug, 1992; Samper, 1995b).

20.6.3.4. Defined component extenders

As discussed most extenders for storage of equine semen are based on either milk or egg yolk and are

Table 20.5. Cream and gel-based extenders in use today.

(a) Cream gel extender. (From Voss and Pickett, 1976; Lawson, 1996.)

Component	Quantity
Gelatin	1.3 g
Distilled water	10.0 ml
Half and half cream	90.0 ml
Penicillin	100,000 IU
Streptomycin	100,000 IU
Polymixin B sulfate	20,000 IU
Osmolarity (mOmol kg^{-1})	280.00
pH	6.52
Add gelatin to distilled water and autoclave for 20 min. Heat cream in boiler at 92–95°C for 10 min and cream to gelatin solution after removing scum from heated cream, to make a total of 100 ml	

(b) Skimmed milk gel extender. (From Voss and Pickett, 1976.)

Component	Quantity
Skimmed milk	100 ml
Gelatin	1.3 g
Add gelatin to skimmed milk and agitate for 1 min. Heat mixture in boiler for 10 min at 92°C, swirling mixture periodically	

Table 20.6. Egg yolk-based extenders.

(a) Dimitropoulous extender. (From De Vires, 1987; Braun et al., 1993; Ijaz and Ducharme, 1995.)

Component	Quantity
Solution A	
Anhydrous glucose	2.0 g
Fructose	2.0 g
Distilled water	100.0 ml
Solution B	
Sodium citrate dehydrate	2.0 g
Glycine	0.94 g
Sulfonilamide	0.35 g
Distilled water	100.0 ml
Egg yolk	20.0 ml
Osmolarity (mOsmol kg^{-1})	280.0
pH	6.9

Make up solution A and B separately, mix 30 ml solution A and 50 ml solution B, to the combined solution add the egg yolk, centrifuge for 20 min at 1200 g. Use the resulting supernatant as the extender.

(b) Glucose–lactose egg yolk extender. (From Martin et al., 1979.)

Component	Quantity
Glucose	30.0 g
Sodium citrate	1.85 g
Sodium EDTA	1.85 g
Sodium bicarbonate	0.6 g
Distilled water	100 ml

Add 50 ml of this solution to 50 ml of 11% lactose solution; supplement with 20% egg yolk and 4% glycerol

then routinely stored at 4–6°C (Varner et al., 1988, 1989; Moran et al., 1992; Aurich, 2005). Work has been most recently carried out into replacing some of the biological products, such as skim milk, with defined proteins, such as phosphocaseinates, in an attempt to only include substances beneficial to sperm, improve standardization and reduce the risk of microbiological contamination. These extenders are known as defined component extenders and reported to be superior to traditional skim milk extenders (Batellier et al., 1998, 2001; Pagl et al., 2006a). As these extenders were not based on milk they have the advantage of allowing storage at room temperature (see section 20.6.4 15°C; Batellier et al., 2000). The most successfully defined component extender to date is INRA 96, which is reported to be appropriate for cooled (4–5°C) and ambient temperature (18–20°C) storage (Table 20.7; Batellier et al., 1998, 2001). Commercial extenders such as EquiPro, based on defined casein and whey proteins, have also been produced and are proving successful both at ambient temperature and cooling (Pagl et al., 2006a; Aurich et al., 2007; Price et al., 2007).

20.6.3.5. Other major components within extenders

The four main types of extender for use with stallion semen have been considered. However, much work has been carried out on the addition of other components to improve, what can be relatively poor and very variable conception rates. One of the major problems encountered in storing semen is the accumulation of metabolic by-products, or waste, including reactive oxygen species (ROS) which subsequently have a toxic effect on sperm (Sanocka and Kurpisz, 2004). In an attempt to counteract this buffers have been used. The most popular in equine semen extenders being TRIS, which in common with several other buffers such as N-tris (hydroxymethyl) methyl-2-aminomethanesulfonic acid (TES), N-2-hydroxyethylpiperazine-N-2-ethanesulfonic acid (HEPES), sodium phosphate, sodium citrate, citric acid, BSA, etc. have been used quite successfully to buffer the accumulation of sperm metabolites over a range of temperatures and pH (Magistrini and Vidamnet, 1992). Citric acid and BSA, as well

Table 20.7. Defined component extender INRA 96. (From Batellier et al., 1998, 2001.)

Component	Quantity
Glucose	67 mM
Hanks salt solution:	
CaCl	0.14 g
KCl	0.40 g
KH$_2$PO$_4$	0.06 g
MgSO$_4$ 7H$_2$O	0.20 g
NaCl	1.25 g
Na$_2$HPO$_4$ 12H$_2$O	0.118 g
NaHCO$_3$	0.35 g
Glucose	13.21 g
Lactose	45.39 g
HEPES	4.76 g
Distilled water	to make up to 1000 ml
Native Phosphocaseinate (NPPC)	27 g l^{-1}
Penicillin	50 IU ml^{-1}
Gentamicin	50 µg ml^{-1}

as superoxide dismutase (SOD) and glutathione peroxidase (GSH-Px), also act as antioxidants, reducing the peroxidation of lipids in the sperm plasma membranes and thereby reducing the detrimental effect of storage on membrane integrity (Kreider *et al.*, 1985; Padilla and Foote, 1991; Kankofer *et al.*, 2005; Pagl *et al.*, 2006b).

Most extenders use glucose as the major source of energy (Katila, 1997). However, other sugars such as sucrose, fructose, pyruvate, lactose and raffinose have been used with some success. In addition to providing a source of energy, a cryoprotectant role for sugars has also been indicated (Arns *et al.*, 1987; Katila, 1997).

20.6.3.6. Antibiotics

Many of the aforementioned extenders include antibiotics. The semen from mammals invariably contains a natural microflora which, under normal conditions, would pose no threat of infection to females mated. The inclusion of antimicrobial agent, such as antibiotics, is necessitated not only by the risk of infection, especially in mares with a compromised uterine defence system, but also by the ideal nature of most extenders, not only for sperm survival, but also for microbial growth. This microbial growth not only presents a risk of infection, but bacteria also compete for substrates within the extender (Evans *et al.*, 1986; Watson, 1988) and have a direct detrimental effect on sperm motility (Pickett *et al.*, 1988; Aurich and Spergser, 2006). Despite all the precautions taken at collection, semen samples are invariably contaminated (Clement *et al.*, 1995). Thorough hygiene procedures during preparation, at collection and during handling do reduce microbial counts but they cannot eliminate them. The use of antibiotics is, therefore, a safety precaution against such contamination and are used whatever the storage method (Lorton *et al.*, 1988a,b). The success of antibiotics is enhanced by cooling the sample to 5°C rather than to just 20°C (Vaillencourt *et al.*, 1993; Price *et al.*, 2007).

The most popular antibiotics, either used in isolation or as a combination, were traditionally, penicillin, streptomycin and polymixin B. These still find favour. More recently ticarcillin, amikacin and gentamicin have been used (Samper, 1995b; Hurtgen, 1997). The presence of both gram-positive and gram-negative organisms in stallion semen, and the recent suggestion that resistance to antibiotics is

developing, has led to the use of new combinations of antibiotics, such as amikacin sulfate, gentamicin sulfate, streptomycin sulfate, sodium or potassium penicillin, ticarcillin disodium and polymixin B sulfate (Varner, 1991; Clement *et al.*, 1995).

Although the use of antibiotics is undoubtedly beneficial their use has been associated with an increase in extender pH and decreased sperm motility (Varner, 1991; Varner *et al.*, 1992; Jasko *et al.*, 1993a). The extent of adverse effects being dependent upon the antibiotics used (Varner *et al.*, 1992; Aurich and Spergser, 2006). Antibiotics, such as gentamicin and amikacin, are known to affect the pH of the extender, necessitating the addition of a buffer, such as sodium bicarbonate, to counteract this acidic effect (Jasko *et al.*, 1993a). This effect on the pH may also curdle the milk in milk-based extenders, so special precautions have to be taken during preparation (as indicated with some of the extenders given in Table 20.4). Antibiotics, such as penicillin, polymixin B sulfate and streptomycin, do not appear to have the same detrimental effect on pH.

Gentamicin sulfate and polymixin B sulfate in particular are reported to depress sperm motility (Jasko *et al.*, 1993a), whereas amikacin sulfate and ticarcillin disodium have no such adverse effect (Jasko *et al.*, 1993a). The exact mechanism by which the toxicity of antibiotics affects cellular mechanisms is unclear. However, it is evident that the significance of the effect increases with inclusion rates and length of storage (Varner, 1991; Clement *et al.*, 1993; Jasko *et al.*, 1993a). The adverse effects limit the applicability of some antibiotics in stallion semen extenders but they may still be used at low concentrations.

20.6.3.7. Removal of seminal plasma

The components and function of seminal plasma have been discussed at some length in Section 2.3. Although it is reported that seminal plasma plays a role in sperm transport and survival in the female tract (Troedsson *et al.*, 2005), as far as semen storage is concerned it appears that some components have a detrimental effect on sperm, possibly causing premature capacitation and reducing their lifespan (Rigby *et al.*, 2001; Pommer *et al.*, 2002; Akcay *et al.*, 2006). This effect is evident as a reduction in motility and subsequent fertility (Webb *et al.*, 1990; Jasko *et al.*, 1992a; Pruitt *et al.*, 1993), but not necessarily as a detrimental effect on sperm morphology (Sanchez et al., 1995). The fraction of seminal plasma responsible for this appears to originate from the seminal vesicles (Varner et al., 1987; Webb et al., 1990), and has been suggested to be due to elevated concentrations of sodium chloride, which effect OP changes, especially during cooling and freezing (Nishikawa, 1975) or a reaction of lipase in seminal plasma specifically with milk-based extenders (Carver and Ball, 2002). Differing levels of sodium chloride and/or lipases may, therefore, account for some of the apparent differences in the ability of different stallions' semen to survive chilling and frozen storage.

The removal of seminal plasma and its replacement with an appropriate extender may potentially allow the requirements of sperm to be met but avoid the disadvantages of seminal plasma. Indeed sperm survival rates increase as inclusion rates of seminal plasma decrease (Palmer et al., 1984) and better longevity at 25°C, 5°C or –196°C is demonstrated in semen from which seminal plasma has been removed (Varner et al., 1987; Webb et al., 1990; Sanchez et al., 1995; Love et al., 2005). However, these results are not supported by all (Pool et al., 1993; Alghamdi et al., 2004). This may in part be due to a particularly adverse reaction between egg yolk and seminal plasma (Bedford et al., 1995a,b). Additionally the composition of seminal plasma varies considerably between stallions (Jasko et al., 1992b; Charneco et al., 1993; Bedford et al., 1995a,b) and therefore, the adverse effects are likely to also vary (Katila, 1997). The ideal may, therefore, be that different extender regimes should be developed for different stallions. Despite the adverse effects of seminal plasma, it may not be beneficial to completely remove it as the inclusion of some has been reported to be advantageous (usually up to 20%; Pruitt et al., 1993; Braun et al., 1994).

There are several methods to remove seminal plasma, the most popular is centrifugation. Centrifugation at 400 g for 9–15 min successfully fractionates a semen sample allowing the supernatant (mainly seminal plasma) to be aspirated off leaving a soft plug of sperm-rich semen in the bottom (Jasko et al., 1992a; Bedford et al., 1995a,b; Heitland et al., 1996). Centrifugation also allows accurate and standard dilution for cool storage and/or freezing and gives a small volume of concentrated samples for efficient storage which are then readily available at standard-dose rates for direct insemination. This is particularly important when considering the freezing of semen where small volumes of 0.5–4 ml are regularly used.

Although centrifugation may be successful, it runs the risk of damage to sperm, manifest as reductions in motility and adverse changes to sperm morphology (Baemgartl et al., 1980). There are a number of ways in which the detrimental effects of centrifugation can be reduced, these include: the addition of a primary extender; minimizing the centrifugal force; and/or minimizing the time of centrifugation; and under layering the semen with a dense, isotonic liquid (Equi Prep, Genus PLC) that provides a protective cushion, reducing physical damage (Revell et al., 1997). However, the extent to which sperm from different stallions are affected by centrifugation is variable (Cochran et al., 1984; Pickett and Amann, 1993).

Once centrifugation has been completed, the pelleted sample is resuspended in an extender by swirling. Standardization of sample concentrations appropriate for insemination is made at this stage. Appropriate dilution rates being calculated in order to obtain the desired number of sperm per insemination dose, $100–800 \times 10^6$ (Hurtgen, 1997).

Alternatively seminal plasma may be removed at semen collection using an open-ended AV. Such an AV allows isolation and collection of only the sperm-rich fraction (Love et al., 1989; Heiskanen et al., 1994b; Kareskoski et al., 2006) which is reported to result in a significantly better sample (sperm motility) after cooled storage, compared to centrifuged unfractionated samples. A further alternative to centrifugation is filtration through a filter such as glass wool sephadex filter. Filtration appears particularly beneficial when freezing semen (Samper et al., 1991), filtered semen showing better motility and conception rates (69.2% and 80.6%) than centrifuged semen (47.6% and 66.6%; Rauterberg, 1994).

Filtration is also used to improve the quality of a semen sample subsequently prepared for cooling or freezing. Finally, the use of a 'swim up' method of separating highly viable sperm from seminal plasma and poor-quality sperm has also been reported to be successful (Casey et al., 1991).

20.6.4. Ambient temperature storage

Cooling semen below 18–20°C exposes them to potential cold shock (see Section 20.6.5.1); hence, storage above this temperature would seem advantageous. Some success has been reported with storing diluted semen at 15–20°C, ideally in the dark, for 12–24 h (Francel et al., 1987; Pickett et al., 1987; Squires et al., 1988; Varner et al., 1989; Varner and Schumacher, 1991). After 24 h, sperm motility declines rapidly compared to that stored at 5°C. It appears, therefore, that for short-term storage (24 h), cooling to 20°C provides a viable method, though for longer-term storage further cooling to 4–6°C is required (Zidane et al., 1991; Love et al., 2002). There are obvious advantages to storage at room temperature especially in places where refrigeration is difficult. However, fluctuations in ambient temperatures are a risk, and ambient temperature is not cool enough to significantly reduce the sperm metabolic rate, and this, therefore, limits the storage time. If storage at ambient temperature is to be extended an alternative method of reducing metabolic rate is required. To this end various agents have been investigated. In cattle, carbon dioxide bubbled though egg yolk extender (illini variable temperature (IVT) extender) successfully reduces the metabolic rate and increases sperm survival over a period of time (Salisbury et al., 1978). Other metabolic inhibitors are included in other extenders, such as Caprogen extender (Shannon, 1972) and Cornell University extender, and successfully allow the storage of bull's semen for up to 3 days at ambient temperature. There are few reports, however, of the use of such metabolic inhibitors with stallion semen, although Caprogen has been used (Province et al., 1985) as has proteinase inhibitors (Katila, 1997) but with only limited success. As discussed in Section 20.6.3.4, defined component extenders have recently been developed and are particularly successful with ambient temperature storage maintaining sperm viability longer that milk-based extenders (for up to 48 h; Batellier et al., 1998, 2000, 2001).

Similarly Price et al. (2007) reported that storage of semen in a defined protein-based extender called EquiPro (Minitub; Pagl et al., 2006a,b) at 15°C gave comparable results to milk-based extenders similarly stored up to 48 h, beyond this the milk-based extender proved superior especially when stored at 5°C. Storage at ambient temperature would have many advantages but as yet results are not consistent enough for it to be anything more than a potential for the future after more research work.

20.6.5. Chilled semen storage

An alternative, and the most popular, method of reducing sperm metabolic rate and hence increas-

ing viability is via cooling (Perry, 1945; Varner, 1986). No particularly sophisticated equipment is required and storage by this method allows adequate time (up to 72 h) for the transportation of semen over reasonably long distances.

As discussed previously the sample should be filtered and seminal plasma possibly removed followed by extension (see Section 20.6.3.7). Numerous extenders have been used, most of which are based on milk, egg yolk or cream/gelatin-based and have been discussed in Section 20.6.3. Chilled semen is normally extended within a range of 1:1 to 1:10 concentrated semen/extender, though a tighter range of 1:2 and 1:4 is normally advised.

Chilled (5°C) semen will survive for 48 h without a significant decline in motility (Malmgren et al., 1994) though some samples have been reported to survive for as long as 96 h (Hughes and Loy, 1970). In general conception rates for cooled semen stored for any period of time in excess of 24 h (Francel et al., 1987; Squires et al., 1988) are lower than those attained for insemination with fresh semen. Conception rates ranging from 50% to 76% are regularly reported for cooled semen in comparison to rates of 60–76% for fresh semen insemination (Van der Holst, 1984; Francel et al., 1987; Jasko et al., 1993b; Heiskanen et al., 1994a). Although chilling semen provides an efficient and successful means of reducing sperm metabolic rate and hence allowing short-term storage, chilling itself has some adverse effects on sperm. These are manifested as a depression in survival rate, motility and conception rates (Braun et al., 1994; Malmgren et al., 1994). These adverse affects of cooling are termed cold shock.

20.6.5.1. Cold shock

Cold shock is the term used to describe the stress response shown by sperm as a reaction to a drop in environmental temperature resulting in damage to the structure and function of the cell (Amann and Pickett, 1987). Damage is evident shortly after the drop in temperature has occurred and is affected by the rate of cooling and final temperature. Cold shock causes membrane alteration which is accompanied by a loss of intracellular components and a reduction in cellular metabolism. It is this damage to the cellular membranes, specifically changes in fluidity and distribution of the phospholipids, that is of most significance and has a carry-over effect on other cellular structures and functions (Hammerstedt *et al.*, 1990; Parks and Lynch, 1992; Aurich, 2005) disrupting membrane function and permeability (Hammerstedt *et al.*, 1990).

Unfortunately many of these changes to membrane configuration are irreversible and subsequent warming of the cooled sperm does not restore the original membrane configuration. Cold-shock damage manifests itself as a decline in cell metabolism, altered membrane permeability, irreversible loss of sperm motility and an increase in the number of dead sperm (Devireddy *et al.*, 2002). As well as motility rates being adversely affected by cooling, the correlation between motility and fertility is also reduced making motility, especially in frozen samples, an even poorer indicator of fertility (Amann and Pickett, 1987; Watson, 1990). Sperm motion characteristics also change with an increase in backward motion due to an over bending of the tail area arising from irreversible changes to the mid-piece and coiling of the tail (Watson, 1990).

The effects of cold shock are more evident when freezing semen, with sperm also undergoing changes similar to those of capacitation and senescence (Ball *et al.*, 1997; Neild *et al.*, 2003). Hence, when cryopreservation is considered in Section 20.6.6.1, cryoprotectants are discussed. For chilled storage the inclusion of components such as egg yolk, milk, glycerol, BSA, polyvinyl alcohol and liposomes in extenders affords some protection to sperm (Pickett and Amann, 1987; Katila, 1997). It is increasingly apparent that the lipoproteins and phospholipids in milk have a significantly beneficial effect in protecting sperm in fresh and cooled storage (Pickett and Amann, 1987) hence the continued popularity and success of milk-based extenders.

20.6.5.2. Cooling rates

The extent of sperm damage due to cold shock is not only dependent upon the temperature drop but also the speed of the drop. In general the faster the rate of cooling, the more severe the damage (Douglas-Hamilton *et al.*, 1984; Varner *et al.*, 1988); however, this also depends on the extender being used.

Several regimes for the rate of cooling have been investigated, from which it is apparent that the rate of temperature drop is most critical over the temperature range of 20–5°C (Kayser *et al.*, 1992; Katila, 1997). Therefore, a cooling regime consisting of rapid cooling from 37°C down to 20°C,

followed by a slower cooling rate of 0.1–0.05°C min^{-1} from 20°C down to 5°C for storage has been suggested. Further work in this area suggests that the critical range can be defined more specifically as 18–8°C, outside this range rapid cooling rates could be resumed (Moran *et al.*, 1992; Katila, 1997). Unfortunately there also appears to be significant inter- and intra-stallion differences in ideal cooling rates and so no hard and fast rule can be proposed, specific cooling regimes for individual stallions would, therefore, be ideal but would pose serious difficulties for commercial work; hence, most practices use the simple protocol of a steady cooling rate of 0.3°C min^{-1} across the whole range of 37–5°C (Douglas-Hamilton *et al.*, 1984). Semen can be cooled by placing in a refrigerator or cooling unit. However, more accurate control over cooling rates can be achieved by computerized cooling systems.

20.6.5.3. Storage temperature

Semen is commonly cooled to between 4°C and 5°C, though, for short-term storage it may be cooled to room temperature (20°C; see Section 20.6.4). Cooling sperm beyond the 20°C risks cold shock; however, despite this the best conception rates are generally obtained with storage at the lower 4–5°C (Pickett *et al.*, 1987; Squires *et al.*, 1988; Varner *et al.*, 1988; Kayser *et al.*, 1992). The benefit of reducing the temperature still further (0°C to −2°C) has been considered but was associated with a greater adverse effect on motility (Moran *et al.*, 1992). Other storage temperatures have been investigated including 10–15°C, which have been suggested to be better than 0°C, 4°C and 5°C (Province *et al.*, 1984; Magistrini *et al.*, 1992).

20.6.5.4. Length of storage

Prolonged exposure to a cooled environment increases the effect of cold shock and though the metabolism of the sperm is depressed by cooling, it does not cease altogether, so eventually, the build-up of toxins and the exhaustion of the nutrient supply results in death. Most studies indicate that insemination of mares with sperm that have been stored at 5°C for up to 24 h will result in acceptable fertilization rates (50–60%; Douglas-Hamilton *et al.*, 1984; Francel *et al.*, 1987; Squires *et al.*, 1988; Varner *et al.*, 1989). Conception rates with storage beyond 24 h at 5°C result in depressed, but still acceptable, fertility rates (Francel *et al.*, 1987; Pickett *et al.*, 1987; Squires *et al.*, 1988).

Although the most consistent results are obtained with cool storage up to 24 h, there are several reports of successful insemination (73–87%) of mares with semen stored up to 96 h (Hughes and Loy, 1969; Van der Holst, 1984; Heiskanen *et al.*, 1994a). This ability of semen to survive at 5°C for prolonged periods of time is used in the development of specialized insulated semen transport containers, the first of which was the Equitainer developed by Douglas-Hamilton *et al.* (1984).

20.6.5.5. Packaging and methods of transporting chilled semen

Sperm may be packaged, for cool storage, in a number of ways including sterilized polyethylene bags (Whirl-Pak), plastic bottles or baby-bottle liners. However, care should be taken as some types of plastic and rubber can be spermotoxic (Broussard *et al.*, 1990; Katila, 1997).

The presence or absence of air (oxygen) during storage is also reported to have an effect, storage in the absence of oxygen resulting in the best motility (Katila, 1997). It seems, therefore, that, for storage at 4°C, air should be excluded and containers well filled prior to storage (Douglas-Hamilton *et al.*, 1984; Magistrini *et al.*, 1992). Additionally storage of semen in tubes on a roller bench (five turns min⁻¹) may prolong survival (Katila, 1997).

If the advantages of chilling semen are to be fully realized then an effective means of storage and transportation needs to be devised. To this end several insulated transport containers have been designed with varying efficacy (Brinsko *et al.*, 2000). The one in most widespread use is the Equitainer (Hamilton-Thorn Research, Danvers, Massachusetts; Fig. 20.7).

The Equitainer is designed to permit controlled cooling of the stored semen sample at 0.3°C min⁻¹ and subsequent maintenance of the sample at 4°C for up to 36 h. The semen should be diluted in a minimum ratio of 1:2 semen/extender and put in the isothermalizer (conductive container). The pre-frozen coolant cans are placed into the bottom of the container with the thermal impedance pad on top followed by the isothermalizer and then the foam rubber protective padding, and the lid is closed. Once closed, the contents of the isothermalizer gradually cool at an initial rate of −0.3°C min⁻¹

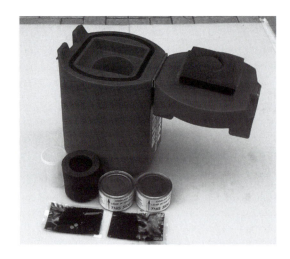

Fig. 20.7. The component parts of the Equitainer. The ballast bags can be seen in the foreground; behind them are the two coolant cans or blocks and to their left the black foam insulating jacket (which acts as a thermal impedance pad and foam rubber protective padding) into which the plastic semen container (isothermalizer) – far left – is placed. (From Hamilton-Thorn Research, Danvers, Massachusetts.)

to 4°C. This container is now a standard form of transporting semen. Other similar disposable containers include the Expecta Foal (Expecta, Parker, Colorado) and Equine Express (MP and J Associates, De Moines, Iowa). The Celle, popular in Germany and the Starstedt, popular in the Netherlands are available (Katila *et al.*, 1997) and work on a similar basis allowing transportation between countries via courier companies, postal services, etc. and in some countries by dedicated semen transport vehicles (Katila, 1997).

20.6.6. Frozen semen

The use of cooled semen is very successful for short-term storage but long-term storage by this method is not possible. In order to realize many of the potential benefits of AI, long-term storage is necessary. This is only really possible by freezing which halts the metabolic processes of the sperm, allowing, in theory, indefinite storage. The discovery by Polge *et al.* (1949) of the cryoprotectant properties of glycerol made cryopreservation possible. As a result, today, the sperm of many species can be stored at −196°C in liquid nitrogen for indefinite periods of time, while still retaining acceptable fertilization rates post thaw. Significant variation in the success rates for frozen semen has been reported. In general pregnancy rates for frozen semen at best reach values approaching those of natural service but may also result in complete failure, despite the protocol for freezing apparently being unchanged (Muller, 1987; Piao and Wang, 1988; Thomassen, 1991; Pickett and Amann, 1993). Pregnancy rates from frozen semen vary enormously particularly with breed, season and between and within stallions (Yi *et al.*, 1983; Magistrini *et al.*, 1987; Vidament *et al.*, 1997) and so the prediction of success is inaccurate. However, it is apparent that in general pregnancy rates from frozen semen are not as high as those expected from using fresh semen, conception rates of 35–60% being acceptable in commercial AI (Piao and Wang, 1988; Thomassen, 1991; Pickett and Amann, 1993; Wockener and Collenbrander, 1993). The biggest problem for commercial AI is this unpredictability of success and in particular the variation between stallions and even between different ejaculates within the same stallion (Loomis, 1993; Torres-Bogino *et al.*, 1995; Vidament *et al.*, 1997), the reason for which is unclear (Aurich *et al.*, 1996).

20.6.6.1. The principles of cryopreservation

Even under ideal conditions, it is inevitable that some damage will occur to sperm during the freezing process. The two main reasons for damage are: internal ice crystal formation resulting in alterations and physical damage to sperm structure; and increases in solute concentrations, possibly to toxic levels, as pure water is withdrawn to form ice both extracellularly and intracellularly (Salisbury *et al.*, 1978; Mazur, 1984). Changes in plasma membrane permeability to calcium has also been demonstrated and is largely manifested as a depression in motility, and possibly acrosome morphology (Wockener and Collenbrander, 1993); hence, motility of sperm is an even poorer indicator of fertility in freeze-thaw samples than it is in fresh or chilled samples (Samper *et al.*, 1991).

Regardless of all these considerations, for cryopreservation to be considered a success the process should enable a sperm to retain its fertilizing capacity post thaw. To achieve this the sperm must retain its ability to produce energy via metabolism; to show progressive motility; to maintain plasma membrane configuration and integrity (in order that it will survive in the female tract and attach to

the oocyte plasma membrane) and to retain enzymes, such as acrosin, within the acrosome (to allow penetration of the ova). Disruption of any of these functions will significantly affect the sperm's ability to achieve fertilization. It is the formation of ice crystals and the resultant movement of water up osmotic gradients that present the greatest risk to the maintenance of these attributes.

During the process of freezing several biophysical changes are evident as the temperature drops from –6°C to –15°C; extracellular ice crystals begin to form from water within the surrounding medium. This ice formation increases the concentration of solutes, such as sugars, salts and proteins. In response to this newly developed OP gradient, and the fact that water within the sperm is slower to form ice crystals than the water in the surrounding medium, water passes out of the sperm, particularly the sperm head, across the semipermeable plasma membrane, as a result the sperm becomes increasingly dehydrated. The rate of efflux of water from the sperm is also dependent upon the speed of temperature drop. The slower the drop, the greater the time allowed for the efflux of water, and hence the greater the dehydration. However, this does reduce the chance of ice crystal formation within the sperm, which can themselves cause considerable physical damage (Amann and Pickett, 1987; Hammerstedt *et al.*, 1990). This advantage, of reducing the likelihood of physical damage, has to be weighed against the greater damage due to increased intracellular dehydration and solute concentration. On the other hand, if the cooling rate is rapid, water has little time to move out of the sperm, across the plasma membrane, and hence large intracellular ice crystals form within the sperm, causing physical damage to cell membranes and components. However, the problems of dehydration and solute concentration are less evident with rapid cooling. The aim, therefore, is to arrive at a compromise between all these factors, this optimum cooling rate, however, is likely to change with the composition of the medium surrounding the sperm, that is with the seminal plasma, and hence the stallion, and with the extender used.

There are two main temperature ranges of concern regarding sperm damage during freezing. These are the period of supercooling (0°C to –5°C) and the formation of ice crystals (–6°C to –15°C). Excessive supercooling (0°C to –5°C) results in rapid ice formation, with the possibility of physical damage. This problem can be overcome by a technique termed 'seeding', which is designed to induce ice formation more gradually over a greater temperature range. However, there is no evidence that seeding a semen sample during the freezing process has any real advantageous effects (Fiser *et al.*, 1991), it is, therefore, not practised. The second area of concern is the formation of ice crystals (–6°C to –15°C) and accompanying change in OP and solute concentrations are previously discussed. In an attempt to overcome this cryoprotectants (or antifreeze agents) are used.

Cryoprotectants may be divided into two types, depending upon their action, either: penetrating, which can penetrate the plasma membrane of the sperm and act intracellularly as well as extracellularly; and non-penetrating that can only act extracellularly. Cryoprotectants act to lower the freezing point of the medium to a temperature much lower than that of water. This in turn reduces the proportion of the medium which is frozen at any one time and hence reduces the effect of low temperature on solute concentrations and hence on OP differences (Amann and Pickett, 1987; Watson and Duncan, 1988). The first cryoprotectant identified was glycerol (Polge *et al.*, 1949) and remains one of the most favoured cryoprotectants. Glycerol is a penetrating cryoprotectant, acting as a solvent and readily taken up by sperm, entering the cell within 1 min of addition to the surrounding medium (Pickett and Amann, 1993). Other penetrating cryoprotectants include dimethyl sulfoxide (DMSO) and propylene glycol (Salisbury *et al.*, 1978). Examples of non-penetrating cryoprotectants include: sugars, phenolic antioxidants, liposomes, egg yolk and milk proteins, detergents such as ethylene diamine tetracetic acid (EDTA), surfactants such as orvus ES paste (OEP), lipids such as phosphatidylcholine (Christanelli *et al.*, 1984; Denniston *et al.*, 1997; Ricker *et al.*, 2006), etc.

It has been evident for sometime that cryoprotectants, both penetrating and non-penetrating, do themselves damage sperm (Demick *et al.*, 1976; Fahy, 1986; Fiser *et al.*, 1991). This may be due to both physical damage, as a result of the changes in OP gradients, and biochemical disruption of cellular components. The adverse effect of cryoprotectants is evident more as a reduction in motility rather than a reduction in fertility hence the particularly poor correlation between motility and fertility rates in post-thaw samples. The use of motility as an indication of viability is not, therefore, a very accurate assessment in freeze-thaw samples. This effect may be due to greater deleterious effect on mitochondria than on the acrosome membrane and

region of the head. The detrimental effects of glycerol on sperm function are more evident in stallion sperm than in other species such as cattle. A reduction in this effect may be achieved by altering the freezing protocol and timing of the addition of glycerol (Section 20.6.6.2).

The protocol for the use of cryoprotectants is ultimately a compromise between the advantageous and detrimental effects of their inclusion. The exact protocol may well ideally vary with individual stallions in order to obtain optimal results. However, such individual tailoring is not practical in a commercial situation and hence further compromise is normally required.

20.6.6.2. *Extenders for use with frozen semen*

Many of the extenders used for the cryopreservation are based upon those used for cool storage (see Section 20.6.3). Often a two-extender protocol is used, a primary extender, for initial dilution which is then aspirated off after centrifugation, prior to the addition of a secondary extender which contains an added cryoprotectant for freezing. Numerous extenders have been used as both primary and/or secondary extenders.

The main function of a primary extender is to maintain sperm motility, but it also acts to protect sperm during the process of centrifugation. There is, therefore, no real requirement for a cryoprotectant in such extenders. Examples of primary extenders are given in Section 20.6.3.

Once centrifuged and the primary extender is aspirated off, the secondary extender for freezing is added. Many of the secondary extenders are also based upon those given for fresh and cooled semen storage and may be similar to the primary extenders used (see Section 20.6.3) but with a cryoprotectant added. These secondary extenders can also be used alone if semen is not centrifuged.

Glycerol and egg yolk-based extenders were among the first to be used for freezing semen (Nishikawa, 1975). Further, more recent work has demonstrated the cryoprotectant nature of many other components including sugars, liposomes, detergents, etc. (Jimenez, 1987; Heitland *et al.*, 1995). Many extenders used for freezing, therefore, contain a mixture of many components in varying ratios. Examples of extenders for freezing include those with added egg yolk, detergents, sugars, glycerols and salts (Table 20.8) and are all commonly used as secondary extenders.

Table 20.8. Examples of secondary extenders used for freezing stallion semen. (From Piao and Wang, 1988; Samper, 1995b.)

(a) Skimmed milk cryopreservation extender. (From Samper, 1995b.)

Component	Quantity
Skimmed milk	2.4 g
Egg yolk	8.0 ml
Sucrose	9.3 g
Glycerol	3.5%
Distilled water	100.0 ml

(b) Detergent (Equex STM, Nova Chemicals Sales, Scituate, Massachusetts)-based cryopreservation extenders. (From Cochran *et al.*, 1984.)

Component	Quantity
Lactose solution (11% weight/volume)	50.0 ml
Glucose EDTA solution (EDTA glucose extender II primary extender)	25.0 ml
Egg yolk	20.0 ml
Glycerol	5.0 ml
Equex STM	0.5 ml

(c) Lactose, mannitol and glucose-based cryopreservation extender. (From Naumenkov and Romankova, 1981, 1983.)

Component	Quantity
Lactose	6.6 g
Mannitol	2.1 g
Glucose	0.7 g
Disodium EDTA	0.15 g
Sodium citrate dehydrate	0.16 g
Sodium bicarbonate	0.015 g
Deionized water	100.0 ml
Egg yolk	2.5 g
Glycerol	3.5 ml

(d) HF-20 extender which may be used as a primary extender (without the 10% glycerol) or a secondary extender (with the 10% glycerol). (From Nishikawa, 1975.)

Component	Quantity
Glucose	5.0 g
Lactose	0.3 g
Raffinose	0.3 g
Sodium citrate	0.15 g
Sodium phosphate	0.05 g
Potassium sodium tartrate	0.05 g
Egg yolk	0.5–2.0 g

Penicillin	25,000 units
Streptomycin	25,000 μg
Deionized water (made up to)	100.0 ml
Glycerol	10%

(e) A simple sugar-based secondary extender. (From Piao and Wang, 1988.)

Component	Quantity
11% sucrose solution	100.0 ml
Skimmed milk	45.0 ml
Egg yolk	16.0 ml
Glycerol	6.0 ml

Glycerol remains the most popular cryoprotectant. However, as mentioned previously, the use of glycerol may itself be detrimental to sperm (Pace and Sullivan, 1975; Demick *et al.*, 1976; Fahy, 1986). Hence, a compromise has to be reached in the concentration of glycerol and the length of time of exposure of sperm to glycerol prior to freezing, in order to maximize its beneficial effects but minimize its toxic effects. The efficiency of glycerol may also be affected by the diluent to which it is added, as well as the stallion. It is now known that 5–10% glycerol with an equilibration time of just a few seconds prior to freezing is adequate for cryoprotection to be achieved (Christanelli *et al.*, 1985; Piao and Wang, 1988; Pickett and Amann, 1993; Burns and Reasner, 1995).

20.6.6.3. Vitrification

It is increasingly evident that the physical damage caused by ice crystals forming in the sperm head at freezing has a significant detrimental effect; to address this, vitrification, prior to freezing, has recently been investigated. Vitrification involves the dehydration of sperm prior to freezing, hence avoiding the formation of ice crystals, followed by very rapid freezing by plunging into liquid nitrogen; thawing should also be very rapid. Dehydration or desiccation is achieved using a non-penetrating cryoprotectant; often sugars such as sucrose, which are added to the extender in high concentrations, significantly increasing OP in the surrounding fluid causing water to be rapidly drawn out of the sperm which then become desiccated. The desiccated sperm are then frozen very rapidly to –196°C for indefinite storage. Though some success has been achieved using vitrification the major problem is the toxic effects of the cryoprotectant prior to freezing hence minimum exposure (a matter of minutes) is essential. Little work has been done on vitrification of equine sperm, but the technique is used successfully in humans and may prove to be a technique for the future (Arav *et al.*, 2002; Kelly *et al.*, 2003; Hossain and Osuamkpe, 2007).

20.6.6.4. Packaging for frozen semen

Several methods are available for the packaging of sperm for freezing. These methods include glass ampoules or vials, polypropylene, polyvinyl or plastic round or flat straws (usually 0.2–4.0 ml in volume; Fig. 20.8), flat aluminium packets (10–15 ml), pellets (0.1–0.2 ml) and macrotubes (10 ml; Loomis *et al.*, 1984; Piao and Wang, 1988; Haard and Haard, 1991).

Initially semen was frozen in glass ampoules or vials with a volume of 1–10 ml. Subsequently the use of straws has become more widespread having the advantage of being smaller in volume and, therefore, taking up less storage space (Merkt *et al.*, 1975). Semen may also be stored in pellets which, due to their size, enable a more rapid decline in temperature to be achieved. Pellets are frozen by placing small drops (approximately 0.1 ml) of concentrated semen into small indentations in a block of solid carbon dioxide or metal plate (Pace and Sullivan, 1975). Today straws are favoured by most and, although some work has suggested an effect of straw volume on subsequent fertility, straws of 0.5–5 ml are most commonly used, each containing adequate sperm so that either one or two straws can

Fig. 20.9. Semen may be conveniently stored indefinitely in liquid nitrogen flasks.

be used per insemination (Wockener and Schuberth, 1993; Heitland *et al.*, 1996). Once frozen regardless of packaging semen is held indefinitely at –196°C in liquid nitrogen flasks (Fig. 20.9).

20.6.6.5. Cooling rates

Traditionally both ampoules and straws were frozen by suspension over, followed by plunging into, liquid nitrogen at –196°C. Most recently it has become evident that rate of cooling is important (Section 20.6.6.1) and that type of storage has a bearing on this. Pelleted semen, for example, ensures a rapid, but rather uncontrolled, drop in temperature. The cooling of straws and aluminium packets can be controlled more easily by initial suspension in racks over liquid nitrogen (varying heights and time of suspension allows some control of cooling rate), followed by plunging into the liq-

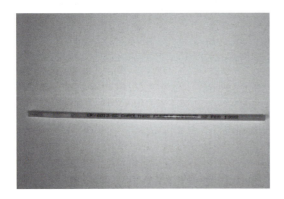

Fig. 20.8. A plastic round straw is one of the most common methods of storing frozen semen. (Photograph courtesy of Julie Baumber and Victor Medina.)

Chapter 20

uid nitrogen. Floating freezing racks into which straws are placed and then floated on liquid nitrogen have also been advocated (Hurtgen, 1997).

The effect of various cooling rates has been investigated and as a result a typical recommended equine cooling curve for a programmable freezer starting at 4°C is 10°C min^{-1} to –10°C, 20°C min^{-1} to –100°C and 60°C min^{-1} to –140°C, this is normally achieved using a computer controlled programmable freezer.

20.6.6.6. Thawing rates and extenders

Thawing rate is known to have an effect on post-thaw quality. The easiest and most commonly used method is to place the packaged semen in a warm water bath. Water bath temperatures of between 4°C and 75°C have been used successfully; however, this depends on the method of storage, volume of semen and conductivity of the packaging, etc. (Muller, 1987; Love *et al.*, 1989; British Equine Veterinary Association, 1991; Borg *et al.*, 1997).

Some protocols for thawing semen involve the addition of warmed thawing extender to aid the thawing process and minimize the effects of OP changes. The extender may also increase the volume of inseminate and help maintain sperm viability until insemination. Thawing extenders may be used for semen stored in pellets, vials or straws, and often contain sugars such as sucrose (Table 20.9; Nishikawa, 1975; Piao and Wang, 1988).

Despite careful adherence to freezing and thawing protocols it is evident that there is considerable variation in success between stallions and even between ejaculates from the same stallion. The reason for this is unclear, ideally it would be possible to identify some marker that indicates the ability of sperm to survive the freezing process, but as yet no such reliable marker has been identified.

20.7. Semen Dilution

The extent of semen dilution depends on the initial concentration of the sample, the motility of the

Table 20.9. A sucrose, milk diluent that may be used as a thawing extender. (From Piao and Wang, 1988.)

Component	Quantity
Sucrose	6.0 g
Powdered skimmed milk	3.4 g
Distilled water	100.0 ml

sperm and method of storage (Magistrini *et al.*, 1987) and, therefore, mean very little as a simple extension ratio. The aim of extension is to store the sample at an appropriate concentration and volume for survival and subsequent insemination, without the need for further treatment. Insemination of diluted semen containing 100 × 10^6 progressively motile sperm per insemination gives acceptable results, but normally 250 × 10^6 sperm per insemination is recommended to allow a margin for error, there appears to be no benefit of insemination more than 500 × 10^6 sperm (Pace and Sullivan, 1975; Householder *et al.*, 1981; Piao and Wang, 1988; Vidament *et al.*, 1997). The optimal sperm concentration for fresh and cooled storage/insemination is considered to be 25–50 × 10^6 sperm ml^{-1} of extender (or extender plus seminal plasma; Varner *et al.*, 1987; Webb *et al.*, 1993). The insemination volume is calculated using the following formula.

Insemination volume (ml) =
number of progressively motile sperm (PMS) required

$$\overline{\text{Number of PMS ml}^{-1}}$$

PMS ml^{-1} = % of PMS × number of sperm ml^{-1}
PMS required = 100–500 × 10^6

A greater number of sperm per insemination (800 × 10^6) is normally advocated for frozen semen, to compensate for loss during freezing sperm (Samper, 1995b). Frozen semen can be inseminated in a very concentrated form direct from straws or a thawing extender can be used and normally results in a sperm concentration of 50–100 × 10^6 ml^{-1}.

20.8. Insemination Volume

In addition to the total number of sperm, volume of inseminate also affects success. Volumes in excess of 100 ml or less than 0.5 ml appear to be detrimental to conception rates (Rowley *et al.*, 1990; Jasko *et al.*, 1992b). Usually the volume of inseminate varies from 10–30 ml for fresh semen, 5–40 ml for chilled and 0.5–5 ml for frozen (British Equine Veterinary Association, 1997). When deciding on dilution rates a happy medium needs to be struck between sperm concentration and inseminate volume (Newcombe *et al.*, 2005).

20.9. Insemination Technique

Due to the cost and organization involved, the oestrous cycle of mares to be covered by AI is invariably manipulated to time ovulation so that the arrival of semen can be planned ahead. There are numerous ways in which this can be achieved (see Section 12.6.2).

20.9.1. Conventional insemination

Mares are conventionally inseminated non-surgically. When using fresh or chilled semen, this should occur, as with natural service, on days 2 and 4 of oestrus and be timed as close as possible to ovulation (Hyland and Bristol, 1979; Voss *et al.*, 1979; Katila *et al.*, 1996; Watson and Nikolakopoulos, 1996). Frozen semen requires better synchrony with ovulation, ideally to within 6 h (Heiskanen *et al.*, 1994a,b; J. Newcombe, Wales, 2007, personal communication). Semen, both diluted and undiluted, is usually deposited into the uterus by means of a plastic (rubber is spermicidal) sterile pipette, with syringe attached or by an insemination gun, guided in through the cervix to the uterus, using the index finger (Figs 20.10 and 20.11). Alternatively, the pipette can be guided in through the cervix as per rectal palpation, the cervix being felt through the rectum wall. Both methods have their merits and disadvantages. Per rectum-guided insemination reduces the risk of contamination of the reproductive tract, no arm is introduced into the vagina and only a relatively small breach of the natural reproductive tract seals occurs, due to the small size of the insemination pipette. However, it is more difficult to locate and manipulate the cervix per rectum. It is largely for this reason that the preferred method of insemination in the mare is guiding the pipette per vagina. Some pipettes have a flexible tip, which allows direction into either of the uterine horns in the belief that deposition of the semen into the horn ipsilateral to the ovulating ovary may improve success rates, though there is no evidence to support this.

Once through the cervix the insemination pipette is pushed into the uterus about 2 cm. When it is in place the semen is slowly expelled by depressing the plunger or the syringe or insemination gun (Fig. 20.11; Davies Morel, 1999).

20.9.2. Low-dose insemination

Recently low-dose insemination methods have been developed to allow the use of smaller semen samples/sperm numbers per insemination. This is potentially important in stallions with poor semen samples, i.e. low in concentration or in volume and in sex-selected semen where, due to the limitations of current technology, only small samples of selected semen are available (Lindsey *et al.*, 2001). It may also have a role in frozen semen where success rates with conventional insemination techniques are much lower and more variable (Samper, 1991; Barbacini *et al.*, 2000); reducing the distance the sperm need to travel through the tract in order to reach the Fallopian tube for fertilization may be beneficial. A progressive reduction in sperm numbers is reported to occur within the uterus (from immediately behind the cervix to the utero-tubular junction area) of many species, including the equine, and that this may reflect some form of selective gradient and so many sperm are lost during passage through the reproductive tract (Parker *et al.*, 1975; Scott *et al.*, 2000). It would, therefore, seem possible that the insemination of a low dose of semen directly into the utero-tubular junction area would bypass this selection mechanism and give equivalent results. This ability to bypass the normal transit through the uterus may also be beneficial in mares with habitual prolonged post-coitum endometritis. There are three main methods employed in low-dose insemination: ultrasound-guided deep uterine insemination; hysteroscopic insemination; and gamete intra-Fallopian tube transfer (GIFT).

Fig. 20.10. In readiness for insemination, the filled syringe is attached to the end of the insemination pipette. (Photograph courtesy of Julie Baumber and Victor Medina.)

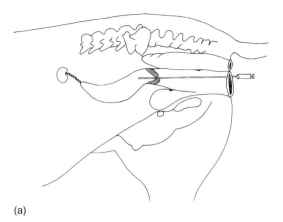

(a)

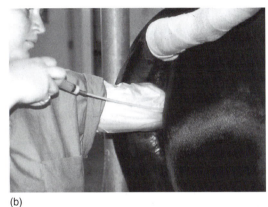

(b)

Fig. 20.11. Artificial Insemination in the mare guided per vagina. A well-lubricated hand is introduced slowly into the vagina along with the insemination pipette. Once in place the plunger of the syringe or insemination gun should be slowly depressed to expel all the semen into the uterus. (Photograph courtesy of Julie Baumber and Victor Medina.)

20.9.2.1. Ultrasound-guided deep intrauterine insemination

Ultrasound-guided intrauterine insemination involves the deposition of as few as 10×10^6 sperm in a small volume of extender (200–1000 µl) directly on to the utero-tubular junction. This is done by introducing the insemination pipette through the cervix as per conventional insemination; however, once the pipette end has passed through the cervix it is visualized and its position monitored by the ultrasonic scanner, the pipette is then pushed further up into the uterine horn to the utero-tubular junction, ipsilateral to (on the same side as) the ovulating ovary, where the semen is deposited (Buchanan *et al.*, 2000). Success

rates have been variable but up to 50% conception rates have been reported (Morris *et al.*, 2000; Morris and Allen 2002; Petersen *et al.*, 2002).

20.9.2.2. Hysteroscopic low-dose insemination

Hysteroscopic (endoscopic or videoendoscopic) low-dose insemination is similar to ultrasound-guided deep intrauterine insemination but uses even lower doses ($1–5 \times 10^6$ sperm) and volumes (10–500 µl; Morris *et al.*, 2000; Alvarenga and Leao 2002; Allen, 2005). The position of the insemination pipette is monitored via an endoscope which is passed through the mare's cervix and up to the tip of the uterine horn so that the utero-tubular junction can be visualized. A long catheter containing the small semen sample is then passed up the working channel of the endoscope until it is at the utero-tubular junction of the horn ipsilateral to (on the same side as) the ovulating ovary, where the semen is then deposited (Morris *et al.*, 2000, 2003; Morris and Allen, 2002). Conception rates are again variable but have been reported to be greater than 60% even at the lowest-dose levels (1×10^6 sperm; Morris *et al.*, 2000; Brinsko *et al.*, 2003; Clulow *et al.*, 2007).

20.9.2.3. Gamete intra-Fallopian tube transfer

GIFT or oviductal insemination (see also Section 21.9.4) involves the surgical placing (via laparoscopy) of sperm directly into the oviduct or Fallopian tube and has been used in attempts to reduce still further the number of sperm required for insemination and to bypass any possible selection, storage and/or channelling function of the utero-tubular junction (Scott *et al.*, 2000). Doses as low as 2×10^5 sperm have been used successfully (Manning *et al.*, 1998; Morris, 2004).

20.10. Conclusion

In many parts of the world equine AI is widespread in its use though the failure of the Thoroughbred industry to recognize and hence register progeny conceived by AI remains a limitation to its use across all breeds. Another major limitation is the relative lack of success and variability with frozen semen. Despite this, it is evident that equine AI is here to stay and will continue to expand, opening up with it exciting opportunities in the selection and breeding of the equine species and reproductive technology.

21 Embryo Transfer in the Mare

21.1. Introduction

The first reported successful equine embryo transfer (ET) was carried out surgically in UK between donkeys and horses (Allen and Rowson, 1972), 2 years later workers in Japan (Oguri and Tsutsumi, 1974) reported the first successful non-surgical transfer and birth of a foal. Research since then has considerably improved the early low success rates and ET is now a commercially successful practice (McKinnon and Squires, 1988; Stout, 2006). Nevertheless, the commercial application of ET in horses has a long way to go before it reaches the sophistication and success of its application in cattle and sheep. One of the major constraints on the development of ET in horses is the continued reluctance of some breed societies, most notably the Thoroughbred, to register foals conceived in this manner and of those that do the vast majority will only allow one foal to be registered per mare per year, restricting many of the potential advantages of ET. ET in horses was first taken on board as a commercial procedure in Argentina where it gained real popularity with breeding high-goal polo ponies. Most recently it has grown in popularity in Europe and North America, helped by the lifting of the restriction on the number of foals per mare per year by the Quarter Horse Breed society.

21.2. Reasons for Using Embryo Transfer

ET may be used for a number of reasons including:

1. To obtain foals from mares that are unable to carry a foal to term or to go through the process of parturition;
2. To obtain foals from older mares without risk;
3. To provide a genetically promising foal with the best maternal environment, both intra- and extra-uterine (maximum milk production);
4. To allow performance mares to breed without interrupting their performance career;
5. To provide embryos for freezing and so provide genetic diversity in the future;
6. To aid in the breeding of exotic equids;
7. To increase the number of foals/mare/lifetime;
8. For biotechnology;
9. To allow cloning, embryo sexing, etc. (Wagoner, 1982; Warren Evans, 1991; Ginther, 1992).

Several concerns have been expressed with regard to the technique, including fear of the economic effect on certain sections of the industry, the possibility of inbreeding and reduction in the genetic pool, the cost of the procedure (East *et al.*, 1999b,c) and how far should techniques such as cloning be allowed to progress.

21.3. Donor and Recipient Mares

The main principle behind ET is the transfer of elite embryos from a genetically superior donor mare, mated to a genetically superior stallion, into a normally, but not necessarily, genetically inferior recipient mare that is reproductively competent. This technique makes use of the fact that the genetic make-up of the mare carrying the foal has no effect whatsoever on the characteristics of that foal. The foal's genetic make-up and, therefore, its characteristics are determined by the mare that produced the ovum and the stallion whose sperm was used to fertilize it. The technique also makes use of the fact that, for the first 16–18 days of its life, the equine embryo is free-living within the uterus and has not yet formed an attachment. Therefore, moving it to another mare's uterus can be carried out with reasonable ease.

In order for ET to be successful, the stage of the uterus into which the embryo is transferred must be synchronized with that of the uterus from which it was collected. This will ensure that the uterine secretions and development match the requirements of the embryo. To achieve this, the oestrous cycles of the donor and recipient mares must be synchronized. This is normally achieved by using exogenous hormone therapy. Further details on the means of synchronizing and timing oestrus and ovulation in the mare are given in Section 12.6.2. Initial evidence suggested that the donor and recipient mares should ovulate within 24 h of each other, but more recent research has reported success if donors ovulate between 3 days before and 1 day after the recipient. The best results, however, are achieved if the recipient mare ovulates 24 h after the donor mare, hence the embryo is placed into a uterus a little behind the stage of the one from which it was removed so compensating for any developmental retardation that may have occurred due to the stress of transfer (McKinnon *et al.*, 1988c).

21.4. Hormonal Treatment of Donor and Recipient

For ET, both the donor and recipient mares are similarly synchronized, normally using prostaglandin F2α (PGF2α), commercially available as Equimate (Palmer and Jousett, 1975). Human chorionic gonadotrophin (hCG) may also be used in combination with PGF2α to further synchronize and hasten ovulation (Allen *et al.*, 1976b; Voss, 1993; see Section 12.6.2.3). Treatment with progesterone for 9–10 days may also be used to synchronize donors and recipients again possibly followed by PGF2α and/or hCG (see Section 12.6.2.1). Whatever protocol is used close ultrasonic monitoring of follicular activity is essential to determine the exact time of ovulation.

21.4.1. The donor mare

In an ideal transfer system, a large number of embryos are collected on one occasion from a single donor, i.e. she is induced to produce many more embryos than she would during her natural oestrous cycle – that is, she is super ovulated. In cattle and sheep super ovulation is quite successful, though results do vary. Equine chorionic gonadotrophin (eCG; also known as pregnant mare serum gonadotrophin (PMSG)) is often used as the super ovulation agent in cattle and sheep; however, it has no effect on mares even at very high doses. The best success until recently was achieved using equine pituitary extract (EPE), known commercially as

Pitropin (Douglas *et al.*, 1974; Lapin and Ginther, 1977; Douglas, 1979). Injected daily, over 7 days, Pitropin increases ovulation rates to an average of 3–4, yielding 1–2 embryos; however, the reaction is not consistent, either between different oestrous cycles in the same mare or between mares. This is likely to be due to the varying concentration of luteinizing hormone (LH) and follicle stimulating hormone (FSH) in the crude preparation; a further disadvantage is the cost of collecting equine pituitaries and extracting and purifying the preparation, plus the limited supply. Numerous protocols have been used to try and improve the response to EPE but variability remains a significant problem (Hofferer *et al.*, 1991; Dippert *et al.*, 1992; Alvarenga *et al.*, 2001). Some success has been obtained using gonadotrophin releasing hormone (GnRH), though this again involves a series of injections (Harrison *et al.*, 1991). Further along these lines, some success has been reported using human menopausal gonadotrophins (hMG; Koene *et al.*, 1990), porcine FSH (Fortune and Kimmich, 1993) and inhibin vaccines. Most recently commercially produced equine FSH has become available and appears promising (Squires and McCue, 2007). Although multiple injections are still required and a greater incidence of anovulatory follicles is reported, the use of equine follicle stimulating hormone (eFSH) can regularly result in 3–4 ovulations per oestrus resulting in on average two embryos (Logan *et al.*, 2007; McCue *et al.*, 2007). This is still a long way off the 6–8 embryos/oestrus obtained, for example, in cattle, but demonstrates that the equine ovary can respond to eFSH and provides a positive start for future research. Even if equine ovaries are able to produce multiple pre-ovulatory follicles, the unique structure of the equine ovary which dictates that ovulation can only occur through the ovulation fossa may present a limitation. Competition is likely to exist between oocytes as they pass through the ovarian stroma towards the ovulation fossa and during actual passage through the ovulation fossa. As such, ovulation of multiple oocytes within a short period of time is unlikely to occur.

A successful super ovulatory agent would not only allow more embryos to be recovered per oestrus but would increase the likelihood of at least one viable embryo per collection. It may also have uses in providing more ova/oestrus in subfertile mares and in those to be mated by a subfertile stallion so again increasing the chances of at least one ova being successfully fertilized. Similarly it may improve conception rates when using frozen semen where fertility is naturally lower. Despite all the work carried out, the lack of a reliable, effective, super ovulation agent remains a major challenge in the development and commercial use of equine ET.

A general routine that may be used to synchronize and attempt to super ovulate a donor mare is given in Table 21.1.

21.4.2. The recipient mare

The ideal recipient mares are multiparous, 5–10 years of age, with a proven breeding record and no history of uterine infection or compromise. Work by Carnevale *et al.* (2000b) suggests that the tone of the recipient's uterus and cervix are also very important selection criteria. Recipient mares need be of no particular genetic merit, as they will in no way affect the genetic make-up of the embryos transferred to them. Ideally they should be larger than the donor, to provide a larger uterus and so maximize fetal development *in utero*, this will have a carry-over effect on birth weight and future post-natal growth and development (East *et al.*, 1999a,c).

Recipients may be treated hormonally in a very similar manner to donors, except no attempt is made to super ovulate them, and they are, of course, not mated. Some people advocate treating the recipient slightly behind the donor in order to be able to put the embryo into a recipient 24 h behind the donor uterus from which it was taken

Table 21.1. A general hormone routine that can be used to time and attempt to super ovulate a donor mare for ET.

Time	Drug to be administered/event
Day 0	PGF2α
Day 6	hCG
Day 10	Oestrus and ovulation in some mares
Day 14	PGF2α and eFSH
Day 15	eFSH
Day 16	eFSH
Day 17	eFSH
Day 18	eFSH, oestrus may start
Day 19	eFSH, oestrus
Day 20	eFSH, oestrus
Day 21	hCG, oestrus
Day 22	Ovulation may occur – covering/AI
Day 24	Ovulation may occur – covering/AI
Day 30	Embryo collection
Day 31	Embryo collection

Table 21.2. An example of hormone regime used to time ovulation in a recipient mare.

Time	Drug to be administered/event
Day 0	PGF2α
Day 6	hCG
Day 10	Oestrus and ovulation in some mares
Day 14	PGF2α
Day 18	Oestrus may start
Day 19	Oestrus
Day 20	Oestrus
Day 21	hCG, oestrus
Day 22	Ovulation may occur
Day 24	Ovulation may occur
Day 30	Embryo transfer
Day 31	Embryo transfer

(see Section 21.3; Allen, 2001, 2005). An example of an exogenous hormone treatment regime used in recipients is given in Table 21.2.

They should be scanned to ensure that they have reacted to the synchronization programme and ovulated. Ovariectomized mares have been used experimentally as recipients. These mares alleviate the need for synchronization and veterinary inspection but, because of the lack of ovarian progesterone, they require artificial progesterone supplementation for the first 120 days of pregnancy (Hinrichs and Kenney, 1988; Squires *et al.*, 1989).

21.5. Embryo Recovery

Once ovulation in the donor is confirmed she is either covered naturally or by artificial insemination (AI; Chapter 21, this volume). The age of embryos recovered varies with the chosen method of recovery. Equine embryos in general are collected at between 4 (morula) and 8 days (blastocyst) of age (Figs 5.2 and 5.3). They can be recovered at a relatively late stage compared to other farm livestock, as equine concepti do not expand into elongated trophoblasts but remain spherical and free-living within the uterus for a prolonged period of time (16–18 days).

21.5.1. Surgical recovery

The initial method of recovery during early work was surgical. This technique is still used by some and recovery rates are similar, or possibly lower, compared with the more popular commercial non-surgical techniques; surgical collection allows younger more robust embryos to be collected. Surgery is carried out under general anaesthetic with the mare presented in a dorsal recumbency with her ventral side (abdomen) uppermost or under sedation with the mare held in stocks. The uterus is exteriorized through a ventral midline incision in the area between the mammary gland and the position of umbilical cord attachment or via a flank incision. The uterine horn is cannulated, with a glass tube and ligated (tied off) near to the uterine body, preventing fluid passing into the body of the uterus. Approximately 50 ml of fluid, often Dulbecco's phosphate-buffered saline, possibly with additional calf serum or oestrous mare serum and penicillin to help prevent infection, is flushed, by means of a blunt-ended needle and attached syringe, from the Fallopian tube towards the uterine horn. As the fluid passes, it takes with it any embryos present and exits via the glass canula to be collected in a warm collecting vessel (Allen and Rowson, 1975; Allen et al., 1977; Castleberry et al., 1980; Imel et al., 1981). Surgical recovery only allows embryos within the Fallopian tube to be recovered, i.e. those younger than 5 days. Recovery rates in the order of 70–77% have been reported (Allen and Rowson, 1975).

21.5.2. Non-surgical recovery

Non-surgical embryo recovery was first used in horses with any consistent success in Texas in 1979 (Douglas, 1979; Vogelsang et al., 1979), though several other researchers had attempted it previously (Oguri and Tsutsumi, 1972; Allen and Rowson, 1975). It has become increasingly popular as there is no need for a general anaesthetic and hence the procedure carries lower risks and is the method of choice for commercial ET. Recovery rates are reported to be as good, if not better, than those for surgical recovery, though it does depend upon the stage of embryo development at collection. Recovery rates of up to 80% have been reported for the collection of 6- to 7-day-old embryos (Oguri and Tsutsumi, 1980). Collection of embryos younger than 5 days old is not possible, as this technique, unlike surgical recovery, cannot flush the mare's fallopian tubes, hence only embryos that are within the uterine horn can be collected.

The techniques for non-surgical recovery are very similar to those used in cattle and remain largely unchanged from when the procedure was first successful. The mare is restrained in stocks, having been prepared and washed as for minimal contamination,

natural covering (see Section 13.5.1). A three-way catheter (French or Rusch foley catheter; Fig. 21.1) is introduced through the cervix of the mare, guided per rectum or by inserting a hand into the vagina and guiding the catheter through the cervix using the index finger (Fig. 21.2).

The catheter is passed as high up into the uterine horn as possible without undue pressure. Once in

Fig. 21.1. A foley catheter used for non-surgical embryo flushing. (Photograph courtesy of INNOVIS. With permission.)

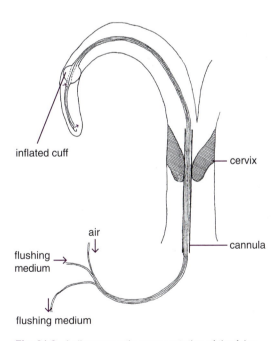

Fig. 21.2. A diagrammatic representation of the foley catheter illustrating the inlet and outlet tubes plus the air inlet for inflating the cuff.

position, the cuff of the catheter is inflated with 15–50 ml of air via the inlet tube, so occluding the base of the uterine horn and, thereby, preventing the escape of flushing medium through the uterus. Fluid, usually Dulbecco's phosphate-buffered saline (PBS) as described for surgical transfer, is then flushed in through the entry catheter up into the top of the uterine horn. The fluid returns, along with any embryos present, via an opening into the outlet tube for collection in a warm collecting vessel (Fig. 21.3). Both horns may be flushed out simultaneously if the inflated cuff is drawn back against the internal os of the cervix (Imel *et al.*, 1980), or independently if the cuff is placed in turn in each horn (Douglas, 1979). The donor is usually flushed two or three times, the volume of fluid used depends upon the position of the inflated cuff. If the whole uterus is flushed then approximately 1 l of fluid per flush for ponies and up to 2 l per flush for horses may be required. Recovery of embryos can be improved by palpation of the uterus, per rectum, at the same time as flushing to dislodge any embryos caught between the endometrial folds (Squires and Seidel, 1995). Oxytocin may also be administered to encourage uterine myometrial activity and hence help in the evacuation of the fluid plus embryo (Jasko, 2002; Hudson and McCue, 2004). Recovery rates vary with method of collection but are also affected by day of recovery, number of ovulations, age of the mare and semen quality. Recovery rates of 50–60% are considered to be acceptable (Squires *et al.*, 1987; Ball *et al.*, 1989).

Both methods of recovery have their advantages and disadvantages. Surgical recovery allows younger (prior to 5 days), and, therefore, more robust embryos to be collected, but at a greater risk to the mare. Non-surgical recovery results in older (day 6 onwards), therefore, less-robust embryos, but with less risk to the mare and with potential for multiple recoveries (Iuliano *et al.*, 1985). The ideal scenario would be one in which early embryos, those prior to day 5, could be collected non-surgically. Indeed work by Robinson *et al.* (1999) suggests that this may be possible via the laparoscopic application of prostaglandin E (PGE) to the external surface of the Fallopian tube ipsilateral to the ovary that has ovulated 4 days previously. Work by Weber *et al.* (1991) demonstrated that the selective passage of fertilized ova through the utero-tubular junction is due to their ability to secrete PGE. Hence, its application by laparoscopy allows the passage of day 4 embryos into the uterus and hence their possible collection via non-surgical means. In reality however, the need to starve a mare for 36 h (a general requirement prior to laparoscopy) deems it not a popular option with competition mares. In non-surgical recovery the next oestrous cycle can be advanced by administering PGF2α at the time of collection, which induces the mare to return to

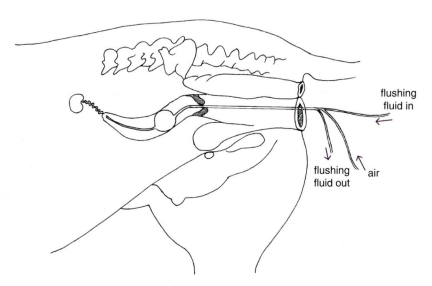

Fig. 21.3. The foley catheter in place ready for flushing and the collection of embryos.

Chapter 21

oestrus 4 days or so later and hence allows another crop of embryos to be collected from that mare within 10–12 days (Griffin *et al.*, 1981). As such, non-surgical recovery of days 7–8 embryos is normally the commercial method of choice.

21.6. Embryo Evaluation

Once collected the flushings are then examined microscopically for embryos; a filter system may be used to aid the search. Due to their relative weight, viable embryos will sink to the bottom of the collecting vessel allowing a significant amount of the flushing medium to be decanted off and so ease identification. Throughout the evaluation process, it is essential that embryos are kept warm (35–38°C) and that all equipment used is also pre-warmed. Embryos are evaluated to assess their viability prior to transfer. Embryos are measured and their stage of development matched to their age. Morphological features such as shape, colour, number and compactness of cells, etc. are noted. Based upon this information, embryos are graded 1–5, 1 being excellent and 5 dead (Betteridge, 1989). Embryos graded 3 or better are normally selected for transfer (Oguri and Tsutsumi, 1972; Allen, 1977). Using grade 1 embryos 15-day pregnancy rates of 70–75% have been reported (Squires *et al.*, 2003).

21.7. Embryo Storage

Recovered embryos are either transferred immediately, or cooled or occasionally frozen for use at a later date. Equine embryos can be stored for up to 24 h at 42°C, i.e. slightly above body temperature, prior to transfer, allowing some limited transportation. Embryos have been successfully transported in ligated rabbit oviducts, providing they are transferred into recipients within 48 h (Allen *et al.*, 1976b). Alternatively, and more normally, embryos can be stored in either physiological saline containing 2% gelatin, or a 1:1 mix of mare's serum: Ringer's solution or modified Dulbecco's PBS (Whittingham, 1971; Oguri and Tsutsumi, 1972, 1974).

21.7.1. Chilled embryo storage

More recently it has been possible to chill embryos and, as such, they can be successfully stored for 24–48 h at 4–5°C (Martin *et al.*, 1991). Storage

mediums include: modified Ham F10 medium which has been previously gassed with 5% CO_2, 5% O_2 and 90% N, but there are also several commercial media such as EmCare and Vigro Holding Plus (A B Technology, Pullman, Washington) that have been used with good success (Squires *et al.*, 2003). Commercially, embryos can then be cooled and stored, as described for semen, in an Equitainer (see Section 20.6.5.5) which acts like a cool box, cooling embryos by −0.3°C min^{-1} down to 5°C, and maintains them at this temperature for in excess of 48 h, allowing reasonable scope for transportation (Allen *et al.*, 1976a; Douglas-Hamilton *et al.*, 1984; Carnevale *et al.*, 1987). Pregnancy rates with cooled embryos stored for 24 h are equivalent to those with fresh transfer (Carney *et al.*, 1991). More recently, in common with semen storage (see Section 20.6.4), cooling and storing at ambient temperature (15–18°C) in Ham's F10 with Hepes Buffer and 04% BSA has been investigated with some success.

21.7.2. Frozen embryo storage

The only means of long-term storage is by cryopreservation (freezing) with or without vitrification. Cryopreservation is the term given to cooling embryos (also sperm (see Section 20.6.6) or oocytes (see Section 21.9.5)) to −196°C, one of the biggest risks of such treatment is the formation of ice crystals within the embryo during the freezing process which causes physical damage in a similar way to that described for sperm (see Section 20.6.6.1). This has been traditionally reduced by the use of a cryoprotectant (or antifreeze; see Section 20.6.6.1) which reduces the freezing temperature and results in the development of ice crystals over a wider temperature range allowing water to pass out of the embryo into the surrounding fluid and so reduce the number of ice crystals within the embryo that can cause damage. Unfortunately cryopreservation is currently not that successful in horses, although it is quite successful and commercially viable in sheep, cattle and goats. The first successful birth of a foal from a frozen embryo was not achieved until 1982, with only one live foal from 14 embryos (Yamamoto *et al.*, 1982). Similar or better success rates have been achieved since, in the region of 20–40% (Czlonkowska *et al.*, 1985; Slade *et al.*, 1985; Farinasso *et al.*, 1989; Squires *et al.*, 1989). Poor success rates have been postulated to be due to a variation in embryo size at freezing. Most of the embryos recovered from mares are via the non-surgical technique, hence are 6–7 days old and so 500–1000 µm in diameter and at the blastocyst stage (Fig. 5.3). Experiments have shown that cryopreservation of equine embryos larger than 250 µm in diameter (early blastocysts approximately day 5) gives relatively poor results. This is confirmed by Skidmore et al. (1990) who achieved pregnancy rates of 40–55% from frozen embryos that were less than 225 µm in diameter. The equine conceptus is unique in developing, at around days 4–5 of pregnancy, an acellular glycoprotein capsule (see Section 5.3.1). It appears that this may impede the passage of cryoprotectants into the conceptus and so reduce success rates. Indeed Legrand et al. (2002) reported that the success of freezing was related to the thickness of the capsule and that if embryos greater than 500 µm in diameter were treated with 0.2% weight to volume trypsin for 15 min prior to addition of glycerol and freezing, 75% of the embryos (three out of four) were viable post thaw and went on to produce pregnancies. The success rates for the transfer of equine morula and early blastocysts which do not as yet have a fully formed capsule are much better. However, recovering such young embryos (prior to day 5), from the mare via the non-surgical technique is near impossible as they have not yet passed into the uterus and so can only be collected surgically, on the other hand, older and larger 6- to 7-day-old blastocysts can be collected more easily non-surgically but due to the presence of the capsule do not freeze well (Boyle et al., 1989).

In order to cryopreserve embryos a cryoprotectant (see Section 20.6.6.1), traditionally glycerol, is required. Other cryoprotectants (1,2 propandiol, ethylene glycol, DMSO) have been tried with limited success (Hochi et al., 1994a; Ferreira et al., 1997; Huhtinen et al., 1997; Bruyas et al., 2000). Embryos, as with sperm (see Section 20.6.6.1), need to be frozen in a slow stepwise fashion. The temperature is initially dropped down quite rapidly to –6° or –7°C and then more gradually dropped through the formation of ice crystals period down to –33°C to –35°C. This is followed by a rapid temperature drop by plunging into liquid nitrogen for storage at –196°C at which temperature storage is presumed to be indefinite (Slade et al., 1985; Skidmore et al., 1990). Prior to transfer they need to be thawed out by a gradual stepwise increase in temperature, with the possible addition of a thawing extender (see Section 20.6.6.6) such as a sucrose solution to aid rehydration and help prevent excessive alterations in osmotic pressure (Hochi et al., 1996; Young et al., 1997).

21.7.2.1. Vitrification

More recently, vitrification (see Section 20.6.6.3; Hochi et al., 1994b, 1995; Young et al., 1997; Oberstein et al., 2001; Carnevale, 2004) has been tried. Vitrification involves the removal of most of the water from within the embryo prior to freezing (desiccation of the embryo) so that no ice crystals can form within the embryo and cause damage. Removal of the intra-embryonic water is brought about by developing a strong osmotic pressure gradient between the embryo and its surrounding fluid. This can be achieved, for example, by placing the embryo in a high-concentrate solution of sugars, often sucrose, which draws the water out of the embryo so dehydrating it before it is then frozen.

As with straightforward freezing vitrification of small (<300 µm) young (<6 days) embryos is most successful. Larger blastocysts again do not store well presumably due to the protective nature of the capsule making the conceptus impermeable and so dehydration difficult. It would seem possible, therefore, that the removal of the capsule with enzymes such as trypsin may allow older embryos to be successfully vitrified prior to freezing.

21.8. Embryo Transfer

The transfer of embryos can, as with collection, be done either surgically or non-surgically.

21.8.1. Surgical embryo transfer

Surgical transfer is not popular today. As with embryo collection, it requires a general anaesthetic. The mare is prepared as for surgical collection and a similar, but smaller, ventral midline incision is made. Just the uterine horn is exteriorized through this incision and a small hole is made at the top of the horn with a blunt needle. A Pasteur pipette or equivalent, containing the embryo, held between two bubbles of air, is introduced through this hole into the uterus and the embryo expelled into the lumen of the uterine horn (Allen, 1982b). A similar, but less traumatic, procedure can be carried out through the flank of the mare while standing restrained in stocks, using a local anaesthetic and general tranquillizer. A small incision is made in the mare's flank, the uterine horn is then exteriorized through this incision and the embryo replaced as described above (Squires *et al.*, 1985).

Both these methods have the advantage that younger embryos, 2–4 days old, may be replaced into the Fallopian tubes, as access to the Fallopian tube as well as the top of the uterine horns is possible. The reported success rates are very variable – 50–90% – though they do tend to be higher than those in non-surgical transfer, especially after cryopreservation (Allen and Rowson, 1975; Imel *et al.*, 1981; Allen, 1982b). Pregnancy rates with cooled embryos are equivalent to fresh transfer (Sertich *et al.*, 1988; Carney *et al.*, 1991).

21.8.2. Laparoscopic embryo transfer

Laparoscopic ET is successful in a number of animals and has been attempted with some success in the mare. A laparoscope inserted in the flank of a mare sedated and held in stocks is used to guide a long needle containing a catheter plus the embryos through the abdominal wall, and into the top of the uterine horn. Similarly a long flexible catheter can be passed down a wide bore needle passed through the anterior wall of the vagina and inserted blind into the uterine horn which is manipulated per rectal palpation (Muller and Cunat, 1993). Though success rates are good both techniques require considerable skill and dexterity.

21.8.3. Non-surgical embryo transfer

Due to the complications of surgical and laparoscopic ET the most popular technique currently

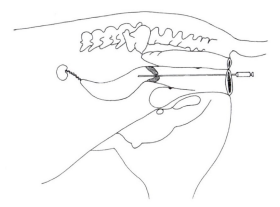

Fig. 21.4. Non-surgical transfer of embryos into a recipient mare.

widely used is transcervical ET very similar to that used in cattle and in AI (see Section 20.9.1; Fig. 20.11, Meira and Henry, 1991). The mare is restrained within stocks, the perineal area thoroughly washed, the transfer gun or catheter with attached syringe, containing the embryo held between two bubbles of air, is passed in through the mare's cervix and into the uterine body. Once in place, the embryo and the associated fluids are expelled into the uterus by slowly depressing the plunger of the syringe (Fig. 21.4; Lagreaux and Palmer, 1989; Jasko, 2002).

Non-surgical transfer is a relatively easy and quick procedure. However, it is less successful than surgical, conception rates of 22–50% and 55–90%, respectively (Douglas, 1982; Squires *et al.*, 1982a; Iuliano *et al.*, 1985; Lagreaux and Palmer, 1989; Allen, 2005). This may be due to the increased risk of introducing low-grade infections into the reproductive tract (Wilsher and Allen, 2004) also possibly due to induced release of PGF2α and/or oxytocin by manual manipulation of the cervix when trying to pass through the catheter, causing early luteolysis (Kask *et al.*, 1997; Handler *et al.*, 2002) and a localized uterine inflammatory response to the technique (Squires *et al.*, 1999). In an attempt to reduce, in particular uterine infection and excessive cervical manipulation, Wilsher and Allen (2004) developed a new technique for the placement of the embryo into the recipient's uterus. Mares are sedated and following a strict hygiene regime a duck-billed vaginal speculum is inserted into the vagina and opened to allow the cervix to be clearly viewed. A pair of elongated, smooth-ended grasping forceps is inserted and the cervical external os grasped and pulled caudally. This straightens the cervical canal, elevates the uterus within the abdomen and straightens the uterine lumen making insertion of the catheter easier and much less traumatic on the reproductive tract. Using this technique, Wilsher and Allen (2004) reported an 85% success rate conventional. Non-surgical transfer only allows the deposition of embryos into the uterine body or the lower part of the uterine horns unless ultrasound guided deep intrauterine and hysteroscopic low-dose insemination techniques (see sections 20.9.2.1 and 20.9.2.2) are used. This technique is, therefore, of limited use with embryos less than 5 days old; for 6- to 8-day-old embryos, pregnancy rates of 40–70% have been reported (Allen and Rowson, 1975; Castleberry *et al.*, 1980; Oguri and Tsutsumi, 1980; Wilson *et al.*, 1987).

The site of embryo deposition in relation to the functional corpora lutea seems to be less crucial in the mare than in the ewe and the cow. This is presumably due to the natural transuterine migration of equine embryos. The medium in which the embryos are transferred also seems to have an effect on pregnancy rates, Dulbecco's PBS being reported to give the best results as a transport medium (Douglas, 1980).

21.9. Embryo Transfer Associated Technologies

Straightforward ET is now a commercially available technique for breeding horses in many parts of the world. Many associated technologies have been developed in other species for example, human, cattle, sheep, pigs, etc. Though the horse in many ways lags behind the sophistication of embryo technology in these species many of the techniques are now being attempted in horses and may present opportunities for future breeding practice.

21.9.1. *In vitro* fertilization (IVF)

IVF is the fertilization of a collected oocyte by a sperm within the laboratory rather than within the mare's tract. Once fertilization and initial development has taken place the embryo is placed into the uterus of a recipient mare. The technique is quite commonly used in humans but is rarely practised in horses, largely due to poor success rates. However, IVF would be of advantage to subfertile mares that have had problems in producing viable embryos for conventional ET, or in the case of stallions with poor semen quality or quantity. To date, unfortunately only two foals have been born from IVF. Oocytes can be collected from pre-ovulatory follicles which are aspirated (see Section 21.9.1.1) in which case they will have matured *in vivo* before collection, these are the oocytes that have been used to produce the few successful IVF foals. Alternatively oocytes can be collected from smaller follicles during dioestrus and then matured *in vitro*, this allows many more oocytes per collection to be gained but they require maturation before use (Hinrichs *et al.*, 2002). Finally, but rarely, oocytes can be collected post mortem and again matured *in vitro*, this is only really of consideration if a mare has died or had to be euthanized suddenly (Dell'Aquila *et al.*, 2000). Unfortunately *in vitro* oocyte maturation is problematic in the mare as is *in vitro* sperm capacitation.

The most successful method of capacitation to date is calcium ionophore A23187 possibly plus heparin (Alm *et al.*, 2001). After such capacitation sperm penetration of the oocyte can be facilitated by exposing the zona pellucida to acid (Li *et al.*, 1995). *In vitro* maturation, both nuclear and cytoplasmic, does occur in some aspirated oocytes stored for 24–30h, especially those that are from follicles >20mm in diameter. Some success has been achieved in *in vitro* oocyte maturation by culturing oocytes in TCM-199, follicular fluid from pre-ovulatory follicles, etc. (Choi *et al.*, 2002). Once fertilized the embryos are developed *in vitro* to the blastocyst stage at which time they can be placed into the uterus of the recipient mare as per conventional ET. Despite all the work, IVF in horses is relatively unsuccessful. In an attempt to overcome some of the problems of IVF, especially those associated with sperm capacitation, intra cytoplasmic sperm injection (ICSI) has been developed (see Section 21.9.2).

21.9.1.1. *Oocyte collection*

Oocyte collection was initially from slaughterhouse material. Ovaries were collected and returned to laboratory where the follicles were aspirated using a needle or catheter and syringe and the fluid was then filtered to isolate the oocyte(s). This is not commonly practised today as a commercial procedure but is occasionally required if a valuable mare dies or has to be euthanized suddenly. Today oocytes are collected from follicles on the ovary of the live mare either at oestrus in which case the follicles will be pre-ovulatory follicles (>30mm in diameter) and so the oocyte will have matured *in vivo*; or alternatively they can be collected during dioestrus, in which case the follicles will be much smaller (varying sizes up to 25mm in diameter) and the oocytes immature and so will need to undergo maturation *in vitro*. Whatever the size of the follicle aspiration is pretty standard and involves transvaginal ultrasound guided follicle aspiration plus possible flushing. A long needle and catheter is passed through the anterior vaginal wall and guided via ultrasound to the follicles to be aspirated. Once the follicle has been located it is punctured with the needle, the catheter is then pushed into the follicle and the fluid, plus hopefully the oocyte, is withdrawn. The attachment of the oocyte to the cumulus oophorus upon which it sits (see Section 1.9.1) in the mare's pre-ovulatory follicle is particularly difficult to disrupt (Bruck *et al.*, 1999). Some workers, therefore, flush the follicle with a small amount of fluid after aspiration in an attempt to improve recovery rates especially in pre-ovulatory oocyte collection. Repeated aspirations and/or flushings are not reported to be detrimental to the future oocyte collection or any subsequent pregnancy (Mari *et al.*, 2005).

21.9.1.2. Embryo splitting

Embryo splitting involves the bisection of an undifferentiated embryo (young morula; see Section 5.3) into a number of potential new individuals that can then be transferred into a number of recipient mares. Commercially such a procedure could compensate for the significant difficulties encountered in super ovulating mares, allowing an increase in the number of embryos per mare. The first two sets of identical twin foals (one set of colts the other fillies) resulting from embryo splitting were reported by Allen and Pashan (1984). The original embryo was collected surgically on days 2–3 post fertilization and pairs of blastomeres from an 8-cell morula were separated and each pair injected into empty pig zona pellucida and imbedded in agar and transferred to sheep oviducts for 3–4 days until the blastocyst stage, when each conceptus was then transferred on to a recipient mare. More recently success has been achieved by collecting morula (6–6.5 days post fertilization) before capsule formation and bisecting them into two demi-embryos. Each embryo is then transferred into one or two recipient mares. Work with embryo bisection and the formation of identical twins which are subsequently transferred into different uteri has underlined the importance played by uterine and placental competence and size in fetal development. Despite identical genetic make-up transfer of one embryo into a mare with a smaller or compromised uterus limits placental size and hence foal birth weight, which may not necessarily be compensated for by accelerated growth before mature size is achieved (Allen, 2005).

21.9.2. Intra cytoplasmic sperm injection (ICSI)

ICSI involves the injection of a single sperm into a collected oocyte, which is usually at metaphase II stage, in order to achieve fertilization. Following fertilization the conceptus is allowed to develop *in vitro* to the morula or early blastocyst before transfer into a recipient mare by standard ET technique. ICSI has been successfully used to produce foals from *in vitro*-matured oocytes (Cochran *et al.*, 1998; McKinnon *et al.*, 2000). The need for dioestrous oocytes to be matured *in vitro* still remains but reasonable success has been achieved by incubation of oocytes prior to, and/or immediately after, ICSI in one or a combination of several media including: follicular fluid from pre-ovulatory follicles; blood serum from oestrous mare, ionomycin, ethanol, thimerosal, inositol, oviductal epithelial cells, TCM-199, fetal fibroblast calls, etc. (Dell'Aquila *et al.*, 1997a,b; Li *et al.*, 2000, 2001; Galli *et al.*, 2007). Culture of ova after ICSI is also important and some of the best results have been obtained by placing the embryos into the oviducts of mares, rabbits or sheep (Galli *et al.*, 2002; Lazzari *et al.*, 2002; Choi *et al.*, 2004). Although results are not that brilliant, an average of 30% develop to the blastocysts stage, live foals have been produced and providing the culture media both before and after ICSI can be perfected ICSI has the potential to provide a means of breeding stallions with very poor semen quality or sex-sorted sperm.

21.9.3. Oocyte transfer

Oocyte transfer is the collection of oocytes from the donor mare rather than an embryo. As discussed in Section 21.9.1.1, oocytes are either matured *in vivo*, i.e. collected from pre-ovulatory follicles or matured *in vitro*, i.e. collected from smaller follicles and matured to metaphase II stage in the laboratory. Once maturation has been achieved the oocyte is transferred on to the fimbrae of the infundibulum of the Fallopian tube of the recipient mare. The recipient mare is then mated either naturally or by AI. Prior to mating the recipient mare has her own pre-ovulatory follicle aspirated to prevent her becoming pregnant with her own foal. Oocyte transfer is of particular use in mares that are also good candidates for IVF, these are often older mares and those that have problems in ovulating or have an incompetent uterus – often due to persistent infections. The advantage over IVF is that fertilization takes place *in vivo* and so the problems associated with sperm capacitation are avoided. The first successful oocyte transfer was reported by McKinnon *et al.* (1988c) but has recently been used more commercially with pre-ovulatory oocytes, i.e. those collected from mares 24–36 h after treatment with hCG in the presence of a >35 mm follicle (Carnevale *et al.*, 2000a,b; Hinrichs *et al.*, 2000). Success rates are much higher with oocytes collected from pre-ovulatory follicles due to the previously mentioned problems of *in vitro* maturation of equine oocytes (Carnevale *et al.*, 2001).

21.9.4. Gamete intra fallopian tube transfer (GIFT)

Oocyte transfer relies upon a natural covering or AI for fertilization, occasionally sperm numbers or the semen quality is so low, due to stallion subfertility or after semen sexing, that the chances of conception are very poor. In this case GIFT is an option as it involves the placing of a low number of sperm plus the oocyte, usually at metaphase II stage, into the Fallopian tube or on to the fimbrae of the infundibulum of the Fallopian tube of a recipient mare (Carnevale, 2004). The same issues with *in vivo* and *in vitro* maturation of oocytes exist and so the best results have been obtained with oocytes collected from a pre-ovulatory follicle. The first successful GIFT foal was reported by Carnevale *et al.* (1999) using *in vivo*-matured oocytes and fresh sperm. Successful GIFT foals have also been reported by Hinrichs *et al.* (2000, 2002), Scott *et al.* (2001) though success with frozen semen is poor (Coutinho da Silva *et al.*, 2002; Squires *et al.*, 2003). The advantage of GIFT over IVF is that fertilization takes place within the most suitable environment, i.e. the Fallopian tube. A similar process to both GIFT and IVF is ZIFT (zygote intra fallopian tube transfer) in which fertilization takes place *in vitro* as per IVF but the fertilized ova (now a zygote) is transferred immediately to the Fallopian tube of the recipient mare instead of allowing initial development to take place *in vitro*. This has not been investigated to date in the mare but is successful in other mammals and so may warrant further consideration.

21.9.5. Oocyte freezing

Much work has been carried out into the freezing of stallion sperm, with some success (see Section 20.6.6). Embryo freezing (see Section 21.7.2) has also been investigated though with little success. An alternative would be to freeze oocytes which could then at a later date be available for fertilization by IVF, GIFT, ZIFT or ICSI. This would be particularly useful for valuable mares that die or have to be euthanized unexpectedly. Some success in oocyte freezing has been reported in other live-stock but work in horses is limited. As with embryo freezing, both cryopreservation and vitrification have been attempted with some success (Hochi *et al.*, 1994a,b; Hurtt *et al.*, 2000; Arav *et al.*, 2002). From the limited work reported to date it appears that freezing mature oocytes and immature oocytes is equally successful but still pregnancy rates with both are poor (20–40%; Maclellan *et al.*, 2002; Squires *et al.*, 2003; Tharasanit *et al.*, 2006).

21.9.6. Cloning

Since the successful cloning and birth of Dolly the sheep, cloning has become a hot topic in reproductive technology work. The horse has not escaped and Woods in 2002 reported the first successful cloned equid, a mule. Cloning or somatic cell nuclear transfer involves the collection of donor oocytes (see Section 21.9.1.1) from the recipient mare, these are then enucleated (the nucleus removed so as to remove their own genetic material) and then fused with diploid cells (the karyoplast) from the desired individual, these are often somatic cells, but skin cells, fetal cells and cumulus cells have also been used (Galli *et al.*, 2003; Vanderwall *et al.*, 2004). Fusion is achieved using either Sendai virus, electrofusion or by direct injection of the cell nucleus into the recipient's oocyte cytoplasm. The transfer of nuclear material, plus activation, normally by an electrical stimulation of the cell, reprogrammes the karyoplast to function as an undifferentiated embryonic stem cell. It then multiplies up like a conventional fertilized ova to form an embryo of identical genetic make-up to the animals from which the original diploid karyoplast cell was taken (Fig. 21.5).

The resulting embryo is then placed into the lumen of the oviduct or uterus (depending on length of time that has been allowed for *in vitro* development) of the recipient mare. As mentioned previously the first cloned equine was a mule, in fact three clones were created from cultured fetal cells (Woods *et al.*, 2003). This was followed very shortly after by the first cloned horse (Galli *et al.*, 2003) which was a clone of an adult skin cell taken from the mare in which the clone was subsequently placed, i.e. the mare was both the donor of the cell and the recipient of the resulting clone and so in essence gave birth to its identical twin. Cloning has a long way to go before it becomes a commercial practice and there are many ethical

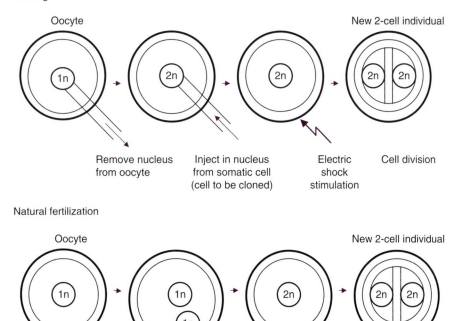

Cloning

Oocyte

New 2-cell individual

1n → 2n → 2n → 2n 2n

Remove nucleus from oocyte — Inject in nucleus from somatic cell (cell to be cloned) — Electric shock stimulation — Cell division

Natural fertilization

Oocyte

New 2-cell individual

1n · 1n → 1n 1n → 2n → 2n 2n

Sperm attaches to ovum — Sperm and ovum nucleus fuse — Fertilization — Cell division

Fig. 21.5. A comparison of the early events in conventional fertilization and cloning.

issues that such techniques raise; however, the great potential for cloning in horses lies in the reproduction of geldings. The majority of top performance horses are geldings and so their superior genetic material is largely lost to subsequent generations; however, if entire (stallion) clones of the gelding can be produced they can then breed 'on behalf of' the original gelding.

21.10. Conclusion

ET in horses has a significant potential for development, especially in the area of cryopreservation, without which its successful commercial application is limited. At present, many breed societies refuse to accept the progeny of ET, or only allow one foal per mare per year to be registered, though this is gradually changing (Bailey *et al.*, 1995). The expansion of ET within the equine industry is dependent upon several factors including, a change in breed registration restrictions; the value of horses; the performance of ET foals; the cost of the procedures; refinement of techniques; and the attitude of the equine industry to its application. However, even within these constraints, ET in horses is a valuable experimental tool. ET may find further application and commercial worth if a reliable super ovulating agent can be found and fetal sexing becomes reliable (Pieppo *et al.*, 1995; Jafer and Flint, 1996; Manz *et al.*, 1998; Stout, 2006). The recent development of a whole host of related technologies, such as IVF, ICSI, GIFT, oocyte freezing, etc. presents exciting opportunities for horse breeding in the future but is likely also to present ethical challenges, especially for techniques such as cloning.

Bibliography

Aanes, W.A. (1969) Surgical repair of third degree perineal laceration and rectovaginal fistula in the mare. *Journal of the American Veterinary Medical Association* 144, 485–490.

Aanes, W.A. (1973) Progress in recto-vaginal surgery. *Proceedings of 19th Annual Meeting of the American Association Equine Practitioners*. Lexington, Kentucky, pp. 225–230.

Aanes, W.A. (1993) Cervical lacerations. In: McKinnon, A.O. and Voss, J.L. (eds) *Equine Reproduction*. Lea and Febiger, Philadelphia, pp. 444–449.

Acland, H.M. (1993) Abortion in mares. In: McKinnon, A.O. and Voss, J.L. (eds) *Equine Reproduction*. Lea and Febiger, Philadelphia, pp. 554–562.

Adams, G.P., Kastelic, J.P., Bergfelt, D.R. and Ginther, O.J. (1987) Effect of uterine inflammation and ultrasonically detected uterine pathology on fertility in the mare. *Journal of Reproduction and Fertility, Supplement* 35, 444–454.

Adams, R. (1993a) Identification of mare and foal at high risk for perinatal problems. In: McKinnon, A.O. and Voss, J.L. (eds) *Equine Reproduction*. Lea and Febiger, Philadelphia, pp. 985–996.

Adams, R. (1993b) Neonatal disease: an overview. In: McKinnon, A.O. and Voss, J.L. (eds) *Equine Reproduction*. Lea and Febiger, Philadelphia, pp. 997–1002.

Adams, W.M. and Wagner, W.C. (1970) Role of corticosteroids in parturition. *Biology of Reproduction* 3, 223–226.

Ainsworth, C.G.V. and Hyland, J.H. (1991) Continuous infusion of gonadotrophin releasing hormone (GnRH) advances the onset of oestrous cycles in Thoroughbred mares on Australian stud farms. *Journal of Reproduction and Fertility, Supplement* 44, 235–240.

Akcay, E., Reilas, T., Andersson, M. and Katilla, T. (2006) Effect of seminal plasma fractions on stallion sperm survival after cooled storage. *Journal of Veterinary Medicine* 53, 481–485.

Alexander, S.L. and Irvine, C.H.G. (1982) Radioimmunoassay and in vitro bioassay of serum LH throughout the equine oestrus cycle. *Journal of Reproduction and Fertility, Supplement* 32, 253–260.

Alexander, S.L. and Irvine, C.H.G. (1991) Control of onset of breeding season in the mare and its artificial regulation by progesterone treatment. *Journal of Reproduction and Fertility, Supplement* 44, 307–318.

Alexander, S.L. and Irvine, C.H.G. (1993) FSH and LH. In: McKinnon, A.O. and Voss, J.L. (eds) *Equine Reproduction*. Lea and Febiger, Philadelphia, pp. 45–56.

Alghamdi, A., Troedsson, M.H., Laschkewitsch, T. and Xue, J.L. (2001) Uterine secretion from mares with post-breeding endometritis alters sperm motion characteristics in vitro. *Theriogenology* 55(4), 1019–1028.

Alghamdi, A.S., Foster, D.N. and Troedsson, M.H.T. (2004) Equine seminal plasma reduces sperm binding to polymorphonuclear neutrophils (PMNs) and improves the fertility of fresh semen inseminated into inflamed uteri. *Reproduction* 127, 593–600.

Allen, W.R. (1969a) A quantitative immunological assay for pregnant mare serum gonadotrophin. *Journal of Endocrinology* 43, 581–591.

Allen, W.R. (1969b) The immunological measurement of pregnant mare serum gonadotrophin. *Journal of Endocrinology* 43, 593–598.

Allen, W.R. (1970) Endocrinology of early pregnancy in the mare. *Equine Veterinary Journal* 2, 64–68.

Allen, W.R. (1974) Palpable development of the conceptus and fetus in Welsh pony mares. *Equine Veterinary Journal* 6, 69–73.

Allen, W.R. (1975) Ovarian changes during early pregnancy in pony mares in relation to PMSG production. *Journal of Reproduction and Fertility, Supplement* 23, 425–428.

Allen, W.R. (1977) Techniques and results in horses. In: Betteridge, K.J. (ed.) *Embryo Transfer in Farm Animals*. Canada Agricultural Monograph, Canada Department of Agriculture, Ottawa 16, pp. 47–49.

Allen, W.R. (1978) Control of ovulation and oestrus in the mare. In: Crighton, D.P., Haynes, H.B., Foxcroft, G.R. and Lamming, G.E. (eds) *Control of ovulation*. Butterworths, London, pp. 453–470.

Allen, W.R. (1979) Evaluation of uterine tube function in pony mares. *Veterinary Record* 105, 364–366.

Allen, W.R. (1982a) Immunological aspects of endometrial cup reaction in horses and donkeys. *Journal of Reproduction and Fertility, Supplement* 31, 57–94.

Allen, W.R. (1982b) Embryo transfer in the horse. In: Adams, C.E. (ed.) *Mammalian Egg Transfer.* CRC Press, Boca Raton, Florida, pp. 135–154.

Allen, W.R. (1987) Endogenous hormonal control of the mare's oestrus cycle. *Proceedings of the Bain-Fallon Memorial Lectures*, Sydney, pp. 2–13.

Allen, W.R. (1991) Investigations into the use of exogenous oxytocin from promoting uterine drainage in mares susceptible to endometritis. *Veterinary Record* 128, 593–594.

Allen, W.R. (1992) The diagnosis and handling of early gestational abnormalities in the mare. *Animal Reproduction Science* 28, 31–38.

Allen, W.R. (1993) Progesterone and the pregnant mare: unanswered chestnuts. *Equine Veterinary Journal, Supplement* 25, 90–91.

Allen, W.R. (2001) Fetomaternal interactions and influences during equine pregnancy. *Reproduction* 121, 513–527.

Allen, W.R. (2005) The development and application of modern reproductive technologies to horse breeding. *Reproduction in Domestic Animals* 40, 310–329.

Allen, W.R. and Goddard, P.J. (1984) Serial investigations of early pregnancy in pony mares using a real time ultrasound scanning. *Equine Veterinary Journal* 15, 509–514.

Allen, W.R. and Hadley, J.C. (1974) Blood progesterone concentrations in pregnant and non-pregnant mares. *Equine Veterinary Journal* 6, 87–93.

Allen, W.R. and Moor, R.M. (1972) The origin of the equine endometrial cup. 1. Production of PMSG by foetal trophoblastic cells. *Journal of Reproduction and Fertility* 29, 313–316.

Allen, W.R. and Pashan, R.L. (1984) Production of monozygotic (identical) horse twins by embryo micromanipulation. *Journal of Reproduction and Fertility* 71, 607–613.

Allen, W.R. and Pycock, J.P. (1989) Current views on the pathogenesis of bacterial endometritis in mares. *Veterinary Record* 125, 241–262.

Allen, W.R. and Rowson, L.E.A. (1972) Transfer of ova between horses and donkeys. *Proceedings of the 7th International Congress on Animal Reproduction and AI.* Munich, Germany, pp. 484–487.

Allen, W.R. and Rowson, L.E.A. (1973) Control of the mare's oestrous cycle by prostaglandins. *Journal of Reproduction Fertility* 33, 539–543.

Allen, W.R. and Rowson, L.E.A. (1975) Surgical and non-surgical egg transfer in horses. *Journal of Reproduction and Fertility, Supplement* 23, 525–530.

Allen, W.R. and Stewart, F. (1993) eCG. In: McKinnon, A.O. and Voss, J.L. (eds) *Equine Reproduction.* Lea and Febiger, Philadelphia, pp. 81–96.

Allen, W.R., Stewart, F., Cooper, N.J., Crowhurst, R.C., Simpson, D.J., McEnry, R.J., Greenwood, R.E.S., Rossdale, P.D. and Ricketts, S.W. (1974) Further studies on the use of synthetic prostaglangin analogues for inducing luteolysis in mares. *Equine Veterinary Journal* 6, 31–35.

Allen, W.R., Bowen, J.H., Frank, C.J., Jeffcote, L.B. and Rossdale, P.D. (1976a) The current position of AI in horse breeding. *Equine Veterinary Journal* 8, 72–74.

Allen, W.R., Stewart, F., Trounson, A.O., Tischner, M. and Bielanski, W. (1976b) Viability of horse embryos after storage and long distance transport in the rabbit. *Journal of Reproduction and Fertility, Supplement* 47, 387–390.

Allen, W.R., Bielanski, W., Cholewinski, G., Tischner, M. and Zwolinski, J. (1977) Blood groups in horses born after double transplantation of embryos. *Bulletin of the Acadamy of the Polish Scientific Service of Science and Biology* 25(11), 757.

Allen, W.R., Kydd, J.H. and Antczak, D.F. (1993) Interspecies and extraspecies equine pregnancies. In: McKinnon, A.O. and Voss, J.L. (eds) *Equine Reproduction.* Lea and Febiger, Philadelphia, pp. 530–553.

Allen, W.R., Mathias, S., Lennard, S.N. and Greenwood, R.E.S. (1995) Serial measurement of periferal oestrogen and progesterone concentrations in oestrus mares to determine optimum mating time and diagnose ovulation. *Equine Veterinary Journal* 27(6), 460–464.

Allen, W.R., Wilsher, S., Morris, L., Crowhurst, J.S., Hillyer, M.H. and Neal, H.N. (2006) Laparoscopic application of PGE2 to re-establish oviducal patency and fertility in infertile mares: a preliminary study. *Equine Veterinary Journal* 38(5), 454–459.

Alm, C.C., Sullivan, J.J. and First, N.L. (1974) Induction of premature parturition by parenteral administration of dexamethasone in the mare. *Journal of the American Veterinary Medical Association* 165, 721–722.

Alm, C.C., Sullivan, J.J. and First, N.L. (1975) The effect of corticosteroid (Dexamethasone) progesterone, estrogen and prostaglandin F2α on gestation length in normal and ovariectomised mares. *Journal of Reproduction and Fertility, Supplement* 23, 637–640.

Alm, H., Torner, H., Blottner, S., Nurnberg, G. and Kanitz, W. (2001) Effect of sperm cryopreservation and treatment with calcium ionophore or heparin on *in vitro* fertilisation of horse oocytes. *Theriogenology* 58, 817–829.

Althouse, G.C. and Hopkins, S.M. (1995) Assessment of boar sperm viability using a combination of two flurophores. *Theriogenology* 43, 595–603.

Alvarenga, M.A. and Leao, K.M. (2002) Hysteroscopic insemination of mares with low number of frozen-thawed spermatozoa selected by Percoll gradient. *Theriogenology* 58, 651.

Alvarenga, M.A., McCue, P.M., Bruemmer, J., Neves Neot, J.R. and Squires, E.L. (2001) Ovarian superstimulatory response and embryo production in mares treated with equine pituitary extract twice daily. *Theriogenology* 56, 879–887.

Amann, R.P. (1981a) A review of the anatomy and physiology of the stallion. *Equine Veterinary Science* 1(3), 83–105.

Amann, R.P. (1981b) Spermatogenesis in the stallion, a review. *Equine Veterinary Science* 1(4), 131–139.

Amann, R.P. (1984) Effects of extender, storage temperature and centrifugation on stallion spermatozoal motility and fertility. *Proceedings of the 10th International Congress on Animal Reproduction and Artificial Insemination.* Urbana, Paper No. 186.

Amann, R.P. (1988) Computerised evaluation of stallion spermatozoa. *Proceedings of the American Association of Equine Practice.* Lexington, Kentucky, pp. 453–473.

Amann, R.P. (1989) Treatment of sperm to predetermine sex. *Theriogenology* 31, 49–60.

Amann, R.P. (1993a) Functional anatomy of the adult male. In: McKinnon, A.O. and Voss, J.L. (eds) *Equine Reproduction.* Lea and Febiger, Philadelphia, pp. 645–657.

Amann, R.P. (1993b) Physiology and endocrinology. In: McKinnon, A.O. and Voss, J.L. (eds) *Equine Reproduction.* Lea and Febiger, Philadelphia, pp. 658–685.

Amann, R.P. and Graham, J.K. (1993) Spermatozoal function. In: McKinnon, A.O. and Voss, J.L. (eds) *Equine Reproduction.* Lea and Febiger, Philadelphia, pp. 715–745.

Amann, R.P. and Pickett, B.W. (1987) Principles of cryopreservation and a review of cryopreservation of stallion spermatozoa. *Equine Veterinary Science* 7, 145–173.

Amann, R.P., Squires, E.L. and Pickett, B.W. (1988) Effects of sample thickness and temperature on spermatozoal motion. *Proceedings of 11th International Congress on Animal Reproduction and AI*, Vol. 3. Dublin, Republic of Ireland, pp. 221a–221c.

Amann, S.F., Threlfall, W.R. and Kline, R.C. (1989) Equine temperature and progesterone fluctuations during estrus and near parturition. *Theriogenology* 31, 1007–1019.

Apter, R.C. and Householder, D.D. (1996) Weaning and weaning management of foals: a review and some recommendations. *Journal of Equine Veterinary Science* 16(10), 428–435.

Arav, A., Yavin, S., Zeron, Y., Natan, D., Dekel, I. and Gacitua, H. (2002) New trends in gamete's cryopreservation. *Molecular and Cell Endocrinology* 187(1–2), 77–81.

Arbeiter, K., Barth, U. and Jochle, W. (1994) Observations on the use of progesterone intravaginally and of desorelin in acyclic mares for induction of ovulation. *Journal of Equine Veterinary Science* 14(1), 21–25.

Argo, C.M., Cox, J.E. and Gray, J.L. (1991) Effect of oral melatonin treatment on the seasonal physiology of pony stallions. *Journal of Reproduction and Fertility, Supplement* 44, 115–125.

Arns, M.J., Webb, G.W., Kreider, J.L., Potter, G.D. and Evans, J.W. (1987) Use of diluent glycolysable sugars to maintain stallion sperm viability when frozen or stored at 37°C and 5°C in bovine serum albumin. *Journal of Reproduction and Fertility, Supplement* 35, 135–141.

Arthur, E.H. (1958) An analysis of the reproductive function of mares based on post mortem examination. *Veterinary Record* 70, 682–686.

Arthur, G.H. and Allen, W.E. (1972) Clinical observations on reproduction on a pony stud. *Equine Veterinary Journal* 4(3), 73–75.

Asbury, A.C. (1984) Uterine defense mechanisms in the mare. The use of intrauterine plasma in the management of endometritis. *Theriogenology* 21, 387–393.

Asbury, A.C. (1987) Infectious and immunological considerations in mare infertility. *Compendulum of Continuing Education for Practicing Veterinary Surgeons* 9, 585–592.

Asbury, A.C. (1990) Large volume uterine lavage in the management of endometritis and acute metritis in the mare. *Compendium of Continuing Education for Practising Veterinary Surgeons* 12, 1477–1479.

Asbury, A.C. (1991) Diseases of the reproductive system. The mare. Examination of the mare. In: Colahan, P.T., Mayhew, I.G., Merritt, A.M. and Moore, J.N. (eds) *Equine Medicine and Surgery*, Vol. 2, 4th edn. American Veterinary Publications, Goleta, California, pp. 949–963.

Asbury, A.C. (1993) Care of the mare after foaling. In: McKinnon, A.O. and Voss, J.L. (eds) *Equine Reproduction.* Lea and Febiger, Philadelphia, pp. 976–980.

Asbury, A.C. and Le Blanc, M.M. (1993) The placenta. In: McKinnon, A.O. and Voss, J.L. (eds) *Equine Reproduction.* Lea and Febiger, Philadelphia, pp. 509–516.

Asbury, A.C. and Lyle, S.K. (1993) Infectious causes of Infertility. In: McKinnon, A.O. and Voss, J.L. (eds) *Equine Reproduction.* Lea and Febiger, Philadelphia, pp. 381–391.

Ashdown, R.R. and Done, S.H. (1987) *Colour Atlas of Veterinary Anatomy. The Horse.* J.B. Lippencott, Gower Medical Publishing, Philadelphia, 852 pp.

Astudillo, C.R., Hajek, G.E. and Diaz, O.H. (1960) Influencia de algunos factores climaticos sobre la duracion de la gestation de yeguas fina sangre de carrera: estudio preliminary (the influence of some climate factors on pregnancy duration in Thoroughbred mares: preliminary account). *Zoolatria* 2, 35/38, 37 (Animal Breeding Abstracts 1962, 30, 2348).

Aurich, C. (2005) Factors affecting the plasma membrane function of cooled-stored stallion spermatozoa. *Animal Reproduction Science* 89, 65–75.

Aurich, C. and Spergser, J. (2006) Influence of genitally pathogenic bacteria and gentamicin on motility and membrane integrity of cooled-stored stallion spermatozoa. *Animal Reproduction Science* 94, 117–120.

Aurich, C., Hoppe, H. and Aurich, J.E. (1995) Role of endogenous opioids for regulation of the oestrous cycle in the horse. *Reproduction in Domestic Animals* 30(4), 188–192.

Aurich, C., Seeber, P. and Muller-Schlosser, F. (2007) Comparison of different extenders with defined protein composition for storage of stallion spermatozoa at 5°C. *Reproduction in Domestic Animals* 42(4), 445–448.

Aurich, J.E., Kuhne, A., Hoppe, H. and Aurich, C. (1996) Seminal plasma affects membrane integrity and motility of equine spermatozoa after cryopreservation. *Theriogenology* 46(5), 791–797.

Aurich, C., Schlote, S., Hoppen, H.-O., Klug, E., Hope, H. and Aurich, J.E. (1994) Effects of opioid antagonist naloxane on release of LH in mares during the anovulatory season. *Journal of Endocrinology* 142, 139–144.

Back, W., Smit, L.D., Schamhardt, H.C. and Barneveld, A. (1999) The influence of different exercise regimens on the development of locomotion in the foal. *Equine Veterinary Journal, Supplement* 31, 106–111.

Bacus, K.L., Ralston, S.L., Noekels, C.F. and McKinnon, A.O. (1990) Effects of transport on early embryonic death in mares. *Journal of Animal Science* 68, 345–351.

Bader, H. (1982) An investigation into sperm migration into the oviducts of the mare. *Journal of Reproduction and Fertility, Supplement* 32, 59–64.

Badi, A.M., O'Bryne, T.N. and Cunningham, E.P. (1981) An analysis of reproductive performance in thoroughbred mares. *Irish Veterinary Journal* 35(1), 1–12.

Baemgartl, C., Bader, H., Drommer, W. and Luning, I. (1980) Ultrastructural alterations of stallion spermatozoa due to semen conservation. *Proceedings of the International Congress of Animal Reproduction and Artificial Insemination* 5, 134–137.

Bailey, M.T., Bott, R.M. and Gimenez, T. (1995) Breed registries regulations on artificial insemination and embryo transfer. *Journal of Equine Veterinary Science* 15(2), 60–61.

Bain, A.M. (1967) The ovaries of the mare during early pregnancy. *Veterinary Record* 80, 229–231.

Baker, C.B., Little, T.V. and McDowell, K.J. (1993) The live foaling rate per cycle in mares. *Equine Veterinary Journal, Supplement* 15, 28–30.

Ball, B.A. (1988) Embryonic loss in mares. *Veterinary Clinics of North America Equine Practice* 4, 263–290.

Ball, B.A. (1993a) Embryonic death in mares. In: McKinnon, A.O. and Voss, J.L. (eds) *Equine Reproduction.* Lea and Febiger, Philadelphia, pp. 517–530.

Ball, B.A. (1993b) Management of twin embryos and foetuses in the mare. In: McKinnon, A.O. and Voss, J.L. (eds) *Equine Reproduction.* Lea and Febiger, Philadelphia, pp. 532–536.

Ball, B.A. and Brinsko, S.P. (1992) Early embryonic loss. A research update. *Modern Horse Breeding* 9(1), 8–9.

Ball, B.A., Altschul, M., McDowell, K.J., Ignotz, G. and Currie, W.B. (1991) Trophoblastic vessicles and maternal recognition of pregnancy in mares. *Journal of Reproduction and Fertility, Supplement* 44, 445–454.

Ball, B.A., Fagnan, M.S. and Dobrinski, V. (1997) Determination of acrosin amidase activity in equine spermatozoa. *Theriogenology* 48(7), 1191–1198.

Ball, B.A., Little, T.V., Weber, J.A. and Woods, G.L. (1989) Survivability of day 4 embryos from young normal mares and aged subfertile mares after transfer to normal recipient mares. *Journal of Reproduction and Fertility* 85, 187–194.

Ball, B.A., Medina, V., Gravance, C.G. and Baumber, J. (2001) Effect of antioxidants on preservation of motility, viability and acrosomal integrity of equine spermatozoa during storage at 5°C. *Theriogenology* 56, 577–589.

Bamberg, E., Choi, H.S., Mostl, E., Wrum, W., Lorin, D. and Arbeiterk, D. (1984) Enzymatic determination of unconjugated oestrogens in faeces for pregnancy diagnosis in mares. *Equine Veterinary Journal* 16, 537–539.

Barbacini, S., Gulden, P., Marchi, V. and Zavaglin, G. (1999) Incidence of embryo loss in mares inseminated before or after ovulation. *Equine Veterinary Education* 11(5), 251–254.

Barbacini, S., Zavaglia, G., Gulden, P., Marchi, V. and Necchi, D. (2000) Retrospective study on the efficacy of hCG in equine artificial insemination programme using frozen semen. *Equine Veterinary Education* 12, 312–317.

Barnes, R.J., Nathanielsz, P.W., Rossdale, P.D., Comline, R.S. and Silver, M. (1975) Plasma progestagens and oestrogens in fetus and mother in late pregnancy. *Journal of Reproduction and Fertility, Supplement* 23, 617–623.

Barnes, R.J., Comline, R.S., Jeffcote, L.B., Mitchell, M.D., Rossdale, P.D. and Silver, M. (1978) Fetal and maternal concentrations of 13,14-dihydro-15-oxo-prostaglandin F in the mare during late pregnancy and at parturition. *Journal of Endocrinology* 78(2), 201–215.

Barnisco, M.J.V. and Potes, N.M. (1987) The effects of teasing on the reproductive cycle in Portugese mares. *Revista Porteguesa Ciencias Veterinarias* 82, 37–43.

Bataille, B., Magistrini, M. and Palmer, E. (1990) Objective determination of sperm motility in frozen-thawed stallion semen. Correlation with fertility. *Quoi de neuf en matiere d'etudes et de recherches de le cheval? 16eme journee d'etude, Paris, 7 Mars 1990.* CEREOPA, Paris, France, pp. 138–141.

Batellier, F., Duchamp, G., Vidament, M., Arnaud, G., Palmer, E. and Magistrini, M. (1998) Delayed insemination is successful with a new extender for storing fresh equine semen at 15°C under aerobic conditions. *Theriogenology* 50, 229–236.

Batellier, F., Gerard, N., Courtens, J.L., Palmer, E. and Magistrini, M. (2000) Preservation of stallion sperm by native phosphocaseinate: a direct or indirect effect. *Journal of Reproduction and Fertility, Supplement* 56, 69–77.

Batellier, F., Vidament, M., Fauquant, J., Duchamp, G., Arnaud, G., Yvon, J.M. and Magistrini, M. (2001) Advances in cooled semen technology. *Animal Reproduction Science* 68, 181–190.

Bazer, F.W., Ott, T.L. and Spencer, T.C. (1994) Pregnancy recognition in ruminants, pigs and horses, signals from the trophoblast. *Theriogenology* 41, 79–94.

Beard, T. and Knight, F. (1992) Developmental orthopaedic disease. In: Robinson, N.E. (ed.) *Current Therapy in Equine Medicine 3.* W.B. Saunders, Philadelphia, pp. 105–166.

Beatty, R.A., Bennett, G.H., Hall, J.G., Hancock, J.L. and Stewart, D.L. (1969) An experiment with heterospermic insemination in cattle. *Journal of Reproduction and Fertility* 19, 491–496.

Beckett, S.D., Hudson, R.S., Walker, D.F., Vachon, R.I. and Reynolds, T.M. (1972) Corpus cavernosum penis pressure and external penile muscle activity during erection in the goat. *Biology of Reproduction* 7(3), 359–364.

Bedford, J.M. and Overstreet, J.W. (1972) A method for objective evaluation of the fertilising ability of spermatozoa irrespective of genetic character. *Journal of Reproduction and Fertility* 31, 407–411.

Bedford, S.J., Jasko, D.J., Graham, J.K., Amann, R.P., Squires, E.L. and Pickett, B.W. (1995a) Use of two freezing extenders to cool stallion spermatozoa to 5°C with and without seminal plasma. *Theriogenology* 43(5), 939–953.

Bedford, S.J., Jasko, D.J., Graham, J.K., Amann, R.P., Squires, E.L. and Pickett, B.W. (1995b) Effect of seminal extenders containing egg yolk and glycerol on motion characteristics and fertility of stallion spermatozoa. *Theriogenology* 43(5), 955–967.

Bell, R.J. and Bristol, F.M. (1991) Equine chorionic gonadotrophin in mares that conceive at foal oestrus. *Journal of Reproduction and Fertility, Supplement* 44, 719–721.

Belling, T.H. (1984) Post ovulatory breeding and related reproductive phenomena in the mare. *Equine Practice* 6, 12–19.

Belonje, C.W.A. (1965) Operation of retroversion of the penis in the stallion. *Journal of the South African Veterinary Medicine Association* 27, 53.

Bennett, D.G. (1987) Diagnosis and treatment of equine bacterial endometritis. *Journal of Equine Veterinary Science* 7, 345–350.

Bennett, W.K., Loch, W.E., Plata-Madrid, H. and Evans, T. (1998) The effects of perphenazine and bromocriptine on follicular dynamics and endocrine profiles in anoestrus pony mares. *Theriogenology* 49, 717–733.

Bergeron, A. and Manjunath, P. (2006) New insights towards understanding the mechanisms of sperm protection by egg yolk and milk. *Molecular Reproduction and Development* 73, 1338–1344.

Bergfelt, D.R. (2000) Anatomy and physiology of the mare. In: Samper, J.C. (ed.) *Equine Breeding Management and Artificial Insemination.* W.B. Saunders, Philadelphia, pp. 141–164.

Bergfeldt, D.R. and Ginther, O.J. (1992) Relationship between circulating concentrations of FSH and follicular waves during early pregnancy in mares. *Journal of Equine Veterinary Science* 12, 274–279.

Bergfeldt, D.R. and Ginther, O.J. (1993) Relationship between FSH surges and follicular waves during the oestrous cycles in mares. *Theriogenology* 39, 781–796.

Bergman, R.V. and Kenney, R.M. (1975) Representation of an uterine biopsy in the mare. *Proceedings of the 21st Annual Meeting of the American Association of Equine Practitioners*, pp. 355–362.

Bermudez, V., Miller, R., Johnson, W., Rosendal, S. and Ruhnke, L. (1987) The prevalence of *Mycoplasma spp.* and their relationship to reproductive performance in selected equine herds in southern Ontario. *Journal of Reproduction and Fertility, Supplement* 35, 671–673.

Berndston, W.E. and Jones, L.S. (1989) Relationship of intratesticular testosterone content to age, spermatogenesis, sertoli cell distribution and germ cell:sertoli cell ratio. *Journal of Reproduction and Fertility* 85, 511–518.

Berndston, W.E., Pickett, B.W. and Nett, T.H. (1974) Reproductive physiology of the stallion and seasonal changes in the testosterone concentrations of peripheral plasma. *Journal of Reproduction and Fertility, Supplement* 39, 115–118.

Bertone, J.J. and Jones, R.L. (1988) Evaluation of a field test kit for determination of serum IgG concentrations in foals. *Journal of Veterinary Internal Medicine* 2, 181–183.

Besognek, B., Hansen, B.S. and Daels, P.F. (1995) Prolactin secretion during the transitional phase and relationships to onset of reproductive season in mares. *Biology of Reproduction, Monograph series* 1, 459–467.

Betsch, J.M., Hunt, P.R., Spalart, M., Evenson, D. and Kenney, R.M. (1991) Effects of chlorhexidene penile washing on stallion semen parameters and sperm chromatin structure assay. *Journal of Reproduction and Fertility, Supplement* 44, 655–656.

Betteridge, K.J. (1989) The structure and function of the equine capsule in relation to embryo manipulation and transfer. *Equine Veterinary Journal, Supplement* 8, 92–100.

Betteridge, K.J. and Mitchell, D. (1975) A surgical technique applied to the study of tubal eggs in the mare. *Journal of Reproduction and Fertility, Suppelment* 23, 519–524.

Betteridge, K.J., Eaglesome, M.D., Mitchell, D., Flood, P.F. and Beriault, R. (1982) Development of horse embryos up to twenty two days after ovulation: observations on fresh specimens. *Journal of Anatomy* 135, 191–209.

Bhattacharya, B.C., Shome, P. and Gunther, A.H. (1977) Successful separation of X and Y spermatozoa in human and bull semen. *International Journal of Fertility* 22, 30–35.

Bielanski, W. (1975) The evaluation of stallion's semen in aspects of fertility control and its use for Artificial Insemination. *Journal of Reproduction and Fertility, Supplement* 23, 19–24.

Binor, Z., Sokoloski, J.E. and Wolf, P.P. (1980) Penetration of the zona free hamster egg by human sperm. *Fertility and Sterility* 33, 321–327.

Bishop, M.W.H., David, J.S.E. and Messervey, A. (1964) Some observations on cryptorchidism in the horse. *Veterinary Record* 76, 1041–1048.

Bishop, M.W.H., David, J.S.E. and Messervy, A. (1966) Cryptorchidism in the stallion. *Proceedings of the Royal Society of Medicine* 59, 769–774.

Blach, E.L., Amann, R.P., Bowen, R.A., Sawyer, H.R. and Hermenet, M.J. (1988) Use of a monoclonal antibody to evaluate integrity of the plasma membrane of stallion sperm. *Gamete Research* 21(3), 233–241.

Blanchard, T.L. (1995) Dystocia and post parturient disease. In: Kobluk, C.N., Ames, T.R. and Giver, R.J. (eds) *The Horse, Diseases and Clincal Management.* W.B. Saunders, Philadelphia, pp. 1021–1027.

Blanchard, T.L. and Varner, D.D. (1993a) Uterine involution and post partum breeding. In: McKinnon, A.O. and Voss, J.L. (eds) *Equine Reproduction.* Lea and Febiger, Philadelphia, pp. 622–625.

Blanchard, T.L. and Varner, D.D. (1993b) Testicular degeneration. In: McKinnon, A.O. and Voss, J.L. (eds) *Equine Reproduction.* Lea and Febiger, Philadelphia, pp. 855–860.

Blanchard, T.L., Elmore, R.G., Youngquist, R.S., Loch, W.E., Hardin, D.K., Bierschwal, C.J., Ganjam, V.K., Balke, J.M., Ellersiek, M.R., Dawson, L.J. and Miner, W.S. (1983) The effects of stanozolol and boldenone undecylenate on scrotal width, testis weight and sperm production in pony stallions. *Theriogenology* 20, 121–131.

Blanchard, T.L., Varner, D.D., Love, C.C., Hurtgen, J.P., Cummings, M.R. and Kenney, R.M. (1987) Use of semen extender containing antibiotic to improve the fertility of a stallion with seminal vesiculitis due to pseudomonas aeroginosa. *Theriogenology* 28, 541–546.

Blanchard, T.L., Thompson, J.A., Brinsko, S.P., Stich, K.L., Wendt, K.M., Varner, D.D. and Rigby, S.L. (2004) Mating mares on foal heat: a 5 year retrospective study. Proceedings of the 50th Annual Convention of the *American Association of Equine Practitioners* pp. 1496–1504.

Blue, M.D. (1981) A cytogenital study of prenatal loss in the mare. *Theriogenology* 15, 295–305.

Bollwein, H., Weber, F., Woschee, I. and Stolla, R. (2004) Transrectal Doppler sonography of uterine and umbilical blood flow during pregnancy in mares. *Theriogenology* 61(2–3), 499–509.

Bone, J.F. (1998) *Animal Anatomy and Physiology*, 3rd edn. Prentice-Hall, New Jersey.

Booth, L.C., Oxender, W.D., Douglas, T.H. and Woodley, S.L. (1980) Estrus, ovulation and serum hormones in mares given prostaglandin F2α, estradiol and gonadotrophin releasing hormone. *American Journal of Veterinary Research* 41, 120–122.

Borg, K., Colenbrander, B., Fazeli, A., Parlevliet, J. and Malmgren, L. (1997) Influence of thawing method on motility, plasma membrane integrity and morphology of frozen-thawed stallion spermatozoa. *Theriogenology* 48(4), 531–536.

Bos, H. and Van der May, G.J.W. (1980) Length of gestation periods for horses and ponies belonging to different breeds. *Livestock Production Science* 7, 181–187.

Bostock, D.E. and Owen, L.N. (1975) *Neoplasms in the Cat, Dog and Horse.* Wolfe Medical Publishers, London.

Bosu, W.T.K. (1982) Ovarian disorders: clinical and morphological observations in 30 mares. *Canadian Veterinary Journal* 23, 6–14.

Bosu, W.T.K. and Smith, C.A. (1993) Ovarian abnormalities. In: McKinnon, A.O. and Voss, J.L. (eds) *Equine Reproduction.* Lea and Febiger, Philadelphia, pp. 397–403.

Bosu, W.T.K., Van Camp, S.D., Miller, R.B. and Owen, R. (1982) Ovarian disorders: clinical and morphological observations in 30 mares. *Canadian Veterinary Journal* 23, 6–14.

Bowen, J.M., Gaboury, C. and Bousquet, D. (1976) Non surgical correction of a uterine torsion in the mare. *Veterinary Record* 99, 495–496.

Bowen, J.M., Niang, P.S., Menard, L., Irvine, D.S. and Moffat, J.B. (1978) Pregnancy without estrus in the mare. *Journal of Equine Medical Surgery* 2(5), 227–232.

Bowen, J.M., Tobin, N. and Simpson, R.B. (1982) Effect of washing on the bacterial flora of the stallion's penis. *Journal of Reproduction and Fertility, Supplement* 32, 41–45.

Bowling, A.T. (1996) *Horse Genetics*, CAB International, Wallingford, UK, 200 pp.

Bowling, A.T. and Hughes, J.P. (1993) Cytogenic abnormalities. In: McKinnon, A.O. and Voss, J.L. (eds) *Equine Reproduction*. Lea and Febiger, Philadelphia, pp. 258–265.

Bowling, A.T., Milton, L. and Hughes, J.P. (1987) An update of chromosomal abnormalities in mares. *Journal of Reproduction and Fertility, Supplement* 35, 149–155.

Boyd, H. (1979) Pregnancy diagnosis. In: Laing, J.A. (ed.) *Fertility and Infertility in Domestic Animals*, 3rd edn. Bailiere, Tindall, London, pp. 36–58.

Boyle, M.S. (1992) Artificial insemination in the horse. *Annules de Zootechnie* 41(3–4), 311–318.

Boyle, M.S., Sanderson, M.W., Skidmore, J.A. and Allen, W.R. (1989) Use of serial progesterone measurements to assess cycle length, time of ovulation and timing of uterine flushes in order to recover equine morulae. *Equine Veterinary Journal, Supplement* 8, 10–13.

Boyle, M.S., Skidmore, J., Zhange, J. and Cox, J.E. (1991) The effects of continuous treatment of stallions with high levels of a potent GnRH analogue. *Journal of Reproduction and Fertility, Supplement* 44, 169–182.

Brachen, F.K. and Wagner, P.C. (1983) Cosmetic surgery for equine pseudohermaphrodism. *Veterinary Medicine of Small Animal Clinics* 78, 879–884.

Bracher, V. (1992) Equine endometritis. PhD thesis. University of Cambridge, Cambridge.

Bracher, V., Mathias, S. and Allen, W.R. (1996) Influence of chronic degenerative endometritis (endometriosis) on placental development in the mare. *Equine Veterinary Journal* 28(3), 180–188.

Braun, J., Oka, A., Sato, K. and Oguri, N. (1993) Effect of extender, seminal plasma and storage temperature on spermatozoal motility in equine semen. *Japanese Journal of Equine Science* 4(1), 25–30.

Braun, J., Torres-Boggino, F., Hochi, S. and Oguri, N. (1994) Effect of seminal plasma on motion characteristics of epididymal and ejaculated stallion spermatozoa during storage at 5°C. *Deutsche Tierarztliche Wochenschrift* 101(8), 319–322.

Breen, V.B. and Bowman, R.T. (1994) Retained placenta: solving a sticky situation. *Modern Horse Breeding* March, 18–20.

Brendemuehl, P.J. and Cross, D.L. (2000) Influence of the dopamine antagonist domperidone on the vernal transition in seasonally anoestrous mares. *Reproduction and Fertility, Supplement* 56, 185–193.

Brewer, B.D., Clement, S.F., Lotz, W.S. and Gronwall, R. (1991) Renal clearance, urinary excretion of endogenous substances, and urinary diagnostic indices in healthy neonatal foals. *Journal of Veterinary Internal Medicine* 5(1), 28–33.

Brewer, R.L. and Kliest, G.J. (1963) Uterine prolapse in the mare. *Journal of the American Veterinary Medical Association* 142, 1118–1120.

Brinsko, S.P. and Varner, D.D. (1992) Artificial insemination and preservation of semen. *Veterinary Clinical Equine Practice* 8, 205–218.

Brinsko, S.P. and Varner, D.D. (1993) Artificial insemination. In: McKinnon, A.O. and Voss, J.L. (eds) *Equine Reproduction*. Lea and Febiger, Philadelphia, London, pp. 790–797.

Brinsko, S.P., Varner, D.D., Blanchard, T.L. and Meyers, S.A. (1990) The effect of post breeding uterine lavage on pregnancy rates in mares. *Theriogenology* 33, 465–475.

Brinsko, S.P., Rowan, K.R., Varner, D.D. and Blanchard, T.L. (2000) Effects of transport container and ambient storage temperature on motion characteristics of equine spermatozoa. *Theriogenology* 53(8), 1641–1655.

Brinsko, S.P., Rigby, S.L., Lindsey, A.C., Blanchard, T.L., Love, C.C. and Varner, D.D. (2003) Pregnancy rates in mares following hysteroscopic or transrectally-guided insemination with low sperm numbers at the utero-tubal papilla. *Theriogenology* 59(3–4), 1001–1009.

Bristol, F. (1981) Studies of oestrus synchronisation in mares. *Proceedings of the Society of Theriogenology 1981*, pp. 258–264.

Bristol, F. (1982) Breeding behaviour of a stallion at pasture with 20 mares in synchronised oestrus. *Journal of Reproductive Fertility, Supplement* 32, 71–77.

Bristol, F. (1986) Estrus synchronisation in mares. In: Morrow, D.A. (ed.) *Current Therapy in Theriogenology*. W.B. Saunders, Philadelphia, pp. 661–664.

Bristol, F. (1987) Fertility of pasture bred mares in synchronised oestrus. *Journal of Reproduction and Fertility, Supplement* 35, 39–43.

Bristol, F. (1993) Synchronization of ovulation. In: McKinnon, A.O. and Voss, J.L. (eds) *Equine Reproduction.* Lea and Febiger, Philadelphia, London, pp. 348–352.

British Equine Veterinary Association (1991) *Codes of Practice for 1) Veterinary Surgeons and 2) Breed Societies In the United Kingdom and Ireland using AI for breeding Equids*, British Equine Veterinary Association, London, 24 pp.

British Equine Veterinary Association (1997) *Equine AI: Course for Technicians.* British Equine Veterinary Association, London.

Britton, J.W. and Howell, C.E. (1943) Physiological and pathological significance of the duration of gestation in the mare. *Journal of the American Veterinary Medical Association* 102, 427–430.

Britton, V. (1998) *Foal to First Ridden*, Crowood Press, Wiltshire, UK, 144 pp.

Brook, D. (1987) Evaluation of a new test kit for estimating the foaling time in the mare. *Equine Practice* 9, 34–36.

Brook, D. (1993) Uterine cytology. In: McKinnon, A.O. and Voss, J.L. (eds) *Equine Reproduction.* Lea and Febiger, Philadelphia, pp. 246–254.

Broussard, J.R., Roussel, J.D., Hibbard, M., Thibodeaux, J.K., Moreau, J.D., Goodeaux, S.D. and Goodeaux, L.L. (1990) The effect of Monojet and Rir-Tite syringes on equine spermatozoa. *Theriogenology* 33, 200.

Brown, C. (1999) Diseases affecting multiple sites. Dourine. In: Colahan, P.T., Merritt, A.M., Moore, J.N. and Mayhew, I.G. (eds) *Equine Medicine and Surgery.* 5th edn. Mosby, St. Louis, pp. 2012–2013.

Brown, J.S. (1984) Surgical repair of the lacerated cervix in the mare. *Theriogenology* 22, 351–359.

Brown, J.S., Varner, D.D., Hinrichs, K. and Kenney, R.M. (1984) Amikacin sulfate in mares: pharmackinetics and body fluid and endometrial concentrations after repeated intramuscular administration. *American Journal of Veterinary Research* 45, 1610–1613.

Brown, M.P., Colahan, P.T. and Hawkins, D.L. (1978) Urethral extension treatment of urine pooling in mares. *Journal of the American Veterinary Medical Association* 173, 1005–1007.

Bruck, I., Greve, T. and Hyttel, P. (1999) Morphology of the oocyte-follicular connection in the mare. *Anatomy and Embryology* 199, 21–28.

Brugmans, A.C. (1997) Investigation on the efficiency of hCG, deslorelin, luprostinol and dinoprost on the termination of ovulation in mares and on pregnancy rates following time interval defined artificial insemination after induction of ovulation with hCG and deslorelin. D.Vet. Med Dissertation Hannover University.

Brun, R., Hecker, H. and Lun, Z.-R. (1998) *Trypanosoma evansi* and *T. equiperdum:* distribution, biology, treatment and phylogenetic relationship. *Veterinary Parasitology* 79, 95–107.

Bruyas, J.F., Sanson, J.P., Battut, I., Fieni, F. and Tainturier, D. (2000) Comparison of the cryoprotectant properties of glycerol and ethylene glycol for early day 6 equine embryos. *Journal of Reproduction and Fertility, Supplement* 56, 549–560.

Bryant-Greenwood, G.D. (1982) Relaxin a new hormone. *Endocrinology Review* 3, 62–90.

Buchanan, B.R., Seidel, G.E. Jr., McCue, P.M., Schenk, J.L., Herickhoff, L.A. and Squires, E.L. (2000) Insemination of mares with low numbers of either unsexed or sexed spermatozoa. *Theriogenology* 53, 1333–1344.

Bunning, M.L., Bowen, R.A., Cropp, C.B., Sullivan, K.G., Davis, B.S., Komar, N., Godsey, M.S., Baker, D., Hettler, D.L., Holmes, D.A., Biggerstaff, B.J. and Mitchell, C.J. (2002) Experimental infection of horses with West Nile virus. *Emerging Infectious Diseases* 8(4), 380–386.

Burkhardt, T. (1947) Transition from anoestrus in the mare and the effects of artificial lighting. *Journal of Agricultural Science Cambridge* 37, 64–68.

Burns, K.A. and Casilla, E.R. (1990) Partial purification and characterisation of an acetylcarnitine hydrolase from bovine epididymal spermatozoa. *Archieves of Biochemistry and Biophysics* 277, 1–7.

Burns, P.J. and Douglas, R.H. (1985) Reproductive hormone concentrations in stallions with breeding problems: case studies. *Journal of Equine Veterinary Science* 5, 40–42.

Burns, P.J. and Reasner, D.S. (1995) Computerized analysis of sperm motion: effects of glycerol concentration on cryopreservation of equine spermatozoa. *Journal of Equine Veterinary Science* 15(9), 377–380.

Burns, P.J., Jaward, M.J., Edmunson, A., Cahill, C., Boucher, J.K., Wilson, E.A. and Douglas, R.H. (1982) Effect of increased photoperiod on hormone concentrations in Thoroughbred stallions. *Journal of Reproduction and Fertility, Supplement* 32, 103–111.

Buss, D.B., Asbury, A.C. and Chevalier, L. (1980) Limitations in equine fetal electrocardiography. *Journal of the American Veterinary Medical Association* 177, 174–176.

Button, C. (1987) Congenital disorders of cardiac blood flow. In: Robinson, N.E. (ed.) *Current Therapy in Equine Medicine*, W.B. Saunders, Philadelphia 2, 167–170.

Camillo, F., Marmorini, P., Romagnoli, S., Vannozzi, J. and Bagliacca, M. (1997) Fertility in the first oestrus compared with fertility at following oestrous cycles in foaling mares and with fertility in non foaling mares. *Journal of Equine Veterinary Science* 17(11), 612–615.

Campbell, M.H.L. and England, G.C.W. (2002) A comparison of the ecbolic efficacy of intravenous and intrauterine oxytocin treatments. *Theriogenology* 58, 473–477.

Campbell, M.L. and England, G.C. (2004) Effect of teasing, mechanical stimulation and the intrauterine infusion of saline on uterine contractions in mares. *Veterinary Record* 155(4), 103–110.

Campbell, M.L. and England, G.C. (2006) Effects of coitus and the artificial insemination of different volumes of fresh semen on uterine contractions in mares. *Veterinary Record* 159(25), 843–849.

Card, C. (2005) Post-breeding inflammation and endometrial cytology in mares. *Theriogenology* 64(3), 580–588.

Card, C.E. (2000) Management of pregnant mares. In: Samper, J.C. (ed.) *Equine Breeding Management and Artificial Insemination.* W.B. Saunders, Philadelphia, pp. 247–266.

Card, C.E. and Hillman, R.D. (1993) Parturition. In: McKinnon, A.O. and Voss, J.L. (eds) *Equine Reproduction.* Lea and Febiger, Philadelphia, pp. 567–574.

Carnevale, E.M. (1998) Folliculogenesis and ovulation. In: Rantanen, N.W. and McKinnon, A.O. (eds) *Equine Diagnostic Ultrasonography.* Williams and Wilkins, Baltimore, Maryland, pp. 201–212.

Carnevale, E.M. (2004) Oocyte transfer and gamete intrafallopian transfer in the mare. *Animal Reproduction Science* 82–83, 617–624.

Carnevale, E.M. and Ginther, O.J. (1992) Relationship of age to uterine function and reproductive efficiency in mares. *Theriogenology* 37, 1101–1105.

Carnevale, E.M. and Ginther, O.J. (1997) Age and pasture effects on vernal transition in mares. *Theriogenology* 47, 1009–1018.

Carnevale, E.M., Squires, E.L. and McKinnon, A.O. (1987) Comparison of Ham F10 with CO_2 or Hepes buffer for storage of equine embryos at 5°C for 24 hrs. *Journal of Animal Science* 65, 1775–1781.

Carnevale, E.M., Griffin, P.G. and Ginther, O.J. (1993) Age associated subfertility before entry of embryos into the uterus in mares. *Equine Veterinary Journal, Supplement* 15, 31–35.

Carnevale, E.M., Alvarenga, M.A., Squires, E.L. and Choi, Y.H. (1999) Use of non cycling mares as recipients for oocyte transfer and GIFT. *Proceedings of the Annual Conference of the Society of Theriogenology,* p. 44.

Carnevale, E.M., Maclellan, L.J., Coutinho da Silva, M., Scott, T.J. and Squires, E.L. (2000a) Comparison of culture and insemination techniques for equine oocyte transfer. *Theriogenology* 54, 981–987.

Carnevale, E.M., Ramirez, R.J., Squires, E.L., Alvarenga, M.A. and McCue, P.M. (2000b) Factors affecting pregnancy rates and early embryonic death after equine embryo transfer. *Theriogenology* 54, 965–979.

Carnevale, E.M., Squires, E.L., Maclellan, L.J., Alvarenga, M.A. and Scott, T.J. (2001) Use of oocyte transfer in a commercial breeding program for mares with various abnormalities. *Journal of the American Veterinary Medicine Association* 218, 87–91.

Carney, N.J., Squires, E.L., Cook, V.M., Seidel, G.E. and Jasko, D.L. (1991) Comparison of pregnancy rates from transfer of fresh versus cooled transported equine embryos. *Theriogenology* 36(1), 23–32.

Caron, J., Barber, S. and Bailey, J. (1985) Equine testicular neoplasia. *Compendium of Continuing Education for Practising Vets* 6, 5296.

Carson, K. and Wood-Gush, D.E.M. (1983a) Behaviour of Thoroughbred foals during nursing. *Equine Veterinary Journal* 15, 257–262.

Carson, K. and Wood-Gush, D.E.M. (1983b) Equine behaviour: I. A review of the literature on social and dam-foal behaviour. II. A review of the literature on feeding, eliminative and resting behaviour. *Applied Animal Ethology* 10(3), 165–190.

Carver, D.A. and Ball, B.A. (2002) Lipase activity in stallion seminal plasma and the effect of lipase on stallion spermatozoa during storage at 5 degrees C. *Theriogenology* 58(8), 1587–1595.

Casey, P.J., Robertson, K.R., Lui, I.K.M., Botta, E.S. and Drobnis, E. (1991) Separation of motile spermatozoa from stallion semen. *Proceedings of the 17th Annual Convention of American Association of Equine Practitioners,* San Francisco, California, pp. 203–210.

Casey, P.J., Hillman, R.B., Robertson, K.R., Yudin, A.I., Lui, I.K. and Drobonis, E. (1993) Validation of an acrosomal stain for equine sperm that differentiates between living and dead sperm. *Journal of Andrology* 14, 282–297.

Cash, R.S.E., Ousey, J.L. and Rossdale, P.D. (1985) Rapid test strip method to assist management of foaling mares. *Equine Veterinary Journal* 17, 61–62.

Caslick, E.A. (1937) The vulva and vulvo-vaginal orifice and its relation to genital tracts of the Thoroughbred mare. *Cornell Veterinarian* 27, 178–187.

Caspo, J., Stefler, J., Martin, T.G., Makray, S. and Cspo-Kiss, Z. (1995) Composition of mare's colostrum and milk. Fat content, fatty acid composition and vitamin content. *International Dairy Journal* 5, 393–402.

Castillo-Olivares, J., Tearle, J.P., Montesso, F., Westcott, D., Kydd, J.H., Davis-Poynter, N.J. and Hannant, D. (2003) Detection of equine arteritis virus (EAV)-specific cytotoxic CD8+ T lymphocyte precursors from EAV-infected ponies. *Journal of General Virology* 84, 1–9.

Castleberry, R.S., Schneider, H.J., Jr. and Griffin, J.L. (1980) Recovery and transfer of equine embryos. *Theriogenology* 13, 90–94.

Castro, T.A.M., Gastal, E.L. and Castro, F.G., Jr. and Augusto, C. (1991) Physical, chemical and biochemical traits of stallion semen. *Anais, IX Congresso Brasileiro de Reproducao Animal*, Belo Horizonte, Brazil, 22 a 26 de Junho de 1991. Vol. II. Belo Horizonte, Brazil; Colegio Brasileiro de Reproducao Animal, p. 443.

Causey, R.C. (2007) Mucus and the mare: how little we know. *Theriogenology* 68(3), 386–394.

Celebi, M. and Demirel, M. (2003) Pregnancy diagnosis in mares by determination of oestradiol-17-β levels in faeces. *Turkish Journal of Veterinary and Animal Sciences* 27, 373–375.

Chan, J.P.W., Huang, T.H., Chuang, S.T., Cheng, F.P., Fung, H.P., Chen, C.L. and Mao, C.L. (2003) Quantitative echotexture analysis for prediction of ovulation in mares. *Journal of Equine Veterinary Science* 23(9), 397–402.

Chandley, A.N., Fletcher, J., Rossdale, P.D., Peace, C.K., Ricketts, S.W., McEnery, R.J., Thorne, J.P., Short, R.V. and Allen, W.R. (1975) Chromosome abnormalities as a cause of infertility in mares. *Journal of Reproduction and Fertility, Supplement* 23, 377–383.

Charneco, R., Pool, K.C. and Arns, M.J. (1993) Influence of vesicular gland rich and vesicular gland poor seminal plasma on the freezability of stallion spermatozoa. *Proceedings of the 13th Conference of the Equine Nutrition and Physiology Symposium*, pp. 385–386.

Chavatte, P. (1991) Maternal behaviour in the horse: theory and practical applications to foal rejection and fostering. *Equine Veterinary Education* 3(4), 215–220.

Chavatte, P., Brown, G., Ousey, J.C., Silver, M., Cotrill, C., Fowden, A.L., McGladdery, A.J. and Rossdale, P.D. (1991) Studies of bone marrow and leucocycte counts in peripheral blood in fetal and newborn foals. *Journal of Reproduction and Fertility, Supplement* 44, 603–608.

Chavatte, P., Holton, D., Ousey, J.C. and Rossdale, P.D. (1997a) Biosynthesis and possible roles of progestogens during equine pregnancy and in the new born foal. *Equine Veterinary Journal, Supplement* 24, 89–95.

Chavatte, P., Rossdale, P.D. and Tait, A.D. (1997b) Corticosteroid synthesis by the equine fetal adrenal. *Biology of Reproduction, Monograph* 1, 13–20.

Chenier, T.S. (2000) Anatomy and physical examination of the stallion. In: Samper, J.C. (ed.) *Equine Breeding Management and Artificial Insemination.* W.B. Saunders, Philadelphia, pp. 1–26.

Chevalier, F. and Palmer, E. (1982) Results of a field trial on the use of ultrasound echography in the mare. *Journal of Reproduction and Fertility, Supplement* 32, 423–430.

Choi, S.J., Anderson, G.B. and Roser, J.F. (1995) Production of estrogen conjugates and free estrogens by the preimplantation equine embryo. *Biology of Reproduction* 52(Supplement 1), 179.

Choi, Y.H., Love, C.C., Love, L.B., Varner, D.D., Brinsko, S. and Hinrichs, K. (2002) Developmental in vitro matured oocytes fertilised by intracytoplasmic sperm injection with fresh or frozen-thawed spermatozoa. *Reproduction* 123, 455–465.

Choi, Y.H., Roasa, L.M., Love, C.C., Varner, D.D., Brinsko, S.P. and Hinrichs, K. (2004) Blastocyst formation rates in vivo and in vitro of in vitro-matured equine oocytes fertilised by intracytoplasmic sperm injection. *Biology of Reproduction* 70(5), 1231–1238.

Christanelli, M.J., Amann, R.P., Squires, E.L. and Pickett, B.W. (1985) Effects of egg yolk and glycerol levels in lactose-EDTA-egg yolk extender and of freezing rate on the motility of frozen-thawed stallion spermatozoa. *Theriogenology* 24(6), 681–686.

Christanelli, M.J., Squires, E.L., Amann, R.P. and Pickett, B.W. (1984) Fertility of stallion semen processed, frozen and thawed by a new procedure. *Theriogenology* 22(1), 39–45.

Christensen, J.W., Zharkikh, T., Ladewig, J. and Yasinetskaya, N. (2002) Social behaviour in stallion groups (*Equus prezewalskii* and *Equus caballus*) kept under natural and domestic conditions. *Applied Animal Behaviour Science* 76, 11–20.

Clay, C.M. and Clay, J.N. (1992) Endocrine and testicular changes with season, artificial photoperiod and peri-pubertal period in stallions. *Veterinary Clinics of North America, Equine Practice* 8(1), 31–56.

Clay, C.M., Squires, E.L., Amann, R.R. and Pickett, B.W. (1987) Influences of season and artificial photoperiod on stallions: testicular size, seminal characteristics and sexual behaviour. *Journal of Animal Science* 64(2), 517–525.

Clayton, H.M. (1978) Ascariasis in foals. *Veterinary Record* 102, 553–556.

Clayton, H.M. (1986) Outbreak of EVA in Alberta Canada. *Journal of Equine Veterinary Science* 7, 101.

Clement, F., Guerin, B., Vidament, M., Diemert, S. and Palmer, E. (1993) Microbial quality of stallion semen. *Pratique Veterinaire Equine* 25(1), 37–43.

Clement, F., Vidament, M. and Guerin, B. (1995) Microbial contamination of stallion semen. *Biology of Reproduction Monograph Equine Reproduction VI* 1, 779–786.

Cleaver, B.D., Grubaugh, W.R., Davis, S.D., Sheerin, P.C., Franklin, K.J. and Sharp, D.C. (1991) Effect of constant light exposure on circulating gonadotrophin levels and hypothalamic gonadotrophin-releasing hormone (GnRH) content in the ovariectomized pony mare. *Journal of Reproduction and Fertility, Supplement* 44, 259–266.

Clulow, J.R., Buss, H., Sieme, H., Rodger, J.A., Cawdell-Smith, A.J., Evans, G., Rath, D., Morris, L.H. and Maxwell, W.M. (2007) Field fertility of sex-sorted and non-sorted frozen-thawed stallion spermatozoa. *Animal Reproduction Science.* Sept 16 (Epub ahead of print)

Cochran, J.D., Amann, R.P., Froman, D.P. and Pickett, G.W. (1984) Effects of centrifugation, glycerol level, cooling to 5°C, freezing rate and thawing rate on the post thaw motility of equine spermatozoa. *Theriogenology* 22, 25–38.

Cochran, R., Meintjes, M., Reggio, B., Hyland, D., Carter, J., Pinto, C., Paccamonti, D. and Godke, R.A. (1998) Live foals produced from sperm-injected oocytes derived from pregnant mares. *Journal of Equine Veterinary Science* 18, 736–740.

Coldrey, C. and Coldrey, V. (1990) *Breaking and Training Young Horses.* Crowood Press, Gipsy Lane, Swindon, Wiltshire, UK, 143 pp.

Coleman, R.J., Mathison, G.W. and Burwash, L. (1999) Growth and condition at weaning of extensively creep fed foals. *Journal of Equine Veterinary Science* 19(1), 45–49.

Colenbrander, B., Puyk, H., Zandee, A.R. and Parlevliet, J. (1992) Evaluation of the stallion for breeding. *Acta Veterinaria Scandinavica, Supplement* 88, 29–37.

Colenbrander, B., Gadella, B.M. and Stout, T.A.E. (2003) The predictive value of semen analysis in the evaluation of stallion fertility. *Reproduction in Domestic Animals* 38(4), 305–311.

Colles, C.M., Parkes, R.D. and May, C.J. (1978) Fetal electrocardiography in the mare. *Equine Veterinary Journal* 10, 32–38.

Collins, A.M. and Buckley, T.C. (1993) Comparison of methods for early pregnancy detection. *Journal of Equine Veterinary Science* 13, 627–630.

Combs, G.B., Leblanc, M.M., Neuwirth, L. and Tran, T.Q. (1996) Effects of prostaglandin F2α, cloprostenol and fenprostenolene on uterine clearance of radiocoloid in the mare. *Theriogenology* 45, 1449–1455.

Comline, R.S., Hall, L.W., Lavelle, R. and Silver, M. (1975) The use of intravascular catheters for long term studies of the mare and fetus. *Journal of Reproduction and Fertility, Supplement* 23, 583–588.

Cooper, M.J. (1981) Prostaglandins in veterinary practice. *In Practice* 3(1), 30–34.

Cooper, W.L. (1979) Clinical aspects of prostaglandins in equine reproduction. *Proceedings of the 2nd Equine Pharmacy Symposium.* American Association of Equine Practitioners, Golden, Colorado, pp. 225–231.

Cornwall, J.C. (1972) Seasonal variation in stallion's semen and puberty in the Quarter horse colt. MSc thesis, Louisiana State University.

Correa, J.R. and Zavos, P.M. (1994) The hypoosmotic swelling test: its employment as an assay to evaluate the functional integrity of frozen-thawed bovine spermatozoa. *Theriogenology* 42, 351–360.

Correa, J.R., Heersche, G. Jr. and Zavos, P.M. (1997) Sperm membrane functional integrity and response of frozen-thawed bovine spermatozoa during the hypoosmotic swelling test incubation at varying temperatures. *Theriogenology* 47, 715–721.

Cottrill, C.M., O'Connor, W.N., Cudd, T. and Rantanen, N.W. (1987) Persistence of foetal circulatory pathways in the new born foal. *Equine Veterinary Journal* 19(3), 252–255.

Cottrill, C.M., Yenho, S. and O'Connor, W.N. (1997) Embryological development of the embryonic heart. *Equine Veterinary Journal, Supplement* 24, 14–18.

Coutinho da Silva, M.A., Carnevale, E.M., Maclellan, K.A., Leao, K.M. and Squires, E.L. (2002) Use of cooled and frozen semen during gamete intrafallopian transfer in mares. *Theriogenology* 58, 763–766.

Couto, M.A. and Hughes, J.P. (1993) Sexually transmitted (venereal diseases of horses). In: McKinnon, A.O. and Voss, J.L. (eds) *Equine Reproduction.* Lea and Febiger, Philadelphia, London, pp. 845–854.

Cox, J.E. (1971) Urine tests for pregnancy in mares. *Veterinary Record* 89, 606–607.

Cox, J.E. (1988) Hernias and ruptures: words to the heat of deeds. *Equine Veterinary Journal* 20, 155–156.

Cox, J.E. (1993a) Developmental abnormalities of the male reproductive tract. In: McKinnon, A.O. and Voss, J.L. (eds) *Equine Reproduction.* Lea and Febiger, Philadelphia, pp. 895–906.

Cox, J.E. (1993b) Cryptorchid castration. In: McKinnon, A.O. and Voss, J.L. (eds) *Equine Reproduction.* Lea and Febiger, Philadelphia, pp. 915–920.

Cox, J.E. and Galina, C.S. (1970) A comparison of the chemical tests for oestrogens used in equine pregnancy diagnoses. *Veterinary Record* 86, 97–100.

Cox, J.E., Williams, J.H., Rowe, P.H. and Smith, J.A. (1973) Testosterone in normal, cryptorchid and castrated male horses. *Equine Veterinary Journal* 5(2), 85–90.

Cox, J.E., Redhead, P.H. and Jawad, N.N.A. (1988) The effect of artificial photoperiod at the end of the breeding season on plasma testosterone concentrations in stallions. *Australian Veterinary Journal* 65, 239–241.

Craig, T.M., Scrutchfield, W.L. and Martin, M.T. (1993) Comparison of prophylactic pyrantel and suppressive invermectin anthelmintic programs in young horses. *Equine Practice* 15(3), 24–29.

Cran, D.G., Johnson, L.A. and Polge, C. (1995) Sex preselection in cattle: a field trial. *Veterinary Record* 136, 495–496.

Crossett, B., Stewart, F. and Allen, W.R. (1995) A unique progesterone dependant equine endometrial protein that associates strongly with embryonic capsule. *Journal of Reproduction and Fertility Abstract Series* 15, 11.

Crouch, J.R.F., Atherton, J.G. and Platt, H. (1972) Venereal transmission of *Klebsiella aerogenes* in a thoroughbred stud farm from a persistently infected stallion. *Veterinary Record* 90, 21–24.

Crowhurst, R.C. (1977) Genital infection in mares. *Veterinary Record* 100, 476–478.

Crowhurst, R.C., Simpson, D.Y., Greenwood, R.E.S. and Ellis, D.R. (1979) Contagious equine metritis. *Veterinary Record* 104, 465.

Cuboni, E. (1934) A rapid pregnancy diagnosis test for mares. *Clinical Veterinerian (Milano)* 57, 85–93.

Cudderford, D. (1996) *Equine Nutrition*, Crosswood Press, Malborough, Wiltshire, UK, 160 pp.

Cullinane, A., Weld, J., Osborne, M., Nelly, M., Mcbride, C. and Walsh, C. (2001) Field studies on equine influenza vaccination regimes in thoroughbred foals and yearlings. *Veterinary Journal* 161(2), 174–185.

Cupps, P.T. (1991) *Reproduction in Domestic Animals*, 4th edn. Academic Press, New York.

Curnow, E.M. (1991) Ultrasonography of the mare's uterus. *Equine Veterinary Education* 3(4), 190–193.

Curran, S. and Ginther, O.J. (1989) Ultrasonic diagnosis of equine fetal sex by location of the genital tubercle. *Journal of Equine Veterinary Science* 9, 77–83.

Czlonkowska, M., Boyle, M.S. and Allen, W.R. (1985) Deep freezing of horse embryos. *Journal of Reproduction and Fertility* 75(2), 485–490.

Daels, P.F. and Hughes, J.P. (1993) The abnormal cycle. In: McKinnon, A.O. and Voss, J.L. (eds) *Equine Reproduction*. Lea and Febiger, Philadelphia, pp. 144–160.

Daels, P.F., Stabenfeldt, G.H., Hughes, J.P., Odensvik, K. and Kindahl, H. (1990) The source of oestrogen in early pregnancy in the mare. *Journal of Reproduction and Fertility, Supplement* 90(1), 55–61.

Daels, P.F., Jorge De Moraes, M., Stabenfeldt, G.H., Hughes, H. and Lesley, B. (1991) The corpus luteum a major source of oestrogen in early pregnancy in the mare. *Journal of Reproduction and Fertility, Supplement* 44, 502–508.

Daels, P.F., Albrecht, B.A. and Mohammed, H.O. (1995) In vitro regulation of luteal function in mares. *Reproduction in Domestic Animals* 30, 211–217.

Daels, P.F., Fatone, B.S., Hansen, B.S. and Concannon, P.W. (2000) Dopamine antagonist-induced reproductive function in anoestrous mares; gonadotrophin secretion and the effects of environmental cues. *Journal of Reproduction and Fertility, Supplement* 56, 173–183.

Darenius, K., Kindahl, H. and Madej, A. (1987) Clinical and endocrine aspects of early foetal death in the mare. *Journal of Reproduction and Fertility, Supplement* 35, 497–498.

Darenius, K., Kindahl, H. and Madej, A. (1988) Clinical and endocrine studies in mares with a known history of repeated conceptus losses. *Theriogenology* 29, 1215–1232.

Dascanio, J. (2000) How to diagnose and treat fungal endometritis. *Proceedings of the 46th Annual American Association of Equine Practitioners*, pp. 316–319.

Davies Morel, M.C.G. (1999) *Equine Artificial Insemination*, CAB International, Wallingford, UK, 406 pp.

Davies Morel, M.C.G. and Gunnarsson, V. (2000) A survey of the fertility of Icelandic stallions. *Animal Reproduction Science* 64, 49–64.

Davies Morel, M.C.G. and Newcombe, J.R. (2008) The efficacy of different hCG dose rates and the effect of hCG treatment on ovarian activity: ovulation, multiple ovulation, pregnancy, multiple pregnancy, synchrony of multiple ovulation in the mare. *Animal Reproduction Science*. Oct 23 (Epub ahead of print)

Davies Morel, M.C.G. and O'Sullivan, J.A.M. (2001) Ovulation rate and distribution in the thoroughbred mare, as determined by ultrasonic scanning: the effect of age. *Animal Reproduction Science* 66, 59–70.

Davies Morel, M.C.G., Newcombe, J.R. and Holland, S.J. (2002) Factors affecting gestation length in the Thoroughbred mare. *Animal Reproduction Science* 74, 175–185.

Davies Morel, M.C.G., Newcombe, J.R. and Swindlehurst, J.C. (2005) The effect of age on multiple ovulation rates, multiple pregnancy rates and embryonic vesicle diameter in the mare. *Theriogenology* 63, 2482–2493.

Dawes, G.S., Fox, H.E. and Leduc, B.M. (1972) Respiratory movements and rapid eye movement sleep in the foetal lamb. *Journal of Physiology* 220(1), 119–143.

Day, F.T. (1939) Some observations on the causes of infertility in horse breeding. *Veterinary Record* 51, 581–582.

Day, F.T. (1940) The stallion and fertility. The technique of sperm collection and insemination. *Veterinary Record* 52, 597–602.

De Coster, R., Cambiaso, C.L. and Masson, P.L. (1980) Immunological diagnosis of pregnancy in the mare by agglutination of latex particles. *Theriogenology* 13, 433.

De la Cueva, F.I., Pujol, M.R., Rigau, T., Bonet, S., Briz, M. and Rodriguez-Gill, J.E. (1997) Resistance to osmotic stress of horse spermatozoa: the role of ionic pumps and their relationship to cryopreservation success. *Theriogenology* 48, 947–968.

Del Campo, M.R., Donoso, M.X., Parrish, J.J. and Ginther, O.J. (1990) In vitro fertilisation of in vitro-matured equine oocytes. *Journal of Equine Veterinary Science* 10, 18–22.

Dell'Aquila, M.E., Cho, Y.S., Minoia, P., Triana, V., Fusco, S., Lacalandra, G.M. and Maritato, F. (1997a) Effects of follicular fluid supplementation of in-vitro maturation medium on the fertilization and development of equine oocytes after in-vitro fertilization or intracytoplasmic sperm injection. *Human Reproduction* 12, 2766–2772.

Dell'Aquila, M.E., Cho, Y.S., Minoia, P., Traina, V., Fusco, S., Lacalandra, G.M. and Maritato, F. (1997b) Intracytoplasmic sperm injection (ICSI) versus conventional IVF on abattoir-derived and in vitro-matured equine oocytes. *Theriogenology* 47, 1139–1156.

Dell'Aquila, M.E., Masterson, M., Maritato, F. and Hinrichs, J. (2000) Chromatin configuration meiotic competence and success of intra-cytoplasmic sperm injection in horse oocytes collected by follicular aspiration or scaping. *Proceedings of the 5th International Symposium on Equine Embryo Transfer, Havemeyer Foundation Monograph Series* 3, pp. 37–38.

Demick, D.S., Voss, J.L. and Pickett, B.W. (1976) Effect of cooling, storage, glycerolization and spermatozoal numbers on equine fertility. *Journal of Animal Science* 43, 633–637.

Denniston, D.J., Graham, J.K., Squires, E.L. and Brinsko, S.P. (1997) The effect of liposomes composed of phosphatidyl-serine and cholesterol on fertility rates using thawed equine spermatozoa. *Journal of Equine Veterinary Science* 17(12), 675–676.

De Ribeaux, M.B. (1994) Weaning wisdom. *Modern Horse Breeding* 11(7), 28–31.

Devireddy, R.V., Swanlund, D.J., Alghamdi, A.S., Duoos, L.A., Troedsson, M.H., Bischof, J.C. and Roberts, K.P. (2002) Measured effect of collection and cooling conditions on the motility and the water transport parameters at subzero temperatures of equine spermatozoa. *Reproduction* 124(5), 643–648.

De Vries, P.J. (1987) Evaluation of the use of fresh, extended, transported stallion semen in the Netherlands. *Journal of Reproduction and Fertility, Supplement* 35, 641.

De Vries, P.J. (1993) Diseases of the testes, penis and related structures. In: McKinnon, A.O. and Voss, J.L. (eds) *Equine Reproduction*. Lea and Febiger, Philadelphia, pp. 878–884.

Diemer, T., Huwe, P., Ludwig, M., Hauck, E.W. and Weidner, W. (2003) Urogenital infection and sperm motility. *Andrologia* 35(5), 283–287.

Dinger, J.E. and Noiles, E.E. (1986a) Effect of controlled exercise on libido in 2 yr old stallions. *Journal of Animal Science* 62(5), 1220–1223.

Dinger, J.E. and Noiles, E.E. (1986b) Prediction of daily sperm output in stallions. *Theriogenology* 26, 61–67.

Dinger, J.F., Noiles, E.E. and Bates, M.L. (1981) Effect of progesterone impregnated vaginal sponges and PMS administration on oestrous synchronisation in mares. *Theriogenology* 16, 231–237.

Dippert, K.D., Hofferer, S., Palmer, E., Jasko, D.J. and Squires, E.L. (1992) Initiation of superovulation in mares 5 or 12 days after ovulation using equine pituitary extract with or without GnRH analogue. *Theriogenology* 38, 695–710.

Doarn, R.T., Threlfall, W.R. and Kline, R. (1985) Umbilical blood flow and effects of premature severance in the neonatal horse. *Proceedings of the 1985 Annual Society of Theriogenology*, pp. 175–178.

Doarn, R.T., Threlfall, W.R. and Kline, R. (1987) Umbilical blood flow and the effects of premature severance in the neonatal horse. *Theriogenology* 28, 789–800.

Doig, P.A. and Waelchi, R.O. (1993) Endometrial biopsy. In: McKinnon, A.O. and Voss, J.L. (eds) *Equine Reproduction*. Lea and Febiger, Philadelphia, pp. 225–233.

Doreau, M. and Boulet, S. (1989) Recent knowledge on mare's milk production. A review. *Livestock Production Science* 22, 213–235.

Doreau, M., Martin-Rosset, W. and Boulet, S. (1988) Energy requirements and the feeding of mares during lactation: a review. *Livestock Production Science* 20(1), 53–68.

Doreau, M., Boulet, S., Bauchart, D., Barlet, J. and Patureau-Mirand, P. (1990) Yield and composition of milk from lactating mares: effect of lactation stage and individual differences. *Journal of Diary Research* 57, 449–454.

Dott, H.M. and Foster, G.C. (1972) A technique for studying the morphology of mammalian spermatozoa which are eosinophilic in differential "live/dead" stain. *Journal of Reproduction and Fertility* 29, 443–445.

Dott, H.M. and Foster, G.C. (1975) Preservation of different staining of spermatozoa by formal citrate. *Journal of Reproduction and Fertility* 45, 57–60.

Douglas, R.H. (1979) Review of induction of superovulation and embryo transfer in the equine. *Theriogenology* 11(1), 33–46.

Douglas, R.H. (1980) Pregnancy rates following non-surgical embryo transfer in the equine. *Journal and Animal Science* 51(Suppl 1), 272.

Douglas, R.H. (1982) Some aspects of equine embryo transfer. *Journal of Reproduction and Fertility* 32, 405–408.

Douglas, R.H. and Ginther, O.J. (1972) Effect of prostaglandin F2α on length of dioestrus in mares. *Research in Prostaglandins* 2, 265–268.

Douglas, R.H. and Ginther, O.J. (1975) Development of the equine fetus and placenta. *Journal of Reproduction and Fertility, Supplement* 23, 503–505.

Douglas, R.H., Nuti, L. and Ginther, O.J. (1974) Induction of ovulation and multiple ovulation in seasonally an-ovulatory mares with equine pituitary fractions. *Theriogenology* 2(6), 133–141.

Douglas-Hamilton, D.H., Osol, R., Osol, G., Driscoll, D. and Noble, H. (1984) A field study of the fertility of transported equine semen. *Theriogenology* 22(3), 291–304.

Dowsett, K.F., Knott, C.M., Woodward, R.A. and Bodero, D.A.V. (1993) Seasonal variation in the oestrus cycle of mares in the subtropics. *Theriogenology* 39, 631–653.

Draincourt, M.A. and Palmer, E. (1982) Seasonal and individual effects on ovarian and endocrine responses of mares to a synchronisation treatment with progestagens impregnated vaginal sponges. *Journal of Reproduction and Fertility* 32, 283–291.

Dunn, H.O., Smiley, D. and Mcentee, K. (1981) Two equine true hermaphrodites with 64XX/64XY and 63X/63XY chimerism. *Cornell Veterinarian* 71, 123–135.

Duren, S.E. and Crandell, K. (2001) The role of vitamins in the growth of horses. In: Pagan, J.O. and Geor, R.J. (eds) *Advances in Equine Nutrition II*. Nottingham University Press, Nottingham, UK, pp. 169–178.

Dyce, K.M., Sack, W.O. and Wensing, C.J.G. (1996) *Textbook of Veterinary Anatomy*, 2nd edn. W.B. Saunders, Philadelphia.

Easley, J. (1993) External perineal conformation. In: Mckinnon, A.O. and Voss, J.L. (eds) *Equine Reproduction*. Lea and Febiger, Philadelphia, pp. 19–26.

East, L.M., Van Saun, R.J. and Vanderwall, D.K. (1999a) A review of the technique and 1997 practitioner-based survey, equine embryo transfer: Part 1: donor recipient selection and preparation. *Equine Practice* 20(8), 16–20.

East, L.M., Van Saun, R.J. and Vanderwall, D.K. (1999b) A review of the technique and 1997 practitioner based survey, equine embryo transfer. Part 2: embryo recovery and transfer techniques. *Equine Practice* 21(1), 8–12.

East, L.M., Van Saun, R.J. and Vanderwall, D.K. (1999c) A review of the technique and 1997 practitioner-based survey, equine embryo transfer: Part 3: client and veterinary economics. *Equine Practice* 21, 16–19.

Eilts, B.E., Scholl, D.T., Paccamonti, D.C., Causey, R., Klimczak, J.C. and Corley, J.R. (1995) Prevalance of endometrial cysts and their effect of fertility. *Biological Reproductive Monographs* 1, 527–532.

Ellendorff, F. and Schams, D. (1988) Characteristics of milk ejection, associated intramammary pressure changes and oxytocin release in the mare. *Journal of Endocrinology* 119, 219–227.

Elliott, F.E., Sherman, J.K., Elliott, E.J. and Sullivan, J.L. (1973) A photographic method of measuring percentage of progressing motile sperm cells using dark field microscopy. *VIIIth International Symposium Zootechnology*, Milan, p. 160.

Enders, A.C. and Lui, I.K.M. (1991) Lodgement of the equine blastocyst in the uterus from fixation through endometrial cup formation. *Journal of Reproduction and Fertility, Supplement* 44, 427.

Enders, A.C., Lantz, K.C., Lui, I.K.M. and Schlafke, S. (1988) Loss of polar trophoblast during differentiation of the blastocyst of the horse. *Journal of Reproduction and Fertility* 83, 447–460.

Enders, A.C., Schlafke, S., Lantz, K.C. and Lui, I.K.M. (1993) Endoderm cells of the yolk sac from day 7 until formation of the definitive yolk sac placenta. *Equine Veterinary Journal, Supplement* 15, 3–9.

England, G.C. and Plummer, J.M. (1993) Hypo-osmotic swelling of dog spermatozoa. *Journal of Reproduction and Fertility, Supplement* 47, 261–270.

Equine Research (1982) *Breeding Management and Foal Development*. Equine Research, Texas, 699 pp.

Ericsson, R.J. and Glass, R.H. (1982) Functional differences between sperm bearing the X- and Y- chromosome. In: Amann, R.P. and Seidel, G.E. (eds) *Prospects for Sexing Mammalian Sperm*. Colorado Association University Press, Boulder, Colorado, pp. 201–211.

Eustace, A.R. (1992) *Explaining Laminitis and Its Prevention*, Laminitis Clinic, Wiltshire, UK, 96 pp.

Evans, H.E. and Sack, W.O. (1973) Prenatal development of domestic and laboratory mammals: growth curv es, external features and selected references. *Anatomy, Histology and Embryology* 2, 11–45.

Evans, K.L., Kasman, L.H., Hughes, J.P., Couto, M. and Lasley, B.L. (1984) Pregnancy diagnosis in the domestic horse through direct urinary estrone conjugate analysis. *Theriogenology* 22, 615–620.

Evans, M.J. and Irvine, C.H.G. (1975) Serum concentrations of FSH, LH and progesterone during the oestrus cycle and early pregnancy in the mare. *Journal of Reproduction and Fertility, Supplement* 23, 193–200.

Evans, M.J. and Irvine, C.H.G. (1976) Measurements of equine follicle stimulating hormone and luteinizing hormone: response of anoestrous mares to gonadotrophin releasing hormone. *Biology of Reproduction* 15, 277–284.

Evans, M.J. and Irvine, C.H.G. (1977) Induction of follicular development, maturation and ovulation by gonadotrophin releasing hormone administration to acyclic mares. *Biology of Reproduction* 16, 452–462.

Evans, M.J. and Irvine, C.H.G. (1979) Induction of follicular development and ovulation in seasonally acyclic mares using gonadotrophin releasing hormones and progesterone. *Journal of Reproduction and Fertility, Supplement* 27, 113–121.

Evans, M.J., Hamer, J.M., Garson, L.M., Graham, C.S., Asbury, A.C. and Irvine, C.H.G. (1986) Clearance of bacteria and non-antigenic markers following intrauterine inoculation into maiden mares. Effect of steroid hormone environment. *Theriogenology* 25, 37–50.

Evans, M.J., Alexander, S.L., Irvine, C.H.G., Livesey, J.H. and Donald, R.A.S. (1991) *In vitro* and *in vivo* studies of equine prolactin secretion throughout the year. *Journal of Reproduction and Fertility, Supplement* 44, 27–35.

Evenson, D.P., Sailer, B.L. and Josh, L.K. (1995) Relationship between stallion sperm deoxyribonucleic acid (DNA), susceptibility to denaturation *in situ* and presence of DNA strand breaks: implications for fertility and embryo viability. *Biology of Reproduction, Monograph Equine Reproduction VI* 1, 655–659.

Fahy, G.M. (1986) The relevance of cryoprotectant 'toxicity' to cryobiology. *Cryobiology* 23, 1–13.

Farinasso, A., De Faria, C., Mariante, A.D. and DeBem, A.R. (1989) Embryo technology applied to the conservation of equids. *Equine Veterinary Journal, Supplement* 8, 84–88.

Farquhar, V.J., McCue, P.M., Vanderwall, D.K. and Squires, E.L. (2000) Efficiency of the GnRH agonist deslorelin acetate for inducing ovulation in mares relative to age of mare and season. *Journal of Equine Veterinary Science* 20, 722–725.

Fay, J.E. and Douglas, R.H. (1987) Changes in thecal and granulosa cell LH and FSH receptor content associated with follicular fluid and peripheral plasma gonadotrophin and steroid hormone concentrations in pre-ovulatory follicles of mares. *Journal of Reproduction and Fertility, Supplement* 37, 169–181.

Fayrer-Hosken, R.A. and Caudle, A.B. (1989) Stallion fertility evaluation: part III. *Equine Practice* 11(4), 30–34.

Fazeli, A.R., Steenweg, W., Bevers, M.M., Broek van den, J., Bracher, V., Parlevliet, J.M. and Collenbrander, B. (1995) Relationships between stallion sperm binding to homologous hemizonae and fertility. *Theriogenology* 44, 751–760.

Ferreira, J.C.P., Meira, C., Papa, F.O., Landin, E., Alvarenga, F.C., Alveranga, M.A. and Buratini, J. (1997) Cryopreservation of equine embryos with glycerol plus sucrose and glycerol plus 1,2-propanediol. *Equine Veterinary Journal, Supplement* 25, 88–93.

First, N.L. and Alm, C. (1977) Dexamethosone induced parturition in pony mares. *Journal of Animal Science* 44, 1072–1075.

Fiser, P.S., Hansen, C., Underhill, K.L. and Shrestha, J.N.B. (1991) The effect of induced ice nucleation (seeding) on the post thaw motility and acrosomal integrity of boar spermatozoa. *Animal Reproduction Science* 24, 293–304.

Fisher, A.T. and Phillips, T.N. (1986) Surgical repair of a ruptured uterus in five mares. *Equine Veterinary Journal* 18, 153–155.

Fitzgerald, B.P. and McManus, C.J. (2000) Photoperiodic signals as determinants of seasonal anestrus in the mare. *Biology of Reproduction* 63, 335–340.

Fitzgerald, B.P., Affleck, K.J., Barrows, S.P., Murdock, W.L., Barker, K.B. and Loy, R.G. (1987) Changes in LH pulse frequency and amplitude in intact mares during the transition into the breeding season. *Journal of Reproduction and Fertility* 79(2), 485–493.

Fitzgerald, B.P., Reedy, S.E., Sessions, D.R., Pwell, D.M. and McManus, C.J. (2002) Potential signals mediating the maintenance of reproductive activity during the non-breeding season of the mare. *Reproduction, Supplement* 59, 115–129.

Fletcher, M.S., Topliff, D.R., Cooper, S.R., Freeman, D.W. and Geisert, R.D. (2000) Influence of age and sex on serum osteocalcin concentrations in horses at weaning and during physical conditioning. *Journal of Equine Veterinary Science* 20(2), 125–126.

Flink, G. (1988) Gonadotrophin secretion and its control. In: Knobil, E. and Neill, J. (eds) *The Physiology of Reproduction*. Raven Press, New York, pp. 1349–1377.

Flint, A.P.F., Ricketts, S.W. and Craig, W.A. (1979) The control of placental steroids synthesis at parturition in domestic animals. *Animal Reproduction Science* 2, 239.

Flipse, R.J., Potton, S. and Almquist, J.O. (1954) Dilutes for bovine semen. III. Effect of lactenin and of lactoperoxidase upon spermatozoan livability. *Journal of Dairy Science* 32, 1205–1211.

Flood, P.F. (1993) Fertilisation, early embryo development and establishment of the placenta. In: McKinnon, A.O. and Voss, J.L. (eds) *Equine Reproduction*. Lea and Febiger, Philadelphia, pp. 473–485.

Flood, P.F., Betteridge, K.J. and Irvine, D.S. (1979a) Oestrogens and androgens in blastocoel fluids and cultures of cells from equine conceptuses of 10–22 days gestation. *Journal of Reproduction and Fertility, Supplement* 27, 413–420.

Flood, P.F., Jong, A. and Betteridge, K.J. (1979b) The location of eggs retained in the oviducts of mares. *Journal of Reproduction and Fertility* 57(2), 291–294.

Flood, P.F., Betteridge, K.J. and Diocee, M.S. (1982) Transmission electron microscopy of horse embryos 3–16 days after ovulation. *Journal of Reproduction and Fertility, Supplement* 32, 319–327.

Forsyth, I.A., Rossdale, P.D. and Thomas, C.R. (1975) Studies on milk composition and lactogenic hormones in the mare. *Journal of Reproduction and Fertility, Supplement* 23, 631–635.

Fortier, G., Vidament, M., DeCraene, F., Ferry, B. and Daels, P.F. (2002) The effect of GnRH antagonist on testosterone secretion, spermatogenesis and viral excretion in EVA-virus excreting stallions. *Theriogenology* 58, 425–427.

Fortune, J.E. and Kimmich, T.L. (1993) Purified pig FSH increases the rate of double ovulations in mares. *Equine Veterinary Journal, Supplement* 15, 95–98.

Fowden, A.L., Mundy, L., Ousey, J.C., McGladdery, A. and Silver, M. (1991) Tissue glycogen and glucose-6-phosphatase levels in fetal and newborn foals. *Journal of Reproduction and Fertility, Supplement* 44, 537–542.

Francavilla, S., Gabriele, A., Romano, R., Ginaroli, L., Ferraretti, A.P. and Francavilla, F. (1994) Sper-zona pellucida binding of human sperm is correlated with immunocytochemical presence of proacrosin and acrosin in the sperm heads but not with proteolytic activity of acrosin. *Fertility and Sterility* 62(6), 1226–1233.

Francel, A.T., Amann, R.P., Squires, E.L. and Pickett, B.W. (1987) Motility and fertility of equine spermatozoa in milk extender after 12 or 24 hours at 20°C. *Theriogenology* 27, 517–525.

Frandson, R.D. and Spurgen, T.L. (1992) *Anatomy and Physiology of Farm Animals*, 5th edn. Lea and Febiger, Philadelphia, 572 pp.

Frape, D. (1989) *Equine Nutrition and Feeding*, 2nd edn. Blackwell Science, Oxford.

Frape, D. (1998) *Equine Nutrition and Feeding*. 2nd edn. Blackwell Science, Oxford, 564 pp.

Frape, D. (2004) *Equine Nutrition and Feeding*, 3rd edn. Blackwell, Oxford, 650 pp.

Fraser, A.F., Keith, N.W. and Hastie, H. (1973) Summarised observations on the ultrasonic detection of pregnancy and foetal life in the mare. *Veterinary Record* 92(1), 20–21.

Fraser, M. (ed.) (1986) *Merk's Veterinary Manual*, 6th edn. Merk, Rathway, New Jersey.

Freedman, L.J., Garcia, M.C. and Ginther, O.J. (1979) Influences of ovaries and photoperiod on reproductive function in the mare. *Journal of Reproduction and Fertility, Supplement* 27, 79–86.

Freeman, D.A., Weber, J.A., Geary, R.T. and Woods, G.L. (1991) Time of embryo transfer through the mare oviduct. *Theriogenology* 36, 823–830.

Freeman, K.P. and Johnston, J.M. (1987) Collaboration of a cytopathologist and practitioner using equine endometrial cytology in a private broodmare practice. *Proceedings of the 13th Annual Convention of the American Association of Equine Practitioners*, pp. 629–639.

Friedman, R., Scott, M., Heath, S.E., Hughes, J.P., Daels, P.F. and Tran, T.Q. (1991) The effects of increased testicular temperature on spermatogenesis in the stallion. *Journal of Reproduction and Fertility, Supplement* 44, 127–134.

Fuchs, A.R., Periyasamy, S., Alexandrova, M. and Soloff, M.S. (1983) Correlation between oxytocin receptor concentration and responsiveness to oxytocin in pregnant rat myometrium. Effects of ovarian steroids. *Endocrinology* 113, 742–749.

Fukuda, T., Kikuchi, M., Kurotaki, T., Oyomada, T., Yoshikawa, W. and Yoshikawa, T. (2001) Age related changes in the testes of horses. *Equine Veterinary Journal* 33(1), 20–25.

Gadella, B.M., Rathi, R., Brouwers, J.F., Stout, T.A. and Colenbrander, B. (2001) Capacitation and the acrosome reaction in equine sperm. *Animal Reproduction Science* 68(3–4), 249–265.

Galli, A., Basetti, M., Balduzzi, D., Martignoni, M., Bornaghi, V. and Maffii, M. (1991) Frozen bovine semen quality and bovine cervical-mucus penetration test. *Theriogenology* 35(4), 837–844.

Galli, C., Crotti, G., Turini, P., Duchi, R., Mari, G., Zavaglia, G., Duchamp, G., Daels, P. and Lazzari, G. (2002) Frozen thawed embryos produced by ovum pick-up of immature oocytes and ICSI are capable to establish pregnancies in the horse. *Theriogenology* 58, 705–708.

Galli, C., Lagutina, I., Crotti, G., Colleoni, S., Turini, P., Ponderato, N., Duchi, R. and Lazzari, G. (2003) Pregnancy: a cloned horse born to its dam twin. *Nature* 424(6949), 635. Erratum in *Nature* 425(6959), 680.

Galli, C., Colleoni, S., Duchi, R., Lagutina, I. and Lazzari, G. (2007) Developmental competence of equine oocytes and embryos obtained by in vitro procedures ranging from in vitro maturation and ICSI to embryo culture, cryopreservation and somatic cell nuclear transfer. *Animal Reproduction Science* 98(1–2), 39–55.

Ganjam, V.K., Kenney, R.M. and Flickinger, G. (1975) Plasma progesterone in cyclic, pregnant and post-partum mares. *Journal of Reproduction and Fertility, Supplement* 23, 441–447.

Ganowiczow, A.M. and Ganowicz, M. (1966) Preliminary observations on the duration of pregnancy in English Thoroughbred mares. *Zesz. probl. Postep. Nauk. roln.* 67, 99–102 (Animal Breeding Abstracts. 39, 2209).

Garcia, M.C., Freedman, L.H. and Ginther, O.J. (1979) Interaction of seasonal and ovarian factors in the regulation of LH and FSH secretion in the mare. *Journal of Reproduction and Fertility, Supplement* 27, 103–111.

Garner, D.L., Pinkel, D., Johnson, L.A. and Pace, M.M. (1986) Assessment of spermatozoal function using dual fluorescent staining and flow-cytometric analyses. *Biology of Reproduction* 34, 127–138.

Gastal, E.L., Augusto, C., Castro, T.A.M.G. and Gastel, M.O. (1991) Relationship between the quality of stallion semen and fertility. *Anais, IX Congresso Brasileiro de Reproducao Animal,* Belo Horizonte, Brazil, 22 a 26 de Junho de 1991. Vol. II. Belo Horizonte, Brazil, Colegio Brasileiro de Reproducao Animal, p. 446.

Gastal, E.L., Bergfelt, D.R., Nogueira, G.P., Gastel, M.O. and Ginther, O.J. (1999) Role of luteinising hormone in follicle deviation based on manipulating progesterone concentrations in the mare. *Biology of Reproduction* 61, 1492–1498.

Gastal, E.L., Gastal, M.O., Beg, M.A. and Ginther, O.J. (2004) Interrelationships among follicles during the common-growth phase of a follicular wave and capacity of individual follicles for dominance in mares. *Reproduction* 128(4), 417–422.

Gastal, M.O., Gastal, E.L., Kot, K. and Ginther, O.J. (1996) Factors related to the time of fixation of the conceptus in mares. *Theriogenology*. 46(7), 1171–1180.

Gentry, L.R., Thompson, D.L. Jr., Gentry, G.T. Jr., Davis, K.A. and Godke, R.A. (2002a) High versus low body condition in mares: interactions with responses to somatotropin, GnRH analog, and dexamethasone. *Journal of Animal Science* 80(12), 3277–3285.

Gentry, L.R., Thompson, D.L. Jr., Gentry, G.T. Jr., Davis, K.A., Godke, R.A. and Cartmill, J.A. (2002b) The relationship between body condition, leptin, and reproductive and hormonal characteristics of mares during the seasonal anovu-latory period. *Journal of Animal Science* 80(10), 2695–2703.

Ghei, J.C., Uppal, P.K. and Yaday, M.P. (1994) Prospects of AI in equines. *Cataur* XI, 1, July.

Gibbs, E.P.Y., Roberts, M.C. and Morris, J.M. (1972) Equine coital exanthema in the U.K. *Equine Veterinary Journal* 4, 74.

Gibbs, P.G., Potter, G.D., Blake, R.W. and McMullan, W.C. (1982) Milk production of quarter Horse mares during 150 days of lactation. *Journal of Animal Science* 54, 496–499.

Gidley-Baird, A.A. and O'Neill, C. (1982) Early pregnancy detection in the mare. *Equine Veterinary Data* 3, 42–45.

Giggers, S., Paccamonti, D.L., Thompson, R.A. and Eilts, B.E. (2001) The effects of ph, osmolarity and urine contamina-tion on equine spermatozoal motility. *Theriogenology* 56(4), 613–622.

Gill, R.J., Potter, G.D., Schelling, G.T., Kreider, J.L. and Boyd, C.L. (1983) Post partum reproductive performance of mares fed various levels of protein. *Proceedings of the 8th Equine Nutrition Physiology Symposium*, Lexington, Kentucky, pp. 311–316.

Gillespie, J.R. (1975) Postnatal lung growth and function in the foal. *Journal of Reproduction and Fertility, Supplement* 23, 667–671.

Ginther, O.J. (1982) Twinning in mares: a review of recent studies. *Journal of Equine Veterinary Science* 2, 127–135.

Ginther, O.J. (1983a) Mobility of the early equine conceptus. *Theriogenology* 19, 603–611.

Ginther, O.J. (1983b) Fixation and orientation of the early equine conceptus. *Theriogenology* 19, 613–623.

Ginther, O.J. (1983c) Sexual behaviour following introduction of a stallion into a group of mares. *Theriogenology* 19, 877.

Ginther, O.J. (1984) Ultrasonic evaluation of the reproductive tract of the mare: the single embryo. *Journal of Equine Veterinary Science* 4, 75–81.

Ginther, O.J. (1985) Embryonic loss in mares: incidence and ultrasonic morphology. *Theriogenology* 24, 73–86.

Ginther, O.J. (1987) Relationship among number of embryos and type of multiple ovulations, number of embryos and type of embryo fixation in mares. *Journal of Equine Veterinary Science* 7, 82–88.

Ginther, O.J. (1989a) Twin embryos in mares: II. Post fixation embryo reduction. *Equine Veterinary Journal* 21, 171–174.

Ginther, O.J. (1989b) Twin embryos in mares: I. From ovulation to fixation. *Equine Veterinary Journal* 21, 166–170.

Ginther, O.J. (1992) *Reproductive Biology of the Mare, Basic and Applied Aspects*. 2nd edn. Equiservices, Cross Plains, Wisconsin, 642 pp.

Ginther, O.J. (1993) Equine foal kinetics: allantoic fluid shifts and uterine horn closures. *Theriogenology* 40, 241–256.

Ginther, O.J. (1995) *Ultrasonic Imaging and Animal Reproduction: Horses*, Book 2, Equiservices, Cross Plains, Wisconsin.

Ginther, O.J. and Bergfelt, D.R. (1988) Embryo reduction before day 11: in mares with twin conceptuses. *Journal of Animal Science* 66(7), 1727–1731.

Ginther, O.J. and Bergfeldt, D.R. (1993) Growth of small follicles and concentrations of FSH during the equine oestrous cycle. *Journal of Reproduction and Fertility* 99, 105–111.

Ginther, O.J. and First, N.L. (1971) Maintenance of the corpus luteum in hysterectomised mares. *American Journal of Veterinary Research* 32(11), 1687–1691.

Ginther, O.J. and Griffin, P.G. (1994) Natural outcome and ultrasonic identification of equine fetal twins. *Theriogenology* 41, 1193–1199.

Ginther, O.J. and Pierson, R.A. (1983) Ultrasonic evaluation of the reproductive tract of the mare: principles, equipment and techniques. *Journal of Equine Veterinary Science* 3, 195–200.

Ginther, O.J. and Pierson, R.A. (1984a) Ultrasonic evaluation of the reproductive tract of the mare: ovaries. *Journal of Equine Veterinary Science* 4, 11–16.

Ginther, O.J. and Pierson, R.A. (1984b) Ultrasonic anatomy of equine ovaries. *Theriogenology* 21, 471–483.

Ginther, O.J. and Pierson, R.A. (1984c) Ultrasonic anatomy and pathology of the equine uterus. *Theriogenology* 21, 505–515.

Ginther, O.J. and Pierson, R.A. (1989) Regular and irregular characteristics of ovulation in the interovulatory interval in the mare. *Journal of Equine Veterinary Science* 9, 4–12.

Ginther, O.J., Douglas, R.H. and Lawrence, J.R. (1982) Twinning in mares: a survey of veterinarians and analysis of theriogenology records. *Theriogenology* 18, 333–347.

Ginther, O.J., Scraba, S.T. and Nuti, R.C. (1983) Pregnancy rates and sexual behaviour under pasture breeding conditions in mares. *Theriogenology* 20, 333–345.

Ginther, O.J., Beg, M.A., Bergfelt, D.R., Donadeu, F.X. and Kot, K. (2001) Follicle selection in monovular species. *Biology of Reproduction* 65, 638–647.

Ginther, O.J., Meira, C., Beg, M.A. and Bergfelt, D.R. (2002) Follicle and endocrine dynamics during experimental follicle deviation in mares. *Biology of Reproduction* 67(3), 862–867.

Ginther, O.J., Beg, M.A., Donadeu, F.X. and Bergfelt, D.R. (2003) Mechanism of follicle deviation in monovular farm species. *Animal Reproduction Science* 78, 239–257.

Ginther, O.J., Gastel, E., Gastel, M. and Beg, M. (2004) Seasonal influence on equine follicle dynamics. *The Journal of Animal Reproduction* 1(1), 31–44.

Glaser, A.L., Chimside, E.D., Horzinek, M.C. and DeVries, A.A.F. (1997) Equine viral arteritis. *Theriogenology* 47, 1275–1295.

Goater, L.E., Maecham, T.N., Gwazdauski, F.C. and Fontenot, J.P. (1981) Effect of dietary energy level in mares during gestation. *Journal of Animal Science* 53(Supplement 1), 295.

Godoi, D.B., Gastel, E.L. and Gastel, M.O. (2002) A comparative study of follicular dynamics between lactating and non-lactating mares: effect of body condition. *Theriogenology* 58, 553–556.

Goff, A.K., Panbrin, D. and Sirois, J. (1987) Oxytocin stimulation of plasma 15-keto-13, 14-dihydro prostaglandin F2α release during the oestrus cycle and early pregnancy in the mare. *Journal of Reproduction and Fertility, Supplement* 35, 253–260.

Golnik, W. (1992) Viruses isolated from fetuses and still born foals. *Equine Infectious Diseases VI: Proceedings of the Sixth International Conference.* RW Publications, Newmarket, UK, p. 314.

Gordon, I. (1997) Introduction to controlled reproduction in horses. In: Gordon, I. (ed.) *Controlled Reproduction in Horses, Deer and Camelids. Controlled Reproduction in Farm Animals Series.* Vol. 4, 1st edn. CAB International, Wallingford, UK, pp. 1–35.

Gradil, C., Yoon, S.Y., Brown, J., He, C., Visconti, P. and Fissore, R. (2006) PLCζ: a marker of fertility for stallions? *Animal Reproduction Science* 94, 23–25.

Graham, E.F., Crabo, B.G. and Pace, M.M. (1978) Current status of semen preservation in the ram, boar and stallion. Proceedings of the XII Biennial Symposium on Animal Reproduction. *Journal of Animal Science, Supplement* 11, 80–119.

Gravance, C.G., Liu, I.K.M., Davis, R.O., Hughes, J.P. and Casey, P.J. (1996) Quantification of normal morphometry of stallion spermatozoa. *Journal of Reproduction and Fertility* 108(1), 41–46.

Green, J. (1993) Feeding the orphan or sick foal. *Equine Veterinary Education* 5(5), 274–275.

Greenhof, G.R. and Kenney, R.M. (1975) Evaluation of the reproductive status of non pregnant mares. *Journal of the American Veterinary Medicine Association* 167, 449–458.

Griffin, P.G. (2000) The breeding soundness examination in the stallion. *Journal of Equine Veterinary Science* 20(3), 168–171.

Griffin, P.G. and Ginther, O.J. (1990) Uterine contractile activity in mares during the estrous cycle and early pregnancy. *Theriogenology* 34(1), 47–56.

Griffin, J.L., Castleberry, R.S. and Schneider, H.S., Jr. (1981) Influence of day of recovery on collection rate in mature cycling mares. *Theriogenology* 15, 106–124.

Griggers, S., Paccamonti, D.L., Thompson, R.A. and Eilts, B.E. (2001) The effects of ph, osmolarity and urine contamination on equine spermatozoal motility. *Theriogenology.* 56(4), 613–622.

Grubaugh, W.R. (1982) The effects of pinealectomy in pony mares. *Journal of Reproduction and Fertility, Supplement* 32, 293–295.

Guerin, M.U. and Wang, X.J. (1994) Environmental temperature has an influence on timing of the first ovulation of seasonal estrus in the mare. *Theriogenology* 42, 1053–1060.

Gygax, A.P., Ganjam, V.K. and Kenney, R.M. (1979) Clinical microbiological and histological changes associated with uterine involution in the mare. *Journal of Reproduction and Fertility, Supplement* 27, 571–578.

Haard, M.C. and Haard, M.G.H. (1991) Successful commercial use of frozen stallion semen abroad. *Journal Reproduction and Fertility, Supplement* 44, 647–648.

Hackeloer, B.J. (1977) The ultrasonic demonstration of follicular development during the normal menstrual cycle and after hormone stimulation. *Proceedings of International Symposium on Recent Advances in Ultrasound Diagnosis*, Dubrovnik, pp. 122–128.

Hafez, E.S.E. and Hafez, B. (2000) *Reproduction in Farm Animals*, 7th edn. Williams and Wilkins, Baltimore, Maryland, 509 pp.

Haffner, J.C., Fecteau, K.A., Held, J.P. and Eiler, H. (1998) Equine retained placenta: technique for and tolerance to umbilical injections of collagenase. *Theriogenology* 49, 711–716.

Hall, J.L. (1981) Relationship between semen quality and human sperm penetration of zona free hamster ova. *Fertility and Sterility* 35, 457–463.

Hallowell, A.L. (1989) The use of ultrasonography in equine reproduction on the breeding farm. *Proceedings of the 2nd Annual Conference Society of Theriogenology* 202–205.

Halnan, C.R.E. (1985) Sex chromosome mosaicism and infertility in mares. *Veterinary Record* 116, 542–543.

Halnan, C.R.E. and Watson, J.I. (1982) Detection of G and C band karyotyping of genome anomalies in horses of different breeds. *Journal of Reproduction and Fertility, Supplement* 32, 626.

Halnan, C.R.E., Watson, J.I. and Pryde, L.C. (1982) Prediction of reproductive performance in horses by karyotype. *Journal of Reproduction and Fertility, Supplement* 32, 627.

Haluska, G.J. and Currie, W.B. (1988) Variation in plasma concentrations of oestradiol 17 Beta and their relationship to those of progesterone. 13,14-dihydro-15-ketoprostaglandin F2α and oxytocin across pregnancy and at parturition in pony mares. *Journal of Reproduction and Fertility* 84, 635–646.

Haluska, G.J. and Wilkins, K. (1989) Predictive utility of the pre-partum temperature changes in the mare. *Equine Veterinary Journal* 21, 116–118.

Haluska, G.J., Lowe, J.E. and Currie, W.B. (1987a) Electromyographic properties of the myometrium correlated with the endocrinology of the pre-partum and post partum periods and parturition in pony mares. *Journal of Reproduction and Fertility, Supplement* 35, 553–564.

Haluska, G.J., Lowe, J.E. and Currie, W.B. (1987b) Electromyographic properties of the myometrium of the pony mare during pregnancy. *Journal of Reproduction and Fertility* 81, 471–478.

Hammerstedt, R.H., Graham, J.K. and Nolan, J.P. (1990) Cryopreservation of mammalian sperm: what we ask them to survive. *Journal of Andrology* 11, 73–88.

Hamon, M., Clarke, S.W., Houghton, E., Fowden, A.L., Silver, M., Rossdale, P.D., Ousey, J.C. and Heap, R.B. (1991) Production of 5αdihydroprogesterone during late pregnancy in the mare. *Journal of Reproduction and Fertility, Supplement* 44, 529–535.

Handler, J., Gomes, T., Waelchli, R.O., Betteridge, K.J. and Raeside, J.I. (2002) Influence of cervical dilation on pregnancy rates and embryonic development in inseminated mares. In: Eavans, M. (ed.) *Equine Reproduction VIII.* Elsevier, New York, pp. 671–674.

Hansen, T.R., Austin, K.J., Perry, D.J., Pru, J.K., Teixeira, M.G. and Johnson, G.A. (1999) Mechanism of action of interferon-tau in the uterus during early pregnancy. *Journal of Reproduction and Ferility, Supplement* 54, 329–339.

Harrison, L.A., Squires, E.L. and McKinnon, A.O. (1991) Comparison of hCG, buserelin and luprostiol for induction of ovulation in cycling mares. *Journal of Equine Veterinary Science* 11, 163–166.

Harrison, R.A.P. and Vickers, S.E. (1990) Use of fluorescent probes to assess membrane integrity in mammalian spermatozoa. *Journal of Reproduction and Fertility* 88, 343–352.

Hayes, K.E.N., Pierson, R.A., Scraba, S.T. and Ginther, O.J. (1985) Effects of oestrous cycle and season on ultrasonic uterine anatomy in mares. *Theriogenology* 24, 465–477.

Heap, R.B. (1972) Role of hormones in pregnancy. In: Austin, C.R. and Short, R.V. (eds) *Reproduction in Mammals. 3. Hormones in Reproduction.* Cambridge University Press, Cambridge, pp. 73–95.

Heap, R.B., Hamon, M. and Allen, W.R. (1982) Studies on oestrogen synthesis by pre-implantation equine conceptus. *Journal of Reproduction and Fertility, Supplement* 32, 343–352.

Heape, W. (1897) The artificial insemination of mammals and subsequent fertility on impregnation of their ova. *Proceedings of the Royal Society of London,* 61, 52.

Heeseman, C.P., Squires, E.L., Webel, S.K., Shideler, R.K. and Pickett, B.W. (1980) The effect of ovarian activity and allyl trenbolone on the oestrous cycle and fertility in mares. *Journal of Animal Science, Supplement* 1(51), 284.

Heidler, B., Parvizi, N., Sauerwein, H., Bruckmaier, R.M., Heintges, U., Aurich, J.E. and Aurich, C. (2003) Effects of lactation on metabolic and reproductive hormones in Lipizzaner mares. *Domestic Animal Endocrinology* 25(1), 47–59.

Heiskanen, M.L., Hilden, L., Hyyppa, S., Kangasniemi, A., Pirhonen, A. and Maenpaa, P.H. (1994a) Freezability and fertility results with centrifuged stallion semen. *Acta Veterinaria Scandinavia* 35(4), 377–382.

Heiskanen, M.L., Huhtinen, M., Pirhonen, A. and Maepaa, P.H. (1994b) Insemination results with slow cooled stallion semen stored for approximately 40 hours. *Acta Veterinarian Scandinavia* 35(3), 257–262.

Heitland, A.V., Jasko, D.J., Graham, J.K., Squires, E.L., Amann, R.P. and Pickett, B.W. (1995) Motility and fertility of stallion spermatozoa cooled and frozen in a modifies skim milk extender containing egg yolk and liposome. *Biology of Reproduction Monograph Equine Reproduction VI* 1, 753–759.

Heitland, A.V., Jasko, D.J., Squires, E.L., Graham, J.K., Pickett, B.W. and Hamilton, C. (1996) Factors affecting motion characteristics of frozen-thawed stallion spermatozoa. *Equine Veterinary Journal* 28(1), 47–53.

Held, J.P. (1987) Retained placenta. In: Robinson, E.N. (ed.) *Current Therapy in Equine Medicine.* 2nd edn. W.B. Saunders, Philadelphia, pp. 547–550.

Heleski, C.R., Shelle, A.C., Neilsen, B.D. and Zanella, A.J. (2002) Influence of housing on weanling horse behaviour and subsequent welfare. *Applied Animal Behaviour Science* 78, 291–302.

Hemberg, E., Lundeheim, N. and Einarsson, S. (2005) Retrospective study on vulvar conformation in relation to endometrial cytology and fertility in Thoroughbred mares. *Journal of Veterinary Medicine* 52, 474–477.

Henderson, K. and Stewart, J. (2000) A dipstick immunoassay to rapidly measure serum oestrone sulphate concentrations in horses. *Reproduction and Fertility Development* 12, 183–189.

Henderson, K. and Stewart, J. (2002) Factors influencing the measurement of oestrone sulphate by dipstick particle capture immunoassay. *Journal of Immunological Methods* 270, 77–84.

Henderson, K., Stevens, S., Bailey, C., Hall, G., Stewart, J. and Wards, R. (1998) Comparison of the merits of measuring equine chorionic gonadotrophin (eCG) and blood and faecal concentrations of oestrone sulphate for determining pregnancy status of miniature horses. *Reproduction, Fertility and Development* 10(5), 441–444.

Henneke, D.R., Potter, G.D., Kreider, J.L. (1984) Body condition during pregnancy and lactation and reproductive efficiency of mares. *Theriogenology* 21, 897–909.

Herd, R.P. (1987a) Anthelmintic and drug resistance. In: Robinson, N.E. (ed.) *Current Therapy in Equine Medicine*. 2nd edn. W.B. Saunders, Philadelphia, pp. 332–334.

Herd, R.P. (1987b) Pasture hygiene. In: Robinson, N.E. (ed.) *Current Therapy in Equine Medicine*. 2nd edn. W.B. Saunders, Philadelphia, pp. 334–336.

Herfen, K., Jager, C. and Wehrend, A. (1999) Genital infection in mares and their clinical significance. *Reproduction in Domestic Animals* 34, 20–21.

Hermenet, M.J., Sawyer, H.R., Pickett, B.W., Amann, R.P., Squires, E.L. and Long, P.L. (1993) Effect of stain, technician, number of spermatozoa evaluated and slide preparation on assessment of spermatozoal viability by light microscopy. *Journal of Equine Veterinary Science* 13(8), 449–455.

Hershman, L. and Douglas, R.H. (1979) The critical period for maternal recognition of pregnancy in mares. *Journal of Reproduction and Fertility, Supplement* 27, 395–401.

Hess, M.B., Parker, N.A., Purswell, B.J. and Dascanio, D.J. (2002) Use of lufenuron as a treatment for fungal endometritis in four mares. *Journal of the American Veterinary Medicine Association* 221(2), 266–267.

Hevia, M.L., Quiles, A.J., Fuentes, F. and Gonzalo, C. (1994) Reproductive performance of Thoroughbred mares in Spain. *Journal of Equine Veterinary Science* 14(2), 89–92.

Hillman, R.B. and Ganjam, V.K. (1979) Hormonal changes in the mare and foal associated with oxytocin induction of parturition. *Journal of Reproduction and Fertility, Supplement* 27, 541–546.

Hillman, R.B. and Lesser, S.A. (1980) Induction of parturition in mares. *Veterinary Clinics of North America, Large Animal Practice* 2(2), 333–334.

Hillman, R.B., Olar, T.T. and Squires, E.L. (1980) Temperature of the artificial vagina and its effect on seminal quality and behavioural characteristics of stallions. *Journal of the American Veterinary Medical Association* 177(8), 720–722.

Hines, K.K., Hodge, S.L., Kreider, J.L., Potter, G.D. and Harms, P.G. (1987) Relationship between body condition and levels of serum luteinising hormone in post partum mares. *Theriogenology* 28, 815–825.

Hinrichs, K. and Hunt, P.R. (1990) Ultrasound as an aid to diagnosis of granulosa cell tumour in the mare. *Equine Veterinary Journal* 22, 99–103.

Hinrichs, K. and Kenney, R.M. (1988) Effect of timing of progesterone administration on pregnancy rate after embryo transfer in ovarectomised mares. *Journal of Reproduction and Fertility, Supplement* 35, 439–443.

Hinrichs, K., Betschart, R.W., McCue, P.M. and Squires, E.L. (2000) Effect of timing of follicle aspiration on pregnancy rate after oocyte transfer in mares. *Journal of Reproduction and Fertility, Supplement* 56, 493–498.

Hinrichs, K., Love, C.C., Brinsko, S.P., Choi, Y.H. and Varner, D.D. (2002) *In vitro* fertilisation of *in vitro* matured equine oocytes: effect of maturation medium, duration of maturation and sperm calcium ionophore treatment, and comparison with rates of fertilisation *in vivo* after oviductal transfer. *Biology of Reproduction* 67, 256–262.

Hintz, H.F. (1980) Growth in the horse. *Stud Manager's Handbook*, Vol. 16. Agriservices Foundation, California, pp. 59–66.

Hintz, H.F. (1983) Feeding programs. In: Robinson, N.E. (ed.) *Current Therapy in Equine Medicine*. 2nd edn. W.B. Saunders, Philadelphia, pp. 412–418.

Hintz, H.F. (1993a) Nutrition of the broodmare. In: McKinnon, A.O. and Voss, J.L. (eds) *Equine Reproduction*. Lea and Febiger, Philadelphia, London, pp. 631–639.

Hintz, H.F. (1993b) Feeding the stallion. In: McKinnon, A.O. and Voss, J.L. (eds) *Equine Reproduction*. Lea and Febiger, Philadelphia, London, pp. 840–842.

Hochereau-de Riviers, M.T., Courtens, J.-L. and Courot, M. (1990) Spermatogenesis in mammals and birds. In: Laming, G.E. (ed.) *Marshall's Physiology of Reproduction, Volume 2, Reproduction in the male*. 4th edn. Churchill Livingstone, London, pp. 106–182.

Hochi, S., Fujimato, T., Choi, Y., Braun, J. and Oguri, N. (1994a) Cryopreservation of equine oocytes by 2-step freezing. *Theriogenology* 42, 1085–1094.

Hochi, S., Fujimato, T., Braun, J. and Oguri, N. (1994b) Pregnancies following transfer of equine embryos cryopreserved by vitrification. *Theriogenology* 42(3), 483–488.

Hochi, S., Fujimato, T. and Oguri, N. (1995) Large equine blastocysts are damaged by vitrification procedures. *Reproduction, Fertilty and Development* 7, 113–117.

Hochi, S., Maruyana, A. and Oguri, N. (1996) Direct transfer of equine blastocysts frozen-thawed in the presence of ethylene glycol and sucrose. *Theriogenology* 46(7), 1217–1224.

Hodge, S.L., Kieider, J.L., Potter, G.D., Harms, P.G. and Fleeger, J.L. (1982) Influence of photoperiod on the pregnant and post partum mare. *American Journal of Veterinary Research* 43, 1752–1755.

Hofferer, S., Duchamp, G. and Palmer, E. (1991) Ovarian response in mares to prolonged treatment with exogenous equine pituitary gonadotrophins. *Journal of Reproduction and Fertility, Supplement* 44, 341–349.

Hoffman, R.M., Konfeld, D.S., Holland, J.R. and Greiwe-Crandell, K.M. (1995) Pre-weaning diet and stall weaning influences on stress response in foals. *Journal of Animal Science* 73, 2922–2930.

Holdstock, N.B., Ousey, J.C. and Rossdale, P.D. (1998) Glomerular filtration rate, effective renal plasma flow, blood pressure and pulse rate in equine neonate during the first 10 days post partum. *Equine Veterinary Journal* 30(4), 335–343.

Holland, J.L., Kronfeld, D.S., Hoffman, R.M. and Harris, P.A. (1997) Weaning stress assessment in mares. *Proceedings of the 15th Equine Nutrition and Physiology Symposium* 14, pp. 219–221.

Holtan, D. (1993) Progestin therapy in mares with pregnancy complication. Necessity and efficacy. *Proceedings of the 39th Annual Association of Equine Practitioners Convention*, San Diego, California, pp. 165–166.

Holtan, D.W. and Silver, M. (1992) Readiness for birth; another piece of the puzzle. *Equine Veterinary Journal* 24(5), 336–337.

Holtan, D.W., Houghton, E., Silver, M., Fowden, A.L., Ousey, J. and Rossdale, P.D. (1991) Plasma progestagens in the mare, fetus and newborn foal. *Journal of Reproduction and Fertility, Supplement* 44, 517–528.

Holtan, D.W., Nett, T.M. and Estergreen, V.L. (1975a) Plasma progestins in pregnant, postpartum and cyclic mares. *Journal of Animal Science* 40, 251–260.

Holtan, D.W., Nett, T.M. and Estergreen, V.L. (1975b) Plasma progestagens in pregnant mares. *Journal of Reproduction and Fertility, Supplement* 23, 419–424.

Holtan, D.W., Douglas, R.H. and Ginther, O.J. (1977) Estrus, ovulation and conception following synchronisation with progestagens, prostaglandin F2α and human CG in pony mares. *Journal Animal Science* 44(3), 431–437.

Holtan, D.W., Squires, E.L., Lapin, D.R. and Ginther, O.J. (1979) Effect of ovariectomy on pregnancy in mares. *Journal of Reproduction and Fertility, Supplement* 27, 457–463.

Honnas, C.M., Spensley, M.S., Laverty, S. and Blanchard, P.C. (1988) Hydramnios causing uterine rupture in a mare. *Journal of American Veterinary Medicine Association* 193, 332–336.

Horse Race Betting Levy Board (2008) *Codes of Practice 2001–2002 on Contagious Equine Metritis, Klebsiella Pneumoniae, Pseudomonas Aeroginosa, Equine Viral Arteritis and Equine Herpes Virus 1*. Horse Race Betting Levy Board, London.

Hossain, A.M. and Osuamkpe, C.O. (2007) Sole use of sucrose in human sperm cryopreservation. *Archives in Andrology* 53(2), 99–103.

Houghton, E., Holton, D., Grainger, L., Voller, B.E., Rossdale, P.D. and Ousey, J.C. (1991) Plasma progestagen concentrations in the normal and dysmature newborn foal. *Journal of Reproduction and Fertility, Supplement* 44, 609–617.

Householder, D.D., Pickett, B.W., Voss, J.L. and Olar, T.T. (1981) Effect of extender, number of spermatozoa and hCG on equine fertility. *Equine Veterinary Science* 1, 9–13.

Howell, C.E. and Rollins, W.C. (1951) Environmental sources of variation in the gestation length of the horse. *Journal of Animal Science* 10, 789–806.

Hudson, J.J. and McCue, P.M. (2004) How to increase recovery rates and transfer success. *Proceedings of the 50th Annual Convention of the American Association of Equine Practitioners*, pp. 406–408.

Huff, A.N., Meacham, T.N. and Wahlberg, M.L. (1985) Feeds and feeding: a review. *Journal of Equine Veterinary Science* 5, 96–108.

Hughes, J.P. (1993) Developmental anomalies of the female reproductive tract. In: McKinnon, A.O. and Voss, J.L. (eds) *Equine Reproduction*. Lea and Febiger, Philadelphia, London, pp. 408–416.

Hughes, J.P. and Loy, R.G. (1969) Investigation on the effect of intrauterine inoculation of *Streptococcus zooepidemicus* in the mare. *Proceedings of the 15th Annual Meeting of the American Association of Equine Practitioners*, pp. 289–292.

Hughes, J.P. and Loy, R.G. (1970) Artificial insemination in the equine. A comparison of natural breeding and AI of mares using semen from six stallions. *Cornell Veterinary* 60, 463–475.

Hughes, J.P. and Loy, R.G. (1975) The relation of infection to infertility in the mare and the stallion. *Equine Veterinary Journal* 7, 155–159.

Hughes, J.P. and Loy, R.G. (1978) Variations in ovulation response associated with the use of prostaglandins to manipulate the lifespan of the normal dioestrus corpus luteum or prolonged corpus luteum in the mare. *Proceedings of the 24th Annual Convention of American Association of Equine Practitioners*, pp. 173–175.

Hughes, J.P. and Stabenfeldt, G.H. (1977) Anestrus in the mare. *Proceedings of the 23rd American Association of Equine Practitioners*, pp. 89–96.

Hughes, J.P., Loy, R.G., Atwood, C., Astbury, A.C. and Burd, H.E. (1966) The occurrence of pseudomonas in the reproductive tract of mares and its effect on fertility. *Cornell Veterinarian* 56(4), 595–610.

Hughes, J.P., Benirschke, K., Kennedy, P.C. and Trommershausen-Smith, A. (1975) Gonadal dysgenesis in the mare. *Journal of Reproduction and Fertility, Supplement* 23, 385–390.

Hughes, J.P., Stabenfeldt, G.H., Kindhal, H., Kennedy, P.C., Edquist, L.E., Nealy, D.P. and Schalm, O.W. (1979) Pyrometra in the mare. *Journal of Reproduction and Fertility, Supplement* 27, 321–329.

Huhtinen, M., Lagneaux, D., Koskinen, E. and Palmer, E. (1997) The effect of sucrose in the thawing solution on the morphology and mobility of frozen equine embryos. *Equine Veterinary Journal, Supplement* 25, 94–97.

Hunt, B., Lein, D.H. and Foote, R.H. (1978) Monitoring of plasma and milk progesterone for evaluation of post partum estrous cycle and early pregnancy in mares. *Journal of the American Veterinary Medical Association* 172(11), 1298–1302.

Hunter, R.H.F. (1990) Gamete lifespan in the mare's genital tract. *Equine Veterinary Journal* 22(6), 378–379.

Hurtgen, J.P. (1987) Stallion genital abnormalities. In: Robinson, N.E. (ed.) *Current Therapy in Equine Medicine, 2.* W.B. Saunders, Philadelphia, pp. 558–562.

Hurtgen, J.P. (1997) Commercial freezing of stallion semen. *Proceedings of the 19th Bain-Fallon Memorial Lectures, Equine Reproduction, Australian Equine Veterinary Association*, Manly, New South Wales, Australia, pp. 17–23.

Hurtgen, J.P. (2000) Breeding management of the warmblood stallion. In: Samper, J.C. (ed.) *Equine Breeding Management and Artificial Insemination.* W.B. Saunders, Philadelphia, pp. 73–80.

Hurtt, A.E., Landin-Alvarenga, F., Seidel, G.E. Jr. and Squires, E.L. (2000) Vitrification of immature and mature equine and bovine oocytes in an ethylene glycol, ficoll and sucrose solution using open-pulled straws. *Theriogenology* 54, 119–128.

Hyland, J.H. and Bristol, F. (1979) Synchronisation of oestrus and timed insemination of mares. *Journal of Reproduction and Fertility, Supplement* 27, 251–255.

Hyland, J.H. and Jeffcote, L.B. (1988) Control of transitional anoestrus in mares by infusion of gonadotrophin releasing hormone. *Theriogenology* 29, 1383–1391.

Hyland, H.J. and Langsford, D.A. (1990) Changes in urinary and plasma oestrone sulphate concentrations after induction of foetal death in mares at 45 days of gestation. *Australian Veterinary Journal* 67, 349–351.

Ijaz, A. and Ducharme, R. (1995) Effects of various extenders and taurine on survival of stallion sperm cooled to 5°C. *Theriogenology* 44(7), 1039–1050.

Imel, K.J., Squires, E.L. and Elsden, R.P. (1980) Embryo recovery and effect of repeated uterine flushing of mares. *Theriogenology* 13, 97–103.

Imel, K.J., Squires, E.L., Elsden, R.P. and Shideler, R.K. (1981) Collection and transfer of equine embryos. *Journal of the American Veterinary Medical Association* 179(10), 987–989.

Irvine, C.H.G. (1984a) Gonadotrophin releasing hormone. *Journal of Equine Veterinary Science* 3, 168–171.

Irvine, C.H.G. (1984b) Hypothyroidism in the foal. *Equine Veterinary Journal* 16, 302–305.

Irvine, C.H.G. and Alexander, S.L. (1991) Effect of sexual arousal on gonadotrophin releasing hormone, luteinising hormone and follicle stimulating hormone secretion in the stallion. *Journal of Reproduction and Fertility, Supplement* 44, 135–143.

Irvine, C.H.G. and Alexander, S.C. (1993a) Secretory patterns and rates of GnRH, FSH and LH revealed by intensive sampling of pituitary venous blood in the luteal phase mare. *Endocrinology* 132, 212–218.

Irvine, C.H.G. and Alexander, S.C. (1993b) GnRH. In: McKinnon, A.O. and Voss, J.L. (eds) *Equine Reproduction.* Lea and Febiger, Philadelphia, pp. 37–45.

Irvine, C.H.G. and Alexander, S.C. (1994) The dynamics of gonadotrophin releasing hormone, LH and FSH secretion during spontaneous ovulatory surge of the mare as revealed by intensive sampling of pituitary venous blood. *Journal of Endocrinology* 140, 283–295.

Irvine, C.H.G., Alexander, S.L. and Turner, J.E. (1986) Seasonal variation in the feedback of sex steroid hormones on serum LH concentration in the male horse. *Journal of Reproduction and Fertility, Supplement* 76, 221.

Ishii, M., Shimamura, T., Utsumi, A., Jitsukawa, T., Endo, M., Fukuda, T. and Yamanoi, T. (2001) Reproductive performance and factors that decrease pregnancy rate in heavy draft horses bred at the Foal Heat. *Journal of Equine Veterinary Science* 21(3), 131–136.

Iuliano, M.F., Squires, E.L. and Cook, V.M. (1985) Effect of age of equine embryos and method of transfer on pregnancy rate. *Journal of Animal Science* 60, 258–263.

Jackson, G., Carson, T., Heath, P. and Cooke, G. (2002) CEMO in a UK stallion. *Veterinary Record* 151, 582–583.

Jackson, P.G.G. (1982) Rupture of the prepubic tendon in a Shire mare. *Veterinary Record* 111(2), 38.

Jacob, R.J., Cohen, D., Bouchey, D., Davis, T. and Borchelt, J. (1988) Molecular pathogenesis of equine coital exanthema: identification of a new equine herpesvirus isolated from lesions reminiscent of coital exanthema in a donkey. In: Powell, D.G. (ed.) *Equine Infectious Diseases V: Proceedings of the 5th International Conference of Equine Infectious Diseases*. University Press of Kentucky, Lexington, Kentucky, pp. 140–146.

Jacobson, N.L. and McGillard, A.D. (1984) The mammary gland and lactation. In: Swenson, M.J. (ed.) *Duke's Physiology of Domestic Animals*, 10th edn. Canstock Publishing Associates, Cornell University Press, Ithaca, New York, pp. 871–891.

Jafer, S.I. and Flint, A.P.F. (1996) Sex selection in mammals – a review. *Theriogenology* 46(2), 191–200.

Jainudeen, M.R. and Hafez, E.S.E. (1993) Reproductive failure in males. In: Hafez, E.S.E. (ed.) *Reproduction in Farm Animals*. 6th edn. Lea and Febiger, Philadelphia, pp. 287–297.

Jasko, D.J. (1992) Evaluation of stallion semen. In: Blanchard, T.L. and Varner, D.D. (eds) *The Veterinary Clinics of North America, Equine Practitioners*. W.B. Saunders, Philadelphia, pp. 129–148.

Jasko, D.J. (2002) Comparison of pregnancy rates following non-surgical transfer of day 8 equine embryos using various transfer devices. In: Evans, M.J. (ed.) *Equine Reproduction VIII Theriogenology* 58, 713–716.

Jasko, D.J., Little, T.V., Smith, K., Lein, D.H. and Foote, R.H. (1988) Objective analysis of stallion sperm motility. *Theriogenology* 30(6), 1159–1167.

Jasko, D.J., Lein, D.H. and Foote, R.H. (1990a) Determination of the relationship between sperm morphologic classifications and fertility in stallions: 66 cases (1987–1988). *Journal of the American Veterinary Association* 197(3), 389–394.

Jasko, D.J., Lein, D.H. and Foote, R.H. (1990b) A comparison of two computer automated semen analysis instruments for the evaluation of sperm motion characteristics in the stallion. *Journal of Andrology* 11, 453–459.

Jasko, D.J., Lein, D.H. and Foote, R.H. (1991) Stallion spermatozoal morphology and its relationship to spermatozoal motility and fertility. *Proceedings of 37th Annual Convention of the American Association of Equine Practitioners*. San Francisco, California, pp. 211–221.

Jasko, D.J., Hathaway, J.A., Schaltenbrand, V.L., Simper, W.D. and Squires, E.L. (1992a) Effect of seminal plasma and egg yolk on motion characteristics of cooled stallion spermatozoa. *Theriogenology* 37, 1241–1252.

Jasko, D.J., Little, T.V., Lein, D.H. and Foote, R.H. (1992b) Comparison of spermatozoal movement and semen characteristics with fertility in stallions: 64 cases (1987–1988). *Journal of the American Veterinary Association* 200(7), 979–985.

Jasko, D.J., Bedford, S.J., Cook, N.L., Mumford, E.C., Squires, E. and Pickett, B.W. (1993a) Effect of antibiotics on motion characteristics of a cooled stallion spermatozoa. *Theriogenology* 40, 885–893.

Jasko, D.J., Moran, D.M., Farlin, M.E., Squires, E.L., Amann, R.P. and Pickett, B.W. (1993b) Pregnancy rates utilising fresh, cooled and frozen-thawed stallion semen. *Proceedings of the 38th Annual Convention of the American Association of Equine Practitioners*, Orlando, Florida, pp. 649–660.

Jeffcote, L.B. (1972) Observations on parturition in crossbred pony mares. *Equine Veterinary Journal* 4, 209–215.

Jeffcote, L.B. (1987) Passive transfer of immunity to foals. In: Robinson, N.E. (ed.) *Current therapy in Equine Medicine*. 2nd edn. W.B. Saunders, Philadelphia, pp. 210–215.

Jeffcote, L.B. and Rossdale, P.D. (1979) A radiographic study of the foetus in late pregnancy and during foaling. *Journal of Reproduction and Fertility, Supplement* 27, 563–569.

Jeffcote, L.B. and Whitwell, K.E. (1973) Twinning as a cause of foetal and neonatal loss in the thoroughbred mare. *Journal of Comparative Pathology* 83, 91–106.

Jeffcote, L.B., Rossdale, P.D. and Leadon, D.P. (1982) Haematological changes in the neonatal period of normal and induced premature foals. *Journal of Reproduction and Fertility, Supplement* 32, 537–544.

Jennes, R. and Sloane, R.G. (1970) Review article. *Dairy Science Abstract* 32, 158–162

Jeyendran, R.S., Van der Ven, H.H., Perez-Pelaez, M., Crabo, B.G. and Zanveld, L.J.D. (1984) Development of an assay to assess the functional integrity of the human sperm membrane and its relationship to other semen characteristics. *Journal of Reproduction and Fertility* 70, 219–228.

Jimenez, C.F. (1987) Effects of Equex STM and equilibration time on the pre-freeze and post thaw motility of equine epididymal spermatozoa. *Theriogenology* 28(6), 773–782.

Jochle, W. (1957) Einflusse auf graviditsdauer und geschlecht bei pferden (Factors affecting gestation period and sex in horses) *Archives Gynak* 190, 122–124 (Animal Breeding Abstracts 26, 598).

Jochle, W. and Trigg, T.E. (1994) Control of ovulation in the mare with ovuplant a short-term release implant (STI) containing GnRH analogue Deslorelin Acetate. Studies from 1990–1994. A review. *Journal of Equine Veterinary Science* 14(12), 632–644.

Jochle, W., Irvine, C.H.G., Alexander, S.L. and Newby, T.J. (1987) Release of LH, FSH and GnRH into pituitary venous blood in mares treated with PGF analogue, luprostiol, during the transition period. *Journal of Reproduction and Fertility, Supplement* 35, 261–267.

Johnson, A.L. (1986a) Serum concentrations of prolactin, thyroxine and triiodothyronine relative to season and the oestrous cycles of the mare. *Journal of Animal Science* 62(4), 1012–1020.

Johnson, A.L. (1986b) Pulsatile administration of gonadotrophin releasing hormone advances ovulation in cycling mare. *Biology of Reproduction* 35, 1123–1130.

Johnson, A.L. (1987a) Gonadotrophin releasing hormone treatment induces follicular growth and ovulation in seasonally anoestrus mares. *Biology of Reproduction* 36(5), 1199–1206.

Johnson, A.L. (1987b) Seasonal and photoperiodic induced changes in serum prolactin and pituitary responsiveness to thyrotrophin-releasing hormone in the mare. *Proceedings of the Society of Experimental Biology and Medicine* 184(11), 118–122.

Johnson, L. (1991a) Spermatogenesis. In: Cupps, D.T. (ed.) *Reproduction in Domestic Animals*. 3rd edn. Academic Press, New York, pp. 173–219.

Johnson, L. (1991b) Seasonal differences in equine spermatogenesis. *Biology of Reproduction* 44, 284–291.

Johnson, L.A. (2000) Sexing mammalian sperm for production of offspring: the state of the art. *Animal Reproduction Science* 60, 93–107.

Johnson, L. and Neaves, W.B. (1981) Age related changes in leidig cell populations, seminiferous tubules and sperm production in stallions. *Biology of Reproduction* 24, 703–712.

Johnson, L. and Nguyen, H.B. (1986) Annual cycle of the Sertoli cell population in adult stallions. *Journal of Reproduction and Fertility* 76, 311–316.

Johnson, L. and Tatum, M.E. (1989) Temporal appearance of seasonal changes in numbers of sertoli cells, leidig cells and germ cells in stallions. *Biology of Reproduction* 40, 994–999.

Johnson, L. and Thompson, D.L. (1983) Age related and seasonal variation in the sertoli cell population, daily sperm production and serum concentration of follicle stimulating hormone, luteinising hormone and testosterone in stallions. *Biology of Reproduction* 29, 777–789.

Johnson, L.A., Flook, J.P. and Hawk, H.W. (1989) Sex preselection in rabbits: live birth from X and Y sperm separated by DNA and cell sorting. *Biology of Reproduction* 41, 199–203.

Johnson, L., Varner, D.D. and Thompson, D.L. (1991) Effect of age and season on the establishment of spermatogenesis in the horse. *Journal of Reproduction and Fertility, Supplement* 44, 87–97.

Johnson, L., Blanchard, T.L., Varner, D.D. and Scrutchfield, W.I. (1997) Factors affecting spermatogenesis in the stallion. *Theriogenology* 48(7), 1199–1216.

Johnson, L.A., Welch, G.R., Rens, W. and Dobrinsky, J.R. (1998) Enhanced cytometric sorting of mammalian X and Y sperm: high speed sorting and orientating nozzle for artificial insemination. *Theriogenology* 49(1), 361.

Jones, C.T. and Rolph, T.P. (1985) Metabolism during fetal life: a functional assessment of metabolic development. *Physiology Review* 65(2), 357–430.

Jones, P.A. (1978) *Educating Horses, from Birth to Riding*. Ring Press Books, Letchworth, UK, p.111.

Jones, W.E. (1995) Visualising the uterus with ultrasound. *Journal of Equine Veterinary Science* 15(7), 302–304.

Jordon, R.M. (1982) Effect of weight loss of gestating mares on subsequent production. *Journal of Animal Science* 55(Supplement 1), 208.

Juzawiak, J.J., Slone, D.E., Santschi, E.M. and Moll, H.D. (1990) Cesarean section in 19 mares. Results and postoperative fertility. *Veterinary Surgery* 19(1), 50–52.

Kainer, R.A. (1993) Reproductive organs of the mare. In: Mckinnon, A.O. and Voss, J.L. (eds) *Equine Reproduction*. Lea and Febiger, Philadelphia, pp. 3–19.

Kamenev, N. (1955) Opyt po primeneniju molocnogo razbavitelja. (The use of milk diluent.) *Konevodstvo,* 25(2), 32–34 (Animal Breeding Abstracts, 23, 1041).

Kankofer, M., Kolm, G., Aurich, J. and Aurich, C. (2005) Activity of glutathione peroxidase, superoxide dismutase and catalase and lipid peroxidation intensity in stallion semen during storage at 5 degrees C. *Theriogenology* 15(63), 1354–1365.

Kareskoski, A.M., Reilas, T., Andersson, M. and Katila, T. (2006) Motility and plasma membrane integrity of spermatozoa in fractionated stallion ejaculates after storage. *Reproduction in Domestic Animals* 41(1), 33–38.

Kask, K., Odensvik, K. and Kindahl, H. (1997) Prostaglandin F release associated with an embryo transfer procedure in the mare. *Equine Veterinary Journal* 29, 286–289.

Kastelic, J.P., Adams, G.P. and Ginther, O.J. (1987) Role of progesterone in mobility, fixation, orientation, and survival of the equine embryonic vesicle. *Theriogenology* 27(4), 655–663.

Katila, T. (1997) Procedures for handling fresh stallion semen. *Theriogenology* 48(7), 1217–1227.

Katila, T., Celebi, M. and Koskinen, E. (1996) Effect of timing of frozen semen insemination on pregnancy rate in mares. *Acta Veterinaria Scandinavica* 37(3), 361–365.

Katila, T., Combes, G.B., Varner, D.D. and Blanchard, T.L. (1997) Comparison of three containers used for the transport of cooled stallion semen. *Theriogenology* 48(7), 1085–1092.

Kayser, J.P., Amann, R.P., Sheideler, R.K., Squires, E.L., Jasko, D.J. and Pickett, B.W. (1992) Effects of linear cooling rates on motion characteristics of stallion spermatozoa. *Theriogenology* 38, 601–614.

Keifer, N.M. (1976) Male pseudohermaphroditism of the testicular feminizing type in a horse. *Equine Veterinary Journal* 8, 38–41.

Keiper, R. and Houpt, K. (1984) Reproduction in feral horses: an eight year study. *American Journal Veterinary Research* 45(5), 991–995.

Kelly, S.M., Buckett, W.M., Abdul-Jalil, A.K. and Tan, S.L. (2003) The cryobiology of assisted reproduction. *Minerva Gynecology* 55(5), 389–398. Review.

Kenney, R.M. (1975a) Clinical fertility evaluation of the stallion. *Proceedings of the 21st American Association of Equine Practitioners*, pp. 336–355.

Kenney, R.M. (1975b) Prognostic value of endometrial biopsy of the mare. *Journal of Reproduction and Fertility, Supplement* 23, 347–348.

Kenney, R.M. (1977) Clinical aspects of endometrium biopsy in fertility evaluation of the mare. *Proceedings of the 23rd Annual Meeting American Association of Equine Practitioners*, pp. 355–362.

Kenney, R.M. (1978) Cyclic and pathological changes of the mare endometrium as detected by biopsy, with a note on early embryonic death. *Journal of the American Veterinary Medical Association* 172(3), 241–262.

Kenney, R.M. (1990) Estimation of stallion fertility: the use of sperm cromatin structure assay, DNA Index and karyotype as adjuncts to traditional tests. *Proceedings of the 5th International Symposium on Equine Reproduction*, pp. 42–43.

Kenney, R.M. and Ganjam, V.K. (1975) Selected pathological changes of the mare's uterus and ovary. *Journal of Reproduction and Fertility, Supplement* 23, 335–339.

Kenney, R.M., Kingston, R.S., Rajamannam, A.H. and Ramburg, C.F. (1971) Stallion semen characteristics for predicting fertility. *Proceedings of the 17th Annual Convention of the American Association of Equine Practitioners*, pp. 53–67.

Kenney, R.M., Bergman, R.V., Cooper, W.L. and Morse, G.W. (1975a) Minimal contamination techniques for breeding mares: techniques and preliminary findings. *Proceedings of 21st Annual Convention of the American Association of Equine Practitioners*, pp. 327–335.

Kenney, R.M., Ganjam, V.K., Cooper, W.R. and Lauderdale, J.W. (1975b) The use of PGF2x – THAM salt in mares in clinical anoestrus. *Journal of Reproduction and Fertility, Supplement* 23, 247–250.

Kenney, R.M., Condon, W.A., Ganjam, J.K. and Channing, C. (1979) Morphological and biochemical correlates of equine ovarian follicles as a function of their state of viability or atresia. *Journal of Reproduction and Fertility, Supplement* 27, 163–171.

Kenney, R.M., Evenson, D.P., Garcia, M.C. and Love, C.C. (1995) Relationships between sperm chromatin structure, motility and morphology of ejaculated sperm and seasonal pregnancy rate. *Biology of Reproduction Monograph Equine Reproduction VI*, 1, 647–653.

Kiley-Worthington, M. and Wood-Gush, D. (1987) Stereotypic behavior. In: Robinson, N.E. (ed.) *Current Therapy in Equine Medicine*. 2nd edn. W.B. Saunders, Philadelphia, pp. 131–134.

Kilmer, D.N., Sharp, D.C., Berhund, L.A., Grubaugh, W., McDowell, K.J. and Peck, L.S. (1982) Melatonin rhythms in pony mares and foals. *Journal of Reproduction and Fertility, Supplement* 32, 303–307.

Kindahl, H., Knudsen, O., Madej, A. and Edquist, L.E. (1982) Progesterone, prostaglandin F2 alpha, PMSG and oestrone sulphate during early pregnancy in the mare. *Journal of Reproduction and Fertility, Supplement* 32, 353–359.

Klingel, H. (1982) Social organisation of feral horses. *Journal of Reproduction and Fertility, Supplement* 32, 89–95.

Klug, E. (1992) Routine artificial applications in the Hannovarian sport horse breeding association. *Animal Reproduction Science* 28, 39–44.

Knottenbelt, D. and Pascoe, R. (2003) Routine stud management procedures. In: Knottenbelt, D.C., LeBlanc, M., Lopate, C.L. and Pascoe, R.R. (eds) *Equine Stud Farm Medicine and Surgery*. W.B. Saunders, Philadelphia, pp. 25–41.

Knottenbelt, D.C., Holdstock, N. and Madigan, J.E. (2004) *Equine Neonatology, Medicine and Surgery*. W.B. Saunders, Philadelphia, 508 pp.

Knowles, J. (1993) *The ABC of Breaking and Schooling Horses*. J.A. Allen, London, 307 pp.

Knudsen, O. and Velle, W. (1961) Ovarian oestrogen levels in the non-pregnant mare. Relationship to histological appearance of the uterus and its clinical status. *Journal of Reproduction and Fertility* 2, 130–137.

Knutti, B., Pycock, J.F., Van Der Weijden, G.C. and Kupfer, U. (2000) The influences of early post breeding uterine lavage on pregnancy rates in mares with intrauterine fluid accumulations after breeding. *Equine Veterinary Education* 12(5), 267–276.

Koene, M.H., Boder, H. and Hoppen, H.O. (1990) Feasibility of using HMG as a superovulatory drug in a commercial embryo transfer programme. *Journal of Reproduction and Fertility, Supplement* 44, 710–711.

Koets, A.P. (1995) The equine endometrial cup reaction: a review. *Veterinary Quaterly* 17, 21–29.

Kohn, C.W., Knight, D., Hueston, W., Jacobs, R. and Reed, S.M. (1989) Cholesterol and serum IgG, IgA and IgM concentrations in Standardbred mares and their foals at parturition. *Journal of the American Veterinary Medical Association* 195, 64–68.

Kolter, L. and Zimmermann, W. (1988) Social behaviour of Przewalski *(Equus przewalskii)* in the Cologne zoo and its consequences for management and housing. *Applied Animal Behaviour Science* 21, 117–124.

Ko, T.H. and Lee, S.E. (1993) Establishment of a biological assay for the fertilising ability of equine spermatozoa. *Korean Journal of Animal Science* 35(3), 175–179.

Kooistra, L.H. and Ginther, O.J. (1975) Effects of photoperiod on reproduction activity and hair in mares. *American Journal of Veterinary Research* 36, 1413–1419.

Kooistra, L.H. and Loy, R.G. (1968) Effects of artificial lighting regimes on reproductive patterns in mares. *Proceedings of the Annual Convention of the Association of Equine Practitioners*, pp. 159–169.

Korenman, S.G. (1998) New insights into erectile dysfunction: a practical approach. *American Journal Medicine* 105, 135–144.

Kosiniak, K. (1975) Characteristics of successive jets of ejaculated semen of stallions. *Journal of Reproduction and Fertility, Supplement* 23, 59–61.

Koskinen, E., Martila, P. and Katila, T. (1997) Effect of 19-norandrostenololylaurate on semen characteristics of colts. *Acta Veterinaria Scandinavica* 38(1), 41–50.

Koterba, A.M. (1990) Physical examination. In: Koterba, A.M., Drummond, W.H. and Kosch, P.C. (eds) *Equine Clinical Neonatology*. Lea and Febiger, Philadelphia, pp. 71–81.

Koterba, A.M. and Kosch, P.C. (1987) Respiratory mechanisms and breathing patterns in the neonatal foal. *Journal of Reproduction and Fertility* 35, 575–586.

Kotilainen, T., Huhtinen, M. and Katila, T. (1994) Sperm induced leukocytosis in the equine uterus. *Theriogenology* 41(3), 629–636.

Kreider, J.L., Tindall, W.C. and Potter, G.D. (1985) Inclusion of bovine serum albumin in semen extenders to enhance maintenance of stallion spermatozoa motility. *Theriogenology* 23, 399–408.

Kubiak, J.R., Crawford, B.H., Squires, E.L., Wrigley, R.H. and Ward, G.M. (1987) The influence of energy intake and percentage of body fat on the reproductive performance of non pregnant mares. *Theriogenology* 28, 587–598.

Kubien, E.M. and Tischner, M. (2002) Reproductive success of a mare with a mosaic karyotype: 64,XX/65,XX, +30. *Equine Veterinary Journal* 34(1), 99–100.

Kuklin, A.D. (1983) Artificial insemination of horses. *Vetereinariya, Moscow, USSR* 7, 57–58.

Kullander, S., Arvidson, G., Ekelung, L. and Astedt, B. (1975) A review of surfactant principles in the foetal physiology of man and animals. *Journal of Reproduction and Fertility, Supplement* 23, 659–661.

Kumi-Diaka, J. and Badtram, G. (1994) Effect of storage on sperm membrane integrity and other functional characteristics of canine spermatozoa: in vitro bioassay for canine semen. *Theriogenology* 41, 1355–1366.

Ladd, P.W. (1985) The male genital system. In: Jubb, K.V.F., Kennedy, P.C. and Palmer, N. (eds) *Pathology of Domestic Animals*, Vol. 3, 3rd edn. Academic Press, New York, pp. 409–459.

Lagares, M.A., Petzoldt, R., Sieme, H. and Klug, E. (2000) Assessing equine sperm-membrane integrity. *Andrologia* 32(3), 163–167.

Lagreaux, D. and Palmer, E. (1989) Are pony and larger mares similar as recipients for non-surgical transfer of day 7 embryos. *Equine Veterinary Journal, Supplement* 8, 64–67.

Lamming, G.E. and Mann, G.E. (1995) Control of endometrial oxytocin receptors and prostaglandin F2∝ production in cows by progesterone and oestradiol. *Journal of Reproduction and Ferility* 103, 69–73.

Lapin, D.R. and Ginther, O.J. (1977) Induction of ovulation and multiple ovulations in seasonally anovulatory and ovulatory mares with equine pituitary extract. *Journal of Animal Science* 44(5), 834–842.

Lasley, B., Ammon, D., Daels, P., Hughes, J., Munro, C. and Stadenfeldt, G. (1990) Estrogen conjugate concentrations in plasma and urine reflect estrogen secretion in non-pregnant and pregnant mare: a review. *Journal of Equine Veterinary Science* 10, 444–448.

Lawson, K.A. (1996) Longevity of stallion semen in various extenders when chilled to 4°C. MSc thesis, University of Wales, Aberystwyth, UK.

Lawson, K.A. and Davies Morel, M.C.G. (1996) The use of mare's milk as a seminal extender for chilled stallion semen. *Warwick Horse Conference February 1996 New Developments in Equine Studies*. Royal Agricultural Society, Kenilworth, UK, pp. 26–39.

Lazzari, G., Crotti, G., Turnin, P., Duchi, R., Mari, G. and Zavaglia, G. (2002) Equine embryos at the compacted morula and blastocyst stage can be obtained by intracytoplasmic sperm injection (ICSI) of in vitro matured oocytes with frozen-thawed spermatozoa from semen of different fertilities. *Theriogenology* 58, 709–712.

Bibliography

Leadon, D.P., Jeffcott, L.B. and Rossdale, P.D. (1984) Mammary secretions in normal spontaneous and induced premature parturition in the mare. *Equine Veterinary Journal* 16(4), 256–259.

Le Blanc, M.M. (1990) Immunologic considerations. In: Koterba, A.M., Drummond, W.H. and Kosch, P.C. (eds) *Equine Clinical Neonatology*. Lea and Febiger, Philadelphia, pp. 275–296.

Le Blanc, M.M. (1991) Reproductive system: the mare. In: Colahan, P.T., Mayhew, I.G., Merritt, A.M. and Moore, J.N. (eds) *Equine Medicine and Surgery*, Vol. II, 4th edn. American Veterinary Publications, Inc. Gdefa, CA. pp. 1148–1216.

Le Blanc, M.M. (1993a) Vaginal examination. In: McKinnon, A.O. and Voss, J.L. (eds) *Equine Reproduction*. Lea and Febiger, Philadelphia, pp. 221–224.

Le Blanc, M.M. (1993b) Endoscopy. In: McKinnon, A.O. and Voss, J.L. (eds) *Equine Reproduction*. Lea and Febiger, Philadelphia, London, pp. 255–257.

Le Blanc, M.M. (1994) Oxytocin. The new wonder drug for treatment of endometritis. *Equine Veterinary Education* 6, 39–43.

Le Blanc, M.M. (1995) Ultrasound of the reproductive tract. In: Kobluk, C.N., Ames, T.R. and Goer, R.J. (eds) *The Horse – Diseases and Clinical Management*, Vol. 2. W.B. Saunders, Philadelphia, pp. 926–935.

Le Blanc, M.M. (1997) Identification and treatment of the compromised foetus. A clinical perspective. *Equine Veterinary Journal, Supplement* 24, 100–103.

Le Blanc, M.M., Lawin, B.I. and Boswell, R. (1986) Relationships among serum immunoglobulin concentrations in foals, colostrol specific gravity and colostrol immunoglobulin concentrations. *Journal of the American Veterinary Medical Association* 186, 57–60.

Le Blanc, M.M., Tran, T. and Widders, P. (1991) Identification and opsonic activity of immunoglobulins recognising *Streptococcus zooepidemicus* antigens in uterine fluids of mares. *Journal of Reproduction and Fertility, Supplement* 44, 289–296.

Le Blanc, M.M., Newrith, L., Maurag, D., Klapstein, E. and Tran, T. (1994) Oxytocin enhances clearance of radiocolloid from the uterine lumen of reproductively normal mares and mare susceptible to endometritis. *Equine Veterinary Journal* 26(4), 279–282.

Le Blanc, M.M., Lopate, C., Knottenbelt, D. and Pascoe, R. (2004) The mare. In: Knottenbelt, D., LeBlanc, M., Lopate, C. and Pascoe, R. (eds) *Equine Stud Farm Medicine and Surgery*. W.B. Saunders, Philadelphia, pp. 113–211.

Lee, H.E. and Morris, L.H.A. (2005) Challenges facing sex preselection of stallion spermatozoa. *Animal Reproduction Science* 89(1–4), 147–157.

Leendertse, I.P., Asbury, A.C., Boening, K.J. and von Saldern, F.C. (1990) Successful management of persistent urination during ejaculation in a Thoroughbred stallion. *Equine Veterinary Education* 2(2), 62–64.

Legrand, E., Bencharif, D., Barrier-Battut, I., Delajarraud, H. and Bruyas, J.-F. (2002) Comparison of pregnancy rates for days 7–8 equine embryos frozen in glycerol with or without previous enzymatic treatment of their capsule. *Theriogenology* 58, 721–723.

Leipold, H.W. (1986) Cryptorchidism in the horse: genetic implications. *Proceedings of the American Association of Equine Practitioners* 31, 579–590.

Leipold, H.W. and Dennis, S.M. (1993) Congenital defects in foals. In: McKinnon, A.O. and Voss, J.L. (eds) *Equine Reproduction*. Lea and Febiger, Philadelphia, pp. 604–613.

Ley, W.B. (1989) Daytime foaling management of the mare. 2. Induction of parturition. *Journal of Equine Veterinary Science* 9, 95–100.

Ley, W.B., Bowen, J.M., Purswell, B.J., Irby, M. and Greive-Candell, K. (1993) The sensitivity, specificity and predictive value of measuring calcium carbonate in mare's prepartum mammary secretions. *Theriogenology* 40, 189–198.

Li, L.Y., Meintjes, M., Graff, K.J., Paul, J.B., Denniston, R.S. and Godke, R.A. (1995) In vitro fertilisation and development of in vitro matured oocytes aspirated from pregnant mares. *Biology of Reproduction, Monographs Series* 1, 309–317.

Li, X., Morris, L.H. and Allen, W.R. (2000) Effects of different activation treatments on fertilization of horse oocytes by intracytoplasmic sperm injection. *Journal of Reproduction and Fertility* 119(2), 253–260.

Li, X., Morris, L.H. and Allen, W.R. (2001) Influence of co-culture during maturation on the developmental potential of equine oocytes fertilized by intracytoplasmic sperm injection (ICSI). *Reproduction* 121(6), 925–932.

Lieberman, R.J. and Bowman, T.R. (1994) Teasing naturally. Strategies to improve your mare's oestrus response. *Modern Horse Breeding* 11, 28–31.

Lieux, P. (1970) Relationship between the appearance of the cervix and the heat cycle in the mare. *Veterinary Medicine Small Animal Clinics* 65, 859–866.

Lieux, P. (1972) Reproductive and genital disease. In: Catcott, E.J. and Smithcors, J.F. (eds) *Equine Medicine and Surgery*, 2nd edn. American Veterinary Publications Book, Wheaton, Illinois, pp. 567–589.

Liggins, G.C. (1979) Initiation of parturition. *British Medical Bulletin* 35(2), 145–150.

Liggins, G.C., Ritterman, J.A. and Forster, C.S. (1979) Foetal maturation related to parturition. *Animal Reproduction Science* 2, 193–199.

Lindsey, A.C., Bruemmer, J.E. and Squires, E.L. (2001) Low dose insemination of mares using non-sorted and sex-sorted sperm. *Animal Reproduction Science* 68(3–4), 279–289.

Lindsey, A.C., Schenk, J.L., Graham, J.K., Bruemmer, J.E. and Squires, E.L. (2002) Hysteroscopic insemination of low numbers of flow sorted fresh and frozen thawed spermatozoa. *Equine Veterinary Journal* 34, 121–127.

Lindsey, A.C., Varner, D.D., Seidel, G.E. Jr., Bruemmer, J.E. and Squires, E.L. (2005) Hysteroscopic or rectally guided, deep uterine insemination of mares with spermatozoa stored 18 h at either 5°C or 15°C prior to flow cytometric sorting. *Animal Reproduction Science* 85, 125–130.

Little, T.V. (1998) Accessory sex glands and internal reproductive tract evaluation. In: Rantanen, N.W. and McKinnon, A.O. (eds) *Equine Diagnostic Ultrasonography.* Williams and Wilkins, Baltimore, Maryland, pp. 271–288.

Little, T.V. and Woods, G.L. (1987) Ultrasonography of accessory sex glands in the stallion. *Journal of Reproduction and Fertility, Supplement* 35, 87–94.

Littlejohn, A. and Ritchie, J.D.S. (1975) Rupture of caecum at parturition. *Journal of the South African Veterinary Association* 46(1), 87.

Lofstedt, R.M. (1988) Control of estrous cycle in the mare. *Veterinary Clinics of North America Equine Practitioners* 4(2), 177–196.

Lofstedt, R.M. (1993) Miscellaneous diseases of pregnancy and parturition. In: McKinnon, A.O. and Voss, J.L. (eds) *Equine Reproduction.* Lea and Febiger, Philadelphia, pp. 596–603.

Lofstedt, R.M. and Newcombe, J.R. (1997) Pregnancy diagnosis and subsequent examinations in mares: when and why. *Equine Veterinary Education* 9(6), 293–294.

Logan, N.L., McCue, P.M., Alonso, M.A. and Squires, E.L. (2007) Evaluation of three equine FSH superovulation protocols in mares. *Animal Reproduction Science* 102(1–2), 48–55.

Lombard, C.W. (1990) Cardio vascular diseases. In: Koterba, A.M., Drummond, W.H. and Kosch, P.C. (eds) *Equine Clinical Neonatology.* Lea and Febiger, Philadelphia, pp. 240–261.

Lombard, C.W., Evans, M., Martin, L. and Tehrani, J. (1984) Blood pressure, electrocardiogram and echocardiogram measurements in the growing pony foal. *Equine Veterinary Journal* 16, 342–347.

Long, S. (1988) Chromosomal anomalies and infertility in the mare. *Equine Veterinary Journal* 20(2), 89–93.

Long, M.T., Ostund, E.N., Porter, M.B. and Crom, R.L. (2002) Equine West Nile Encephalitis: epidemiological and clinical review for practitioners. *American Association of Equine Practitioners* 48, 1–6.

Loomis, P.R. (1993) Factors affecting the success of AI with cooled, transported semen, *Proceedings of 38th Annual Convention of American Association of Equine Practitioners* 2, pp. 629–647.

Loomis, P.R., Amann, R.P., Squires, E.L. and Pickett, B.W. (1984) Fertility of unfrozen and frozen stallion spermatozoa extended in EDTA-lactose-egg yolk and packaged in straws. *Journal of Animal Science* 56, 687–693.

Lopate, C., Le Blanc, M., Pascoe, R. and Knottenbelt, D. (2003) Parturition. In: Knottenbelt, D., Le Blanc, M., Lopate, C. and Pascoe, R.R. (eds) *Equine Stud Farm Medicine and Surgery.* W.B. Saunders, Philadelphia, pp. 269–324.

Lopate, C., LeBlanc, M. and Knottenbelt, D. (2004) The stallion. In: Knottenbelt, D., LeBlanc, M., Lopate, C. and Pascoe, R. (eds) *Equine Stud Farm Medicine and Surgery.* W.B. Saunders, Philadelphia, pp. 43–113.

Lorch, J. (1998) *Foal to Full Grown.* David and Charles, Newton Abbott, Devon, UK, 191 pp.

Lorton, S.P., Sullivan, J.J., Bean, B., Kaprotl, M., Kellgren, M. and Marshall, C. (1988a) A new antibiotic combination for frozen bovine semen. 2. Evaluation of seminal quality. *Theriogenology* 29, 593–607.

Lorton, S.P., Sullivan, J.J., Bean, B., Kaprotl, M., Kellgren, M. and Marshall, C. (1988b) A new antibiotic combination for frozen bovine semen. 3. Evaluation of fertility. *Theriogenology* 29, 609–614.

Love, C.C. (1992) Semen collection techniques. *Veterinary Clinics of North America, Equine Practice* 8(1), 111–128.

Love, C.C., Loch, W.L., Bristol, F., Garcia, M.C. and Kenney, R.M. (1989) Comparison of pregnancy rates achieved with frozen semen using two packaging methods. *Theriogenology* 31, 613–622.

Love, C.C., Garcia, M.C., Riera, F.R. and Kenney, R.M. (1991) Evaluation of measurements taken by ultrasonography and calipur to estimate testicular volume and predict daily sperm output in the stallion. *Journal of Reproduction and Fertility, Supplement* 44, 99–105.

Love, C.C., Thompson, J.A., Lowry, V.K. and Varner, D.D. (2002) Effect of storage time and temperature on stallion sperm DNA and fertility. *Theriogenology* 57, 1135–1142.

Love, C.C., Thompson, J.A., Brinsko, S.P., Rigby, S.L., Blanchard, T.L., Lowry, V.K. and Varner, D.D. (2003) Relationship between stallion sperm motility and viability as detected by two fluorescence staining techniques using flow cytometry. *Theriogenology* 60(6), 1127–1138.

Love, C.C., Brinsko, S.P., Rigby, S.L., Thompson, J.A., Blanchard, T.L. and Varner, D.D. (2005) Relationship of seminal plasma level and extender type to sperm motility and DNA integrity. *Theriogenology* 63(6), 1584–1591.

Lowe, J.N. (2001) Diagnosis and management of urospermia in a commercial Thoroughbred stallion. *Equine Veterinary Education* 13(1), 4–7.

Lowis, T.C. and Hyland, J.H. (1991) Analysis of post partum fertility in mares on a Thoroughbred stud in South Victoria. *Australian Veterinary Journal* 68(9), 304–306.

Loy, R.G. (1980) Characteristics of post partum reproduction in mares. *Veterinary Clinics of North America, Large Animal Practice* 2, 345–358.

Loy, R.G. and Swann, S.M. (1966) Effects of exogenous progestagens on reproductive phenomena in mares. *Journal of Animal Science* 25, 821–825.

Loy, R.G., Buell, J.R., Stevenson, W. and Hamm, D. (1979) Sources of variation in response intervals after prostaglandin treatment in mares with functional corpus lutea. *Journal of Reproduction and Fertility, Supplement* 27, 229–235.

Loy, R.G., Pernstein, R., O'Conna, D. and Douglas, R.H. (1981) Control of ovulation in cycling mares with ovarian steroids and prostaglandin. *Theriogenology* 15, 191–197.

Lubbeke, M., Klug, E., Hoppen, H.O. and Jochle, W. (1994) Attempts to synchronise estrus and ovulation in mares using progesterone (CIDR-B) and GnRH analogue deslorelin. *Reproduction in Domestic Animals* 29, 305–314.

Lui, I.K.M., Lantz, K.C. and Schlafke, S. (1990) Clinical observations of oviductal masses. *Proceedings of the 36th Annual Conference for the American Association of Equine Practitioners*, Lexington, Kentucky, pp. 41–50.

MacDonald, A.A., Fowden, A.L., Silver, M., Ousey, J. and Rossdale, P.D. (1988) The foramen ovale of the foetal and neonatal foal. *Equine Veterinary Journal* 20, 255–260.

MacDonald, A.A. and Fowden, A.L. (1997) Microscopic anatomy of the ungulate placenta. *Equine Veterinary Journal, Supplement* 24, 7–13.

Machida, N., Yasuda, J., Too, K. and Kudo, N. (1998) A morphological study on the obliteration of the dustus arteriosus in the horse. *Equine Veterinary Journal* 20(4), 249–254.

Maclellan, L.J.M., Carnevale, E.M., Coutinho da Silva, M.A., Scoggin, C.F., Bruemmer, J.E. and Squires, E.L. (2002) Pregnancies from vitrified equine opocytes collected from superstimulated and non-stimulated mares. *Theriogenology* 5, 911–919.

Macpherson, M.L. and Reimer, J.M. (2000) Twin reduction in the mare: current options. *Animal Reproduction Science* 60–61, 233–244.

Madigan, J.E. (1990) Management of the newborn foal. *Proceedings of the 36th Annual Convention of the American Association of Equine Practitioners*, pp. 99–116.

Madill, S. (2002) Reproductive considerations: mare and stallion. *Veterinary Clinics of North America Equine Practice* 18(3), 591–619. Review.

Madill, S., Troedsson, M.H.T., Alexander, S.L., Shand, N., Santaschi, E.M. and Irvine, C.H.G. (2000) Simultaneous recording of pituitary oxytocin secretion and myometrial activity in oestrus mares exposed to various breeding stimuli. *Journal of Reproduction and Fertility, Supplement* 56, 351–361.

Madsen, M. and Christensen, P. (1995) Bacterial flora of semen collected from Danish warmblood stallions by artificial vagina. *Acta Veterinaria Scandinavia* 36(1), 1–7.

Magistrini, M. and Vidamnet, M. (1992) Artificial insemination in horses. *Recueil de Medicine Veterinaire Special: Reproduction des Equides* 168(11–12), 959–967.

Magistrini, M., Chanteloube, P. and Palmer, E. (1987) Influence of season and frequency of ejaculation on the production of stallion semen for freezing. *Journal of Reproduction and Fertility, Supplement* 35, 127–133.

Magistrini, M., Couty, I. and Palmer, E. (1992) Interactions between sperm packaging, gas environment, temperature and diluent on fresh stallion sperm survival. *Acta Veterinaria Scandinavia, Supplement* 88, 97–110.

Magistrini, M., Guitton, E., Levern, Y., Nicolle, J.C., Vidament, M., Kerboeuf, D. and Palmer, E. (1997) New staining methods for sperm evaluation estimated by microscopy and flow cytometry. *Theriogenology* 48, 1229–1235.

Mahon, G.A.T. and Cunningham, E.P. (1982) Inbreeding and the inheritance of fertility in the Thoroughbred mare. *Livestock Production Science* 9, 743–754.

Makinen, A., Katila, T., Anderson, M. and Gustavsson, I. (2000) Two sterile stallions with XXY syndrome. *Equine Veterinary Journal* 32(4), 358–360.

Malacarne, M., Martuzzi, F., Summer, A. and Mariani, P. (2002) Review: protein and fat composition of mare's milk: some nutritional remarks with reference to human and cow's milk. *International Dairy Journal* 12, 869–877.

Malinowski, K., Halquist, N.A., Heylar, L., Sherman, A.R. and Scanes, C.G. (1990) Effect of different separation protocols between mares and foals on plasma cortisol and cell mediated immune response. *Proceedings of the 11th Equine Nutrition and Physiology Symposium* 10(5), 363–368.

Malmgren, L. (1992) Ultrasonography: a new diagnostic tool in stallions with genital tract infection? *Acta Veterinaria Scandinavia, Supplement* 88, 91–94.

Malmgren, L. (1997) Assessing the quality of raw semen: a review. *Theriogenology* 48, 523–530.

Malmgren, L., Kamp, B., Wockener, A., Boyle, M. and Colenbrander, B. (1994) Motility, velocity and acrosome integrity of equine spermatozoa stored under different conditions. *Reproduction in Domestic Animals* 29(7), 469–476.

Mann, T. (1975) Biochemistry of stallion semen. *Journal of Reproduction and Fertility, Supplement* 23, 25–28.

Manning, S.T., Bowman, P.A., Fraser, L.M. and Card, C.E. (1998) Development of hysteroscopic insemination of the uterine tube in the mare. *Proceedings of Annual Meeting Society for Theriogenology*, Baltimore, Maryland, pp. 84–85.

Manz, E., Vogel, T., Glatzel, P. and Schmidtke, J. (1998) Identification of equine Y chromosome specific to gene locus (ETSPY) with potential in preimplantation sex diagnosis. *Theriogenology* 49(1), 364–370.

Mari, G., Castagnetti, C. and Belluzzi, S. (2002) Equine fetal sex determination using a single ultrasonic examination under farm conditions. *Theriogenology* 58, 1237–1243.

Mari, G., Iancono, E., Merlo, B. and Castagnetti, C. (2004) Reduction of twin pregnancy in the mare by transvaginal ultrasound-guided aspiration. *Reproduction in Domestic Animals* 39(6), 434–439.

Mari, G., Barbara, M., Eleonora, I. and Stefano, B. (2005) Fertility in the mare after repeated transvaginal ultrasound-guided aspirations. *Animal Reproduction Science* 88(3–4), 299–308.

Martin, J.C., Klug, E. and Gunzel, A.R. (1979) Centrifugation of stallion semen and its storage in large volume straws. *Journal of Reproduction and Fertility* 27, 47–51.

Martin, R.G., McMeniman, N.P. and Dowsett, K.F. (1991) Effects of a protein deficient diet and urea supplementation on lactating mares. *Journal of Reproduction and Fertility, Supplement* 44, 543–550.

Martin, R.G., McMeniman, N.P. and Dowsett, K.F. (1992) Milk and water intakes of foals suckling grazing mares. *Equine Veterinary Journal* 24, 295–299.

Mather, E.C., Refsal, K.R., Gustafisson, B.K., Seguin, B.E. and Whitmore, H.C. (1979) The use of fiber-optic techniques in clinical diagnosis and visual assessment of experimental intrauterine therapy in mares. *Journal of Reproduction and Fertility, Supplement* 25, 293–297.

Mathews, R.G., Rophia, R.T. and Butterfield, R.M. (1967) The phenomenon of foal heat in mares. *Australian Veterinary Journal* 43, 579–585.

Mayhew, I.G. (1990) Neurological aspects of urospermia in the horse. *Equine Veterinary Education* 2(2), 68–69.

Mazur, P. (1984) Freezing of living cells. Mechanisms and implications. *American Journal of Physiology* 247, C125–C142.

McAllister, R.A. and Sack, W.O. (1990) Identification of anatomical features of the equine clitoris as potential growth sites for *Taylorella equigenitalis*. *Journal of American Veterinary Medicine Association* 196(12), 1965–1966.

McBride, S.D. and Hemmings, A. (2005) Altered mesoaccumbens and nigro-striatal dopamine physiology is associated with stereotypy development in a non-rodent species. *Behaviour Brain Research* 159(1), 113–118.

McCall, C.A., Potter, G.D. and Kreidel, J.L. (1985) Locomotor, vocal and other behavioural responses to varying methods of weaning foals. *Applied Animal Behaviour Science* 14(1), 27–35.

McCall, C.A., Potter, G.D., Kreidel, J.L. and Jenkins, W.L. (1987) Physiological responses in foals weaned by abrupt and gradual methods. *Journal of Equine Veterinary Science* 7(6), 368–374.

McCaughey, W.J., Hanna, J. and O'Brien, J.J. (1973) A comparison of three laboratory tests for pregnancy diagnosis in the mare. *Equine Veterinary Journal* 5, 94–95.

McClure, C.C. (1993) The immune system. In: McKinnon, A.O. and Voss, J.L. (eds) *Equine Reproduction*. Lea and Febiger, Philadelphia, pp. 1003–1016.

McCue, P.M. (1993) Lactation. In: McKinnon, A.O. and Voss, J.L. (eds) *Equine Reproduction*. Lea and Febiger, Philadelphia, pp. 588–595.

McCue, P.M., Farquhar, V.J., Carnevale, E.M. and Squires, E.L. (2002) Removal of deslorelin (Ovuplant) implant 48 h after administration results in normal interovulatory intervals in mares. *Theriogenology* 58(5), 865–870.

McCue, P.M., LeBlanc, M.M. and Squires, E.L. (2007) eFSH in clinical equine practice. Review. *Theriogenology* 68(3), 429–433.

McDonnell, S.M. (1992) Normal and abnormal sexual behaviour. *Veterinary Clinics of North America Equine Practitioners* 8, 71–89.

McDonnell, S.M. (2000a) Reproductive behaviour of stallion and mares: comparison of free-running and domestic in-hand breeding. *Animal Reproduction Science* 60–61, 211–219.

McDonnell, S.M. (2000b) Stallion sexual behaviour. In: Samper, J.C. (ed.) *Equine Breeding Management and Artificial Insemination*. W.B. Saunders, Philadelphia, pp. 53–61.

McDonnell, S.M. and Love, C.C. (1990) Manual stimulated collection of semen from stallions. Training time, sexual behaviour and semen. *Theriogenology* 33, 1201–1210.

McDonnell, S.M. and Love, C.C. (1991) Xylazine induced ex-copulatory ejaculation in stallions. *Theriogenology* 36, 73–76.

McDonnell, S.M. and Murray, S.C. (1995) Bachelor and Harem stallion behaviour and endocrinology. *Biology of Reproduction Monographs* 1, 577–590.

McDonnell, S.M. and Turner, R.M.O. (1994) Post-thaw motility and longevity of motility of impramine-induced ejaculate of pony stallions. *Theriogenology* 42(3), 475–481.

McDonnell, S.M., Garcia, M.C., Blanchard, T.L. and Kenney, R.M. (1986) Evaluation of androgenized mares as an estrus detection aid. *Theriogenology* 26, 261–266.

McDonnell, S.M., Hinrichs, K., Cooper, W.L. and Kenney, R.M. (1988) Use of an androgenised mare as an aid in detection of oestrus in mares. *Theriogenology* 30(3), 547–553.

McDonnell, S.M., Pozor, M.A., Beech, J. and Sweeney, R.W. (1991) Use of manual stimulation for the collection of semen from an atactic stallion unable to mount. *Journal of the American Veterinary Medical Association* 199(6), 753–754.

McDowell, K.J., Sharp, D.C., Grubaugh, W., Thatcher, W.W. and Wilcox, C.J. (1988) Restricted conceptus mobility results in failure of pregnancy maintenance in mares. *Biology of Reproduction* 39(2), 340–348.

McEntee, K. (1970) The male genital system. In: Jubb, K.V.F. and Kennedy, P.C. (eds) *Pathology of Domestic Animals.* 2nd edn. Academic Press, New York, pp. 450–454.

McGreevy, P.D., Cripps, P.J., French, N.P., Green, L.E. and Nicol, C.J. (1995) Management factors associated with stereotypic and redirected behavior in the thoroughbred horse. *Equine Veterinary Journal* 27(2), 86–91.

McGuire, T.C. and Crawford, T.B. (1973) Passive immunity in the foal: measurement of immunoglobulin classes and specific antibodies. *American Journal of Veterinary Research* 34, 1299–1303.

McKinnon, A.O. (1987a) Ultrasound evaluation of the mare's reproductive tract. Part 1. *Compendium of Continuing Education for Veterinary Practitioners* 9, 336–345.

McKinnon, A.O. (1987b) Ultrasound evaluation of the mare's reproductive tract. Part 11. *Compendium of Continuing Education for Veterinary Practitioners* 9, 472–482.

McKinnon, A.O. (1987c) Diagnostic ultrasonography of uterine pathology in the mare. *Proceedings of the 33rd Annual Meeting of the American Association of Equine Practitioners,* pp. 605–622.

McKinnon, A.O. (1993) Pregnancy diagnosis. In: McKinnon, A.O. and Voss, J.L. (eds) *Equine Reproduction.* Lea and Febiger, Philadelphia, pp. 501–508.

McKinnon, A.O. (1998) Uterine pathology. In: Rantanen, N.W. and McKinnon, A.O. (eds) *Equine Diagnostic Ultrasonography.* Williams and Wilkins, Baltimore, Maryland, pp. 181–200.

McKinnon, A.O. and Carnevale, E.M. (1993) Ultrasonography. In: McKinnon, A.O. and Voss, J.L. (eds) *Equine Reproduction.* Lea and Febiger, Philadelphia, London, pp. 211–220.

McKinnon, A.O. and Squires, E.L. (1988) Equine E.T. *Equine Practice* 4, 305–333.

McKinnon, A.O. and Voss, J.L. (1993) Breeding the problem mare. In: McKinnon, A.O. and Voss, J.L. (eds) *Equine Reproduction.* Lea and Febiger, Philadelphia, pp. 369–378.

McKinnon, A.O., Lacham-Kaplan, O. and Trounson, A.O. (2000) Pregnancies produced from fertile and infertile stallions by intracytoplasmic sperm injection (ICSI) of single frozen/thawed spermatozoa into in-vivo matured oocytes. *Journal of Reproduction and Fertility, Supplement* 56, 513–517.

McKinnon, A.O., Squires, E.L., Harrison, L.A., Blach, E.L. and Shideler, R.K. (1988a) Ultrasonic studies on the reproductive tracts of mares after parturition: Effect of involution and uterine fluid on pregnancy rates in mares with normal and delayed post partum ovulatory cycles. *Journal of the American Veterinary Medical Association* 192, 350–353.

McKinnon, A.O., Voss, J.L., Trotter, G.W., Pickett, B.W., Shideler, R.K. and Squires, E.L. (1988b) Hemospermia of the stallion. *Equine Practice* 10(9), 17–23.

McKinnon, A.O., Carnevale, E.M., Squires, E.L., Voss, J.L. and Seidel, G.E. Jr. (1988c) Heterogenous and xenogenous fertilisation of in vitro matured equine oocytes. *Journal of Equine Veterinary Science* 8, 143–147.

McKinnon, A.O., Voss, J.L., Squires, E.L. and Carnevale, E.M. (1993) Diagnostic ultrasonography. In: McKinnon, A.O. and Voss, J.L. (eds) *Equine Reproduction.* Lea and Febiger, Philadelphia, London, pp. 266–302.

Meager, D.M. (1978) Granulosa cell tumours in the mare. A review of 78 cases. *Proceedings of the 23rd Annual Meeting of the American Association of Equine Practitioners.* pp. 133–139.

Meinert, C., Silva, J.F.S., Kroetz, I., Klug, E., Trigg, T.E., Hoppen, H.O. and Jochle, W. (1993) Advancing the time of ovulation in the mare with a short term implant releasing the GnRH analogue deslorelin. *Equine Veterinary Journal* 25, 65–68.

Meira, C. and Henry, M. (1991) Evaluation of two non-surgical equine embryo transfer methods. *Journal of Reproduction and Fertility, Supplement* 44, 712–713.

Melrose, R.A., Walker, R.F. and Douglas, R.H. (1990) Dopamine in the cerebral spinal fluid of prepubertal and adult horses. *Brain and Behavioural Evolution* 5, 98–106.

Mepham, B. (1987) *Physiology of Lactation.* Open University Press, Milton Keynes, 207 pp.

Merchenthaler, I., Setalo, G., Csontos, C., Petruz, P., Flerko, B. and Negro-wiler, A. (1989) Combined retrograde tracing and immunocytochemical identification of luteinising hormone-releasing hormone and somatostatin containing neurons projecting into the median eminence of the rat. *Endocrinology* 125, 2812–2820.

Merkt, H., Klug, E. and Jochle, W. (2000) Reproduction management in the German Thoroughbred industry. *Journal of Equine Veterinary Science* 20(12), 822–825, 867–868.

Merkt, H., Klug, E., Krause, D. and Bader, H. (1975) Results of long-term storage of stallion semen frozen by pellet method. *Journal of Reproduction and Fertility, Supplement* 23, 105–106.

Metcalf, E.S. (2001) The role of international transport of equine semen on disease transmission. *Animal Reproduction Science* 3(68), 229–237.

Meuten, D.J. and Rendano, V. (1978) Hypertrophic osteopathy in a mare with dysgerminoma. *Equine Medical Surgery* 2, 445–450.

Meyers, P.J. (1997) Control and synchronisation of oestrous cycles and ovulation. In: Youngquist, R.S. (ed.) *Current Therapy in Large Animal Theriogenology.* W.B. Saunders, Philadelphia, pp. 96–102.

Meyers, S.A., Liu, I.K.M., Overstreet, J.W. and Drobnis, E.Z. (1995) Induction of acrosome reactions in stallion sperm by equine zona pellucida, porcine zona pellucida and progesterone. *Biology of Reproduction Monograph Equine Reproduction VI* 1, 739–744.

Meyers, S.A., Liu, I.K.M., Overstreet, J.W., Vadas, S. and Drobnis, E.Z. (1996) Zona pellucida binding and zona induced acrosome reactions in horse spermatozoa: comparisons between fertile and sub fertile stallions. *Theriogenology* 46(7), 1277–1288.

Mihailov, N.I. (1956) Primenenie moloka v kacestve razbavitelja semini seljskohozjaistvennyh zivotnyh (Milk as a diluent for livestock semen). *Zivotnovodstvo,* 4, 74 (*Animal Breeding Abstracts* 24, 953).

Miller, A. and Woods, G.L. (1988) Diagnosis and twin correction of twin pregnancy in the mare. *Veterinary Clinics of North America Equine Practice* 4, 215–220.

Mills, D. and Nankervis, K. (1999) *Equine Behaviour, Principles and Practice.* Blackwell Science, Oxford, 232 pp.

Mitchell, D. (1973) Detection of foetal circulation in the mare and cow by Doppler ultrasound. *Veterinary Record* 93, 365–368.

Mitchell, D. and Betteridge, K.J. (1973) Persistence of endometrial cups and serum gonadotrophin following abortion in the mare. *Proceedings of the 7th International Congress for Animal Reproduction and AI.* Munich 1, pp. 567–570.

Monin, T. (1972) Vaginoplasty: a surgical treatment for urine pooling in the mare. *Proceedings of the 18th Annual Congress of the American Association of Equine Practitioners,* pp. 99–110.

Moran, D.M., Jasko, D.J., Squires, E.L. and Amann, R.P. (1992) Determination of temperature and cooling rate which induce cold-shock in stallion spermatozoa. *Theriogenology* 38, 999–1012.

Morgenthal, J.L. and Van Niekerk, C.H. (1991) Plasma progestagen levels in normal mares with luteal deficiency during early pregnancy and in twinning habitual aborters. *Journal of Reproduction and Fertility, Supplement* 44, 728–729.

Morris, L.H. (2004) Low dose insemination in the mare: an update. *Animal Reproduction Science* 82–83, 625–632.

Morris, L.H.A. (2005) Challenges facing sex preselection of stallion spermatozoa. *Animal Reproduction Science* 89(1–4), 147–157.

Morris, L.H.A. and Allen, W.R. (2002) An overview of low dose insemination in the mare. *Reproduction in Domestic Animals* 37, 206–210.

Morris, L.H.A., Hunter, R.H.F. and Allen, W.R. (2000) Hysteroscopic insemination of small numbers of spermatozoa at the uterotubal junction of preovulatory mares. *Journal of Reproduction and Fertility* 118, 95–100.

Morris, L.H.A., Tiplady, C. and Allen, W.R. (2003) Pregnancy rates in mares after a single fixed time hysteroscopic insemination of low numbers of frozen-thawed spermatozoa onto the uterotubal junction. *Equine Veterinary Journal* 35, 197–201.

Morris, R.P., Rich, G.A., Ralston, S.L., Squires, E.L. and Pickett, B.W. (1987) Follicular activity in transitional mares as affected by body condition and dietary energy. *Proceedings of the 10th Equine Nutrition and Physiology Symposium,* pp. 93–99.

Morse, M.J. and Whitmore, W.F. (1986) Neoplasms of the testis. In: Walsh, P.C., Gittes, R.F., Perlmutter, A.D. and Stomey, T.A. (eds) *Cambell's Urology,* Vol. 2, 5th edn. W.B. Saunders, Philadelphia, pp. 1535–1575.

Morehead, J.P., Colon, J.L. and Blanchard, T.L. (2001) Clinical experience with native GnRH therapy to hasten follicular development and first ovulation of the breeding season. *Journal of Equine Veterinary Science* 21, 54–84.

Moss, G.E., Estergreen, V.L., Becker, S.R. and Grant, B.D. (1979) The source of 5 pregnanes that occur during gestation in mares. *Journal of Reproduction and Fertility, Supplement* 27, 511–519.

Muller, Z. (1987) Practicalities of insemination of mares with deep frozen semen. *Journal of Reproduction and Fertility, Supplement* 35, 121–125.

Muller, Z. and Cunat, L. (1993) Special surgical transfers of horse embryos. *Equine Veterinary Journal, Supplement* 15, 113–115.

Mumford, E.L., Squires, E.L., Peterson, K.D., Nett, T.M. and Jasko, D.J. (1994) Effect of various doses of Gonadotrophin releasing hormone analogue on induction of ovulation in anostrus mares. *Journal of Animal Science* 72, 178–183.

Mumford, E.L., Squires, E.L., Jochle, E., Harrison, L.A., Nett, T.M. and Trigg, T.E. (1995) Use of deslorelin short term implants to induce ovulation in cycling mares during 3 consecutive estrous cycles. *Animal Reproduction Science* 39, 129–140.

Mumford, J.A., Hannant, D. and Jessett, D.M. (1996) Abortigenic and neurological disease caused by experimental infection with equine herpes virus. In: *Equine Infectious Diseases VII, Proceedings of the 7th International Conference.* R.W. Publications, Newmarket, UK, pp. 261–275.

Munro, C.D., Renton, J.P. and Butcher, R.A. (1979) The control of oestrous behaviour in the mare. *Journal of Reproduction and Fertility, Supplement* 27, 217–227.

Naden, J., Amann, R.R. and Squires, E.L. (1990) Testicular growth, hormone concentrations, seminal characteristics and sexual behaviour in stallions. *Journal of Reproduction and Fertility, Supplement* 88, 167–176.

Nagy, P., Guillaume, D. and Daels, P. (2000) Seasonality in mares. *Animal Reproduction Science* 60/61, 245–262.

Nash, J.G. Jr., Voss, J.L. and Squires, E.L. (1980) Urination during ejaculation in a stallion. *Journal of the American Veterinary Medicine Association* 176, 224–227.

Nathanielsz, P.W., Rossdale, P.D., Silver, M. and Comline, R.S. (1975) Studies on foetal, neonatal and maternal cortisol metabolism in the mare. *Journal of Reproduction and Fertility, Supplement* 23, 625–630.

Nathanielsz, P.W., Grussani, D.A. and Wu, W.X. (1997) Stimulation of the switch in myometrial activity from contractures to contractions in the pregnant sheep and non human primate. *Equine Veterinary Journal, Supplement* 24, 83–88.

National Research Council (1989) *Nutrient Requirements for Horses.* 5th edn. Revised. National Academy Press, National Academy of Sciences, National Research Council, Washington, DC, 100 pp.

National Research Council (2007) *Nutrient Requirements for Horses.* 6th edn. Revised. National Academy Press, National Academy of Sciences, National Research Council, Washington, DC, 315 pp.

Naumenkov, A. and Romankova, N. (1981) An improved semen diluent. *Konevodstvoi Konnyi Sport* 4, 34, *Animal Breeding Abstracts* 49, 6207.

Naumenkov, A. and Romankova, N. (1983) Improving diluent composition and handling for stallion sperm. *Nauchnye Trudy. Vsesoyuznyi Nauchno Issledovatel'skii Institut Konevodstva,* 38–47. *Animal Breeding Abstracts* 51, 6370.

Naylor, J.M. and Bell, R.J. (1987) Feeding the sick or orphaned foal. In: Robinson, N.E. (ed.) *Current Therapy in Equine Medicine.* 2nd edn. W.B. Saunders, Philadelphia, pp. 205–209.

Neely, D.P. (1983) Evaluation and therapy of genital disease in the mare. In: Hughes, J.P. (ed.) *Equine Reproduction.* Hoffman-LaRoche, Nutley, New Jersey, pp. 40–56.

Neely, D.P., Kindahl, H., Stabenfeldt, G.H., Edquist, L.-E. and Hughes, J.P. (1979) Prostaglandin release patterns in the mare: physiological, pathophysiological and therapeutic responses. *Journal of Reproduction and Fertility, Supplement* 27, 181–189.

Neild, D.M., Chaves, M.D., Flores, M., Mora, N., Beconi, M. and Aguero, A. (1999) Hyposmotic test in equine sperm. *Theriogenology* 51, 721–727.

Neild, D.M., Gadella, B.M., Chaves, M.G., Miragaya, M.H., Colenbrander, B. and Aguero, A. (2003) Membrane changes during different stages of a freeze-thaw protocol for equine cryopreservation. *Theriogenology* 59(8), 1693–1705.

Neild, D.M., Brouwers, J.F., Colenbrander, B., Aguero, A. and Gadella, B.M. (2005a) Lipid peroxidase formation in relation to membrane stability of fresh and frozen thawed stallion spermatozoa. *Molecular Reproductive Development* 72(2), 230–238.

Neild, D.M., Gadella, B.M., Aguero, A., Stout, T.A. and Colenbrander, B. (2005b) Capacitation, acrosome function and chromatin structure in stallion sperm. *Animal Reproduction Science* 89(1–4), 47–56.

Nequin, L.G., King, S.S., Johnson, A.L., Gow, G.M. and Ferreira-Dias, G.M. (1993) Prolactin may play a role in stimulating the equine ovary during the spring reproductive transition. *Equine Veterinary Science* 13, 631–635.

Nett, T.M. (1993a) Estrogens. In: McKinnon, A.O. and Voss, J.L. (eds) *Equine Reproduction.* Lea and Febiger, Philadelphia, pp. 65–68.

Nett, T.M. (1993b) Reproductive peptide and protein hormones. In: McKinnon, A.O. and Voss, J.L. (eds) *Equine Reproduction.* Lea and Febiger, Philadelphia, pp. 109–114.

Nett, T.M. (1993c) Reproductive endocrine function testing in stallions. In: McKinnon, A.O. and Voss, J.L. (eds) *Equine Reproduction.* Lea and Febiger, Philadelphia, pp. 821–824.

Nett, T.M., Holtan, D.W. and Estergreen, V.L. (1973) Plasma estrogens in pregnant mares. *Journal of Animal Science* 37(4), 962–970.

Nett, T.M., Holtan, D.W. and Estergreen, V.L. (1975) Oestrogens, LH, PMSG and prolactin in serum of pregnant mares. *Journal of Reproduction and Fertility, Supplement* 23, 457–462.

Neuschaefer, A., Bracher, V. and Allen, W.R. (1991) Prolactin secretion in lactating mares before and after treatment with bromocryptine. *Journal of Reproduction and Fertility, Supplement* 44, 551–559.

Newcombe, J.R. (1994) Conception in a mare to a single mating 7 days before ovulation. *Equine Veterinary Education* 6, 27–28.

Newcombe, J.R. (1995) Incidence of multiple ovulation and multiple pregnancy in mares. *Veterinary Record* 137, 121–123.

Newcombe, J.R. (2000) Embryonic loss and abnormalities of early pregnancy. *Equine Veterinary Education* 12(2), 88–101.

Newcombe, J.R. and Wilson, M.C. (1997) The use of progesterone releasing intravaginal devices to induce estrus and ovulation in anoestrus standardbred mares in Australia. *Equine Practice* 19(6), 13–21.

Newcombe, J.R., Lichtwark, S. and Wilson, M.C. (2005) Case report: the effect of sperm number, concentration, and volume of insemination dose chilled, stored and transported semen on pregnancy rate in Standardbred mares. *Journal of Equine Veterinary Science* 25(12), 1–6.

Nicol, C.J., Davidson, H.P., Harris, P.A., Waters, A.J. and Wilson, A.D. (2002) Study of crib-biting and gastric inflammation and ulceration in young horses. *Veterinary Record* 151(22), 658–662.

Nicol, C.J., Badnell-Waters, A.J., Bice, R., Kelland, A., Wilson, A.D. and Harris, P.A. (2005) The effects of diet and weaning method on the behaviour of young horses. *Applied Animal Behaviour Science* 95(3–4), 205–221.

Nie, G.J., Johnson, K.E. and Wenzel, J.G.W. (2001) Use of glass ball to suppress behavioural estrus in mares. *Proceedings of American Association of Equine Practitioners* 47, 246–252.

Nie, G.J., Johnson, K.E., Braden, T.D. and Wenzel, J.G.W. (2003) Use of intra-uterine glass ball protocol to extend luteal function in mares. *Journal of Equine Veterinary Science* 23(6), 266–272.

Nikolakopoulos, E. and Watson, E.D. (2000) Effect of infusion volume and sperm numbers on persistence of uterine inflammation in mares. *Equine Veterinary Journal* 32(2), 164–166.

Nikolakopoulos, E., Kindahl, H., Gilbert, C.L., Goode, J. and Watson, E.D. (2000a) Release of oxytocin and prostaglandin F2 alpha around teasing, natural service and associated events in the mare. *Animal Reproduction Science* 63, 89–99.

Nikolakopoulos, E., Kindahl, H. and Watson, E.D. (2000b) Oxytocin and PGF2 alpha release in mares resistant and susceptible to persistent mating-induced endometritis. *Journal of Reproduction and Fertility, Supplement* 56, 363–372.

Nishikawa, Y. (1975) Studies on the preservation of raw and frozen semen. *Journal of Reproduction and Fertility, Supplement* 23, 99–104.

Norris, H.Y., Taylor, W.B. and Garner, F.M. (1968) Equine ovarian granular tumours. *Veterinary Record* 82, 419–420.

Oba, E., Bicudo, S.D., Pimentel, S.L., Lopes, R.S., Simonetti, F. and Hunziker, R.A. (1993) Quantitative and qualitative evaluation of stallion semen. *Revista Brasileira de Reproducao Animal* 17(1–2), 57–74.

Oberstein, N., O'Donovan, M.K., Bruemmer, J.E., Seidel, G.E. Jr., Carnevale, E.M. and Squires, E.L. (2001) Cryopreservation of equine embryos by open pull straw, cryoloop, or conventional slow cooling methods. *Theriogenology* 55, 607–613.

Oetjen, M. (1988) Use of different acrosome stains for evaluating the quality of fresh and frozen semen. Thesis, Tierarztliche Hochschule Hannover, German Federal Republic.

Oftedal, O.T., Hints, H.F. and Schryver, H.F. (1983) Lactation in the horse: milk composition and intake by foals. *Journal of Nutrition* 113, 2096–2106.

O'Grady, S. (1995) Umbilical care in foals. *Journal of Equine Veterinary Science* 15, 12–14.

Oguri, N. and Tsutsumi, Y. (1972) Non surgical recovery of equine eggs and an attempt at non surgical embryo transfer in the horse. *Journal of Reproduction and Fertility* 31, 187–195.

Oguri, N. and Tsutsumi, Y. (1974) Non-surgical egg transfer in mares. *Journal of Reproduction and Fertility, Supplement* 41, 313–317.

Oguri, N. and Tsutsumi, Y. (1980) No surgical transfer of equine embryos. *Archives of Andrology* 5, 108.

Onuma, H. and Ohnami, Y. (1975) Retention of tubal eggs in mares. *Journal of Reproduction and Fertility, Supplement* 23, 507–511.

Oriol, J.G. (1994) The equine capsule. *Equine Veterinary Journal* 26(3), 184–186.

Oriol, J.G., Sharom, F.J. and Betteridge, K.J. (1993) Developmentally regulated changes in the glycoproteins of the equine embryonic capsule. *Journal of Reproduction and Fertility* 99, 653–664.

Osborne, V.E. (1966) An analysis of the pattern of ovulation as it occurs in the reproductive cycle of the mare in Australia. *Australian Veterinary Journal* 42, 149–154.

Osborne, V.E. (1975) Factors influencing foaling percentages in Australian mares. *Journal of Reproduction and Fertility, Supplement* 23, 477–483.

Ott, E.A. (2001) Energy, protein and amino acids requirements for growth of young horses. In: Pagan, J.D. and Guer, R.J. (eds) *Advances in Equine Nutrition II*. Nottingham University Press, Nottingham, UK, pp. 153–159.

Ott, E.A. and Asquith, R.L. (1986) Influence of level of feeding and the nutrient content of the concentrate on growth and development of yearling horses. *Journal of Animal Science* 62, 290–299.

Ousey, J.C. (1997) Thermoregulation and energy requirement of the newborn foal, with reference to prematurity. *Equine Veterinary Journal, Supplement* 24, 104–108.

Ousey, J.C., Dudan, F. and Rossdale, P.D. (1984) Preliminary studies of mammary secretions in the mare to assess foetal readiness for birth. *Equine Veterinary Journal* 16, 259–263.

Ousey, J.C., McArthur, A.J. and Rossdale, P.D. (1991) Metabolic changes in Thoroughbred and pony foals during the first 24 hours post partum. *Journal of Reproduction and Fertility, Supplement* 44, 561–570.

Ousey, J.C., Rossdale, P.D., Fowden, A.L., Palmer, L., Turnbull, C. and Allen, W.R. (2004) Effects of manipulating intra-uterine growth on post natal adrenocortical development and other parameters of maturity in neonatal foals. *Equine Veterinary Journal* 36(7), 616–621.

Overstreet, J.W., Yanagimachi, R., Katz, D.F., Hayashi, K. and Hanson, F.W. (1980) Penetration of human spermatozoa into the human zona pellucida and the zona-free hamster egg: a study of fertile donors and infertile patients. *Fertility and Sterility* 33, 534–542.

Oxender, W.D., Noden, P.A. and Hafs, H.D. (1977) Estrus, ovulation and serum progesterone, estradiol and LH concentrations in mares after increased photoperiod during winter. *American Journal of Veterinary Research* 38, 203–207.

Pace, M.M. and Sullivan, J.J. (1975) Effect of timing of insemination, numbers of spermatozoa and extender components on the pregnancy rate in mares inseminated with frozen stallion semen. *Journal of Reproduction and Fertility, Supplement* 23, 115–121.

Padilla, A.W. and Foote, R.H. (1991) Extender and centrifugation effects on the motility patterns of slow cooled spermatozoa. *Journal of Animal Science* 60, 3308–3313.

Padilla, A.W., Tobback, C. and Foote, R.H. (1991) Penetration of frozen-thawed, zona-free hamster oocytes by fresh and slow-cooled stallion spermatozoa. *Journal of Reproduction and Fertility, Supplement* 44, 207–212.

Pagl, R., Aurich, J.E., Muller-Schlosser, F., Kankofer, M. and Aurich, C. (2006a) Comparison of an extender containing defined milk protein fractions with a skim milk-based extender for storage of equine semen at 5°C. *Theriogenology* 66, 1115–1122.

Pagl, R., Aurich, C. and Kanofer, M. (2006b) Anti-oxidative status and semen quality during cooled storage in stallions. *Journal of Medical Anatomy, Physiology, Pathology and Clinical Medicine* 53(9), 486–489.

Pagan, J.D. and Hintz, H.F. (1986) Composition of milk from pony mares fed various levels of digestible energy. *Cornell Veterinarian* 76(2), 139–148.

Palmer, B. and Jousset, B. (1975) Synchronisation of oestrus in mares with prostaglandin analogue and hCG. *Journal of Reproduction and Fertility, Supplement* 23, 269–274.

Palmer, E. (1979) Reproductive management of mares without detection of oestrus. *Journal of Reproduction and Fertility, Supplement* 27, 263–270.

Palmer, E., Draincourt, M.A. and Ortavant, R. (1982) Photoperiodic stimulation of the mare during winter anoestrus. *Journal of Reproduction and Fertility* 32, 275–282.

Palmer, E., Domerg, D., Fauquenot, A. and de Sainte-Marie, T. (1984) Artificial insemination of mares: results of five years of research and practical experience. In: *Le cheval. Reproduction, selection, alimentation, exploitation.* Institut National de la Recherche Agronomique, Paris, France, pp. 133–147.

Palmer, E., Draincourt, M.A. and Chevalier, F. (1985) Breeding without oestrus detection. Synchronisation of oestrus. *Bulletin des Groupements Techniques Veterinaires* 1, 41–50.

Palmer, E. and Quelier, P. (1988) Uses of LHRH and analogues in the mare. In: *Proceedings of the 11th International Congress on Animal Reproduction and AI.* Elsevier, Amsterdam, pp. 338–346.

Panchal, M.T., Gujarati, M.L. and Kavani, F.S. (1995) Study of some of the reproductive traits of kathi mares in Gujarat state. *Indian Journal of Animal Reproduction* 16, 47–49.

Pantke, P., Hyland, J., Galloway, D.B., Maclean, A.A. and Hoppen, H.O. (1991) Changes in luteinising hormone bioactivity associated with gonadotrophin pulses in the cycling mare. *Journal of Reproduction and Fertility, Supplement* 44, 13–20.

Pantke, P., Hyland, J.H., Galloway, D.B., Liu, D.Y. and Baker, H.W.G. (1992) Development of a zona pellucida sperm binding assay for the assessment of stallion fertility. *Australian Equine Veterinarian* 10(2), 91.

Pantke, P., Hyland, J.H., Galloway, D.B., Liu, D.Y. and Baker, H.W.G. (1995) Development of a zona pellucida sperm binding assay for the assessment of stallion fertility. *Biology of Reproduction Monograph Equine Reproduction VI* 1, 681–687.

Papa, F.O., Alvarenga, M.A., Lopes, M.D. and Campos Filho, E.P. (1990) Infertility of autoimmune origins in a stallion. *Equine Veterinary Journal* 22, 145–146.

Papaioannou, K.Z., Murphy, R.P., Monks, R.S., Hynes, N., Ryan, M.P., Boland, M.P. and Roche, J.F. (1997) Assessment of viability and mitochondrial function of equine spermatozoa using double staining and flow cytometry. *Theriogenology* 48, 299–312.

Parker, W.G., Sullivan, J.J. and First, N.L. (1975) Sperm transport and distribution in the mare. *Journal of Reproduction and Fertility, Supplement* 23, 63–66.

Parkes, R.D. and Colles, C.M. (1977) Fetal electrocardiography in the mare as a practical aid to diagnosing singleton and twin pregnancies. *Veterinary Record* 100, 25–26.

Parks, J.E. and Lynch, D.V. (1992) Lipid composition and thermotropic phase behaviour of boar, bull, stallion and rooster sperm membrane. *Cryobiology* 29(2), 255–266.

Parlevilet, J.M. (2000) Pre-seasonal breeding evaluation of the stallion. *Pferdeheilkunde* 15(6), 523–528.

Parlevliet, J.M. and Colenbrander, B. (1999) Prediction of first season fertility in three year old Dutch Warmbloods with prebreeding assessment of morphologically live sperm. *Equine Veterinary Journal* 31(3), 248–251.

Parlevliet, J.M. and Samper, J.C. (2000) Disease transmission through semen. In: Samper, J.C. (ed.) *Equine Breeding Management and Artificial Insemination.* W.B. Saunders, Philadelphia, pp. 133–140.

Parlevliet, J.M., Bleumink-Pluym, N.M.C., Houwers, D.J., Remmen, J.L.A.M., Sluijter, F.J.H. and Colenbrander, B. (1997) Epidemiologic aspects of Taylorella equigenitalis. *Theriogenology* 47, 1169–1177.

Pascoe, R.R. (1979a) Observations of the length of declination of the vulva and its relation to fertility in the mare. *Journal of Reproduction and Fertility, Supplement* 27, 299–305.

Pascoe, R.R. (1979b) Rupture of the utero-ovarian on middle uterine artery in the mare at parturition. *Veterinary Record* 104(4), 77.

Pascoe, D.R. (1995) Effects of adding autologous plasma to an intrauterine antibiotic therapy after breeding on pregnancy rates in mares. *Biology of Reproduction Monograph* 1, 539–543.

Pascoe, R.R. and Bagust, T.J. (1975) Coital exanthema in stallions. *Journal of Reproduction and Fertility, Supplement* 23, 147–150.

Pascoe, R.R., Spradbrow, P.B. and Bagust, T.Y. (1968) Equine coital exanthema. *Australian Veterinary Journal* 44, 485–490.

Pascoe, R.R., Spradbrow, P.B. and Bagust, T.Y. (1969) An equine genital infection resembling coital exanthema associated with a virus. *Australian Veterinary Journal* 45, 166.

Pascoe, R.R., Meagher, D.M. and Wheat, J.D. (1981) Surgical management of uterine torsion in the mare, a review of 26 cases. *Journal of American Veterinary Medical Association* 179, 351–356.

Pascoe, R.R., Pascoe, R.R. and Hughes, J.P. (1987a) Management of twin conceptuses by manual embryonic reduction: comparison of two techniques and three hormone treatments. *American Journal of Veterinary Research* 48, 1594–1599.

Pascoe, R.R., Pascoe, D.R. and Wilson, M.C. (1987b) Influence of follicular status on twinning rate in mares. *Journal of Reproduction and Fertility, Supplement* 35, 183–189.

Pashan, R.L. (1980) Low doses of oxytocin can induce foaling at term. *Equine Veterinary Journal* 12, 85–89.

Pashan, R.L. (1982) Oxytocin the induction agent of choice? *Journal of Reproduction and Fertility, Supplement* 32, 645–650.

Pashan, R.L. (1984) Maternal and foetal endocrinology during late pregnancy and parturition in the mare. *Equine Veterinary Journal* 16, 233–238.

Pashan, R.L. and Allen, R. (1979a) The role of the fetal gonads and placenta in steroid production, maintenance of pregnancy and parturition. *Journal of Reproduction and Fertility, Supplement* 27, 499–509.

Pashan, R.L. and Allen, W.R. (1979b) Endocrine changes after foetal gonadectomy and during normal and induced parturition in the mare. *Animal Reproduction Science* 2, 271–280.

Patel, J. and Lofstedt, R.M. (1986) Uterine rupture in a mare. *Journal of the American Veterinary Medical Association* 189, 806–807.

Pattison, M.L., Chen, C.L. and King, S.L. (1972) Determination of LH and estradiol-17β surge with reference to the time of ovulation in mares. *Biology of Reproduction* 7, 136–140.

Pattle, R.E., Rossdale, P.D., Schock, C. and Creasey, J.M. (1975) The developments of the lung and its surfactant in the foal and other species. *Journal of Reproduction and Fertility, Supplement* 23, 651–657.

Peaker, M., Rossdale, P.D., Forsyth, I.A. and Falk, M. (1979) Changes in mammary development and composition of secretion during late pregnancy in the mare. *Journal of Reproduction and Fertility, Supplement* 27, 555–561.

Pearson, H. and Weaver, B.M. (1978) Priapism after sedation, neuroleptanalgesia and anaesthesia in the horse. *Equine Veterinary Journal* 10, 85–90.

Pearson, R.C., Hallowell, A.C., Bayley, W.M., Torbeck, R.L. and Perryman, L.E. (1984) Times of appearance and disappearance of colostral IgG in the mare. *American Journal of Veterinary Research* 45, 186–190.

Perez, C.C., Rodriguez, I., Mota, J., Dorado, J., Hidalgo, M., Felipe, M. and Sanz, J. (2003) Gestation length in Cathusian Spanish mares. *Livestock Production Science* 82, 181–187.

Perkins, N. (1996) Equine reproductive ultrasonography. *Veterinary Continuing Education* No. 172. Massey University, New Zealand, pp. 129–146.

Perkins, N.R. and Grimmett, J.B. (2001) Pregnancy and twinning rates in Thoroughbred mares following administration of human chorionic gonadotrophin (hCG). *New Zealand Veterinary Journal* 49, 94–100.

Perkins, N.R. and Threlfall, W.R. (1993) Mastitis in the mare. *Equine Veterinary Education* 5, 192–194.

Perry, E.J. (1945) Historical. In: Perry, E.J. (ed.) *The Artificial Insemination of Farm Animals.* Rutgers University Press, New Brunswick, New Jersey, pp. 3–8.

Perry, E.J. (1968) *The Artificial Insemination of Farm Animals,* 4th edn. Rutgers University Press, New Brunswick, New Jersey.

Pesch, S., Bostedt, H., Failing, K. and Bergmann, M. (2006) Advanced fertility diagnosis in stallion semen using transmission electron microscopy. *Animal Reproduction Science* 91, 285–298.

Petersen, M.M., Wessel, M.T., Scott, M.A., Liu, I.K.M. and Ball, B.A. (2002) Embryo recovery rates in mares after deep intrauterine insemination with low numbers of cryopreserved spermatozoa. In: Evans, M.J. (ed.) *Equine Reproduction VIII. Theriogenology*, Vol. 58. Lea and Febiger, Philadelphia, pp. 663–666.

Philpott, M. (1993) The danger of disease transmission by artificial insemination and embryo transfer. *British Veterinary Journal* 149, 339–369.

Piao, S. and Wang, Y. (1988) A study on the technique of freezing concentrated semen of horses (donkeys) and the effect of insemination. *Proceedings of the International Congress for Animal Reproduction and Artificial Insemination*, 3, pp. 286a–286c.

Pickett, B.W. and Amann, R.P. (1987) Extension and storage of stallion spermatozoa. A review. *Equine Veterinary Science* 7, 289–302.

Pickett, B.W. and Back, D.G. (1973) *Proceedings for the Preparation, Collection, Evaluation and Inseminating of Stallion Semen*. Bulletin 935, Agricultural Experimental Station Animal Reproduction Laboratory General Service, Colorado State University, Fort Collins, Colorado.

Pickerel, T.M., Cromwell-Davis, S.L., Cundle, A.B. and Estep, D.Q. (1993) Sexual preferences of mares *(Equus Caballus)* for individual stallions. *Applied Animal Behaviour Science* 38, 1–13.

Pickett, B.W. (1993a) Collection and evaluation of stallion semen for artificial insemination. In: McKinnon, A.O. and Voss, J.L. (eds) *Equine Reproduction*. Lea and Febiger, Philadelphia, London, pp. 705–714.

Pickett, B.W. (1993b) Seminal extenders and cooled semen. In: McKinnon, A.O. and Voss, J.L. (eds) *Equine Reproduction*. Lea and Febiger, Philadelphia, London, pp. 746–754.

Pickett, B.W. (1993c) Reproductive evaluation of the stallion. In: McKinnon, A.O. and Voss, J.L. (eds) *Equine Reproduction*. Lea and Febiger, Philadelphia, London, pp. 755–768.

Pickett, B.W. (1993d) Sexual behaviour. In: McKinnon, A.O. and Voss, J.L. (eds) *Equine Reproduction*. Lea and Febiger, Philadelphia, London, pp. 809–820.

Pickett, B.W. and Amann, R.P. (1993) Cryopreservation of semen. In: McKinnon, A.O. and Voss, J.L. (eds) *Equine Reproduction*. Lea and Febiger, Philadelphia, London, pp. 769–789.

Pickett, B.W. and Back, D.G. (1973) *Procedures for preparation, collection, evaluation and inseminating of stallion semen*. Bulletin 935 of the Agricultural Experimental Station Animal Reproduction Laboratory General Service Colorado State University.

Pickett, B.W. and Shiner, K.A. (1994) Recent developments in AI in horse. *Livestock Production Science* 40, 31–36.

Pickett, B.W. and Voss, J.L. (1972) Reproductive management of stallions. *Proceedings of the 18th Annual Convention of the American Association of Equine Practitioners*, San Francisco, California, 18, 501–531.

Pickett, B.W. and Voss, J.C. (1975a) The effect of semen extenders on mare fertility. *Journal of Reproduction and Fertility, Supplement* 23, 95–98.

Pickett, B.W. and Voss, J.L. (1975b) Abnormalities of mating behaviour in domestic stallions. *Journal of Reproduction and Fertility, Supplement* 23, 129–134.

Pickett, B.W., Faulkner, L.C. and Sutherland, T.N. (1970) Effect of month and stallion on seminal characteristics and sexual behaviour. *Journal of Animal Science* 31, 713.

Pickett, B.W., Faulkner, L.C. and Voss, J.L. (1975a) Effect of season on some characteristics of stallion semen. *Journal of Reproduction and Fertility, Supplement* 23, 25.

Pickett, B.W., Sullivan, J.J. and Seidel, G.E. Jr. (1975b). Effect of centrifugation and seminal plasma on motility and fertility of stallions and bull spermatozoa. *Fertility and Sterility* 26, 167–174.

Pickett, B.W., Sullivan, J.J. and Seidel, G.E. Jr. (1975c) Reproductive physiology of the stallion. V. Effect of frequency of ejaculation on seminal characteristics and spermatozoal output. *Journal of Animal Science* 40, 917–923.

Pickett, B.W., Voss, J.L., Squires, E.L. and Amann, R.P. (1981) Management of the stallion for maximum reproductive efficiency. *Animal Reproduction Laboratory General Series Bulletin No.1005*, Colorado State University, Fort Collins, Colorado.

Pickett, B.W., Neil, J.R. and Squires, E.L. (1985) The effect of ejaculation frequency on stallion sperm output. *Proceedings of the 9th Equine Nutrition and Physiology Symposium*, pp. 290–295.

Pickett, B.W., Squires, E.L. and McKinnon, A.O. (1987) Procedures for collection, evaluation and utilisation of stallion semen for AI. *Animal Reproduction Laboratory Bulletin No. 03*. Colorado State University, Fort Collins, Colorado.

Pickett, B.W., Voss, J.L., Bowen, R.A., Squires, E.L. and McKinnon, A.O. (1988) Seminal characteristics and total scrotal width (TSW) of normal and abnormal stallions. *Proceedings of the 33rd Annual Convention of the American Association of Equine Practitioners*, pp. 485–518.

Pickett, B.W., Amann, R.P., McKinnon, A.O., Squires, E.L. and Voss, J.L. (1989) Management of the stallion for maximum reproductive efficiency II. *Bulletin of the Colorado State University Agriculture Experimental Station Animal Reproduction Laboratory General Series Bulletin No. 05*. Colorado State University, Fort Collins, Colorado, pp. 121–125.

Pickett, B.W., Voss, J.L., Squires, E.L., Vanderwall, D.K., McCue, P.M. and Bruemmer, J.E. (2000) Collection, preparation and insemination of stallion semen. *Bulletin No 10 Animal Reproduction and Biotechnology Laboratory*. Colorado State University, Fort Collins, Colorado, pp. 1–15.

Pieppo, J., Huntinen, M. and Kotilainen, T. (1995) Sex diagnosis of equine preimplantation embryos using the polymerase chain reaction. *Theriogenology* 44(5), 619–627.

Pierson, R.A. (1993) Folliculogenesis and ovulation. In: McKinnon, A.O. and Voss, J.L. (eds) *Equine Reproduction*. Lea and Febiger, Philadelphia, pp. 161–171.

Pierson, R.A. and Ginther, O.J. (1985a) Ultrasonic evaluation of the corpus luteum of the mare. *Theriogenology* 23, 795–806.

Pierson, R.A. and Ginther, O.J. (1985b) Ultrasonic evaluation of the pre-ovulatory follicle in the mare. *Theriogenology* 24, 359–368.

Piquette, G.N., Kenney, R.M., Sertich, P.L., Yamoto, M. and Hsueh, A.J.W. (1990) Equine granulosa theca cell tumours express inhibinα and βA subunit messenger ribonucleic acids and proteins. *Biology of Reproduction* 43, 1050–1057.

Pitra, C., Schafer, W. and Jewgenow, K. (1985) Quantitative measurement of the fertilising capacity of deep-frozen stallion semen by means of the hamster egg penetration test. *Monatschefte fur Veterinarmedizin* 40(7), 235–237.

Plata-Madrid, H., Youngquist, R.S., Murphy, C.N., Bennett-Wimbrush, K., Braun, W.F. and Loch, W.E. (1994) Ultrasonographic characteristics of the follicular and uterine dynamics in Belgium mares. *Journal of Equine Veterinary Science* 14, 421–423.

Platt, H. (1975) Infection of the horse fetus. *Journal of Reproduction and Fertility, Supplement* 23, 605–610.

Platt, H., Atherton, J.G. and Simpson, D.Y. (1978) The experimental infection of ponies with contagious equine metritis. *Equine Veterinary Journal* 10, 153–158.

Plewinska-Wierzobska, D. and Bielanski, W. (1970) The methods of evaluation the speed and sort of movement of spermatozoa. *Medycyna Weterynaryjna* 26, 237–250.

Polge, C., Smith, A.U. and Parkes, A.S. (1949) Revival of spermatozoa after vitrification and dehydration at low temperature. *Nature* 164, 666.

Pommer, A.C., Linfor, J.J. and Meyers, S.A. (2002) Capacitation and acrosomal exocytosis are enhanced by incubation of stallion spermatozoa in a commercial semen extender. *Theriogenology* 57(5), 1493–1501.

Pool, K.C., Charneco, R. and Arns, M.J. (1993) The influence of seminal plasma from fractionated ejaculation on the cold storage of equine spermatozoa. *Proceedings of the 13th Conference of the Equine Nutrition and Physiology Symposium*, pp. 395–396.

Potter, J.T., Kreider, J.L., Potter, G.D., Forrest, D.W., Jenkins, W.L. and Evans, J.W. (1987) Embryo survival during early gestation in energy deprived mares. *Journal of Reproduction and Fertility, Supplement* 35, 715–716.

Pouret, E.J.M. (1982) Surgical techniques for the correction of pneumo and uro vagina. *Equine Veterinary Journal* 14, 249–250.

Power, S.G.A. and Challis, R.G. (1987) Steroid production by dispersed cells from fetal membranes and intrauterine tissue of sheep. *Journal of Reproduction and Fertility* 81(1), 65–76.

Pozor, M. and McDonnell, S. (2002) Ultrasonographic measurement of accessory sex glands, ampullae and urethra of normal stallions of various size and type. *Theriogenology* 58, 1425–1430.

Pozor, M. and McDonnell, S. (2004) Colour Doppler ultrasound evaluation of testicular blood flow in stallions. *Theriogenology* 61, 799–810.

Pozor, M.A., McDonnell, S.M., Kenney, R.M. and Tischner, M. (1991) GnRH facilitates the copulatory behaviour in geldings treated with testosterone. *Journal of Reproduction and Fertility, Supplement* 44, 666–667.

Price, S., Aurich, J., Davies-Morel, M. and Aurich, C. (2007) Effects of oxygen exposure and gentamicin on stallion semen stored at 5°C and 15°C. *Reproduction in Domestic Animals* 43 (3), 261–266.

Price, S.B.P. (2008) Effects of storage conditions on cooled-stored stallion semen. MSc thesis, University of Wales, Aberystwyth, UK.

Prickett, M.E. (1966) Pathology of the equine ovary. *Proceedings of the 12th Annual Convention of the Annual Association of Equine Practitioners*, pp. 145–150.

Province, C.A., Amann, R.P., Pickett, B.W. and Squires, E.L. (1984) Extenders for preservation of canine and equine spermatozoa at 5°C. *Theriogenology* 22, 409–415.

Province, C.A., Squires, E.L., Pickett, B.W. and Amann, R.P. (1985) Cooling rates storage temperature and fertility of extended equine spermatozoa. *Theriogenology* 23, 925–934.

Pruitt, J.A., Arns, M.J. and Pool, K.C. (1993) Seminal plasma influences recovery of equine spermatozoa following *in vitro* culture (37°C) and cold storage (5°C). *Theriogenology* 39, 291.

Pugh, D.G. (1985) Equine ovarian tumours. *Compendium of Continuing Education, Practice Veterinarian* 7, 710–715.

Pugh, D.G. and Schumacher, J. (1990) Management of the broodmare. *Proceedings of the 36th Annual Convention of the American Association of Equine Practitioners*, pp. 61–78.

Purvis, A.D. (1972) Electric induction of labour and parturition in the mare. *Proceedings of the18th Annual Convention American Association Equine Practice*, San Francisco, California, pp. 113–119.

Pycock, J.F. (1994) A new approach to treatment of endometritis. *Equine Veterinary Education* 6(1), 36–38.

Pycock, J.F. (2000) Breeding management of the problem mare. In: Samper, J.C. (ed.) *Equine Breeding Management and Artificial Insemination*. W.B. Saunders, Philadelphia, pp. 195–228.

Pycock, J.F. and Newcombe, J.R. (1996) Assessment of the effects of three treatments to remove intrauterine fluid on pregnancy rates in the mare. *Veterinary Record* 138(14), 320–323.

Quinn, G.C. and Woodford, N.S. (2005) Infertility due to a uterine leiomyoma in a Thoroughbred mare: clinical findings, treatment and outcome. *Equine Veterinary Education* 17(3), 150–152.

Raeside, J.I. (1969) The isolation of oestrone sulfate and oestradiol 18B sulfate from stallion testis. *Canadian Journal of Biochemistry* 47, 811–816.

Raeside, J.I., Liptrap, R.M., McDonnell, W.N. and Milne, E.J. (1979) A precursor role for dihydroepiandrosterone DHA in feto-placental unit for oestrogen formation in the mare. *Journal of Reproduction and Fertility, Supplement* 27, 493–497.

Raeside, J.I., Ryan, P.L. and Lucas, Z. (1991) A method for the measurement of oestrone sulphate in faeces in feral mares. *Journal of Reproduction and Fertility, Supplement* 44, 638.

Rahaley, R.S., Gordon, B.J., Leipold, H.W. and Peter, J.E. (1983) Sertoli cell tumour in a horse. *Equine Veterinary Journal* 15(1), 68–70.

Ralston, S.L. (1997) Feeding the rapidly growing foal. *Journal of Equine Veterinary Science* 17(12), 634–636.

Ralston, S.L., Rich, G.A., Jackson, S. and Squires, E.L. (1986) The effect of vitamin A supplementation on sexual characteristics and vitamin A absorption in stallions. *Journal of Equine Veterinary Science* 6(4), 203–207.

Rambegs, B.P., Stout, T.A. and Rijkenhuizen, A.B. (2003) Ovarian granulosa cell tumours adherent to other abdominal organs; surgical removal from 2 warmblood mares. *Equine Veterinary Journal* 35(6), 627–632.

Rantanen, N.W. and Kinkaid, B. (1989) Ultrasound guided fetal cardiac puncture: a method of twin reduction. *Proceedings of the American Association of Equine Practitioners* 34, 173–179.

Rathor, S.S. (1989) Equine cital exanthema in thoroughbreds. *Journal of Equine Veterinary Science* 9, 34.

Raub, R.H., Jackson, S.G. and Baker, J.P. (1989) The effect of exercise on bone growth and development in weanling horses. *Journal of Animal Science* 67, 2508–2514.

Rauterberg, H. (1994) Use of glasswool sephadex filtration for the collection of fresh semen from horses. Laboratory studies and field trials. Thesis, Tierarztliche Hochschule, Hannover, Germany, 105 pp.

Reichart, M., Lederman, H., Hareven, D., Keden, P. and Bartoov, B. (1993) Human sperm acrosin activity with relation to semen parameters and acrosomal ultrastructure. *Andrologia* 25(2), 59–66.

Reef, V.B. (1993) Diagnostic ultrasonography of the foal's abdomen. In: McKinnon, A.O. and Voss, J.L. (eds) *Equine Reproduction*. Lea and Febiger, Philadelphia, pp. 1088–1094.

Reef, V.B. (1998) Fetal ultrasonography. In: Reef, V.B. (ed.) *Equine Diagnostic Ultrasound*. W.B. Saunders, Philadelphia, pp. 425–445.

Reef, V., Vaala, W. and Worth, L. (1995) Ultrasonographic evaluation of the fetus and intrauterine environment in healthy mares during gestation. *Veterinarian Radio Ultrasound* 1995, 256–258.

Reifenrath, H. (1994) Use of L4 leucocyte absorption membrane filtration in AI in horses, using fresh or frozen semen. Thesis, Tierarztliche Hochschule Hannover, German Federal Republic.

Reinfenrath, H., Jensen, A., Sieme, H. and Klug, E. (1997) Ureteroscopic catheterisation of the vesicular glands in the stallion. *Reproduction in Domestic Animals* 32, 47–49.

Revell, S.G. (1997) A sport horse for the future. *Proceedings of the British Society for Animal Science Equine Conference*, Cambridge, July 1997.

Revell, S.G. and Mrode, R.A. (1994) An osmotic resistance test for bovine semen. *Animal Reproduction Science* 36, 77–86.

Ricker, J.V., Linfor, J.J., Delfino, W.J., Kysar, P., Scholtz, E.L., Tablin, F., Crowe, J.H., Ball, B.A. and Meyers, S.A. (2006) Equine sperm membrane phase behaviour: the effects of lipid-based cryoprotectants. *Biology of Reproduction* 74(2), 359–365.

Ricketts, S.W. (1975a) The technique and clinical application of endometrial biopsy in the mare. *Equine Veterinary Journal* 7, 102–108.

Ricketts, S.W. (1975b) Endometrial biopsy as a guide to diagnosis of endometrial pathology in the mare. *Journal of Reproduction and Fertility, Supplement* 23, 341–345.

Ricketts, S.W. (1978) Histological and histopathological studies of the endometrium of the mare. Fellowship thesis, Royal College of Veterinary Surgeons, London.

Ricketts, S.W. (1993) Evaluation of stallion semen. *Equine Veterinary Education* 5(5), 232–237.

Ricketts, S.W. and Alonso, S. (1991) The effect of age and parity on the development of equine chorionic endometrial disease. *Equine Veterinary Journal* 23, 189–192.

Ricketts, S.W. and Mackintosh, M.E. (1987) Role of anaerobic bacteria in equine endometritis. *Journal of Reproduction and Fertility, Supplement* 35, 343–351.

Ricketts, S.W., Young, A. and Medici, E.B. (1993) Uterine and clitoral cultures. In: McKinnon, A.O. and Voss, J.L. (eds) *Equine Reproduction.* Lea and Febiger, Philadelphia, London, pp. 234–245.

Ricketts, S.W., Barrelet, A. and Whitwell, K.E. (2003) Equine abortion. *Equine Veterinary Education* 6, 18–21.

Rigby, S.L., Brinsko, S.P., Cochran, M., Blanchard, T.L., Love, C.C. and Varner, D.D. (2001) Advances in cooled semen technologies: seminal plasma and semen extender. *Animal Reproduction Science* 68(3–4), 171–180.

Roberts, M.C. (1975) The development and distribution of mucosal enzymes in the small intestine of the foetus and young foal. *Journal of Reproduction and Fertility, Supplement* 23, 717–723.

Roberts, S.J. (1971) Artificial insemination in horses. In: Roberts, S.J. (ed.) *Veterinary Obstetrics and Genital diseases (Theriogenology)*, 2nd edn. Comstock, Ithaca, New York, pp. 740–743.

Roberts, S.J. (1986a) Infertility in the mare. In: Roberts, S.J. (ed.) *Veterinary Obstetrics and Genital Diseases (Theriogenology)*, 3rd edn. Edwards Brothers, North Pomfret, Vermont, pp. 581–635.

Roberts, S.J. (1986b) Infertility in male animals (andrology). In: Roberts, S.J. (ed.) *Veterinary Obstetrics and Genital Diseases (Theriogenology)*, 3rd edn. Edwards Brothers, North Pomfret, Vermont, pp. 752–893.

Roberts, S.J. (1986c) Gestation and pregnancy diagnosis in the mare. In: Morrow, D.A. (ed.) *Current Therapy in Theriogenology.* W.B. Saunders, Philadelphia, pp. 670–678.

Roberts, S.J. and Myhre, G. (1983) A review of twinning in horses and the possible therapeutic value of supplemental progesterone to prevent abortion of equine twin fetuses in latter half of the gestation period. *Cornell Veterinarian* 73, 257–264.

Roberts, S.M. (1993) Ocular disorders. In: McKinnon, A.O. and Voss, J.L. (eds) *Equine Reproduction.* Lea and Febiger, Philadelphia, pp. 1076–1087.

Robinson, J.A., Allen, G.K., Green, E.M., Fales, W.H., Loch, W.E. and Wilkerson, G. (1993) A prospective study of septicaemia in colostrums deprived foals. *Equine Veterinary Journal* 25(3), 214–219.

Robinson, S.J., Neal, H. and Allen, W.R. (1999) Modulation of oviductal transport in the mare by local application of prostaglandin E2. *Journal of Reproduction and Fertility, Supplement* 56, 587–592.

Roger, J.F. and Hughes, J.P. (1991) Prolonged pulsatile administration of gonadotrophin releasing hormone (GnRH) to fertile stallions. *Journal of Reproduction and Fertility, Supplement* 44, 155–168.

Roger, J.F., Kiefer, B.L., Evans, J.W., Neely, D.P. and Pacheco, C.A. (1979) The development of antibodies to human CG following its repeated injection in the cycling mare. *Journal of Reproduction and Fertility, Supplement* 27, 173–179.

Rogers, C.W., Gee, E.K. and Faram, T.L. (2004) The effect of two different weaning procedures on the growth of pasture reared Thoroughbred foals in New Zealand. *New Zealand Veterinary Journal* 52(6), 401–403.

Rollins, W.C. and Howell, C.E. (1951) Genetic sources of variation in gestation length of the horse. *Journal of Animal Science* 10, 797–805.

Rooney, J.R. (1964) Internal haemorrhage related to gestation in the mare. *Cornell Veterinary* 54, 11–19.

Rooney, J.R. (1970) *Autopsy of the Horse, Technique and Interpretation.* Williams and Wilkins, Baltimore, Maryland.

Rophia, R.T., Mathews, R.G., Butterfield, R.M., Moss, G.E. and McFalden, W.J. (1969) The duration of pregnancy in Thoroughbred mares. *Veterinary Record* 84, 552–555.

Rose, R.J. (1988) Cardiorespiratory adaptations in neonatal foals. *Equine Veterinary Journal, Supplement* 5, 11–13.

Roser, A.J. (1995) Endocrine profiles in fertile, subfertile and infertile stallions. Testicular response to human chorionic gonadotrophin in infertile stallions. *Biology of Reproduction, Monograph, Equine Reproduction, VI* 1, 661–669.

Roser, J.F. (1997) Endocrine basis for testicular function in the stallion. *Theriogenology* 48(5), 883–892.

Roser, J.F., O'Sullivan, J., Evans, J.W., Swedlow, J. and Papkoff, H. (1987) Episodic release of prolactin in the mare. *Journal of Reproduction and Fertility, Supplement* 35, 687–688.

Rossdale, P.D. (1967) Clinical studies on the newborn thoroughbred foal. 1. Perinatal behaviour. *British Veterinary Journal* 123, 470–481.

Rossdale, P.D. (1968) Clinical studies on the newborn Thoroughbred foal. III. Thermal stability. *British Veterinary Journal* 124(12), 18–22.

Rossdale, P.D. (1987) Twin pregnancy. In: Robinson, N.E. (ed.) *Current Therapy in Equine Medicine*, 2nd edn. W.B. Saunders, Philadelphia, pp. 532–533.

Rossdale, P.D. and Jeffcote, L.B. (1975) Problems encountered during induced foaling in pony mares. *Veterinary Record* 97, 371–372.

Rossdale, P.D. and Mahafrey, L.W. (1958) Parturition in the Thoroughbred mare with particular reference to blood deprivation in the new born. *Veterinary Record* 70, 142.

Rossdale, P.D. and Ricketts, S.W. (1980) *Equine Stud Farm Medicine*, 2nd edn. Balliere Tindall, London.

Rossdale, P.D. and Short, R.V. (1967) The time of foaling of Thoroughbred mares. *Journal of Reproduction and Fertility, Supplement* 13, 341–343.

Rossdale, P.D., Silver, M., Comline, R.S., Hall, L.W. and Nathanielsz, P.W. (1973) Plasma cortisol in the foal during the late foetal and early neonatal period. *Research in Veterinary Science* 15(3), 395–397.

Rossdale, P.D., Pashan, R.L. and Jeffcote, L.B. (1979a) The use of prostaglandin analogue (fluprostenol) to induce foaling. *Journal of Reproduction and Fertility, Supplement* 27, 521–529.

Rossdale, P.D., Hunt, M.D.N., Peace, C.K., Hopes, R., Ricketts, S.W. and Wingfield-Digby, N.Y. (1979b) C.E.M.: the case of A.I. *Veterinary Record* 104, 536.

Rossdale, P.D., Ousey, J.C., Silver, M. and Fowden, A.L. (1984) Studies on equine prematurity, guidelines for assessment of foal maturity. *Equine Veterinary Journal* 16, 300–302.

Rossdale, P.D., Ousey, J.C., Cottrill, C.M., Chavatte, P., Allen, W.R. and McGladdery, A.J. (1991) Effects of placental pathology on maternal plasma progestagen and mammary secretion calcium concentrations and on neonatal adrenocortical function in the horse. *Journal of Reproduction and Fertility, Supplement* 44, 579–590.

Rossdale, P.D., McGladdery, A.J., Ousey, J., Holdstock, N., Grainger, C. and Houghton, E. (1992) Increase in plasma progestagen concentrations in the mare after foetal injection with CRH, ACTH or beta methasone in late gestation. *Equine Veterinary Journal* 24(5), 347–350.

Rossdale, P.D., Ousey, J.C. and Chavatte, P. (1997) Readiness for birth: an endocrinological duet between fetal foal and mare. *Equine Veterinary Journal, Supplement* 24, 96–99.

Rota, A., Furzi, C., Panzani, D. and Camillo, F. (2004) Studies on motility of cooled stallion spermatozoa. *Reproduction in Domestic Animals* 39(2), 103–109.

Rouse, B.T. and Ingram, D.G. (1970) The total protein and immunoglobulin profile of equine colostrums and milk. *Immunology* 19, 901–907.

Rowley, M.S., Squires, E.L. and Pickett, B.W. (1990) Effect of insemination volume on embryo recovery in mares. *Equine Veterinary Science* 10, 298–300.

Roy, S.K. and Greenwald, G.S. (1987) In vitro steroidogenesis by primary to antral follicles in the hamster during the periovulatory period. Effects of follicle stimulating hormone, luteinising hormone and prolactin. *Biology of Reproduction* 37(1), 39–46.

Rudak, E., Jacobs, P. and Yanagimachi, R. (1978) Direct analysis of chromosome constitution of human spermatozoa. *Nature (London)* 174, 911–913.

Rutten, D.R., Chaffaux, S., Valon, M., Deletang, F. and De Haas, V. (1986) Progesterone therapy in mares with abnormal oestrus cycles. *Veterinary Record* 119, 569–571.

Saar, L.T. and Getty, R. (1975) Equine lymphatic system. In: Getty, R. (ed.) *Sisson and Grossman's The Anatomy of Domestic Animals*, Vol. 1. W.B. Saunders, Philadelphia, pp. 625–627.

Saastamoinen, M.T., Lahdekorpi, M. and Hyppa, S. (1990) Copper and zinc levels in the diet of pregnant and lactating mares. *Proceedings of the 41st Annual Meeting of the European Association for Animal Production*, pp. 1–9.

Sack, W.O. (1991) Isolated male organs. *Rooney's Guide to the Dissection of the Horse*, 6th edn. Veterinary Text books, Ithaca, New York, pp. 75–78.

Salisbury, G.W., Van Denmark, N.L. and Lodge, J.R. (1978) Part 2. The storage and the planting. *Physiology of Reproduction and AI of Cattle*, 2nd edn. W.H. Freeman, San Francisco, California, pp. 187–578.

Samper, J.C. (1991) Relationship between the fertility of fresh and frozen stallion semen and semen quality. *Journal of Reproduction and Fertility, Supplement* 44, 107–114.

Samper, J.C. (1995a) Diseases of the male system. In: Kobluk, C.N., Ames, T.R. and Goer, R.J. (eds) *The Horse, Diseases and Clinical Management*, Vol. 2. W.B. Saunders, Philadelphia, pp. 937–972.

Samper, J.C. (1995b) Stallion semen cryopreservation: male factors affecting pregnancy rates. *Proceedings of the Society for Theriogenology*, San Antonio, Texas, pp. 160–165.

Samper, J.C. (1997) Reproductive anatomy and physiology of breeding stallion. In: Youngquist, R.S. (ed.) *Current Therapy in Large Animal Theriogenology*. W.B. Saunders, Philadelphia, pp. 3–12.

Samper, J.C. (2000) *Equine Breeding Management and Artificial Insemination*. W.B. Saunders, Philadelphia, 306 pp.

Samper, J.C. and Crabo, B.G. (1988) Filtration of capacitated spermatozoa through filters containing glass wool and/or Sephadex. *Proceedings of the 11th International Congress on Animal Production and Artificial Insemination*, June 26–30, Vol. 3, University College Dublin, Republic of Ireland, paper 294.

Samper, J.C. and Tibary, A. (2006) Disease transmission in horses. *Theriogenology* 66, 551–559.

Samper, J.C., Loseth, K.J. and Crabo, B.G. (1988) Evaluation of horses spermatozoa with Sephadex filtration using three extenders and three dilutions. *Proceedings of the 11th International Congress on Animal Production and Artificial Insemination*, June 26–30, Vol. 3, University College Dublin, Republic of Ireland, Brief communications, 294.

Samper, J.C., Behnke, E.J., Byers, A.P., Hunter, A.G. and Crabo, B.G. (1989) *In vitro* capacitation of stallion spermatozoa in calcium-free Tyrode's medium and penetration of zona-free hamster eggs. *Theriogenology* 31(4), 875–884.

Samper, J.C., Hellander, J.C. and Crabo, B.G. (1991) The relationship between the fertility of fresh and frozen stallion semen and semen quality. *Journal of Reproduction and Fertility, Supplement* 44, 107–114.

Samper, J.C., Jensen, S., Sergeant, J. and Estrada, A. (2002) Timing of induction of ovulation in mares treated with Ovuplant or Chorulon. *Journal of Equine Veterinary Science* 22, 320–323.

Sanchez, R.A., von Frey, G.W. and de los Reyers, S.M. (1995) Effect of semen diluents and seminal plasma on the preservation of refrigerated stallion semen. *Veterinaria Argentina* 12(113), 172–178.

Sanocka, D. and Kurpisz, M. (2004) Reactive oxygen species and sperm cells. *Reproductive Biology and Endocrinology* 2, 12.

Savage, N.C. and Liptrap, R.M. (1987) Induction of ovulation in cyclic mares by administration of a synthetic prostaglandin fenprostalene, during estrus. *Journal of Reproduction and Fertility, Supplement* 35, 239–243.

Scheffrahn, N.S., Wiseman, B.S.,Vincent, D.L.,Harrison, P.C. and Kesler, D.J. (1980) Ovulation control in pony mares during early spring using progestins, PGF2x, hCG and GnRH. *Journal of Animal Science, Supplement* 1(51), 325.

Scherbarth, R., Pozvari, M., Heilkenbrinker, T. and Mumme, J. (1994) Genital microbial flora of the stallion – microbiological examination of presecretion samples between 1972 and 1991. *Deutsche Tierarztliche Wochenschrift* 101(1), 18–22.

Schneider, R.K., Milne, D.W. and Kohn, C.W. (1982) Acquired inguinal hernia in the horse: a review of 27 cases. *Journal of the American Veterinary Medicine Association* 180, 317–332.

Schryver, H.F., Ofledal, O.T., Williams, J., Soderholm, L.V. and Hintz, H.F. (1986) Lactation in the horse: the mineral composition of mare's milk. *Journal of Nutrition* 116, 2142–2147.

Schuler, G. (1998) Indirect pregnancy diagnosis in the mare: determination of oestrone sulphate in blood and urine. *Praktische Tierarzt* 79(1), 43–49.

Schumacher, J. and Varner, D.D. (1993) Neoplasia of the stallion's reproductive tract. In: McKinnon, A.O. and Voss, J.L. (eds) *Equine Reproduction*. Lea and Febiger, Philadelphia, pp. 871–878.

Schumacher, J., Varner, D.D., Schmitz, D.G. and Blanchard, T.L. (1995) Urethral defects in geldings with hematuria and stallions with haemospermia. *Veterinary Surgery* 24(3), 250–254.

Schutzer, W.E. and Holton, D.W. (1995) Novel progestin metabolism by the equine utero-fetal-placental unit. *Biology of Reproduction* 52(Supplement 1), 188.

Schwab, C.A. (1990) Prolactin and progesterone concentrations during early pregnancy and relationship to pregnancy loss prior to day 45. *Journal of Equine Veterinary Science* 10(4), 280–283.

Scott, M.A., Liu, I.K.M., Overstreet, J.W. and Enders, A.C. (2000) The structural morphology and epithelial association of spermatozoa at the utero-tubal junction: a descriptive study of equine spermatozoa in situ using scanning electron microscopy. *Journal of Reproduction and Fertility, Supplement* 56, 415–421.

Scott, T.J., Carnevale, E.M., Maclellan, L.J., Scoggin, C.F. and Squires, E.L. (2001) Embryo development rates after transfer of oocytes matured *in vivo, in vitro,* or within oviducts of mares. *Theriogenology* 55, 705–715.

Scraba, S.T. and Ginther, O.J. (1985) Effect of lighting programs on the ovulatory season in mares. *Theriogenology* 24, 607–679.

Seamens, M.C., Roser, J.F., Linford, R.L., Liu, I.K.M. and Hughes, J.P. (1991) Gonadotrophin and steroid concentrations in jugular and testicular venous plasma in stallions before and after GnRH injection. *Journal of Reproduction and Fertility, Supplement* 44, 57–67.

Searle, D., Dart, A.J., Dart, C.M. and Hodgson, D.R. (1999) Equine castration: review of anatomy, approaches, techniques and complications in normal, cryptorchid and monorchid horses. *Australian Veterinary Journal* 77(7), 428–471.

Seidel, G.E. Jr., Herickhoff, L.A., Schenk, J.L., Doyle, S.P. and Green, R.D. (1998) Artificial insemination of heifers with cooled, unfrozen sexed semen. *Theriogenology* 49(1), 365.

Seltzer, K.L., Divers, T.J., Vaala, W.E., Byars, T.D. and Rubin, J.L. (1993) The urinary system. In: McKinnon, A.O. and Voss, J.L. (eds) *Equine Reproduction*. Lea and Febiger, Philadelphia, pp. 1030–1040.

Senger, P.L. (1999) *Pathways to Pregnancy and Parturition*, 1st revised edn. Current Conceptions, Pullman, Washington, 281 pp.

Sertich, P.L. (1993) Cervical problems in the mare. In: McKinnon, A.O. and Voss, J.L. (eds) *Equine Reproduction*. Lea and Febiger, Philadelphia, London, pp. 404–407.

Sertich, P.L. (1998) Ultrasonography of the genital tract of the mare. In: Reef, V.B. (ed.) *Equine Diagnostic Ultrasound*. W.B. Saunders, Philadelphia, pp. 405–424.

Sertich, P.L., Love, L.B., Hodgson, M.R. and Kenny, R.M. (1988) 24 hour cooled storage of equine embryos. *Theriogenology* 30(5), 947–952.

Setchell, B.P. (1991) Male reproductive organs and semen. In: Cupps, P.T. (ed.) *Reproduction in Domestic Animals*, 4th edn. Academic Press, London, pp. 221–249.

Shannon, P. (1972) The effect of egg yolk level and dose rate on conception rate of semen diluted in caprogen. *Proceedings of the 7th International Congress of Animal Reproduction and AI*, Munich, pp. 279–280.

Sharma, R., Hogg, J. and Bromham, D. (1993) Is spermatozoan acrosin a predictor of fertilisation and embryo quality in the human? *Fertility and Sterility* 60(5), 881–887.

Sharp, D.C. (1980) Environmental influences on reproduction in horses. In: Hughes, J. (ed.) *Veterinary Clinics of North America: Large Animal Practice.* W.B. Saunders, Philadelphia, pp. 207–273.

Sharp, D.C. (1993) Maternal recognition of pregnancy. In: McKinnon, A.O. and Voss, J.L. (eds) *Equine Reproduction.* Lea and Febiger, Philadelphia, pp. 473–485.

Sharp, D.C. and Clever, B.D. (1993) Melatonin. In: McKinnon, A.O. and Voss, J.L. (eds) *Equine Reproduction.* Lea and Febiger, Philadelphia, pp. 100–108.

Shaw, E.B., Houpt, K.A. and Holmes, D.F. (1988) Body temperature and behaviour of mares during the last two weeks of pregnancy. *Equine Veterinary Journal* 20, 199–200.

Shaw, F.D. and Morton, H. (1980) The immunological approach to pregnancy diagnosis: a review. *Veterinary Record* 106, 268–270.

Shideler, R.K. (1993a) History. In: McKinnon, A.O. and Voss, J.L. (eds) *Equine Reproduction.* Lea and Febiger, Philadelphia, London, pp. 196–198.

Shideler, R.K. (1993b) External examination. In: McKinnon, A.O. and Voss, J.L. (eds) *Equine Reproduction.* Lea and Febiger, Philadelphia, London, pp. 199–203.

Shideler, R.K. (1993c) Rectal palpation. In: McKinnon, A.O. and Voss, J.L. (eds) *Equine Reproduction.* Lea and Febiger, Philadelphia, pp. 204–210.

Shideler, R.K. (1993d) The prefoaling period. In: McKinnon, A.O. and Voss, J.L. (eds) *Equine Reproduction.* Lea and Febiger, Philadelphia, pp. 955–963.

Shoemaker, C.F., Squires, E.L. and Shideler, R.K. (1989) Safety of altrenogest in pregnant mares and on health and development of offspring. *Journal of Equine Veterinary Science* 9, 67–72.

Silva, P.F. and Gadella, B.M. (2006) Detection of damage in mammalian sperm cells. *Theriogenology* 65(5), 958–978.

Silva, L.A., Gastal, E.L., Beg, M.A. and Ginther, O.J. (2005) Changes in vascular perfusion of the endometrium in association with changes in location of the embryonic vesicle in mares. *Biology of Reproduction* 72(3), 755–761.

Silver, M. (1990) Prenatal maturation, the timing of birth and how it may be regulated in domestic animals. *Experimental Physiology* 75(3), 285–307.

Silver, M. and Fowden, A.L. (1994) Prepartum adrenocortical maturation in the fetal foal: responses to ACTH $_{1-24}$. *Journal of Endocrinology* 142, 417–425.

Silver, M., Steven, D.H. and Comline, R.S. (1973) Placental exchange and morphology in ruminants and mares. In: Comline, R.S., Cross, K.W., Davies, G.S. and Nathanielsz, P.W. (eds) *Foetal and Neonatal Physiology.* Cambridge University Press, Cambridge, pp. 245–262.

Silver, M., Barnes, R.J., Comline, R.S., Fowden, A.L., Clover, L. and Mitchell, M.D. (1979) Prostaglandins in maternal of foetal plasma and in allantoic fluid during the second half of gestation in the mare. *Journal of Reproduction and Fertility, Supplement* 27, 531–539.

Silver, M., Ousey, J.C., Dudan, F.E., Fowden, A.L., Knox, J., Cash, R.S. and Rossdale, P.D. (1984) Studies on equine prematurity 2: Post natal adrenocortical activity in relation to plasma adrenocorticotrophic hormone and catecholamine levels in term and premature foals. *Equine Veterinary Journal* 16(4), 278–286.

Sinnemaa, L., Jarvimaa, T., Lehmonen, N., Makela, O., Reilas, T., Sankari, S. and Katila, T. (2003) Effect of insemination volume on uterine contractions and inflammatory response and on elimination of semen in the mare's uterus – scintigraphic and ultrasonographic studies. *Theriogenology* 60(4), 727–733.

Sirosis, J., Ball, B.A. and Fortune, J.E. (1989) Patterns of growth and regression of ovarian follicles during the oestrous cycle and after hemiovarectomy in mares. *Equine Veterinary Journal, Supplement* 18, 43–48.

Sissener, T.R., Squires, E.L. and Clay, C.M. (1996) Differential suppression of endometrial prostaglandin F_2 alpha by the equine conceptus. *Theriogenology* 45, 541–546.

Sisson, S. (1975) Female genital organs. In: Getty, R. (ed.) *Sisson and Grossman's The Anatomy of the Domestic Animal,* Vol. 1, 5th edn. W.B. Saunders, Philadelphia, pp. 542–549.

Sist, M.D. (1987) Fecal oestrone sulphate assay for pregnancy. *Veterinary Medicine* 82, 1036–1043.

Skidmore, J.A., Boyle, M.S. and Allen, W.R. (1990) A comparison of two different methods of freezing horse embryos. *Journal of Reproduction and Fertility, Supplement* 44, 714–716.

Slade, N.P., Takeda, T., Squires, E.L., Elsden, R.P. and Seidel, G.E. Jr. (1985) A new procedure for the cryopreservation of equine embryos. *Theriogenology* 24, 45–58.

Slusher, S.H. (1997) Infertility and diseases of the reproductive tract in stallions. In: Youngquist, R.S. (ed.) *Current Therapy in Large Animal Theriogenology.* W.B. Saunders, Philadelphia, pp. 16–23.

Smith, J.A. (1973) The occurrence of larvae of strongylus edentatus in the testicles of stallions. *Veterinary Record* 93, 604–606.

Smolders, E.A.A., Van Der Veen, N.G. and Van Polanen, A. (1990) Composition of horse milk during the suckling period. *Livestock Production Science* 25, 163–171.

Spensley, M.S. and Markel, M.D. (1993) Management of rectal tears. In: McKinnon, A.O. and Voss, J.L. (eds) *Equine Reproduction*. Lea and Febiger, Philadelphia, pp. 464–472.

Spincemaille, J., Bouters, R., Vanderplassche, M. and Bante, P. (1975) Some aspects of endometrial cup formation and PMSG production. *Journal of Reproduction and Fertilty, Supplement* 23, 415–418.

Sprouse, R.F., Garner, H.E. and Lager, K. (1989) Protection of ponies from heterologous and homologous endotoxin challenges via Salmonella typhimurium bacterin-toxoid. *Equine Practice* 11, 34–40.

Squires, E.L. (1993a) Progesterone. In: McKinnon, A.O. and Voss, J.L. (eds) *Equine Reproduction*. Lea and Febiger, Philadelphia, pp. 57–64.

Squires, E.L. (1993b) Progestin. In: McKinnon, A.O. and Voss, J.L. (eds) *Equine Reproduction*. Lea and Febiger, Philadelphia, London, pp. 311–318.

Squires, E.L. (1993c) Endocrinology of pregnancy. In: McKinnon, A.O. and Voss, J.L. (eds) *Equine Reproduction*. Lea and Febiger, Philadelphia, pp. 495–500.

Squires, E.L. (1993d) Estrus detection. In: McKinnon, A.O. and Voss, J.L. (eds) *Equine Reproduction*. Lea and Febiger, Philadelphia, pp. 186–195.

Squires, E.L. (1993e) Embryo transfer. In: McKinnon, A.O. and Voss, J.L. (eds) *Equine Reproduction*. Lea and Febiger, Philadelphia, pp. 357–367.

Squires, E.L. and Ginther, O.J. (1975) Follicular and luteal development in pregnant mares. *Journal of Reproduction and Fertility, Supplement* 23, 429–433.

Squires, E.L. and McCue, P.M. (2007) Superovulation in mares. *Animal Reproduction Science* 99(1–2), 1–8.

Squires, E.L. and Seidel, G.E. (1995) *Collection and Transfer of Equine Embryos*. Colorado State University, Fort Collins, Colorado, pp. 11–16.

Squires, E.L., Stevens, W.B., McGlothin, D.E. and Pickett, V.W. (1979a) Effect of an oral progestagen the oestrous cycle and fertility of mares. *Journal of Animal Science* 49, 729–735.

Squires, E.L., Pickett, B.W. and Amann, R.P. (1979b) Effect of successive ejaculation on stallion seminal characteristics. *Journal of Reproduction and Fertility, Supplement* 27, 7–12.

Squires, E.L., Webel, S.K., Shideler, R.K. and Voss, J.L. (1981a) A review on the use of altrenogest for the broodmare. *Proceedings of the 26th Annual Convention of the American Association of Equine Practitioners*, pp. 221–231.

Squires, E.L., McGlothlin, D.E., Bowen, R.A., Berndtson, W.E. and Pickett, B.W. (1981b) The use of antibiotics in stallion semen for the control of *Klebsiella pneumoniae* and *Pseudomonas aeroginosa*. *Equine Veterinary Science* 1, 43–48.

Squires, E.L., Todter, G.E., Berndtson, W.E. and Pickett, B.W. (1982a) Effect of anabolic steroids on reproductive function of young stallions. *Journal of Animal Science* 54, 576–582.

Squires, E.L., Tunel, K.J., Iuliano, M.F. and Shidler, R.K. (1982b) Factors affecting reproductive efficiency in an equine transfer programme. *Journal of Reproduction and Fertility, Supplement* 32, 409–414.

Squires, E.L., Heesemann, C.P., Webel, S.K., Shideler, R.K. and Voss, J.L. (1983a) Relationship of altrenogest to ovarian activity, hormone concentrations and fertility in mares. *Journal of Animal Science* 56, 901–910.

Squires, E.L., Voss, J.L. and Villahoz, M.D. (1983b) Immunological methods for pregnancy detection in mares. *Proceedings of the 28th Annual Convention of the American Association of Equine Practitioners*, pp. 45–51.

Squires, E.L., Garcia, R.H. and Ginther, O.J. (1985) Factors affecting the success of equine embryo transfer. *Equine Veterinary Journal, Supplement* 3, 92–95.

Squires, E.L., McClain, M.G., Ginther, O.J. and McKinnon, A.O. (1987) Spontaneous multiple ovulation in the mare and its effect on the incidence of twin embryo collections. *Therionology* 28, 609–614.

Squires, E.L., McKinnon, A.O. and Shideler, R.K. (1988) Use of ultrasonography in reproductive management of mares. *Theriogenology* 29, 55–70.

Squires, E.L., Seidel, G.E. Jr. and McKinnon, A.O. (1989) Transfer of cryopreserved equine embryos to progestin treated ovarectomised mares. *Equine Veterinary Journal, Supplement* 8, 89–95.

Squires, E.L., Badzinski, S.L., Amann, R.P., McCue, P.M. and Nett, T.M. (1997) Effects of altrenogest on scrotal width, seminal characteristics, concentration of LH and testosterone and sexual behaviour of stallions. *Theriogenology* 48(2), 313–328.

Squires, E.L., McCue, P.M. and Vanderwall, D. (1999) The current status of equine embryo transfer. *Theriogenology* 51, 91–100.

Squires, E.L., Carnevale, E.M., McCue, J.E. and Bruemmer, J.E. (2003) Embryo technologies in the horse. *Theriogenology* 59, 151–170.

Stabenfeldt, G.H. (1979) Clinical findings, pathological changes and endocrinological secretory patterns in mares with ovarian tumours. *Journal of Reproduction and Fertility, Supplement* 27, 277–285.

Stabenfeldt, G.H., Hughes, J.P., Kennedy, P.C., Meagher, D.M. and Neely, D.P. (1979) Clinical findings, pathological changes and endocrinological secretory patterns in mares with ovarian tumours. *Journal of Reproduction and Fertility, Supplement* 27, 277–285.

Stabenfeldt, G.H., Daels, P.F., Munro, C.J., Kindahl, H., Hughes, J.P. and Lasley, B. (1991) An oestrogen conjugate enzyme immunoassay for monitoring pregnancy in the mare: limitations of the assay between days 40 and 70 of gestation. *Journal of Reproduction and Fertility, Supplement* 44, 37–44.

Stainer, W.B., Akers, R.M., Williams, C.A., Kronfeld, D.S. and Harris, P.A. (2001) Plasma insulin-like growth factor-I (IGF-I) in growing Thoroughbred foals fed a fat and fiber versus a sugar and a starch supplement. *Proceedings of the 17th Equine Nutrition and Physiology Symposium*, The University of Kentucky, Lexington, 31 May–2 June 2001, pp. 176–177.

Stanton, M.B., Steiner, J.V. and Pugh, D.G. (2004) Endometrial cysts in the mare. *Journal of Equine Veterinary Science* 24, 14–19.

Starbuck, G.R., Stout, T.A.E., Lamming, G.E., Allem, W.R. and Flint, A.P.R. (1998) Endometrial oxytocin receptor and uterine prostaglandin secretion in mares during the oestrous cycle and early pregnancy. *Journal of Reproduction and Ferility* 113, 173–179.

Stashak, T.S. (1993) Inguinal hernia. In: McKinnon, A.O. and Voss, J.L. (eds) *Equine Reproduction*. Lea and Febiger, Philadelphia, London, pp. 925–932.

Stashak, T.S. and Vandeplassche, M. (1993) Cesarean section. In: McKinnon, A.O. and Voss, J.L. (eds) *Equine Reproduction*. Lea and Febiger, Philadelphia, pp. 437–443.

Stecco, R., Paccamonti, D., Gutjahr, S., Pinto, C.R.F. and Eilts, B. (2003) Day of cycle affects changes in equine intrauterine pressure in response to teasing. *Theriogenology* 60(4), 727–733.

Steiner, J.N. (2000) Breeding management of the Thoroughbred stallion. In: Samper, J.C. (ed.) *Equine Breeding Management and Artificial Insemination*. W.B. Saunders, Philadelphia, pp. 67–72.

Stevenson, K.R., Parkinson, T.J. and Wathes, D.C. (1991) Measurements of oxytocin concentration in plasma and ovarian extracts during the oestrus cycle of mares. *Journal of Reproduction and Fertility, Supplement* 93, 437–441.

Stewart, D.R., Stabenfeldt, G.H., Hughes, J.P. and Meagher, D.M. (1982a) Determination of the source of elaxir. *Biology of Reproduction* 27, 17–25.

Stewart, D.R., Stabenfeldt, G.H. and Hughes, J.P. (1982b) Relaxin activity in foaling mares. *Journal of Reproduction and Fertility, Supplement* 32, 603–609.

Stewart, F., Allen, W.R. and Moor, R.M. (1977) Influence of foetal genotype on FSH:LH ratio of PMSG. *Journal of Endocrinology* 73, 415–425.

Stewart, F., Charleston, B., Crossett, B., Baker, P.J. and Allen, W.R. (1995) A novel uterine protein that associates with the embryonic capsule in equids. *Journal of Reproduction and Fertility* 105, 65–70.

Stewart, J.H., Rose, R.J. and Barko, A.M. (1984) Respiratory studies in foals from birth to seven days old. *Equine Veterinary Journal* 16, 323–328.

Stone, R. (1994) Timing of mating in relation to ovulation to achieve maximum reproductive efficiency in the horse. *Equine Veterinary Education* 6, 29–31.

Stoneham, S.J. (1991) Failure of passive transfer of colostral immunity in the foal. *Equine Veterinary Education* 3, 43–44.

Stout, T.A. (2005) Modulating reproductive activity in stallions: a review. *Animal Reproduction Science* 89(1–4), 93–103.

Stout, T.A. (2006) Equine embryo transfer: review of developing potential. *Equine Veterinary Journal* 38(5), 467–478.

Stout, T.A. and Allen, W.R. (2001) Role of prostaglandins in intrauterine migration of the equine conceptus. *Reproduction* 121(5), 771–775.

Stout, T.A. and Colenbrander, B. (2004) Suppressing reproductive activity in horses using GnRH vaccines, antagonists or agonists. *Animal Reproduction Science* 82–83, 633–643.

Stout, T.A.E., Lamming, G.E. and Allen, W.R. (2000) Oxytocin and its endometrial receptor are integral to luteolysis in the cycling mare. *Journal of Reproduction and Fertility, Supplement* 56, 281–287.

Stradaioli, G., Chiacchiarini, P., Monaci, M., Verini Supplizi, A., Martion, G. and Piermati, C. (1995) Reproductive characteristics and seminal plasma carnitine concentration in maiden maremmano stallions. *Proceedings of the 46th Annual Meeting of the European Association for Animal Production*, Prague, 4–7 September.

Strzemienski, P.J., Sertich, P.L., Varner, D.D. and Kenney, R.M. (1987) Evaluation of cellulose acetate/nitrate filters for the study of stallion sperm motility. *Journal of Reproduction and Fertility, Supplement* 35, 33–38.

Sukalic, M., Herak, M. and Ljubesic, J. (1982) The first use of imported deep frozen stallion semen in Yugoslavia. *Veterinarski Glasnik* 36(12), 1027–1032.

Sullivan, J.J. and Pickett, B.W. (1975) Influence of ejaculation frequency of stallions on characteristics of semen and output of spermatozoa. *Journal of Reproduction and Fertility, Supplement* 23, 29–34.

Sullivan, J.J., Turner, P.C., Self, L.C., Gutteridge, H.B. and Bartell, D.E. (1975) Survey of reproductive efficiency in the quarter-horse and Thoroughbred. *Journal of Reproduction and Fertility, Supplement* 23, 315–318.

Sundberg, J.P., Burnstein, T., Page, E.H., Kirkham, W.W. and Robinson, F.R. (1977) Neoplasms of equidae. *Journal of the American Veterinary Medical Association* 170, 150–152.

Sutton, E.I., Bowland, J.P. and Rattcliff, W.D. (1977) Influence of level of energy and nutrient intake by mares on reproductive performance and on blood serum composition of the mares and foals. *Canadian Journal of Animal Science* 57, 551–558.

Swerczek, T.W. (1975) Immature germ cells in the semen of Thoroughbreds. *Journal of Reproduction and Fertility, Supplement* 23, 135–137.

Swierstra, E.E., Pickett, B.W. and Gebauer, M.R. (1975) Spermatogenesis and duration of transit of spermatozoa through the excurrent ducts of stallions. *Journal of Reproduction and Fertility, Supplement* 23, 53–57.

Swinker, A.M., Squires, E.L., Mumford, E.L., Knowles, J.E. and Kniffen, D.M. (1993) Effect of body weight and body condition score on follicular development and ovulation in mares treated with GnRH analogue. *Journal of Equine Veterinary Science* 13, 519–520.

Tainturier, D.J. (1981) Bacteriology of *Hemophilus equigenitalis*. *Journal of Clinical Microbiology* 14, 355.

Takagi, M., Nishimura, K., Oguri, N., Ohnuma, K., Ito, K., Takahashi, J., Yasuda, Y., Miyazawa, K. and Sato, K. (1998) Measurement of early pregnancy factor activity for monitoring the viability of the equine embryo. *Theriogenology* 50, 255–262.

Tanaka, Y., Nagamine, N., Nambo, Y., Nagata, S., Nagaoka, K., Tsunoda, N., Taniyama, H., Yoshihara, T., Oikawa, M., Watanbe, G. and Taya, K. (2000) Ovarian secretion of inhibin in mares. *Journal of Reproduction and Fertility, Supplement* 56, 239–245.

Terqui, M. and Palmer, E. (1979) Oestrogen pattern during pregnancy in the mare. *Journal of Reproduction and Fertility, Supplement* 27, 441–446.

Tetzke, T.A., Ismail, S., Mikuckis, G. and Evans, J.W. (1987) Patterns of oxytocin secretion during the oestrous cycle of the mare. *Journal of Reproduction and Fertility, Supplement* 35, 245–252.

Teuscher, C., Kenney, R.M., Cummings, M.R. and Catten, M. (1994) Identification of two stallion sperm specific proteins and their autoantibody response. *Equine Veterinary Journal* 26(2), 148–151.

Tharasanit, T., Colenbrander, B. and Stout, T.A. (2006) Effect of maturation stage at cryopreservation on post-thaw cytoskeleton quality and fertilizability of equine oocytes. *Molecular Reproductive Development*. 73(5), 627–637.

Thomas, P.G.A. and Ball, B.A. (1996) Cytofluorescent assay to quantify adhesion of equine spermatozoa to oviduct epithelial cells in vitro. *Molecular Reproduction and Development* 43(1), 55–61.

Thomas, P.G.A., Ignotz, G.G., Ball, B.A., Brinsko, S.P. and Currie, W.B. (1995) Effect of coculture with stallion spermatozoa on de novo protein synthesis and secretion by equine oviduct epithelial cells. *American Journal of Veterinary Research* 56(12), 1657–1662.

Thomassen, R. (1991) Use of frozen semen for artificial insemination in mares, results in 1990. *Norsk Veterinaertidsskrift* 103(3), 213–216.

Thompson, D.L. Jr. (1992) Reproductive physiology of stallions and jacks. In: Warren Evans, J. (ed.) *Horse Breeding and Management*. Elsevier, Amsterdam, pp. 237–261.

Thompson, D.L. (1994) Breeding management of stallions: breeding soundness evaluations. *Journal of Equine Veterinary Science* 14(1), 19–20.

Thompson, D.L. and Honey, P.G. (1984) Active immunisation of prepubertal colts against estrogens, hormonal and testicular response after puberty. *Journal of Animal Science* 51, 189–196.

Thompson, D.L. and Johnson, L. (1987) Effects of age, season and active immunisation against oestrogen on serum prolactin concentrations in stallions. *Domestic Animal Endocrinology* 4, 17–22.

Thompson, D.L. Jr., Pickett, B.W., Berndtson, W.E., Voss, J.L. and Nett, T.H. (1977) Reproductive physiology of the stallion. VIII. Artificial photoperiod, collection interval and seminal characteristics, sexual behaviour and concentration of LH and testosterone in serum. *Journal of Animal Science* 44(4), 656–664.

Thompson, D.L. Jr., Pickett, B.W., Squires, E.L. and Amman, R.P. (1979) Testicular measurements and reproductive characteristics in stallions. *Journal of Reproduction and Fertility, Supplement* 27, 13–17.

Thompson, D.L. Jr., Pickett, B.W., Squires, E.L. and Nett, T.M. (1980) Sexual behaviour, seminal pH and accessory sex gland weights in geldings administered testosterone and/or estradiol 17β. *Journal of Animal Science* 51, 1358–1366.

Thompson, D.L., Weist, J.J. and Nett, T.M. (1986) Measurement of equine prolactin with an equine-canine radioimmunoassay, seasonal affects of prolactin response to thyrotrophin releasing hormone. *Domestic Animal Endocrinology* 3, 247–251.

Thompson, K.N. (1995) Skeletal growth of weanling and yearling Thoroughbred horses. *Journal of Animal Science* 73, 2513–2517.

Thompson, K.N., Jackson, S.G. and Baker, J.P. (1988) The influence of high planes of nutrition on skeletal growth and development of weanling horses. *Journal of Animal Science* 66, 2459–2467.

Thomson, C.H., Thompson, D.L. Jr., Kincaid, L.A. and Nadal, M.R. (1996) Prolactin involvement with increase in seminal volume after sexual stimulation in stallions. *Journal of Animal Science* 74(10), 2468–2472.

Threlfall, W.R. (1979). Antibiotic infusion of the uterus of the mare. *Proceedings of the Society of Theriogenology*, pp. 45–50.

Threlfall, W.R. (1993) Retained placenta. In: McKinnon, A.O. and Voss, J.L. (eds) *Equine Reproduction*. Lea and Febiger, Philadelphia, pp. 614–621.

Threlfall, W.R. and Carleton, C.L. (1996) Mare's genital tract. In: Traub-Dargatz, J.L. and Brown, C.M. (eds) *Equine Endoscopy*. 2nd edn. Mosby, St. Louis, Missouri, pp. 204–217.

Threlfall, W.R., Carelton, C.L., Robertson, J., Rosol, T. and Gabel, A. (1990) Recurrent torsion of the spermatic chord and scrotal testis in a stallion. *Journal of American Veterinary Association* 196, 1641–1643.

Timoney, P.J. and McCollum, W.H. (1987) Equine viral arteritis. *Canadian Veterinary Journal* 28, 693–695.

Timoney, P.J. and McCollum, W.H. (1988) Equine viral arteritis: epidemiology and control. *Equine Veterinary Science* 8, 54–59.

Timoney, P.J. and McCollum, W.H. (1997) Equine viral arteritis: essential facts about the disease. *Proceedings of the 43rd Annual American Association of Equine Practitioners Convention* 43, 189–195.

Timoney, P.J., McCollum, W.H., Murphy, T.W., Roberts, A.W., Willard, J.G. and Caswell, G.D. (1988) The carrier state in equine arteritis virus infection in the stallion with specific emphasis on the venereal mode of virus transmission. *Journal of Reproduction and Fertility, Supplement* 35, 95–102.

Tischner, M., Kosiniak, K. and Bielanski, W. (1974) Analysis of ejaculation in stallions. *Journal of Reproduction and Fertility* 41, 329–335.

Toal, R. (1996) *Ultrasound for Practitioners*. PSI, Effingham, 211 pp.

Torbeck, R.L. (1986) Diagnostic ultrasound in equine reproduction. *Veterinary Clinics of North America Large Animal Practice* 2, 227–252.

Torres-Bogino, F., Sato, K., Oka, A., Kamo, Y., Hochi, S.I., Oguri, N. and Braun, J. (1995) Relationship among seminal characteristics, fertility and suitability for semen preservation in draft stallions. *Journal of Veterinary Medical Science* 57(2), 225–229.

Traub-Dargatz, J.L. (1993a) Post natal care of the foal. In: McKinnon, A.O. and Voss, J.L. (eds) *Equine Reproduction*. Lea and Febiger, Philadelphia, pp. 981–984.

Traub-Dargatz, J.L. (1993b) Disorders of the digestive tract. In: McKinnon, A.O. and Voss, J.L. (eds) *Equine Reproduction*. Lea and Febiger, Philadelphia, pp. 1023–1029.

Troedsson, M.H., Desvousges, A., Alghamdi, A.S., Dahma, B., Dow, C.A., Hayna, J., Valesco, R., Collahan, P.T., Macpherson, M.L., Pozor, M. and Buhi, W.C. (2005) Components in seminal plasma regulating sperm transport and elimination. *Animal Reproduction Science* 89(1–4), 171–186.

Troedsson, M.H.T. (1999) Uterine clearance and resistance to persistent endometritis in the mare. *Theriogenology* 52, 461–471.

Tucker, H.A. and Wetterman, R.P. (1976) Effects of ambient temperature and relative humidity on serum prolactin and growth hormone in heifers. *Proceedings of the Society for Experimental Biology and Medicine* 151, 623–630.

Tucker, K.E., Henderson, K.A. and Duby, R.T. (1991) In vitro steroidogenesis by granulosa cells from equine preovulatory follicles. *Journal of Reproduction and Fertility, Supplement* 44, 45–55.

Turner, A.S. and McIlwraith, C.W. (1982) Umbilical herniorrhaphy in the foal. In: Turner, A.S. and McIlwraith, C.W. (eds) *Techniques in Large Animal Surgery*. Lea and Febiger, Philadelphia, pp. 254–259.

Turner, R.M. (1998) Ultrasonography of genital tract of stallion. In: Reef, V.B. (ed.) *Equine Diagnostic Ultrasound*. W.B. Saunders, Philadelphia, pp. 446–479.

Turner, R.M., McDonnell, S.M., Feit, E.M., Grogan, E.H. and Foglia, R. (2006) Real-time ultrasound measure of the fetal eye (vitreous body) for prediction of parturition date in small ponies. *Theriogenology* 66(2), 331–337.

Tyznik, W.J. (1972) Nutrition and disease. In: Catcott, E.J. and Smithcors, J.R. (eds) *Equine Medicine and Surgery*. 2nd edn. American Veterinary Publications, Santa Barbara, California, pp. 239–250.

Ullrey, D.E., Struthers, R.D., Hendricks, D.G. and Brent, B.E. (1966) Composition of mare's milk. *Journal of Animal Science* 25, 217–222.

Umphenor, M.W., Sprinkle, T.A. and Murphy, H.Q. (1993) Natural service. In: Mckinnon, A.O. and Voss, J.L. (eds) *Equine Reproduction*. Lea and Febiger, Philadelphia, pp. 798–820.

Vaala, W.E. (1993) The cardiac and respiratory systems. In: McKinnon, A.O. and Voss, J.L. (eds) *Equine Reproduction*. Lea and Febiger, Philadelphia, pp. 1041–1059.

Vaillencourt, D., Fretz, P. and Orr, J.P. (1979) Seminoma in the horse. Report of 2 cases. *Journal of Equine Medical Surgery* 3, 213–218.

Vaillencourt, D., Gucy, P. and Higgins, R. (1993) The effectiveness of gentamicin or polymixin B for the control of bacterial growth in equine semen stored at 20°C or 5°C for up to fourty eight hours. *Canadian Journal of Veterinary Research* 57(4), 277–280.

Van Buiten, A., Zhang, J. and Boyle, M.S. (1989) Integrity of plasma membrane of stallion spermatozoa before and after freezing. *Journal of Reproduction and Fertility* 4, 18–22.

Van Buiten, A., Van der Broek, J., Schukken, Y.H. and Colenbrander, B. (1999) Validation of non-return rates as a parameter for stallion fertility. *Livestock Production Science* 60, 13–19.

Van Camp, S.D. (1993) Uterine abnormalities. In: McKinnon, A.O. and Voss, J.L. (eds) *Equine Reproduction*. Lea and Febiger, Philadelphia, London, pp. 392–396.

Vandeplassche, M. (1975) Uterine prolapse in the mare. *Veterinary Record* 97, 19.

Vandeplassche, M. (1980) Obstetrician's views of the physiology of equine parturition and dystocia. *Equine Veterinary Journal* 12, 45–49.

Vandeplassche, M. (1987) The pathogenesis of dystocia and fetal malformation in the horse. *Journal of Reproduction and Ferility, Supplement* 35, 547–552.

Vandeplassche, M. (1993) Dystocia. In: McKinnon, A.O. and Voss, J.L. (eds) *Equine Reproduction*. Lea and Febiger, Philadelphia, pp. 578–587.

Vandeplassche, M. and Henry, M. (1977) Salpingitis in the mare. *Proceedings of the 23rd Annual Convention of the American Association of Equine Practitioners*, pp. 123–129.

Vandeplassche, M., Spincmaille, J. and Bouters, R. (1971) Aetiology, pathogenesis and treatment of retained placenta in the mare. *Equine Veterinary Journal* 3(4), 144–137.

Vandeplassche, M., Spincmaille, J. and Bouters, R. (1976) Dropsy of the fetal sac in mares: induced and spontaneous abortion. *Veterinary Record* 99, 67–69.

Vandeplassche, M., Bouters, R. and Spincemaille, J. (1979a) Caesarian section in the mare. *Proceedings of the American Association of Equine Practitioners* 25, 75–80.

Vandeplassche, M., Henry, M. and Coryn, M. (1979b) The mature mid-cycle follicle in the mare. *Journal of Reproduction and Fertility, Supplement* 27, 157–162.

Van der Holst, W. (1984) Stallion semen production in AI programs in the Netherlands. In: Courot, M. (ed.) *The Male in Farm Animal Production*. Martinus Nijhoff, Boston, Massachusetts, pp. 195–201.

Van der Veldon, M. (1988) Surgical treatment of acquired inguinal hernia in the horse: a review of 39 cases. *Equine Veterinary Journal* 20, 173–177.

Vanderwall, D.K. and Woods, G.L. (2003) Effect on fertility of uterine lavage performed immediately prior to insemination in mares. *Journal of the American Veterinary Association* 222(8), 1108–1110.

Vanderwall, D.K., Woods, G.L., Freeman, D.A., Weber, A., Roch, R.W. and Tester, D.F. (1993) Ovarian follicles, ovulations and progesterone concentrations in aged versus young mares. *Theriogenology* 40, 21–32.

Vanderwall, D.K., Juergens, T.D. and Woods, G.L. (2001) Reproductive performance of commercial broodmares after induction of ovulation either hCG or ovuplant (Deslorelin). *Journal of Equine Veterinary Science* 21, 539–542.1

Vanderwall, D.K., Woods, G.L., Aston, K.I., Bunch, T.D., Meerdo, L.N. and White, K.L. (2004) Cloned horse pregnancies produced using adult cumulus cells. *Reproduction Fertilization and Development* 16(7), 675–679.

Van Dierendonck, M. and Goodwin, G. (2005) Human animal relationship. *Animals in Philosophy and Science* 4(2), 25–35.

Van Duijn, C. Jr. and Hendrikse, J. (1968) Rational analysis of seminal characteristics of stallions in relation to fertility. Instituut woor Veeteelkundig Onderzoek "Schoonoord" Report B97, Pretoria (Cited in Pickett, 1993a).

Van Furth, R., Cohn, Z.A., Hirsch, J.G., Humphrey, J.H., Spector, W.G. and Langevoot, H.L. (1972) The mononuclear phagocyte system: a new classification of macrophages, monocytes and their cell lines. *Bulletin of the World Health Organisation* 47(5), 651–658.

Van Huffel, X.M., Varner, D.D., Hinrichs, K., Garcia, M.C., Stremienski, P.J. and Kenney, R.M. (1985) Photomicrographic evaluation of stallion spermatozoal motility characteristics. *American Journal of Veterinary Research* 46(6), 1272–1275.

Van Maanen, C., Bruin, G., de Boer-Luijtze, E., Smolders, G. and de Boer, G.F. (1992) Interference of maternal antibodies with the immune response of foals after vaccination against equine influenza. *Veterinary Quaterly* 14(1), 13–17.

Van Niekerk, C.H. (1965) Early embryonic resorption in mares. A preliminary study. *Journal of the South African Veterinary Medical Association* 36, 61–70.

Van Niekerk, C.H. and Allen, W.E. (1975) Early embryonic development in the horse. *Journal of Reproduction and Fertility, Supplement* 23, 495–498.

Van Nierkerk, C.H. and Morgenthal, J.C. (1976) Plasma progesterone and oestrogen concentrations during induction of parturition with flumethasone and prostaglandin. *Proceedings of the 8th International Congress on Animal Reproduction and AI*. Krackow 3, 386–389.

Van Niekerk, C.H. and Morgenthal, J.C. (1982) Foetal loss and the effect of stress on plasma progesterone levels in pregnant Thoroughbred mares. *Journal of Reproduction and Fertility, Supplement* 32, 453–457.

Van Niekerk, C.H. and Van Heerden, J.S. (1972) Nutritional and ovarian activity of mares early in the breeding season. *Journal of the South African Veterinary Medicine Association* 43(4), 351–360.

Van Niekerk, C.H. and Van Heerden, J.S. (1997) The effect of dietary protein on reproduction in the mare. III. Ovarian and uterine changes during the anovulatory season, transitional and ovulatory periods in the non-pregnant mare. *Journal of the South African Veterinary Association* 68, 86–92.

Van Niekerk, C.H., Coughborough, R.I. and Doms, H.W.H. (1973) Progesterone treatment of mares with abnormal oestrus cycle early in the breeding season. *Journal of the South African Veterinary Association* 44, 37–45.

Varner, D.D. (1983) Equine perinatal care, part 1. Prenatal care of the dam. *Compendium of Continuing Education Practical Vet* 5, S356–S362.

Varner, D.D. (1986) Collection and preservation of stallion spermatozoa. *Proceedings of the Annual Meeting (1986) of the Society of Theriogenology*, pp. 13–33.

Varner, D.D. (1991) Composition of seminal extenders and its effect on motility of equine spermatozoa. *Proceedings of the Annual Meeting of the Society for Theriogenology* 1991, 146–150.

Varner, D.D. and Schumacher, J. (1991) Diseases of the reproductive system: the stallion. In: Colahan, P.T., Mayhew, I.G., Merritt, A.M. and Moore, J.N. (eds) *Equine Medicine and Surgery*, Vol. 2, 4th edn. American Veterinary Publication Incorporated, California, pp. 847–948.

Varner, D.D. and Schumacher, J. (1999) Diseases of the scrotum. In: Colahan, P.T., Merritt, A.M. and Moore, J.N. (eds) *Equine Medicine and Surgery*, 5th edn. Mosby, Philadelphia, pp. 1034–1035.

Varner, D.D., Blanchard, T.L., Love, C.L., Garcia, M.C. and Kenney, R.M. (1987) Effects of sperm fractionation and dilution ratio on equine spermatozoal motility parameters. *Theriogenology* 28, 709–723.

Varner, D.D., Blanchard, T.L., Love, C.L., Garcia, M.C. and Kenney, R.M. (1988) Effects of cooling rate and storage temperature on equine spermatozoal motility parameters. *Theriogenology* 29, 1043–1054.

Varner, D.D., Blanchard, T.L., Meyers, P.S. and Meyers, S.A. (1989) Fertilizing capacity of equine spermatozoa stored for 24 hours at 5°C or 20°C. *Theriogenology* 32, 515–525.

Varner, D.D., Schumacher, J., Blanchard, T.L. and Johnson, L. (1991) *Diseases and Management of Breeding Stallions*. American Veterinary Publications, Goleta, California.

Varner, D.D., McIntosh, A.L., Forrest, D.W., Blanchard, T.L. and Johnson, L. (1992) Potassium penicillin G, amikacin sulphate or a combination in seminal extenders for stallions: effects on spermatozoal motility. *Proceedings of the International Congress of Animal Reproduction and AI* 12, 1496–1498.

Varner, D.D., Taylor, T.S. and Blanchard, T.L. (1993) Seminal vesiculitis. In: McKinnon, A.O. and Voss, J.L. (eds) *Equine Reproduction*. Lea and Febiger, Philadelphia, pp. 861–863.

Vasey, J.R. (1993) Uterine torsion. In: McKinnon, A.O. and Voss, J.L. (eds) *Equine Reproduction*. Lea and Febiger, Philadelphia, pp. 456–460.

Vaughan, J.T. (1993) Penis and prepuce. In: McKinnon, A.O. and Voss, J.L. (eds) *Equine Reproduction*. Lea and Febiger, Philadelphia, pp. 885–894.

Vernon, M.W., Strauss, S., Simonelli, M., Zavy, M.T. and Sharp, D.C. (1979) Specific PGF2x binding by the corpus luteum of the pregnant and non-pregnant mare. *Journal of Reproduction and Fertility, Supplement* 27, 421–429.

Veronesi, M.C., Battocchio, M., Faustini, M., Gandini, M. and Cairoli, F. (2003) Relationship between pharmacological induction of estrous and/or ovulation and twin pregnancy in the Thoroughbred mare. *Domestic Animal Endocrinology* 25, 133–140.

Vidament, M., Dupere, A.M., Julienne, P., Evain, A., Noue, P. and Palmer, E. (1997) Equine frozen semen: freezability and fertility field results. *Theriogenology* 48(6), 907–917.

Villani, C., Sighieri, C., Tedeschi, D., Gazzano, G., Sampieri, G.U., Ducci, M. and Martelli, F. (2000) Serial progesterone measurements in early pregnancy diagnosis in the mare. *Selezione Vetinaria, Supplement* s287–s291.

Vivette, S.L., Reimers, T.J. and Knook, L. (1990) Skeletal diseases in a hypothyroid foal. *Journal of the American Veterinary Association* 197, 1635–1638.

Vogelsang, M.M., Kraemer, D.C., Potter, G.D. and Scott, G.G. (1987) Fine structure of the follicular oocyte of the horse. *Journal of Reproduction and Fertility, Supplement* 35, 157–167.

Vogelsang, S.G., Sorensen, A.M. Jr., Potter, G.D., Burns, S.J. and Dreamer, D.C. (1979) Fertility of donor mares following non-surgical collection of embryos. *Journal of Reproduction and Fertility, Supplement* 27, 383–386.

Volkmann, D.H., Botschinger, H.J. and Schulman, M.L. (1995) The effect of prostaglandin E₂ on the cervices of dioestrus and pre-partum mares. *Reproduction in Domestic Animals* 30, 240–244.

Voss, J.L. (1993) Human chorionic gonadotrophin. In: McKinnon, A.O. and Voss, J.L. (eds) *Equine Reproduction*. Lea and Febiger, Philadelphia, pp. 325–328.

Voss, J.L. and McKinnon, A.O. (1993) Hemospermia and urospermia. In: McKinnon, A.O. and Voss, J.L. (eds) *Equine Reproduction*. Lea and Febiger, Philadelphia, pp. 864–870.

Voss, J.L. and Pickett, B.W. (1975) The effect of rectal palpation on the fertility of cyclic mares. *Journal of Reproduction and Fertility, Supplement* 23, 285–290.

Voss, J.L. and Pickett, B.W. (1976) *Reproductive Management of Broodmare*. Animal Reproduction Laboratory General Series Bulletin No. 961. Colorado State University, Fort Collins, Colorado.

Voss, J.L., Sullivan, J.J., Pickett, B.W., Parker, W.G., Burwash, L.D. and Larson, L.L. (1975) The effect of hCG on the duration of oestrus, ovulation time and fertility in mares. *Journal of Reproduction and Fertility, Supplement* 23, 297–301.

Voss, J.L., Wallace, R.A., Squires, E.L., Pickett, B.W. and Shideler, R.K. (1979) Effects of synchronisation and frequency of insemination on fertility. *Journal of Reproduction and Fertility, Supplement* 27, 257–261.

Waelchi, R.O., Gerber, D., Volkmann, D.H. and Betteridge, K.J. (1996) Changes in the osmolarity of the equine blastocyst fluid between days 11 and 25 of pregnancy. *Theriogenology* 45, 290.

Wagoner, D.M. (1982) *Breeding Management and Foal Development*. Equine Research, Texas, pp. 351–358.

Walt, M.L., Stabenfeldt, G.H., Hughes, J.P., Neely, D.P. and Bradbury, R. (1979) Development of the equine ovary and ovulation fossa. *Journal of Reproduction and Fertility, Supplement* 27, 471–477.

Walter, I., Handler, J., Reifinger, M. and Aurich, C. (2001) Association of endometrosis in horses with differentiation of periglandular myofibroblasts and changes of extracellular matrix proteins. *Reproduction* 121, 581–586.

Warren, L.K., Lawrence, L.M., Griffin, A.S., Parker, A.L., Barnes, T. and Wright, D. (1998a) Effect of weaning on foal growth and radiographic density. *Journal of Equine Veterinary Science* 18(5), 335–342.

Warren, L.K., Lawrence, L.M., Parker, A.L., Barnes, T. and Griffin, A.S. (1998b) The effect of weaning age on foal growth and radiographic bone density. *Proceedings of 16th Equine Nutrition and Physiology Symposium* 18(5), 335–340.

Warren Evans, J. (1991) *Horse Breeding and Management*. World Animal Science, Elsevier Scientific, Amsterdam, 325 pp.

Warszawsky, L.F., Parker, W.G., First, N.L. and Ginther, O.J. (1972) Gross changes of internal genitalia during the estrous cycle in the mare. *American Journal of Veterinary Research* 33(1), 19–26.

Waters, A.J., Nicol, C.J. and French, N.P. (2002) Factors influencing the development of stereotypic and redirected behaviours in young horses: findings of a four year prospective epidemiological study. *Equine Veterinary Journal* 34(6), 572–579.

Watson, E.D. (1988) Uterine defence mechanisms in mares resistant and susceptible to persistent endometritis. *Proceedings of the 34th American Association of Equine Practitioners*, pp. 279–387.

Watson, E.D. (1997) Fertility problems in stallions. *In Practice* 19(5), 260–269.

Watson, E.D. (2000) Post-breeding endometritis. *Animal Reproduction Science* 60–61, 221–232.

Watson, E.D. and Nikolakopoulos, E. (1996) Sperm longevity in the mare's uterus. *Journal of Equine Veterinary Science* 16(9), 390–392.

Watson, E.D., Clarke, C.J., Else, R.W. and Dixon, P.M. (1994a) Testicular degeneration in three stallions. *Equine Veterinary Journal* 26(6), 507–510.

Watson, E.D., McDonnell, A.M. and Cuddeford, D. (1994b) Characteristics of cyclicity in maiden thoroughbred mares in the United Kingdom. *Veterinary Record* 135, 104–106.

Watson, E.D., Heald, M., Tsigos, A., Leask, R., Steele, M., Groome, N.P. and Riley, S.C. (2002) Plasma FSH, inhibin A and inhibin isoforms containing pro- and -αC during winter anoestrus, spring transition and the breeding season in mares. *Reproduction* 123, 535–542.

Watson, P.F. (1990) AI and the preservation of semen. In: Lamming, G.E. (ed.) *Marshall's Physiology of Reproduction, Volume 2, Male Reproduction*. Churchill Livingston, London, pp. 747–869.

Watson, P.F. and Duncan, A.G. (1988) Effect of salt concentration and unfrozen water fraction on the viability of slowly frozen ram spermatozoa. *Cryobiology* 25, 131–142.

Webb, G.W., Arns, M.J. and Pool, K.C. (1993) Spermatozoa concentration influences the recovery of progressively motile spermatozoa and the number of inseminates shipped in conventional containers. *Journal of Equine Veterinary Science* 13, 486–489.

Webb, R.L., Evans, J.W., Arns, M.J., Webb, G.W., Taylor, T.S. and Potter, G.D. (1990) Effects of vesiculectomy on stallion spermatozoa. *Journal of Equine Veterinary Science* 10(3), 218–223.

Weber, J.A. and Woods, G.L. (1993) Ultrasonic measurements of stallion accessory glands and excurrent ducts during seminal emission and ejaculation. *Biology of Reproduction* 49(2), 267–273.

Weber, J.A., Geary, R.T. and Woods, G.L. (1990) Changes in accessory sex glands of stallions after sexual preparation and ejaculation. *Journal of the American Veterinary Medical Association* 186(7), 1084–1089.

Weber, J.A., Freeman, D.A., Vanderwall, D.K. and Woods, G.L. (1991) Prostaglandin E$_2$ hastens oviductal transport of equine embryos. *Biology of Reproduction* 45, 544–546.

Weedman, P.J., King, S.S., Newmann, K.R. and Nequin, L.G. (1993) Comparison of circulating oestradiol 17β and folliculogenesis during the breeding season, autumn transition and anestrus in the mare. *Journal of Equine Veterinary Science* 13, 502–505.

Welch, G.R. and Johnson, L.A. (1999) Sex preselection: laboratory validation of the sperm sex ratio of flow sorted X- and Y- sperm by sort reanalysis for DNA. *Theriogenology* 52, 1343–1352.

Welsch, B.B. (1993) The neurologic system. In: McKinnon, A.O. and Voss, J.L. (eds) *Equine Reproduction*. Lea and Febiger, Philadelphia, pp. 1017–1022.

Wesson, J.A. and Ginther, O.J. (1981) Influence of season and age on reproductive activity in pony mares on the basis of a slaughterhouse survey. *Journal of Animal Science* 52(1), 119–129.

Wheat, J.D. and Meager, D.M. (1972) Uterine torsion and rupture in mares. *Journal of the American Veterinary Medical Association* 160, 881–884.

Whitmore, H.L., Wentworth, B.C. and Ginther, O.J. (1973) Circulating concentrations of luteinising hormone during estrous cycles of mares as determined by radioimmunoassay. *American Journal of Veterinary Research* 34, 631–636.

Whittingham, D.G. (1971) Survival of mouse embryo after freezing and thawing. *Nature London* 233, 125–126.

Wide, M. and Wide, L. (1963) Diagnosis of pregnancy in mares by an immunological method. *Nature London* 198, 1017–1019.

Wilhelm, K.M., Graham, J.K. and Squires, E.L. (1996) Comparison of the fertility of cryopreserved stallion spermatozoa with sperm motion analyses, flow cytometry evaluation and zona-free hamster oocyte penetration. *Theriogenology* 46(4), 559–578.

Wilsher, S. and Allen, W.R. (2004) Special article, an improved method for nonsurgical embryo transfer in the mare. *Equine Veterinary Education* 16(1), 39–44.

Wilson, C.G., Downie, C.R., Hughes, J.P. and Roser, J.F. (1990) Effects of repeated hCG injection on reproductive efficiency in mares. *Journal of Equine Veterinary Science* 10, 301–308.

Wilson, G.L. (1985) Diagnostic and therapeutic hysteroscopy for endometrial cysts in mares. *Veterinary Medicine* 8, 59–62.

Wilson, J.H. (1987) Eastern equine encephalomyelitis. In: Robinson, N.E. (ed.) *Current Therapy on Equine Medicine*, 2nd edn. W.B. Saunders, Philadelphia, pp. 345–347.

Wilson, J.M., Rowley, M.B., Rowley, W.K., Smith, H.A., Webb, R.L. and Tolleson, D.R. (1987) Success of non surgical transfer of equine embryos in post partum mares. *Theriogenology* 27, 295.

Windsor, D.P., Evans, G. and White, I.G. (1993) Sex predetermination by separation of X and Y chromosome-bearing sperm: a review. *Reproduction, Fertility and Development* 5, 155–171.

Winskill, L.C., Warren, N.K. and Young, R.J. (1996) The effect of a foraging device (a modified Edinburgh football) on the behaviour of the stabled horse. *Applied Animal Behaviour Science* 48, 25–35.

Witherspoon, D.M. (1972) Technique and evaluation of uterine curettage. *Proceedings of the 18th Annual Convention of the American Association of Equine Practitioners*, pp. 51–60.

Witherspoon, D.M. (1975) The site of ovulation in the mare. *Journal of Reproduction and Fertility, Supplement* 23, 329–331.

Witherspoon, D.M. (1984) Vaccination against equine herpesvirus 1 and equine influenza infection. *Veterinary Record* 115, 363.

Wockener, A. and Collenbrander, B. (1993) Liquid storage and freezing of semen from New Forest and Welsh pony stallions. *Deutsche Tierarztliche Wochenschrift* 100(3), 125–126.

Wockener, A. and Schuberth, H.J. (1993) Freezing of maiden stallion semen – motility and morphology findings in sperm cells assessed by various staining methods including a monoclonal antibody with reactivity against an antigen in the acrosomal ground substance. *Reproduction in Domestic Animals* 28(6), 265–272.

Wood, J.L., Chirnside, E.D., Mumford, J.A. and Higgins, A.J. (1995) First recorded outbreak of equine viral arteritis in the United Kingdom. *Veterinary Record* 136, 381–385.

Woods, G.L. (1989) Pregnancy loss. A major cause of infertility in the mare. *Equine Practice* 29–32.

Woods, G.L., White, K.L., Vanderwall, D.K., Aston, K.I., Bunch, T.D. and Campbell, K.D. (2002) Cloned mule pregnancies produced using nuclear transfer. *Theriogenology* 58, 779–782.

Woods, G.L., White, K.L., Vanderwall, D.K., Li, G.P., Aston, K.I., Bunch, T.D., Meerdo, L.N. and Pate, P.L. (2003) A mule cloned from fetal cells by nuclear transfer. *Science* 301, 1063–1065.

Woods, J., Berfelt, D.R. and Ginther, O.J. (1990) Effects of time of insemination relative to ovulation on pregnancy rate embryonic loss rate in mares. *Equine Veterinary Journal* 22, 410–415.

Worthy, K., Escreet, R., Renton, J.P., Eckersall, P.D., Douglas, T.A. and Flint, D.J. (1986) Plasma prolactin concentrations and cyclic activity in pony mares during parturition and early lactation. *Journal of Reproduction and Fertility* 77(2), 569–574.

Worthy, K., Colquhoun, K., Escreet, R., Dunlop, M., Renton, J.P. and Douglas, T.A. (1987) Plasma prolactin concentrations in non- pregnant mares at different times of the year and in relation to events in the cycle. *Journal of Reproduction and Fertility, Supplement* 35, 269–276.

Wright, J.G. (1963) The surgery of the inguinal canal in animals. *Veterinary Record* 75, 1352–1363.

Wright, P.J. (1980) Serum sperm agglutinins and semen quality in the bull. *Australian Veterinary Journal* 56(1), 10–13.

Wu, W. and Nathanielsz, P.W. (1994) Changes in oxytocin receptor messenger RNA in the endometrium, myometrium, mesometrium and cervix of sheep in late gestation and during spontaneous and cortisol induced labour. *Journal of the Society of Gynaecological Investigation* 1(3), 191–196.

Yanagimachi, R., Yanagimachi, H. and Roger, B.J. (1976) The use of a zona-free animal ova as a test system for the assessment of the fertilising capacity of human spermatozoa. *Biology of Reproduction* 15, 471–472.

Yamamoto, K., Yasuda, J. and Too, K. (1992) Arrhythmias in newborn Thoroughbred foals. *Equine Veterinary Journal* 23, 169–173.

Yamamoto, Y., Oguri, N., Tsutsumi, Y. and Hachinohe, Y. (1982) Experiments in freezing and storage of equine embryos. *Journal of Reproduction and Fertility, Supplement* 2, 399–403.

Yates, D.J. and Whitacre, M.D. (1988) Equine artificial insemination. *Veterinary Clinics of North America. Equine Practice* 4(2), 291–304.

Yi, L.H., Li, Y.J. and Hao, B.Z. (1983) A study on the conception rate of frozen horse semen. *Chinese Journal of Animal Science* 3, 2–4.

Young, C.A., Squires, E.L., Seidel, G.E., Kato, H. and McCue, P.M. (1997) Cryopreservation procedures for day 7–8 equine embryos. *Equine Veterinary Journal, Supplement* 25, 98–102.

Yurdaydin, N., Tekin, N., Gulyuz, F., Aksu, A. and Klug, E. (1993) Field trials of oestrus synchronisation and artificial insemination results in Arab broodmare herd in the National Stud Hasire/Eskisehir (Turkey). *Deutsche Tierarztiche Wochenschrift* 100(11), 432–434.

Zafracas, A.M. (1975) Candida infection of the genital tract in thoroughbred mares. *Journal of Reproduction and Fertility, Supplement* 23, 349.

Zafracas, A.M. (1994) The equines in Greece nowadays. *Bulletin of Veterinary Medical Society* 45(4), 333–338.

Zavos, P.M. and Gregory, G.W. (1987) Employment of the hyposmotic swelling (HOS) test to assess the integrity of the equine sperm membrane. *Journal of Andrology* 8, 25.

Zavy, M.T., Vernon, M.W., Sharp, D.C. and Bazer, F.W. (1984) Endocrine aspects of early pregnancy in pony mares. A comparison of uterine luminal and peripheral plasma levels of steroids during the oestrous cycle and early pregnancy. *Endocrinology* 115, 214–219.

Zerbe, H., Engelke, F., Klug, E., Schoon, H.-A. and Leibold, W. (2004) Degenerative endometrial changes do not change the functional capacity of immigrating uterine neutrophils in mares. *Reproduction in Domestic Animals* 39, 94–98.

Zhang, J., Boyle, M.S., Smith, C.A. and Moore, H.D.M. (1990a) Acrosome reaction of stallion spermatozoa evaluated with monoclonal antibody and zona-free hamster eggs. *Molecular Reproduction and Development* 27(2), 152–158.

Zhang, J., Rickett, S.J. and Tanner, S.J. (1990b) Antisperm antibodies in the semen of a stallion following testicular trauma. *Equine Veterinary Journal* 22, 138–141.

Zidane, N., Vaillancourt, D., Guay, P., Poitras, P. and Bigras-Poulin, M. (1991) Fertility of fresh equine semen preserved for up to 48 hours. *Journal of Reproduction and Fertility, Supplement* 44, 644.

Index